Microbial Ecology of Aerial Plant Surfaces

Microbial Ecology of Aerial Plant Surfaces

Edited by

M.J. Bailey
Centre for Ecology and Hydrology, Mansfield Road, Oxford, UK

A.K. Lilley
Centre for Ecology and Hydrology, Mansfield Road, Oxford, UK

T.M. Timms-Wilson
Centre for Ecology and Hydrology, Mansfield Road, Oxford, UK

and

P.T.N. Spencer-Phillips
School of Biosciences, University of the West of England, Coldharbour Lane, Bristol, UK.

www.cabi.org

CABI is a trading name of CAB International

CABI Head Office
Nosworthy Way
Wallingford
Oxfordshire OX10 8DE
UK

Tel: +44 (0)1491 832111
Fax: +44 (0)1491 833508
E-mail: cabi@cabi.org
Website: www.cabi.org

CABI North American Office
875 Massachusetts Avenue
7th Floor
Cambridge, MA 02139
USA

Tel: +1 617 395 4056
Fax: +1 617 354 6875
E-mail: cabi-nao@cabi.org

A catalogue record for this book is available from the British Library, London, UK.

A catalogue record for this book is available from the Library of Congress, Washington, DC.

ISBN-10: 1 84593 061 4
ISBN-13: 978 1 84593 061 5

Printed and bound in the UK from copy supplied by the editors by Athenaeum Press, Gateshead.

Contents

Preface viii
Contributors xiv

Section I: Biodiversity and Population Genetics of Phyllosphere Communities

1. **Phyllosphere Microbiology: A Perspective** 1
S. Lindow

2. **Microbial Diversity in the Phyllosphere and Rhizosphere of Field Grown Crop Plants: Microbial Specialisation at the Plant Surface** 21
T.M. Timms-Wilson, K. Smalla, T.I. Goodall, A. Houlden, V. Gallego and M.J. Bailey

3. **Diversity, Scale and Variation of Endophytic Fungi in Leaves of Tropical Plants** 37
P. Bayman

4. **Microorganisms in the Phyllosphere of Temperate Forest Ecosystems in a Changing Environment** 51
T. Müller, K. Strobel and A. Ulrich

Section II: Spatial Distribution and Biofilms

5. **Bacterial Biofilm Formation, Adaptation and Fitness** 67
S. Rice and S. Kjelleberg

6. **Bacterial Assemblages on Plant Surfaces** 83
J.-M. Monier

7. **The Role of Plant Genetics in Determining Above- and Below-ground Microbial Communities** 107
J.A. Schweitzer, J.K. Bailey, R.K. Bangert, S.C. Hart and T.G. Whitham

8. **A Survey of A-L Biofilm Formation and Cellulose Expression Amongst Soil and Plant-Associated *Pseudomonas* Isolates** 121
A.J. Spiers, D.L. Arnold, C.D. Moon and T.M. Timms-Wilson

Section III: Biological Control and Pathogenicity

9. **Biological Control of Plant Diseases by Phyllosphere Applied Biological Control Agents** 133
B.J. Jacobsen

10. **Ecophysiology of Biocontrol Agents for Improved Competence in the Phyllosphere** 149
N. Magan

11. **Compost Teas: Alternative Approaches to the Biological Control of Plant Diseases** 165
W.F. Mahaffee and S.J. Scheuerell

Section IV: Gene Expression and Phyllosphere Genomics

12. **Molecular Interactions at the Leaf Surface: *Xanthomonas* and its Host** 181
M. Wilson, W. Moss, Y. Zhang and J. Jones

13. ***Erwiniae*: Genomics and the Secret Life of a Plant Pathogen** 191
I.K. Toth, L. Moleleki, L. Pritchard, H. Liu, S. Humphris, L. Hyman, G.W. Axelsen, M.B. Brurberg, M. Ravensdale, E. Gilroy and P.R.J. Birch

14. **Host-Pathogen Interactions of Relevance to the Phyllosphere** 201
G.W. Sundin

Section V: Leaf Colonisation and Dispersal

15. **Effects of Endophytes on Colonisation by Leaf Surface Microbiota** 209
R.W.S. Weber and H. Anke

16. **Plant Control of Phyllosphere Diversity: Genotype Interactions with Ultraviolet-B Radiation** 223
A.E. Stapleton and S.J. Simmons

17. **Population Growth and the Landscape Ecology of Microbes on Leaf Surfaces** 239
J.H. Andrews

18. **What DNA Microarrays Can Tell Us About Bacterial Diversity: A New Light on an Old Question** 251
G.L. Andersen, Y.M. Piceno, T.Z. DeSantis and E.L. Brodie

Section VI: Aerobiology and Plant Surface Microbiology

19. **Human Pathogens and the Health Threat of the Phyllosphere** 269
M.T. Brandl

20. **Post-harvest Spoilage of Wheat Grains: Malodour Formation and the Infection Process** 287
P.T.N. Spencer-Phillips, A. Wallington, W. Köckenberger, H.E. Gunson and N.M. Ratcliffe

21. **Atmospheric Composition and the Phyllosphere: The Role of Foliar Surfaces in Regulating Biogeochemical Cycles** 304
D. Fowler, J.N. Cape, J. Muller, M. Coyle, M.A. Sutton, R. Smith and C. Jeffree

Index 316

The colour plate section is inserted after p.238

PREFACE

In July 2005, the Centre of Ecology and Hydrology, Oxford (CEH-Oxford) hosted an International Symposium on the Microbiology of Aerial Plant Surfaces at St Catherine's College, Oxford. This was the eighth in this series of meetings which have been held every five years since 1970. The symposia bring together scientists studying the biology and ecology of the microbiology of the surfaces of above ground, aerial portion of vascular plants (stem, leaves, fruit, flowers, etc.) collectively known as the phylloplane. The presentations, discussions and the following chapters developed from this meeting, Phyllosphere 2005, highlight both the value of this highly diverse habitat to research in microbiology and the importance of this research to plant health and ecosystem functions.

The abundance of life in the phyllosphere is matched by the habitat range that plants occupy in both terrestrial and aquatic environments. Plant leaves provide the greatest surface area on the planet, tolerating geographic and climatic extremes that can fluctuate on a daily cycle from sub-zero night time temperatures to leaf surface temperatures that exceed 50°C in direct sunlight. Plants are found on over 90% of the approximately 2×10^8 km^2 of terrestrial surface of the planet. The total surface area of leaves for terrestrial plants may approach one billion square kilometres (1×10^9 km^2). Microorganisms (bacteria, fungi, archaea, protists) that have adapted to life in the phyllosphere must exhibit a range of phenotypic characteristics to mitigate against the effect of these physical parameters. These are perhaps greater than those experienced, for example, by soil or rhizosphere bacteria where temperatures, moisture content and nutrient availability are fairly constant when compared to the phyllosphere. Resource limitation, in respect to nutrient supply and water availability, are common selective factors that dictate the range and functional capacity of microbial life at plant surfaces. To understand more about the wider context of the microbiology of plant associated habitats, we would recommend that you also draw on the published proceedings of the other symposia on phyllosphere microbiology and of the chapters that follow.

The surface and interior of aerial parts of plants, including flowers, fruits, stems and leaves represent the phyllosphere. Specialised microbial colonists, phytopathogens, spoilage and plant protecting organisms as well as periodic immigrants have all been described for this diverse habitat. As plants represent one of the most important features of our landscape and our primary food source it is somewhat surprising that only a limited number of detailed investigations have been conducted that describe their above-ground microbiology. This is in stark contrast to the attention that soil rhizosphere systems receive. Pertinent investigations reveal a diverse and specialised microbial community that is distinct from the rhizosphere. Bacteria are by far the most numerically abundant colonisers of the phyllosphere, and typical community densities in the order of 2×10^7 cells per cm^2 of leaf surface have been recorded, although these numbers can vary from as little as 10s to in excess of 1×10^{12} in arid and senescing leaves respectively. Densities vary over the plant surface, in leaf buds or on young emerging leaves densities may be two orders of magnitude greater than those estimated in established mature leaves. These differences can be explained in part by the increased availability of water as these regions are often protected or collect moisture derived from rainfall or dew. In addition, nutrient loss from young, developing leaves provides suitable carbon (C) and nitrogen (N) sources for colonising bacteria which can form complex mixed assemblages or biofilms that facilitate survival. These assemblages also represent suites of significant bacterial activity and horizontal gene transfer. The source of these primary colonists remains somewhat obscure but in studies of developing seedlings the bacteria identified are often typical of those also found in the spermosphere and in rhizosphere soils. As plants develop, the phyllosphere community develops and becomes distinct, both in terms of relative abundance and complexity, from that found in the below-ground environment. Bacteria are often the focus of phyllosphere study as they are reported to be the numerically most abundant of isolates recorded using both isolation and culture independent methods. These general observations however are not intended to imply that yeasts and other fungi have a less important role.

Section I of this book concerns the **biodiversity and population genetics of phyllosphere communities**. The aerial portion of plants supports a diversity of microorganisms, including bacteria, fungi and archaea. As the following chapters reveal, there has been a major advance in the study of leaf surface organisms. The development of molecular tools has allowed the accurate description of the microbial diversity of many environments based on gene analyses and comparison. Specific measures of gene expression are reported, not only based on the assessment of messenger RNA expression but also through the analysis of protein synthesis using antibody assays or the direct assessment of function. Furthermore with the use of reporter genes that fluoresce, emit light or produce an enzyme that catabolises the production of colorimetric substrates, it is now possible to record how individual cells perceive and respond to their environment.

Methods based on the polymerase chain reaction (PCR) include amplified ribosomal DNA restriction analysis (ARDRA), terminal restriction fragment length polymorphism (T-RFLP), and length heterogeneity or LH-PCR which are all based on enzymic digestion of PCR derived amplimers with restriction endonuclease enzymes. Denaturing gradient gel electrophoretic (DGGE) analysis of the products of PCR amplified 16S and 18S rRNA genes has been applied to describe the total diversity of phyllosphere communities. Unsurprisingly this reveals that many previously described bacterial genera are indeed abundant, but interestingly only a few novel taxa were detected.

Some studies are beginning to integrate a variety of sampling methods to produce more integrated views of the phyllosphere microbial community structure. In Chapter 2, Timms-Wilson and colleagues report on the use of a variety of methods to assay changes in bacterial and fungal community structure and function on three field grown crop plants through a growing season. These methods include standard culture based and non-culture based analyses, including use of selective plate counts on microbiological agars for bacteria and fungi, community level physiological profiling (CLPP) with BIOLOG plates containing a range of sole carbon sources to analyse the 'metabolic potential' of responsive bacterial communities and PCR-DGGE of 16S and 18S ribosomal RNA genes to assess "total" diversity for bacteria and fungi.

Progress in the development of appropriate and representative sampling methods continues to play a major role in improving our understanding of microbial diversity. In Chapter 3, Bayman elegantly describes how different sampling strategies affect the diversity recorded in tropical and temperate phyllosphere bacterial and fungal communities. The chapter demonstrates considerable leaf-to-leaf variation in diversity and between adjacent epiphyte and endophyte biotas. Of particular note is the observation that a reduction in the size of the leaf fragment sampled leads to substantial increases in the diversity of fungi detected. Bayman proposes, where practical, that the selection of an optimal sample size of leaf fragment, for example one that contains a single colony, may avoid the inhibitory effects of different populations on one another.

Phyllosphere biodiversity studies are moving from investigation of populations and communities to experimentation on their role in global ecology. The phyllosphere of the temperate forest tree canopy is a substantial habitat which can influence regional and global ecosystem cycling. Environmental changes in climate, CO_2 levels, UV radiation and air pollutants all affect the phyllosphere microbial populations which in turn can exhibit altered growth and activity. In Chapter 4, Müller and colleagues report the effects of environmental change on the ecosystems of the temperate forest phyllosphere. This habitat is the site of microbial activity transforming organic matter and affecting nutrient cycling. The authors consider this activity, the effects of air pollution and insect invasions on forest ecosystems. They show that phylloplane microbial community activity may be enhanced by invasive insect species, which may grow better when plants are stressed by atmospheric pollutants. Such combined responses can lead to elevated nutrient cycling in the canopies of forest trees with higher organic matter and energy flow from the canopy to the forest floor.

Section II of this book concerns the **spatial distribution and biofilm structures of microbes on the phylloplane**. Technological advances have facilitated the direct surface imaging of the distribution and activity of key groups. These methods have facilitated real progress in the understanding of bacterial biofilm formation in many habitats. In Chapter 5, Rice and Kjelleberg provide a comprehensive review of the progress in methods, models and the understanding of the formation of bacterial biofilms. They stress the roles of signalling and biofilm formation in bacterial adaptation and fitness. On the phylloplane, these are important factors in the interaction of bacteria with plants and microbial eukaryotes.

Studies of bacterial phyllosphere communities have often shown that many bacteria are present as solitary cells, although the majority are highly aggregated in large assemblages embedded in an exopolymeric matrix. In Chapter 6, Monier reports on studies of these high density multi-population aggregates by integrating molecular methods and light microscopy. These studies have confirmed that bacterial phylloplane activity shows a variety of density-dependent traits where leaf surface colonisation involves a high level of spatial segregation. Understanding of interactions at the micro scale is providing important biological and ecological insights.

Moving to a new level of spatial scale and interaction, in Chapter 7, Schweitzer and colleagues examine the role that plant genetics play in determining both phyllosphere and rhizosphere microbial communities. It is proposed that the genetics of a host plant are an important common factor that can impact diverse communities which provides a plant-related basis to ecological and evolutionary processes.

Bacterial biofilms in soil and plant habitats use a variety of extracellular polysaccharides (EPS) initially for attachment and then to provide a matrix for colonisation. In Chapter, 8 Spiers and colleagues survey biofilm formation and cellulose expression amongst soil and plant-associated pseudomonads. Partially acetylated cellulose has been implicated in biofilm formation and successful phytosphere colonisation by the plant growth promoting rhizobacteria, PGPR *Pseudomonas fluorescens* SBW25.

Section III of this book concerns **biological control and pathogenicity**. Plant protection from pathogens using biological control is today a common component in crop protection. In Chapter 9, Jacobsen reviews the use of bacteria and fungi against a variety of plant diseases and pests. Successful biological control agents (BCAs) need to be produced with appropriate formulation, shelf-life and reproducibility of performance in the phyllosphere. Magan's Chapter 10 considers the use of physiological manipulation of the growth of fungal BCAs. Specific growth and formulation of inoculum has the potential to enhance the synthesis of nutrient reserves which can be exploited to provided improved tolerance to environmental stresses and conserved BCA capacity.

Alternatives to synthetic pesticides and biological control agents produced commercially are being sought in many areas of plant production. Compost teas are being extensively used in urban, horticultural and agriculture settings for their fertility and disease control properties. In Chapter 11, Mahaffee and Scheuerell review this interesting field which they explain is in need of more scientific investigation.

Section IV concerns **gene expression and phyllosphere genomics**. A variety of reporter methods are now available for the *in situ* study of microbes and their activities. These have been applied to understand the importance of cell-cell signalling and the response of assemblages of microorganisms to environmental cues. For example, sets of genes are being identified that are only expressed by bacteria as they colonise or interact with the leaf. These sets of genes or operons may in many cases be unique to phyllosphere specialists and provide the insight we need to understand the mechanistic basis and ecology of organisms that have adapted to life in this extreme habitat. Molecular and genomic methods have also allowed detailed assessments of the distribution and availability of nutrients (C and N sources) to be mapped and recorded as well as the importance of local water and pH gradients on the metabolic activity of leaf surface bacteria and fungi. These approaches, combined with total community assessments, have been used similarly to test the ability of immigrants, and the range of suitable local conditions needed, to successfully colonise developing plants or exchange genetic information by horizontal gene transfer.

While much research has focused on the "resistant" reaction between pathogen and host, in Chapter 12, Wilson and co-workers consider the less understood early stages of the "susceptible" reaction wherein a virulent pathogen invades a susceptible host. They have used the bacterial pathogen *Xanthomonas euvesicatoria* to study the molecular interactions between the pathogen and its plant host. They stress the importance of this stage which includes: pathogen recognition of the host plant; host recognition of a potential pathogen; elicitors and pathogen-associated molecular patterns; and the early, non-specific defence responses, recently likened to the "innate immunity" observed in mammalian systems.

Whole genome sequencing is opening up new avenues to study the adaptation and activity of important environmental bacteria. The potato pathogen *Erwinia carotovora* subsp. *atroseptica* (Eca) was the first plant pathogenic enterobacterium to be sequenced fully. A considered overview is provided by Toth and colleagues in Chapter 13 who study the developments and insight provided by recent detailed genome sequence analysis and the resulting experimental approaches that they illuminate. These approaches reveal the complex interactions and diversity of genes potentially involved in the non-pathogenic component of the bacterium life cycle as well as worked examples that demonstrate the presence of genes responsible for the expression of phytotoxins, especially those previously unknown in this group of organisms. Comparative analyses are used to suggest that these genes may have been acquired from other plant-associated organisms in the environment via horizontal transfer.

While substantial progress has been made in our knowledge of the host-pathogen interactions of the bacterial pathogen *Pseudomonas syringae,* understanding of the relevance of these in the phyllosphere milieu is still limited. In Chapter 14, Sundin reports the contribution being made by genomic studies to assessing the impact of environmental stress and host genotype on gene expression in pathogens. These studies are revealing interesting effects of stress on global gene expression in the plant pathogen.

Section V of the book concerns **leaf colonisation and dispersal**. In Chapter 15 on the effects of fungal endophytes on colonisation by leaf surface microbiota, Weber and Anke present evidence that *Phoma medicaginis* synthesises brefeldin A for defence against phylloplane organisms attempting to gain entry into the plant from the outside. In another approach to phyllosphere colonisation, Stapleton and Simmons show that plant genotype and ultraviolet-b (UV-B) radiation interact to affect plant influences on phyllosphere diversity. They manipulated the genotype of maize plants (the two parents of a commonly used mapping population and the F1 hybrid) and found, in field experiments, that plant genotype is more important than UV-B in explaining patterns of rDNA microbial diversity on maize leaves.

The observed diversity and activity of microbes on plants is the outcome and sum of many activities and interactions taking place at the micro, and other, scales. The significance of heterogeneity, spatial configuration, scale and other factors used in landscape ecology for microbial ecology are reviewed by Andrews in Chapter 17. Using Geographic Information Systems (GIS) software, data have been processed and models linking pattern and process tested for interpreting microbial colonisation. The processes investigated at a fine spatial scale have now been used to understand emergent patterns at a coarser spatial scale. This has important implications for phyllosphere community ecology.

With an interest in assaying the effects of perturbations on natural ecosystems, Andersen and colleagues stress in Chapter 18 the importance of defining detailed changes in the microbial community structure and how influencing factors shift population dynamics and the relative abundance of component taxa. What is particularly important here is the authors' development and application of gene-based microarrays that allow comprehensive assessments of diversity within samples. Surely this is the future for accuracy and detail, where sensitivities are as good as if not better than any other method. They report on the development a microarray system used to accurately measure the key microbial components in air, water, and soil environments. The application to air samples highlights the role of ambient temperature, temporal variation, local bacterial reservoirs and meteorological conditions in affecting bacterial deposition.

Section VI concerns **aerobiology and plant surface microbiology**. Whilst many technological advances are helping us to understand plant microbe interactions, they also contribute to a number of broader fields. These range from testing basic ecological theory of niche competition, dispersal and biogeography, to their value in the biogeochemical cycling of C, N and sulphur (S), and the microbial response and role in measuring the impact of pollution and climate change in ecosystem function and nutrient turnover. These latter attributes have allowed considerable insight to the ecology of a select number of species, and have contributed to a more detailed functional understanding of the population genetics and genetic diversity of leaf bacteria needed to identify the key adaptive traits they possess for life in the phyllosphere.

Although enteropathogenic bacteria are usually associated with contaminated meat or dairy products, survey data and epidemiological studies of foodborne outbreaks have provided strong evidence for the presence of these bacteria on fresh fruits and vegetables, with links to epidemics of enteric disease. In Chapter, 19 Brandl reports investigations into *Salmonella enterica* colonisation of the phyllosphere of cilantro plants (*Coriandrum sativum*), and its resultant interactions with the indigenous phyllosphere microbiota. Using population studies, fluorescence microscopy, and digital image analysis, the effects of various biotic and abiotic factors on the fitness of *S. enterica* in the phyllosphere are reported. The interaction of *S. enterica* with plant pathogens such as *Pseudomonas viridiflava* and *Erwinia chrysanthemi,* and with the epiphytic fungus *Aspergillus niger,* are all considered along with the changing microarray-based gene expression profiles of *S. enterica* during these interactions.

A more traditional cause of post-harvest spoilage is that caused by phyllosphere fungi. These can seriously reduce marketable yield. Spencer-Phillips *et al.* illustrate how an understanding of the fundamental biology of interactions that start in the phyllosphere and other aerial plant surfaces can lead to improvements in the quality and safety of the food produced. They report the use of light microscopy and environmental scanning electron microscopy to investigate fungal colonisation of wheat grains, and facilitate interpretation of nuclear magnetic resonance micro-imaging of the distribution of water within the grains.

These data are interpreted in relation to the formation of malodours, which are a significant reason for rejection of wheat in commercial grain intake laboratories.

Plant-atmosphere exchange of reactive gases is the focus of Chapter 21 by Fowler and colleagues, who demonstrate that reactive gases including O_3, SO_2, HNO_3 and HCl are deposited primarily on the external surfaces of vegetation. This chapter elaborates on the role of plant surfaces in regulating the rates of deposition and biogeochemical cycles of these gases.

The phyllosphere is a special environment that deserves greater attention. It is accessible, ubiquitous and the keeper of many interesting microbial functions. The aerial plant surface is providing a habitat for the direct study of how the plasticity of the genome, the diversity of mixed communities and the complexity of aggregates maintains and defines the diversity of functional microbial communities. The essays in this book highlight some of the best research seeking to answer these questions. In the first chapter, Lindow provides a detailed overview of phyllosphere biology to date and eloquently considers the major questions that remain unanswered. He draws our attention to unresolved issues in phyllosphere microbiology that go beyond the areas that have been addressed successfully in new and innovative research. These questions extend from the general 'what is the nature of plant surfaces' to the very specific 'how dry is a leaf'. Key questions remaining to be answered include who is present on plants, how do they get there, and what drives community development? Further questions regard the nature of these communities; which populations are epiphytes, where are they located, what kinds of interactions are occurring, are epiphytes strictly commensalistic on plants and do they engage in multi-trophic level interactions?

Acknowledgements

We are especially grateful to Angela Morrison for her editorial assistance. Thanks are also due to the following organisations who sponsored Phyllosphere 2005, the Eighth International Symposium on the Microbiology of Aerial Plant Surfaces held at St Catherine's College, Oxford, United Kingdom, 24-27 July 2005: British Mycological Society, British Society for Plant Pathology, Common Agricultural Bureau International, Centre for Ecology and Hydrology and the Society for General Microbiology.

M.J. Bailey
A.K. Lilley
T.M. Timms-Wilson
P.T.N. Spencer-Phillips

Contributors

G.L. Andersen, Center for Environmental Biotechnology, Lawrence Berkeley National Laboratory, 1 Cyclotron Road, Berkeley, CA 94720, USA

J.H. Andrews, Plant Pathology Department, 1630 Linden Dr, University of Wisconsin, Madison, WI 53706, USA

H. Anke, Institute of Biotechnology and Drug Research (IBWF), Erwin-Schroedinger-Str. 56, 67663 Kaiserslautern, Germany

D.L. Arnold, School of Biosciences, Faculty of Applied Sciences, University of the West of England, Coldharbour Lane, Bristol, BS16 1QY, UK

G.W. Axelsen, Department of Chemistry, Biotechnology and Food Science, Norwegian University of Life Sciences, Chr. M. Falsensvei 1, PO Box 5003, 1432 As, Norway

J.K. Bailey, Department of Biological Sciences, Northern Arizona University, Box 5640, Flagstaff, AZ-86011, USA

M.J. Bailey, Centre for Ecology and Hydrology, Mansfield Road, Oxford, OX1 3SR, UK

R.K. Bangert, Department of Biological Sciences, Northern Arizona University, Box 5640, Flagstaff, AZ-86011, USA

P. Bayman, Departamento de Biologia, Universidad de Puerto Rico, Rio Piedras, PO Box 23360, San Juan, PR 00931, USA

P.R.J. Birch, Scottish Crop Research Institute, Invergowrie, Dundee, DD2 5DA, Scotland

M.T. Brandl, Produce Safety and Microbiology, USDA/ARS, WRRC, 800 Buchanan Str., Albany, CA 94710, USA

E.L. Brodie, Center for Environmental Biotechnology, Lawrence Berkeley National Laboratory, 1 Cyclotron Road, Berkeley, CA 94720, USA

M.B. Brurberg, Department of Biotechnological Sciences, Agricultural University of Norway, Norway

J.N. Cape, Centre for Ecology and Hydrology, Bush Estate, Penicuik, Midlothian, EH26 0QB, Scotland

C. Coyle, Centre for Ecology and Hydrology, Bush Estate, Penicuik, Midlothian, EH26 0QB, Scotland

T.Z. DeSantis, Center for Environmental Biotechnology, Lawrence Berkeley National Laboratory, 1 Cyclotron Road, Berkeley, CA 94720, USA

D. Fowler, Centre for Ecology and Hydrology, Bush Estate, Penicuik, Midlothian, EH26 0QB, Scotland

V. Gallego, Department of Microbiology and Parasitology, Faculty of Pharmacy, University of Seville, 41012, Seville, Spain

E. Gilroy, Scottish Crop Research Institute, Invergowrie, Dundee, DD2 5DA, Scotland

T.I. Goodall, Centre for Ecology and Hydrology, Mansfield Road, Oxford, OX1 3SR, UK

H.E. Gunson, School of Biosciences, Faculty of Applied Sciences, University of the West of England, Coldharbour Lane, Bristol BS16 1QY, UK

S.C. Hart, School of Forestry, Northern Arizona University, Box 15018, Flagstaff AZ-86011, USA

A. Houlden, Department of Animal and Plant Sciences, The University of Sheffield, Western Bank, Sheffield, South Yorkshire S10 2TN, UK

S. Humphris, Scottish Crop Research Institute, Invergowrie, Dundee, DD2 5DA, Scotland

L. Hyman, Scottish Crop Research Institute, Invergowrie, Dundee, DD2 5DA, Scotland

B.J. Jacobsen, Department of Plant Sciences and Plant Pathology, Montana State University, 119 AgBiosciences Bldg Bozeman, MT 59717-3150 USA
C. Jeffree, Institute of Molecular Plant Sciences, University of Edinburgh, King's Buildings, Mayfield Road, Edinburgh, EH9 3JH, UK
J. Jones, Department of Plant Pathology, University of Florida, Gainesville, FL 32611, USA
S. Kjelleberg, The Centre for Marine Biofouling and Bio-Innovation and The School of Biotechnology and Biomolecular Sciences, The University of New South Wales, Sydney, NSW 2052, Australia
W. Köckenberger, Sir Peter Mansfield Magnetic Resonance Centre, School of Physics and Astronomy, University of Nottingham, University Park, Nottingham, NG7 2RD, UK
S. Lindow, University of California, Department of Plant and Microbial Biology, 111 Koshland Hall, Berkeley, CA 94720-3102, USA
H. Liu, Scottish Crop Research Institute, Invergowrie, Dundee, DD2 5DA, Scotland
N. Magan, Applied Mycology Group, Institute of BioScience and Technology, Cranfield University, Silsoe, Bedford, MK45 4DT, UK
W. Mahaffee, United States Department of Agriculture, Agriculture Research Service, Horticulture Crops Research Laboratory, 3420 NW Orchard Ave, Corvallis, Oregon 97330, USA
L. Moleleki, Scottish Crop Research Institute, Invergowrie, Dundee, DD2 5DA, Scotland
J.-M. Monier, Laboratoire D'Ecologie Microbienne, UMR 5557, Université Claude Bernard, Lyon, F-69622, France
C.D. Moon, Department of Plant Sciences, University of Oxford, South Parks Road, Oxford, OX1 3RB, UK
W. Moss, Department of Entomology and Plant Pathology, Auburn University, Auburn, AL 36830, USA
T. Müller, Leibniz-Centre for Agricultural Landscape Research (ZALF) e.V. Müncheberg, Institute of Landscape Matter Dynamics, Eberswalder Str. 84, D-15374 Müncheberg, Germany
J. Muller, Centre for Ecology and Hydrology, Bush Estate, Penicuik, Midlothian, EH26 0QB, Scotland
Y.M. Piceno, Center for Environmental Biotechnology, Lawrence Berkeley National Laboratory, 1 Cyclotron Road, Berkeley, CA 94720, USA
L. Pritchard, Scottish Crop Research Institute, Invergowrie, Dundee, DD2 5DA, Scotland
N.M. Ratcliffe, School of Biosciences, Faculty of Applied Sciences, University of the West of England, Coldharbour Lane, Bristol BS16 1QY, UK
M. Ravensdale, Scottish Crop Research Institute, Invergowrie, Dundee, DD2 5DA, Scotland
S. Rice, The Centre for Marine Biofouling and Bio-Innovation and The School of Biotechnology and Biomolecular Sciences, The University of New South Wales, Sydney, NSW 2052, Australia
S.J. Scheuerell, Evergreen State College, 2700 Evergreen Parkway NW, Olympia, Washington 98505, USA
J. Schweitzer, School of Forestry, Northern Arizona University, Box 15018, Flagstaff AZ-86011, USA
S.J. Simmons, Department of Mathematics and Applied Statistics, University of North Carolina at Wilmington, Wilmington, NC 28403, USA

K. Smalla, Institute for Plant Virology, Microbiology and Biosafety, Federal Biological Research Centre for Agriculture and Forestry, Messeweg 11/12, D-38104 Braunschweig, Germany
R. Smith, Centre for Ecology and Hydrology, Bush Estate, Penicuik, Midlothian, EH26 0QB, Scotland
P.T.N. Spencer-Phillips, School of Biosciences, Faculty of Applied Sciences, University of the West of England, Coldharbour Lane, Bristol, BS16 1QY, UK
A. Spiers, Department of Plant Sciences, University of Oxford, South Parks Road, Oxford, OX1 3RB, UK
A.E. Stapleton, Department of Biology and Marine Biology, University of North Carolina at Wilmington, Wilmington, NC 28403, USA
K. Strobel, Bayreuth Institute for Terrestrial Ecosystem Research, University of Bayreuth, D-95440, Germany
G.W. Sundin, Department of Plant Pathology, Center for Microbial Ecology, Michigan State University, 103 Center for Integrated Plant Systems, East Lansing, MI 48824-1311, USA
M.A. Sutton, Centre for Ecology and Hydrology, Bush Estate, Penicuik, Midlothian, EH26 0QB, Scotland
T.M. Timms-Wilson, Centre for Ecology and Hydrology, Mansfield Road, Oxford, OX1 3SR, UK
I.K. Toth, Scottish Crop Research Institute, Invergowrie, Dundee, DD2 5DA, Scotland
A. Ulrich, Leibniz-Centre for Agricultural Landscape Research (ZALF) e.V. Müncheberg, Institute of Landscape Matter Dynamics, Eberswalder Str. 84, D-15374 Müncheberg, Germany
A. Wallington, School of Biosciences, Faculty of Applied Sciences, University of the West of England, Coldharbour Lane, Bristol, BS16 1QY, UK
R.W. Weber, Department of Biotechnology, University of Kaiserslautern, Erwin-Schroedinger-Str. 56, 67663 Kaiserslautern, Germany
T.G. Whitham, Department of Biological Sciences, Northern Arizona University, Box 5640, Flagstaff, AZ-86011, USA
M. Wilson, Biology Department, The Colorado College, Colorado Springs, CO 80903, USA
Y. Zhang, Department of Plant Pathology, University of Florida, Gainesville, FL 32611, USA

8th International Symposium on the Microbiology of Aerial Plant Surfaces
St Catherine's College, University of Oxford, UK

01. Jacobsen, B.
02. Pedersen, C.
03. Greenberg, J.
04. Marin-Ceveda, V.
05. Limm, E.
06. Dulla, G.
07. Fuentes, L.
08. Toussaint, V.
09. He, J.
10. Preston, G.
11. Paul, P.
12. Timms-Wilson, T.
13. Morrison, A.
14. Karamanoli, K
15. Lee, S.
16. Elad, Y.
17. Derridj, S.
18. Harling, R.
19. McGrath, M.
20. Jeger, M.
21. Stapleton, A.
22. Smalla, K.
23. Fowler, D.
24. Vorholt, J.
25. Rice, S.
26. Andersen, G.
27. Lindow, S.
28. Meakin, H.
29. Cotton, A.
30. Magan, N.
31. Andrews, J.
32. Sundin, G.
33. De Vicente, A.
34. Perez-Garcia, A.
35. Weber, R.
36. Spiers, A.
37. Spencer-Phillips, P.
38. Amey, R.
39. Jackson, R.
40. Pertot, I.
41. Houlden, A.
42. Bailey, M.
43. Mahaffee, W.
44. Schweitzer, J.
45. Lilley, A.
46. Monier, J.- M.
47. Ferrari, A.
48. Turner, S.
49. Spear, R.
50. Krsek, M.
51. Yoshida, S.
52. Braun, S.
53. Leveau, J.
54. Allitt, U.
55. Mueller, T.
56. Bayman, P.
57. Last, P.
58. Last, F.

Chapter 1

Phyllosphere Microbiology: A Perspective

S. Lindow
University of California, Department of Plant and Microbial Biology,
111 Koshland Hall, Berkeley, CA 94720-3102, USA
icelab@socrates.berkeley.edu

Introduction

The field of phyllosphere microbiology is a large one, encompassing such diverse topics as the chemical and physical environment of a diversity of different types of plant surfaces, a description of a wide diversity of inhabitants of plant surfaces, the interactions that such organisms have with the plant and with each other, as well as many other related topics. Because the field is so broad, it is impossible for any one overview to be comprehensive. Furthermore, the biases of the author will certainly dictate the emphasis on certain topics. In that respect, this overview will also reveal the biases of this author. It will thus be somewhat bacterial-centric given both the predominant experiences of this author with this group of organisms as well as the prominence of bacterial studies at the expense of the studies of other microbial groups. Several excellent reviews of the microbiology of the phyllosphere or related topics have appeared in recent years (Beattie and Lindow, 1995; Jacques and Morris, 1995; Kinkel, 1997; Johnson and Stockwell, 1998; Beattie and Lindow, 1999; Andrews and Harris, 2000; Hirano and Upper, 2000; Lindow and Leveau, 2002; Lindow and Brandl, 2003) in which more detailed discussions of various aspects of phyllosphere microbiology are provided. This overview will address several different aspects of the field that have been illustrated recently in the literature, as well as address the major questions that remain unanswered in the field. By pointing out the many unresolved issues in phyllosphere microbiology, this overview will perhaps highlight more questions than the areas that have been addressed successfully in new and innovative research.

What is the Nature of Plant Surfaces?

Given that there is a wide variety of different types of plants and plant tissues that harbour microbial colonists, it is difficult to provide an unambiguous description of the nature of the plant surfaces on which they reside. Most studies of microbial colonisation of plants have emphasised the leaf surface habitat (Beattie and Lindow, 1995; Jacques and Morris, 1995;

Kinkel, 1997; Beattie and Lindow, 1999; Andrews and Harris, 2000; Hirano and Upper, 2000; Lindow and Brandl, 2003). The microbiology of flowers has also been studied extensively, largely because flowers are the site of infection by serious plant pathogens such as *Erwinia amylovora* as well as by other bacterial and fungal pathogens (Johnson and Stockwell, 1998; Nuclo *et al.*, 1998; Stockwell, *et al.*, 1999; Johnson *et al.*, 2000; Ngugi *et al.*, 2002; Lindow and Suslow, 2003; Thomson and Gouk, 2003; Walcott *et al.*, 2003; Ngugi and Scherm, 2004; Pusey and Curry, 2004). A few conceptual reviews of the nature of the leaf surface and how it affects the chemical environment of the leaf have appeared (Schönherr and Baur, 1996; Reiderer and Schreiber, 2001; Schreiber *et al.*, 2001). The conceptual models of the leaf surface environment are perhaps more numerous than direct studies of plant surfaces, however.

Ultraviolet Radiation

Given that aerial plant surfaces are generally exposed directly to ultraviolet radiation, there has been much conjecture of its significance to the composition of microbial communities. The role of UV irradiation on microbial communities has been reviewed recently by Sundin (2002). It has long been presumed that the occurrence of UV radiation would select against UV-sensitive microbes on plants. Indeed, it has long been assumed that the relatively high frequency of occurrence of pigmented bacteria and fungi on plants is a result of selection for such species that are tolerant of UV (Stout, 1960; Ayers *et al.*, 1996; Sundin, 2002). Work by Sundin and Jacobs (1999) did, in fact, reveal that pink, orange and yellow-pigmented bacteria were more tolerant of UV-B than non-pigmented bacteria. There have been several reports in recent years that have altered the flux of UV radiation of different wavelengths to plants under natural settings, either with supplemental UV emission or by using UV-absorbing filters to determine the extent to which UV flux and the abundance or composition of microbial communities were influenced by UV. When oak trees were exposed to supplemental UV-B radiation the yeasts *Aureobasidium pullulans* and *Sporobolomyces roseus* were significantly reduced, especially on adaxial leaf surfaces, compared to trees with ambient UV conditions (Newsham *et al.*, 1997). Several groups have now shown that reducing the flux of UV to plants using filters can alter the composition of the bacterial microbiota (Sundin and Jacobs, 1999; Jacobs and Sundin, 2001; Kadivar and Stapleton, 2003). Such studies, when combined with the observation that bacterial populations are often larger on the abaxial surfaces than on the adaxial surfaces (Sundin and Jacobs, 1999) both support the idea that UV is a strong selection agent on plants. The mechanisms for UV resistance in phyllosphere bacteria, particularly *Pseudomonas syringae,* have been investigated thoroughly and are reviewed by Sundin (2002). Mutagenic DNA repair by *rulAB* and its homologs seem to be common in epiphytes, enabling them to tolerate UV even though it is likely to increase the rate of mutation in such strains (Sundin, 2002; Zhang and Sundin, 2004). Surprisingly, such an elevated mutation rate was not associated with reduced fitness of a *P. syringae* strain after exposure to UV (Zhang and Sundin, 2004).

The fact that most aerial plant surfaces are part of a complex, three-dimensional plant canopy consisting of overlapping leaves and stems has not been considered adequately in relation to UV exposure of epiphytes until recently. Most discussions of UV exposure to microbes on plants have assumed the worst-case scenario, that of a microbe on a planar leaf exposed perpendicular to the sun. In fact, the microclimate of a plant canopy provides a complex assortment of surfaces, only a few of which have this exceptional UV exposure. It has been shown that such microclimate effects can have a great influence on growth and survival of bacteria on plants; increased shading is associated with higher populations of most bacteria studied (Giesler *et al.,* 2000). Because microbes on leaves within a plant canopy will be exposed to a combination of both direct and diffuse UV light, it is important to know UV flux at various points within the canopy to understand if UV is really a stressing agent or not. Shulski *et al.,* (2004) have recently reported such measurements using various V sensors placed in alfalfa (*Medicago sativa*) and tall fescue (*Festuca arundinacea*) canopies. They reported that, as expected, UV-B transmittance decreased greatly, and in a non-linear fashion with increasing leaf area index. In canopies of both pant species the transmission of UV-B was less than 10% at leaf area indices of greater than about five (Shulski *et al.,* 2004). Such measures will enable, for the first time, others to predict UV flux into other types of canopies since leaf area index is readily measured, unlike UV flux. In another exciting development, Yuen and coworkers (Yuen *et al.,* 2002) developed a small UV-B biodosimeter comprising of small pads containing cells of the DNA repair-deficient *E. coli* stain CSR06 which could be placed at different locations and inclinations within plant canopies. This UV-hypersensitive strain produced accurate estimates of UV flux at small spatial scales enabling them to account for the striking variation in UV microclimate within plant canopies. Such a tool should prove valuable in further studies of UV exposure to microbes on plants under realistic conditions and better defining the "microclimate" of plants.

Water Availability on Plants

A conspicuous feature of aerial plant surfaces is the fact that free moisture is quite transient. Plants are frequently wetted by rain or dew, but such wetness often disappears within a few hours, leaving a surface that appears dry to the naked eye. The disappearance of free moisture on plants is often associated with death of epiphytic bacteria that had multiplied on wet leaves (O'Brien and Lindow, 1989; Lindow *et al.,* 1993; Beattie and Lindow, 1994 a, b). The question that has remained unanswered, however, is "How dry is a leaf?". Given that a plant surface is irregular at small spatial scales (Jeffre, 1996; Beattie and Marcell, 2002; Marcell and Beattie, 2002) it has remained unclear whether all parts of a leaf would be exposed to the same degree of desiccation stress, and to what degree the presence of a laminar boundary layer on plants, that might be expected to be hydrated due to transpiration water loss, might temper water stresses to cells on plant surfaces. A breakthrough in this field of study was reported recently by Axtell and Beattie (2002). They described a *proU-gfp* transcriptional fusion that could be introduced into *E. coli, P. syringae* and *Pantoea agglomerans* to yield GFP fluorescence that is inversely related to water availability to these bacteria both in culture and on plants. Following introduction of the *P. agglomerans* biosensor onto bean leaves, the cells detected, on average, a water potential equivalent to that imposed by about 55 mM NaCl (Axtell and Beattie, 2002). Surprisingly, these initial studies did not find evidence for high variability in water potential on leaves, nor the extremely low water potentials that might have been expected. Such a tool should provide new insights into the physical phyllosphere environment when applied to plants under field conditions.

Role of Nutrients in Numbers and Spatial Distribution of Microbes on Leaves

Considerable evidence supports the concept that the abundance of nutrients on plant surfaces dictates the population size of various microbes. Depending on the system studied, carbon compounds alone, or both carbon and nitrogen compounds were shown to be the limiting factors for bacterial and yeast populations on leaves (Bashi and Fokkema, 1977; Dik *et al.,* 1991; Wilson and Lindow, 1994a, b). Because the cuticle is considered to be a major barrier to diffusion of polar compounds such as sugars and amino acids (Schönherr and Baur, 1996), the amounts of intrinsic nutrients available to epiphytes from the plant part on which the microbe resides has long been considered to be relatively low. Very few studies of the amount and composition of nutrients on plants have been performed. In a recent report, Mercier and Lindow (2000) noted that there was a range of about 0.5 to 12 μg of sugars on the surface of individual snap bean (*Phaseolus vulgaris*) leaves (average of about 10 μg/leaf). Furthermore, the log-normal distribution of sugar abundance on bean leaves was predictive of the distribution of population sizes of *Pseudomonas fluorescens* attained after it was inoculated onto these leaves (Mercier and Lindow, 2000). This group also found a direct correlation between the total amount of sugar on leaves of different plant species before bacterial inoculation and the maximum populations of *P. fluorescens* attained after inoculation, suggesting that under these greenhouse conditions leaf surface sugars limited bacterial abundance on a range of different plant species (Mercier and Lindow, 2000). Such studies, however, beg the question as to whether on other plant species and under field conditions compounds other than sugars might limit microbial population size, or whether exogenous nutrients such as from pollen are major sources of nutrients. Both pollen and aphid honeydew have been suggested to be important sources of nutrients for various microbes at certain times of the year (Fokkema, 1971; Ercolani, 1991; Dik *et al.,* 1992). Such exogenous sources of nutrients are likely to be temporally and spatially variable, contributing to the large variations in microbial populations between and within leaves (Kinkel, 1997; Kinkel *et al.,* 2000). Much more work is needed to determine if the nutrient abundance and composition of bean leaves is at all representative of other plant species and of the relative importance of intrinsic versus exogenous nutrients in phyllosphere populations. Since most studies have been performed with well-fertilised crop plants, it might be possible that non-fertilised plants might have lower relative rates of leakage of nitrogenous compounds and hence have a more nitrogen-limited leaf surface.

Recent studies using nutritional biosensors have provided information on the spatial localisation of nutrients on leaves, giving new insight into the processes leading to microbial colonisation of plants. The molecular biosensors developed for different sugars have revealed much about the chemical environment of the phyllosphere at the small spatial scales of relevance to microbes, and suggest that nutrient availability on leaves is highly spatially heterogeneous. Whole cell bacterial biosensors responsive to fructose, consisting of *Erwinia herbicola* cells harbouring short-half-life variants of a *gfp* reporter gene fused to a fructose-responsive *fru*B promoter, have provided considerable insight into the process of consumption of nutrients by bacteria on plants (Leveau and Lindow, 2001). This report noted that while nearly all fructose bioreporter cells were engaged in consumption of fructose (as evidenced by GFP fluorescence) within 1 h after inoculation onto plants, this dropped to less than 1% of the cells within 24 h, suggesting a highly heterogeneous availability of nutrients to individual cells and that most cells had access to only very small amounts of sugar. Furthermore, direct *in planta* examination of the bioreporter cells on leaves revealed that those cells that continued to consume sugars were usually in localised sites on the plant, and not randomly dispersed across the leaf. Similar findings of limited and spatially heterogeneous sucrose availability detected with a *gfp*-based biosensor for sucrose was reported by Miller *et al.* (2001). This suggests that most immigrants to a leaf will find themselves in a nutritional "desert" with only small amounts of nutrients, while the lucky immigrant may encounter "oases" of relatively abundant nutrients.

Recent quantitative microscopy analysis of leaves has revealed that bacterial populations are not randomly distributed across plant surfaces, but instead, are spatially aggregated (Morris *et al.,* 1997; Monier and Lindow, 2004). Cells of *P. syringae* inoculated uniformly onto leaves in a greenhouse, as well as natural populations of bacteria on field conditions, exhibited such an aggregated pattern of colonisation on leaves. For example, the majority of cells of *P. syringae* were found in aggregates of at least 100 cells in size a week after inoculation onto bean leaves (Monier and Lindow, 2004). It seems likely that these bacterial aggregates formed from single immigrant cells at the discrete sites of continuing nutrient release from the plant as noted in studies using *gfp*-based nutritional biosensors, but further work will be required to link aggregate development with nutritional availability.

The presence of both bacterial aggregates and apparently localised nutrient release on leaves raises questions as to the heterogeneous nature of leaf surfaces and what factors might enable such heterogeneity to persist. Several other recent findings also suggest that leaf surfaces have spatially heterogeneous chemical and/or physical microenvironments. For example, the expression of *ipdC*, encoding indole pyruvate decarboxylase, one of the initial enzymes required for indole acetic acid production in *Erwinia herbicola*, is induced under conditions of low water availability in culture (Brandl and Lindow, 1997). The expression of *ipdC* in *E. herbicola* cells on plants was spatially heterogeneous, and many cells did not express this gene whilst on plants while closely adjacent cells did (Brandl and Lindow, 2001). These results suggest that the chemical and/or physical environment of the leaf was only locally conducive to expression of *ipdC*. Likewise, only about 10% of the cells of *P. syringae* expressed a Fe^{+3}-regulated gene involved in pyoverdine siderophore production when cells were inoculated onto bean leaves (Joyner and Lindow, 2000). This result suggests strongly that Fe^{+3} is abundant over most areas of the leaf but was in low amounts in about 10% of the sites that supported the growth of *P. syringae*. While these studies all suggest a spatially heterogeneous leaf surface environment, it is unclear what processes could maintain such heterogeneity. Such variation suggests that when leaves are wetted, diffusion is somewhat limited across the leaf; if diffusion was not limited then it would be expected that a uniform chemical environment would be found on plants. If leaves are subject to limited diffusion does this imply that some parts of the leaf are not fully wetted when the leaf is macroscopically “wet”? Because of the high surface tension of water it seems possible that the small-scale surface irregularities associated with cuticular waxes and other plant surface features might prevent small crevices from being wetted. This issue needs to be addressed better since it appears to be central to our understanding of leaf surface colonisation. No clear relationship between the structure of cuticular surfaces and bacterial colonisation was seen in plant mutants with altered cuticular waxes (Beattie and Marcell, 2002), suggesting that complex chemical as well as physical factors play a role in surface colonisation, and perhaps diffusion. There have been several reports of the ability of bacteria to alter the wetability of leaf surfaces (Bunster *et al.,* 1989; Ercolani, 1994; Schreiber, 1996; Knoll and Schreiber, 1998, 2000). Alteration of cuticular waxes or production of biosurfactants both might be expected to increase diffusion across the leaf, enabling microbes to access more distant nutrient sources. This issue seems central to our understanding of microbial colonisation and deserves more study.

What is Present on Plants?

While there have been many anecdotal descriptions of the microbial colonists of plants, there have been few systematic studies of the diversity of these communities. Studies of the composition of bacterial communities on leaves have been numerous but rather limited in scope. It is generally believed that populations of culturable aerobic bacteria on leaves are dominated by a few genera. A few exhaustive studies of the variations in the microbial community of leaves over multiple time and space scales have provided important detailed knowledge about the identity and the ecology of bacterial leaf inhabitants (Dickinson *et al.*, 1975; Ercolani, 1978, 1991; Thompson *et al.*, 1995). Ercolani (1991) made an extensive inventory of culturable aerobic bacteria isolated from the surface of olive leaves over six growing seasons and reported distinct bacterial community structures on leaves of the same age at a given time of the growing season over the entire sampling period. Thompson *et al.* (1995) analysed 1,236 bacterial strains from immature, mature and senescent leaves of field grown sugar beet over a complete growing season. They identified 78 species, and 37 named and 12 unnamed genera of bacteria. Most importantly, like Ercolani (1991), they found distinct patterns of microbial colonisation at different times of the year, with bacterial community diversity being lowest during the warmest and driest months of the season, and highest during the cooler and rainy months. Coincidentally, in both of the above studies, communities on young leaves were composed of a greater number of taxa than those of old leaves. Thus specific natural environments of the phyllosphere apparently select for the presence of specific genotypes within the leaf bacterial community. This is supported further by the finding that the acquisition by *P. fluorescens* of plasmids that are indigenous to the leaf microbiota, coincided with a specific maturation stage of the plant over two consecutive years (Lilley *et al.*, 2002). This indicated that traits carried on these plasmids conferred variable selective fitness to specific plasmid-bacterial host combinations during the growing season, possibly in response to changing conditions in the phyllosphere habitat (Bailey *et al.*, 2002).

Despite the numerous descriptions of phyllosphere microbial communities, virtually all have been restricted to microbes that are culturable. To date, only two studies have used culture-independent methods to describe epiphytic bacterial communities. Yang *et al.* (2001) and Kadivar and Stapleton (2003) both used denaturing gradient gel electrophoresis (DGGE) to differentiate PCR-amplified 16R rRNA sequences from DNA isolated from bacteria recovered directly from leaves. In both studies, a sub-set of the isolated 16S rRNA clones were sequenced to provide insight into the unculturable microbiota. Kadivar and Stapleton (2003) not only found that DGGE analysis could reveal differences in communities with differential exposure to UV irradiation, but that several bacterial species not previously described on leaves were present. Yang *et al.* (2001) used such techniques to more clearly show that the culturable bacterial microbiota differed considerably from the total microbiota determined by DGGE, suggesting the presence of many unculturable bacteria. This later study of citrus microbiota was done in summer months in southern California when culturable microbial populations on citrus are known to be low (Lindow and Andersen, 1996), leaving open the possibility that many of the bacteria recovered from citrus leaves by DGGE were not active colonists of the leaves and represented soil microbes which had accumulated during the summer months. More culture-independent studies of the microbiota on leaves are highly justified since it should provide insight into the nature of the leaf surface as well as to the types of interactions that might be expected among the epiphytic microbiota. Since viable but non-cultural bacteria have most commonly been associated with environments that are nutrient-limited rather than simply being stressful (Wilson and Lindow, 1992; 1999), it will be informative to learn of how commonly unculturable microbes are found on plants. If plants do not provide an environment that is as oligotrophic as commonly assumed, then a relatively high proportion of epiphytes might be expected to be culturable.

It has long been assumed that microbes commonly found on plants are far more fit as epiphytes than those microbes that are not usually found on plants. The presence of chromogenic bacteria on leaves but not in soil or on roots has often been used as evidence that epiphytic bacteria are distinct from those in soil, etc., but this begs the question as to whether soil-borne microbes could exhibit epiphytic fitness. Very few studies have addressed this issue. O'Brien and Lindow (1989) found that while both plant associated bacterial species such as *P. syringae,* and bacteria such as *E. coli, Salmonella* and *Rhizobium* that are not commonly found on aerial surfaces of plants, could achieve high population sizes on plants that were kept moist after inoculation, the latter could not survive desiccation stresses on plants. Similarly, Brandl and Mandrell (2002) found that *Salmonella enterica* was less fit than *Pantoea agglomerans* and *Pseudomonas chlororaphis*, two common colonists of plants, especially at lower incubation temperatures. More of these studies are needed to define better how distinctive the common epiphytes are from other microbes. Such studies, when coupled with physiological and genomic investigations should help define traits involved with epiphytic fitness.

How do Microbes get to Plants and What Drives Community Development?

Before one considers the processes that contribute to the inoculation and multiplication of microbes on plant tissues, it is important to recognise the many differences in the types of habitats that different plant parts provide. Vegetative plant tissues, whether they emerge from seeds, vegetative buds, or flower buds generally are without significant microbial colonisation (Kinkel, 1997; Hirano and Upper, 2000). Plants thus usually develop a larger and more complex microbial community as plant parts age. As such, island biogeography models are thought to be applicable to the development of microbial communities on plants (Andrews *et al.*, 1987; Andrews and Harris, 2000). Plant parts such as flowers, which are very ephemeral, however, almost certainly are subject to quantitatively different processes that affect colonisation when compared to long-lived leaves.

Conceptually, there are four major processes that can influence the size and composition of microbial populations on leaves. Microbial population sizes on a leaf can increase by both multiplication on that leaf as well as by immigration of microbes from other leaves. In contrast, both the death of microbial cells and their migration from a leaf can contribute to decreases in population size. It appears that most workers have assumed that growth and death of microbes are the predominant processes that determine population sizes. These processes are obviously much easier to study than immigration and emigration since they can be done in isolation in the laboratory or greenhouse. For example, the multiplication of bacteria on plants can be studied easily by inoculating plants that are isolated from other plants in incubation chambers; increases in population size are directly attributable to multiplication of the bacteria. In contrast, studies of the importance of immigration of bacteria to the population sizes of bacteria on a leaf require a source of immigrant bacteria, and that conditions facilitating immigration be maintained. Since the conditions which favour the immigration of bacteria to plants are not yet well understood, such studies must be done under field conditions. They therefore face the complications of variable environmental conditions which make the experiments difficult to reproduce. For this reason, most information relevant to understanding processes that occur on plants are inferences made from simple field or laboratory observations. Most observations have not been sufficiently detailed to partition and elucidate the many factors that can influence epiphytic microbial populations. To date, there have been few studies designed to understand the processes that contribute to epiphytic microbial populations. Extensive reviews of the factors that address the broad processes affecting microbial community development have appeared (Lindow, 1996; Andrews and Harris, 2000; Kinkel, 2000). While a few excellent and detailed studies have addressed the processes such as immigration that lead to the development of fungal populations on plants (Kinkel *et al.*, 1989; Kinkel, 1991), relatively little work has been done in this area. Thus I will emphasise how this lack of information on the role of immigration on the development of epiphytic microbial populations hinders our understanding of community development.

Role of Immigration in Establishing Epiphytic Microbial Communities

Immigration of microbes to a leaf is coupled strongly to their emigration from another leaf since most epiphytes have leaves as their primary habitat. Thus, for microbes to immigrate to a leaf they must first emigrate from another plant having epiphytic bacterial populations. The processes of emigration of microbes from leaves by rain splash and aerosol particle movement has been studied extensively and reviewed previously (Lindow, 1996). Immigration to a leaf may occur via several modes of transportation. Many bacteria can be transported to a leaf via rain splash. While quantitatively important in releasing bacteria from plants, rain deposits a large percentage of the bacteria released from plants onto the soil and hence this is not a particularly important mechanism of inoculation of new plants (Lindow, 1996). A number of phytopathogenic bacteria can also be transferred from infected plants to healthy plants by insect vectors. While there are some bacterial pathogens that are disseminated due to intimate associations of the pathogen with the insect vector, most often bacteria are transmitted via insects which are contaminated during their foraging or nectar collecting activities (Lindow, 1996). This latter phenomenon is most well studied in the case of the vectoring of *Erwinia amylovora* from cankers or infected flowers of pear and apple trees to newly opened flowers where infection can occur. For this reason, insect vectoring of phyllosphere microorganisms has generally been considered to be a relatively inefficient and uncommon phenomenon. Recently, however, detailed work by Hirano and Upper has shown that the common epiphyte, *P. syringae,* commonly may be moved from one leaf to another via nocturnal insects (Hirano *et al.,* 1995; Hirano and Upper, 2000). While few bacteria could be transferred by insects trapped when leaves were dry, *P. syringae* as well as other bacteria were commonly transferred onto Petri dishes by insects such as *Glischrochilus quadrisignatus* when they were collected from leaves wet with dew in the early morning (Hirano and Upper, 2000). Not only does it seem likely that insects would more likely entrain phyllosphere microbes that were dispersed in a liquid film on plants, they might also be more readily deposited onto, and survive on other wet leaves that insects might visit. This exciting observation warrants further study to determine if it is a common but under-appreciated means of dispersal of phyllosphere microbes.

Aerosols that contain phyllosphere microbes can be produced under both wet and dry conditions, and deposition of aerosols has often been considered of primary important in the immigration of inoculum to plants (Lindow, 1996). While the processes of aerosol generation, dispersion and deposition onto plants is relatively well studied, the quantitative contribution of inoculum to plants via this route compared to that via insect transfer, etc. is not well understood.

Immigration is significant for populations of epiphytic bacteria in at least two different ways: 1) immigrant cells are a source of inoculum of bacteria that subsequently multiply on leaves, and 2) immigrant cells contribute directly to the population size of leaves on plants on which little growth is possible. The role of immigrant cells as inoculum of plant pathogens that subsequently proliferate on plants has been examined most closely. A popular model of the microbiology of leaves is based on "island biogeography" theory

(Andrews *et al.,* 1987). In this model, the population structure of living beings on an island is strongly influenced by the nature and size of immigrants to the islands that are isolated from each other and are initially devoid of colonists (Andrews *et al.,* 1987). The relatively low population size of microbes on newly emerged tissues as discussed above is often apparently due to a lack of inoculum of appropriate epiphytic bacterial strains. Adjacent leaves often harbour quantitatively and qualitatively different resident microbiota (Hirano *et al.,* 1995). Similarly, the diversity of bacteria on a given leaf may be either very high or very limited (Hirano and Upper, 2000). Therefore the colonisation of leaves may be a rather stochastic process involving repeated (but limited) introduction of genotypes of bacteria capable of colonisation of leaves. Occurrence of a particular species (such as a plant pathogen) on a particular leaf will be dependent on its successful immigration (at least once) from another source. This has led to the interest in identifying "collateral hosts", plants that can support epiphytic growth of the pathogen but not lesion development, which can serve as a source of immigrant inoculum of plant pathogenic bacteria (Hirano and Upper, 2000). Upper *et al.* (2003) have shown that even relatively small amounts of immigrant inoculum are sufficient to establish *P. syringae* on plants, especially if the plants to which the inoculum has moved were previously relatively uncolonised by other microbes. Such results clearly are in support of the island biogeography model of plant colonisation. Further, such a process could apply equally well to non-pathogenic microbes of all types. The time of arrival of immigrant inoculum relative to plant phenological stage and hence level of prior inoculation by other potential epiphytic colonists thus seems to dictate the success of the inoculation process, with initial colonists having a much higher likelihood of contributing to microbial community development. Unfortunately, while these few studies suggest that immigration processes are important in dictating epiphytic communities, this phenomenon requires much more study to understand all of the processes involved.

While immigration has been recognised as contributing qualitatively to the composition of phyllosphere bacterial populations, there has been little appreciation of its contribution to the size of these populations. Immigration may be relatively more important in determining population size on some plant species than on others. A wide range of bacterial population sizes occurs on plants, even after leaves have been inoculated with epiphytic strains and allowed to incubate under moist conditions conducive for growth for long periods of time (Lindow *et al.,* 1978; Lindow, 1982; O'Brien and Lindow, 1989). While maximum population sizes for some plants such as bean and cucumber exceed 10^7 cells/cm^2, some plant species, such as citrus species and conifers, support less than about 10^3 cells/cm^2 (Lindow *et al.,* 1978; Lindow, 1982; O'Brien and Lindow, 1989; Lindow and Andersen, 1996). The rate of deposition of bacteria in landscapes that include an abundance of plants having high epiphytic bacterial populations has been estimated to be about 10^3 cells/100 cm^2 /day (Hirano *et al.,* 1995; Lindow, 1996). This rate of deposition would yield a cumulative immigrant population of about 10^4 cells per month on an average sized leaf (such as a bean leaf). If we assume that most leaves in nature are subject to immigration of about 10^4 total cells per month, then immigrant cells could account for a substantial fraction of the bacterial population observed on plants having a relatively low population size, but only a very small proportion on plants like bean and cucumber having relatively high population sizes. For example, navel orange leaves in California seldom harbour more than about 10^5 cells (Lindow, 1982; Lindow and Andersen, 1996). Since these leaves are long-lived and have been shown to be exposed to an average of about 10^4 immigrant cells per month (Lindow, 1996) most of the cells on their leaves may be attributable to immigration. Further support for such a conjecture is obtained from examining the growth of genetically marked bacterial strains on citrus leaves after inoculation. Little growth of common epiphytic bacteria such as *P. syringae*, *E. herbicola* and *P. fluorescens* was observed (Lindow and Andersen, 1996). Thus, many plant species having small epiphytic bacterial population sizes may not support the epiphytic growth of bacteria. Instead, such plant species may simply harbour "casual" occupants (immigrants) that did not arise by growth on the plants. As shown by Jacques *et al.* (1995), immigration may be important in determining the epiphytic population size on emerging leaves of plant species that harbour large populations on their mature leaves.

Immigration-limited Colonisation of Plants

Some communities, including those of some fungal epiphytes, respond directly to enhancements in recruitment or immigration; by definition, these communities are said to be immigration limited (Kinkel *et al.,* 1989; Hughes, 1990; Kinkel, 1997). For example, Kinkel *et al.* (1989) found that *Aureobasidium pullulans* population sizes on apple leaves in the field to which large numbers of this species had been inoculated were significantly larger than population sizes on leaves that had not been inoculated, even after substantial apparent growth of this species on the leaves. This direct response in population size to an increase in the numbers of immigrants indicated that *A. pullulans* is immigration-limited (Kinkel *et al.,* 1989). Similarly, Jacques *et al.* (1995) found that population sizes could be increased by up to 100-fold when the exposure of emerging leaves to air was increased, particularly later in the growing season when the abundant epiphytic bacterial populations that existed on adjacent plants could apparently serve as a source of immigrant inoculum. Lindow and Suslow (2003) also found that total bacterial population sizes were much higher on flowers of pear trees inoculated with *P. fluorescens* several weeks earlier than on uninoculated flowers. While this strain exhibited substantial multiplication on inoculated flowers to achieve similarly high population sizes on treated flowers and become the dominant bacterial strain, uninoculated flowers remained poorly colonised by any bacteria (Lindow and Suslow, 2003). Such findings raise important questions regarding the fitness and host specificity of immigrant inoculum. Given the relatively high flux of total bacteria that arrive at a plant each day, it would seem that if strains that were capable of exploiting the new plant habitat, then immigration limitation would not be observed. The fact that plant parts, can remain “undercolonised” for extended periods of time suggests that many of the immigrant cells either are poor epiphytes of any plant or are specialised to certain plants. More work is needed to distinguish such scenarios and should shed much light on the behaviour of plant-associated microbes. This work would, for example, help answer questions as to whether plants are a source of the large numbers of airborne microbes, and whether plants are the ultimate “sink” for these airborne microbes. Unfortunately, few other studies have addressed this immigration of fungal or bacterial species to and from leaves.

Where are Microbes Located in the Phyllosphere and What is an Epiphyte?

Even after much study, there is still considerable uncertainty about where microbes are located in and on plants and what constitutes an epiphyte. Epiphytic populations of bacteria are normally described functionally as those bacteria that can be washed off plants (Hirano and Upper, 2000). While very convenient and supportive of ecological studies, this definition begs the question of where such bacteria are located, and what we should call microbes that are not readily washed from plants. While most of the bacterial cells washed from plants probably were located on the surface of the plant, not all such bacteria may be removed, and some which might have been "within" the plant such as in the apoplast might also be removed by washing (Hirano and Upper, 2000). Surface sterilants such as UV irradiation and hydrogen peroxide have been used to kill 'surface' bacteria, leaving more "internal" bacteria to be recovered from treated plants (Wilson *et al.,* 1999; Hirano and Upper, 2000). Most of these studies have been conducted under laboratory conditions and have revealed that most of the cells of plant pathogens such as *P. syringae* on susceptible hosts (75 to 95%) are external to the plant, while non-pathogenic strains of *P. syringae* and non-pathogens were almost exclusively found on the surface of plants (Wilson *et al.,* 1999; Hirano and Upper, 2000). The little data available would thus suggest that only pathogenic bacteria have the ability to colonise the "interior" of plants to a significant extent. Not all plant pathogenic bacteria, however, may have the same propensity to enter plants. While fewer than 10% of the cells of *P. syringae* pv. *syringae* strain B728a survived surface sterilisation after inoculation onto susceptible bean plants (Wilson *et al.,* 1999), a majority of cells of *P. syringae* pv. *tomato* DC3000 were found within susceptible tomato (Boureau *et al.,* 2002). Such findings have led Beattie and Lindow (1999) to suggest that there is probably a continuum of bacterial strains on plants, with some nearly exclusively external to the plant while others are more internally located. Unfortunately, such a proposal is based on very limited data, most of which have come from laboratory and greenhouse studies which may not be reflective of field conditions. A recent study by Sabaratnam and Beattie (2003) has addressed this issue for the first time under field conditions. Interestingly, they found that a majority of the cells of *P. syringae* on bean plants were not subject to killing by surface sterilants and hence were "interior" to the susceptible bean plants, while less than about 1% of the population was internal to maize leaves, a non-host for this strain. Likewise, less than about 1% of the cells of *P. agglomerans* were internal to either bean or maize in these field studies (Sabaratnam and Beattie, 2003). More studies are needed to clarify our understanding of the location of microbes on and in plants. This would help clarify the nature of "endophytes" that are commonly described in plants (Hallmann *et al.,* 1997; Kobayashi and Palumbo, 2000). These endophytes are usually defined as those microbes that survive surface sterilisation and include a large number of different species, each of which is usually found in small numbers ($<10^3$ to 10^4 cells/g) (Hallmann *et al.,* 1997; Kobayashi and Palumbo, 2000). Since, with the exception of bacteria closely related to plant pathogenic species, endophytic bacteria seldom are found in high numbers in plants, it is tempting to speculate that they represent occasional introductions into a plant and, unlike plant pathogens, lack the ability to exploit the plant interior after introduction. Thus, for example, do the few cells of *P. agglomerans* found within maize in studies such as by Sabaratnam and Beattie (2003) indicate that this species is an endophyte, when far more cells are found on the surface of the plant? Clearly the relationships between epiphytes and endophytes need to be clarified.

What Kinds of Interactions Occur on Plants?

Are Epiphytes Strictly Commensalistic on Plants?

It is generally assumed that except for pathogens, other residents of plants generally have a commensalistic relationship with plants. Webster's 1966 dictionary defines a commensal as "one living with, on, or in another, without injury to either". The ability of plant-associated microbes to injure a plant in the absence of obvious pathogenicity has not been addressed widely. There are, however, several phenotypes that epiphytic bacteria possess that might logically alter the plant in a way that might be somewhat injurious, but providing benefit to the bacteria. For example, many strains of *P. syringae* produce the phytotoxins syringomycin or coronatine, which are considered virulence factors in compatible plant interactions leading to disease (Bender *et al.,* 1999). The toxin affects ion transport across the plant cell plasma membrane by the formation of ion channels, and this ion flux leads to the release of metabolites from the plant cells and ultimately to cell lysis (Hutchinson *et al.*, 1995). However, significant levels of pore-forming activity in plant cells were detected at concentrations of syringomycin much lower than that required for measurable cell lysis (Hutchinson *et al.*, 1995). Non-pathogenic strains of *P. syringae* may synthesise syringomycin in amounts that are insufficient to cause cell necrosis and disease, but high enough to trigger the low release of plant metabolites. Might the production of this compound on plants alter the nutritional environment of the leaf surface? In addition to its effect on plant cells, syringomycin acts as a potent biosurfactant. As noted earlier, many epiphytes produce other biosurfactants that might alter the cuticular surface, thereby stimulating nutrient diffusion to the surface of the plant and enabling more efficient usage of those nutrients already present on leaves (Lindow and Brandl, 2003). Schreiber *et al.,* (2005) have recently described epiphytic bacteria that do, in fact, alter leaf surface permeability. While the mechanism of this effect is unknown and such strains are not pathogenic *per se*, they might be considered to injure the plant by facilitating loss of water and perhaps nutrients.

In contrast to the production of biosurfactants and syringomycin, which appear to be restricted to certain groups of bacteria, the biosynthesis of the plant growth regulator indole-3-acetic acid (IAA) is widespread among bacterial colonisers of the phyllosphere (Fett *et al.,* 1987; Glickmann *et al.,* 1988; Lindow *et al.,* 1998). Although bacterial production of auxin is a major pathogenicity determinant in many bacteria-inducing hyperplasia on plants, its role in other pathogenic and non-pathogenic plant-associated bacteria has not been well studied. Because IAA is involved in many aspects of plant development, it is of great interest that microbes that colonise plant surfaces have the ability to synthesise an auxin that is identical to that found in plants. IAA production was shown to benefit the epiphytic colonisation of pear and bean by *E. herbicola* (Brandl and Lindow, 1998). IAA promotes cell wall loosening at very low concentrations (Vanderhoff and Dute, 1981), and exogenously applied auxin stimulates the release of saccharides from the plant cell wall (Fry, 1989). Because bacteria are frequently nutrient-limited on plants, it seems possible that IAA-producing strains are more fit epiphytically due to enhanced nutrient availability caused by increased leakage of saccharides from plant cells in their vicinity. It seems likely that other microbes that alter the plant to change the surface environment to one that is more conducive to their colonisation will be found.

Do Multi-trophic Level Interactions Occur on Plants?

While protists are considered to be important predators of bacteria in soil and aquatic habits, their role in the phyllosphere has received very little attention. Since protozoans are generally considered to reside in aquatic habitats, and since free moisture is generally transient on plant surfaces, protists have been assumed to play little role in regulating bacterial numbers on plants. At least one protozoan has been shown to have a wide distribution on plants. *Colpoda cucullus* was found on 88% of the small herbaceous plants examined as well as on the bark of over 98% of the trees surveyed (Mueller and Mueller, 1970). This species was also commonly found in soil and other terrestrial habitats (Mueller and Mueller, 1970). It was found to encyst quite rapidly upon drying, as well as to undergo excystment rapidly upon rehydration (Mueller and Mueller, 1970). It is perhaps because of these traits that it may be well suited to the transient moisture conditions of plants. Except for this limited survey of the occurrence of this protist on plants, there was no information available about its potential to prey on bacteria on plant surfaces. A preliminary assessment was therefore made by Christina Chen, an undergraduate major in Microbial Biology at the University of California, Berkeley, of the ability of *C. cucullus* to consume cells of *P. syringae* on bean plants. Bean plants were inoculated with: 1) about 10^6 cells/ml of *P. syringae* B728a alone; 2) *P. syringae* mixed with about 1000 cells/ml of *C. cucullus* which had been grown on a diet of E. coli cells, or 3) *P. syringae* mixed with *C. cucullus* cells which were killed by exposure to 10 µg/ml cycloheximide. Cell suspensions were sprayed onto bean plants and population sizes of *P. syringae* B728a measured by dilution plating of leaf washings on Kings B medium containing 100 ug/ml rifampicin and 100 µg/ml cycloheximide. While population sizes of *P. syringae* increased rapidly with time when inoculated alone, population sizes of this bacterium exhibited little net growth and were about 100-fold lower than on plants inoculated with *P. syringae* alone after 72 h incubation when co-inoculated with *P. syringae* and *C. cucullus* (Fig. 1.1). *P. syringae* populations on plants inoculated also with killed *C. cucullus* were nearly as large as on plants inoculated with *P. syringae* alone (Fig. 1.1). These preliminary results suggest that protozoa such as *C. cucullus* are capable of consuming large numbers of bacteria on moist plants. Further work will be required to determine how rapidly they respond to fluctuating moisture conditions on plants and whether they can consume bacteria at reasonable rates under field conditions. Clearly this is an important and understudied area of phyllosphere microbiology.

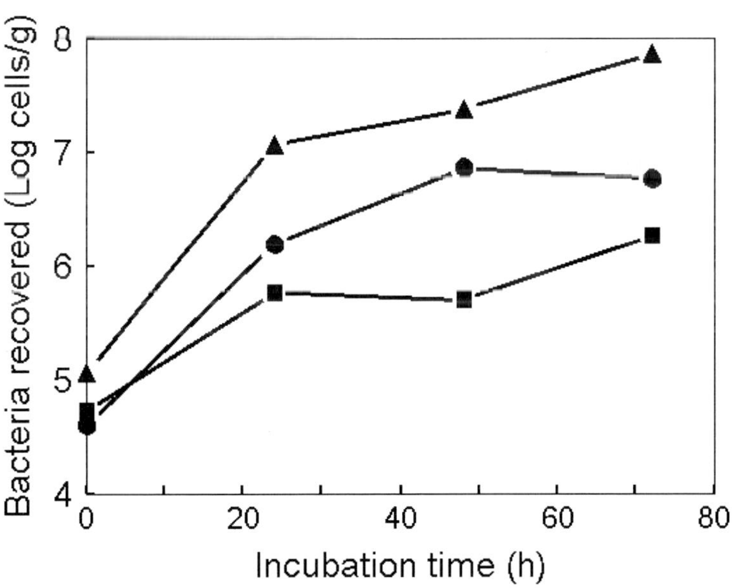

Fig. 1.1. Population size of *Pseudomonas syringae* B728a on bean leaves at various times after inoculation alone (triangles) or when inoculated with live *Colpoda cucullus* (squares) or *C. cucullus* that had been killed with cycloheximide (circles).

Bacteria and fungi both commonly occur together on healthy leaf surfaces. Many fungi have the potential to infect plants and withdraw nutrients directly from the plant interior, rather than relying on diffusion of nutrients to the plant surface. Many fungi also have large spores that support extensive proliferation of hyphae on plant surfaces after spores germinate after immigration to plants. While nutrients of plant origin that are present on leaves of many plant species are often present in low amounts (Mercier and Lindow, 2000), bacteria might be able to achieve higher population sizes if they were to parasitize other organisms such as fungi on plant surfaces rather than relying only on release of nutrients from the pant interior directly onto the phyllosphere. Fungi growing on a leaf surface thus represent a large potential source of nutrients for epiphytic bacteria. A potential novel, antagonistic interaction between the epiphytic bacterium *P. syringae* and filamentous fungi, mediated by a previously uncharacterised *P. syringae* protein, has been described (Wichmann and Lindow, 2004). This bacterial protein, termed Phc, has high similarity to Het-c from *Neurospora crassa*, a protein known to control fungal heterokaryon incompatibility (HI). HI is a programmed cell death pathway utilised by filamentous fungi during vegetative growth. Homologues of Het-c are also found in a wide variety of plant-associated fungi. It is hypothesized that by co-opting HI in filamentous fungi, *P. syringae* can kill fungal cells in order to obtain nutrients on leaves. Transformation of *N. crassa* with *phc* yields transformants with HI phenotypes indistinguishable from HI phenotypes seen after transformation with the appropriate fungal *het-c* control (Wichmann and Lindow, 2004). Further examination of the bacterial-fungal interaction has found that *P. syringae* can colonise *N. crassa* hyphae extensively, as well as use *N. crassa* as a nutrient source under nutrient limited conditions. In addition, the upstream regulatory region of *phc* contains a putative type III promoter element, suggesting that Phc could be delivered directly into fungal hypha (Wichmann and Lindow, 2004). This phenomenon thus appears to represent a new form of prokaryotic-eukaryotic interaction. While further work will be required to determine the fitness advantage conferred by *phc* and the extent to which *P. syringae* interacts with various fungal associates on leaves, these data should stimulate work to determine if parasitic interactions between bacteria and fungi might be common on plants.

Cell-Cell Communication of Microbes on Plants and Cross-Talk Among Epiphytes

The aggregated structure of *P. syringae* cells on plants, as discussed earlier, strongly affects their behaviour on plants. After several cycles of wet and dry conditions, most cells of *P. syringae* that were not in relatively large aggregates (> 100 cells) on leaves had succumbed to desiccation stress (Monier and Lindow, 2003). Such results suggest that cells within aggregates on leaves were more stress tolerant because they differ physiologically from more solitary cells, prompting an examination of the basis and consequences of cell density-dependent gene expression in *P. syringae.* Such a density-dependent behaviour would also suggest that there are ways that cells communicate their local cell density, and that there would be opportunities for bacteria to interact in ways that might exploit such communication.

There is growing appreciation that traits important to a variety of bacterial behaviours are regulated in a cell density-dependent manner by a process known as quorum sensing (Von Bodman *et al.,* 2003). In this process, bacteria assess their local population density by detecting the production of a diffusible signal molecule, which accumulates as population densities increase. *P. syringae* and several other gram-negative bacteria use members of a family of N-acyl homoserine lactones (AHL) as the signal molecule, expressing virulence factors and secondary metabolites that mediate a successful colonisation of a host only upon local accumulation of AHL. Initial characterisation of quorum sensing in *P. syringae* demonstrated that the production of the cell density-dependent AHL signal, 3-oxo-hexanoyl-homoserine lactone (3-oxo-C6-HSL), requires the expression of the AHL synthase gene, ahlI, and the AHL regulator gene, ahlR as well as a second regulator, AefR (Quiñones *et al.,* 2004). Furthermore, mutants deficient in production of the AHL signal molecule were less fit epiphytically than the parental strain (Quiñones *et al.,* 2004).

Quorum sensing controls expression of several traits that contribute to the epiphytic fitness and virulence of *P. syringae* (Quiñones *et al.,* 2005). The production of extracellular polysaccharides, resistance to reactive oxygen species such as hydrogen peroxide, and ability to produce macerated disease symptoms in inoculated bean pods were all positively regulated by the quorum sensing system in *P. syringae* (Quiñones *et al.,* 2005). In contrast swarming motility is suppressed by quorum sensing in *P. syringae* (Quiñones *et al.,* 2005). The various behaviours exhibited by *P. syringae* in a cell density-dependent fashion are sensible when one considers the nature of the leaf surface and the process of immigration of inoculum to a plant. Upon immigration of single cells of *P. syringae* to the leaf surface, the likelihood of their landing on or near an area of abundant nutrients conducive to growth is low due to the heterogeneity of the leaf surface. It thus would benefit the cell to move about on the leaf either to a site where invasion of the leaf could occur or to a surface nutrient oasis. Such solitary cells would not be in a "quorum-sensing state" and hence would not be repressed for motility. A cell fortunate enough to find such a site would proliferate and form an aggregate; further movement away from the site into the surrounding nutritional desert would be expected to be detrimental to the bacterium. Upon reaching a quorum in such an aggregate, motility would be suppressed. Subsequent EPS production would further hamper movement of the cells, and other stress tolerance traits might be expressed to enable the survival of the assemblage. Given that many plant-associated bacteria produce AHL signal molecules (Cha *et al.,* 1998), it seems likely that at least some of the traits that affect their epiphytic fitness are also controlled in a density-dependent fashion as in the case of *P. syringae.* Much more work will be needed to understand how commonly bacteria exhibit such quorum-sensing dependent behaviours.

The diverse microbial communities on plants may affect quorum sensing of the epiphytic inhabitants via chemical cross-talk. Preliminary studies have found that as many as 18% of culturable epiphytic bacteria from leaves produce small diffusible molecules that can interfere with quorum sensing in *P. syringae* (Dulla *et al.,* 2004). About 7% of these bacterial epiphytes produce the same 3-oxo-C6-AHL, often in amounts far higher than *P. syringae.* Given that quorum-sensing controls both epiphytic fitness and virulence of *P. syringae,* its confusion by interference with signals by other community members may lead to a new paradigm in biological control of plant pathogens. Preliminary results have revealed that, compared to bean plants inoculated with *P. syringae* alone, the number and size of lesions formed on plants co-inoculated with *P. syringae* and AHL-producing strains were smaller. Conversely, co-inoculation of bean plants with *P. syringae* and quorum-sensing inhibiting strains increased the size and numbers of lesions. These results suggest strongly that bacterial epiphytes can alter the signal molecule abundance in their vicinity, thereby altering the behaviour of neighbouring bacterial strains whose behaviour is controlled by such signals. While more work is needed to determine the ubiquity and magnitude of effects of chemical cross-talk, such communication may prove to be an important, if previously unrecognised, phenomenon that affects the behaviour of microbes in the phyllosphere.

References

Andrews, J.H. and Harris, R.F. (2000) The ecology and biogeography of microorganisms on plant surfaces. *Annual Review of Phytopathology* 38, 145-180.

Andrews, J.H., Kinkel, L.L., Berbee, F.M. and Nordheim, E.V. (1987) Fungi, leaves, and the theory of island biogeography. *Microbial Ecol*ogy 14, 277-290.

Axtell, C.A. and Beattie, G.A. (2002) Construction and characterization of a *proU-gfp* transcriptional fusion that measures water availability in a microbial habitat. *Applied and Environmental Microbiology* 68, 4604-4612.

Ayers, P.G., Gunasekera, T.S., Rasanayagam, M.S. and Paul, N.D. (1996) Effects of UV-B radiation (280-320 nm) on foliar saprotrophs and pathogens. In: Gadd, G.M. and Frankland, J.C. (ed.) *Fungi and Environmental Change*. Cambridge University Press, Cambridge, UK. pp. 32-50.

Bailey, M.J., Rainey, P.B., Zhang, X.-X. and Lilley, A.K. (2002) Population dynamics, gene transfer and gene expression in plasmids, the role of the horizontal gene pool in local adaptation at the plant surface. In: Lindow, S.E., Hecht-Poinar, E.I. and Elliott, V.J. (eds) *Phyllosphere Microbiology*. APS Press, St. Paul, Minnesota, USA. pp. 173-192.

Bashi, E. and Fokemma, N.J. (1977) Environmental factors limiting growth of *Sporobolomyces roseus* on wheat leaves. *Transactions of the British Mycological Society* 68, 17-25.

Beattie, G.A. and Lindow, S.E. (1994a) Survival, growth, and localization of epiphytic fitness mutants of *Pseudomonas syringae* on leaves. *Applied and Environmental Microbiology* 60, 3790-3798.

Beattie, G.A. and Lindow, S.E. (1994b) Comparison of the behavior of epiphytic fitness mutants of *Pseudomonas syringae* under controlled and field conditions. *Applied and Environmental Microbiology* 60, 3799-3808.

Beattie, G.A. and Lindow, S.E. (1995) The secret life of foliar bacterial pathogens on leaves. *Annual Review of Phytopathology* 33, 145-172.

Beattie, G.A. and Lindow, S.E. (1999) Bacterial colonization of leaves: a spectrum of strategies. *Phytopathology* 89, 353-359.

Beattie, G.A. and Marcell, L.M. (2002) Effect of alterations in cuticular wax biosynthesis on the physicochemical properties and topography of maize leaf surfaces. *Plant Cell and Environment* 25, 1-16.

Bender, C.L. Alarcon-Chaidez, F. and Gross, D.C. (1999) *Pseudomonas syringae* phytotoxins: mode of action, regulation, and biosynthesis by peptide and polyketide synthetases. *Microbiology and Molecular Biology Reviews* 63, 266-92.

Boureau, T., Routtu, J., Roine, E., Taira, S. and Romantschuk, M. (2002) Localization of *hrpA*-induced *Pseudomonas syringae* pv. tomato DC3000 in infected tomato leaves. *Molecular Plant Pathology* 3, 451-460.

Brandl, M.T. and Lindow, S.E. (1997) Environmental signals modulate the expression of an indole-3-acetic acid biosynthetic gene in *Erwinia herbicola*. *Molecular Plant-Microbe Interactions* 10, 499-505.

Brandl, M.T. and Lindow, S.E. (1998) Contribution of indole-3-acetic acid production to the epiphytic fitness of *Erwinia herbicola*. *Applied and Environmental Microbiology* 64, 3256-63.

Brandl, M.T. and Mandrell, R.E. (2002) Fitness of *Salmonella enterica* serovar Thompson in the cilantro phyllosphere. *Applied and Environmental Microbiology* 68, 3614-3621.

Brandl, M.T., Quinones, B. and Lindow, S.E. (2001) Heterogeneous transcription of an indoleacetic acid biosynthetic gene in *Erwinia herbicola* on plant surfaces. *Proceedings of the National Academy of Sciences, USA* 98, 3454-3459.

Bunster, L., Fokkema, N.J. and Schippers, B. (1989) Effect of surface activity of *Pseudomonas* spp. on leaf wettability. *Applied and Environmental Microbiology* 55, 1340-1345.

Cha, C., Gao, P., Chen, Y.-C., Shaw, P.D. and Farrand, S.K. (1998) Production of acyl-homoserine lactone quorum-sensing signals by gram-negative plant-associated bacteria. *Molecular Plant-Microbe Interactions* 11, 1119-1129.

Dickinson, C.H., Austin, B. and Goodfellow, M. (1975) Quantitative and qualitative studies of phylloplane bacteria from *Lolium perenne*. *Journal of General Microbiology* 91, 157-166.

Dik, A.J., Fokkema, N.J. and Van Pelt, J.A. (1991) Consumption of aphid honeydew, a wheat yield reduction factor, by phyllosphere yeasts under field conditions. *Netherlands Journal of Plant Pathology* 97, 209-232.

Dik, A.J., Fokkema, N.J. and Van Pelt, A. (1992) Influence of climatic and nutritional factors on yeast population dynamics in the phyllosphere of wheat. *Microbial Ecology* 23, 41-52.

Dulla, G., Quinones, B. and Lindow, S.E. (2004) Quorum sensing contributes to virulence and epiphytic fitness of *Pseudomonas syringae*. *Phytopathology* 94, S27.

Ercolani, G.L. (1978) *Pseudomonas savastanoi* and other bacteria colonizing the surface of olive leaves in the field. *Journal of General Microbiology* 109, 245-257.

Ercolani, G.L. (1991a) Distribution of epiphytic bacteria on olive leaves and the influence of leaf age and sampling time. *Microbial Ecology* 21, 35-48.

Ercolani, G.L. (1991b) Factor analysis of fluctuations in profiling characteristics of the olive epiphyllous bacteria. *Annali di Microbiologia Ed Enzimologia* 41, 17-26.

Ercolani, G.L. (1994) Changes in the wettability of olive leaves colonized by epiphytic bacteria. *Annali di Microbiologia Ed Enzimologia* 44, 85-95.

Ercolani, G.L. (1995) Factor analysis of *Pseudomonas syringae* pv. *savastanoi* on the phylloplane of the olive. *Microbial Ecology* 11, 41-49.

Fett, W.F., Osman, S.F. and Dunn, M.F. (1987) Auxin production by plant pathogenic pseudomonads and xanthomonads. *Applied and Environmental Microbiology* 53, 1839-1845.

Fokkema, N.J. (1971) The effect of pollen in the phyllosphere of rye on colonization by saprophytic fungi and infection by *Helminthosporium sativum* and other leaf pathogens. *Netherlands Journal of Plant Pathology* 77, 3-60.

Fry, S.C. (1989) Cellulases, hemicelluloses and auxin-stimulated growth: a possible relationship. *Physiologia Plantarum* 75, 532-536.

Giesler, L.J., Yuen, G.Y. and Horst, G.L. (2000) Canopy microenvironments and applied bacterial population dynamics in shaded tall fescue. *Crop Science* 40, 1325-1332.

Glickmann, E., Gardan, L., Jacquet, S., Hussain, S., Elasri, M., Petit, A. and Dessaux Y. (1988) Auxin production is a common feature of most pathovars of *Pseudomonas syringae*. *Molecular Plant-Microbe Interactions* 11, 156-162.

Hallmann, J., Quadt-Hallmann, A., Mahaffee, W.F. and Kloepper, J.W. (1997) Bacterial endophytes in agricultural crops. *Canadian Journal of Microbiology* 43, 895-914.

Hirano, S.S. and Upper, C.D. (2000) Bacteria in the leaf ecosystem with emphasis on *Pseudomonas syringae*: a pathogen, ice nucleus, and epiphyte. *Microbiology and Molecular Biology Reviews* 64, 624-653.

Hirano, S.S., Rouse, D.I., Clayton, M.K. and Upper, C.D., (1995) *Pseudomonas syringae* pv. *syringae* and bacterial brown spot of bean: a study of epiphytic phytopathogenic bacteria and associated disease. *Plant Disease* 79,1085-1093.

Hughes, T.P. (1990) Recruitment limitation, mortality, and population regulation in open systems: a case study. *Ecology* 71, 12-20.

Hutchison, M.L., Tester, M.A. and Gross, D.C. (1995) Role of biosurfactant and ion channel-forming activities of syringomycin in transmembrane ion flux: a model for the mechanism of action in the plant-pathogen interaction. *Molecular Plant-Microbe Interactions* 8, 610-20.

Jacobs, J.L. and Sundin, G.W. (2001) Effect of solar UV-B radiation on a phyllosphere bacterial community. *Applied and Environmental Microbiology* 67, 5488-5496.

Jacques, M.-A., Kinkel, L.L. and Morris, C.E. (1995) Population sizes, immigration and growth of epiphytic bacteria on leaves of different ages and positions of field-growth endive (*Cichorium endiva* var. *latifolia*). *Applied and Environmental Microbiology* 61, 899-906.

Jacques, M.-A. and Morris, C.E. (1995) A review of issues related to the quantification of bacteria from the phyllosphere. *FEMS Microbiology Ecology* 18, 1-14.

Jeffree, C.E. (1996) Structure and ontogeny of plant cuticles. In: Kerstiens, G. (ed.) *Plant Cuticles; an Integrated Functional Approach.* BIOS Scientific Publishers, Oxford, UK. pp. 33-82.

Johnson, K.B. and Stockwell, V.O. (1998) Management of fire blight: a case study in microbial ecology. *Annual Review of Phytopathology* 36, 227-248.

Johnson, K.B., Stockwell, V.O., Sawyer, T.L. and Sugar, D. (2000) Assessment of environmental factors influencing growth and spread of *Pantoea agglomerans* on and among blossoms of pear and apple. *Phytopathology* 90, 1285-1294.

Joyner, D.C. and Lindow, S.E. (2000). Heterogeneity of iron bioavailability on plants assessed with a whole-cell GFP-based bacterial biosensor. *Microbiology (Reading)* 146, 2435-2445.

Kadivar, H. and Stapleton, A.E. (2003). Ultraviolet radiation alters maize phyllosphere bacterial diversity. *Microbial Ecology* 45, 353-361.

Kinkel, L.L. (1991) Fungal community dynamics. In: Andrews, J.H. and Hirano, S.S (eds) *Microbial Ecology of Leaves*. Springer-Verlag, New York, USA. pp. 253-270.

Kinkel, L.L. (1997) Microbial population dynamics on leaves. *Annual Review of Phytopathology* 35, 327-347.

Kinkel, L.L., Andrews, J.H. and Nordheim, E.V. (1989) Fungal immigration dynamics and community development on apple leaves, *Microbial Ecology* 18, 45-58.

Kinkel, L.L., Wilson, M. and Lindow, S.E. (2000) Biological and physical factors influencing variability in epiphytic bacterial population size. *Microbial. Ecology* 39, 1-11.

Kobayashi, D.Y. and Palumbo, J.D. (2000) Bacterial endophytes and their effects on plants and uses in agriculture. In: Bacon, C.W. and White, J.F. Jr. (eds) *Microbial Endophytes*. Marcel Dekker, Inc., New York, USA. pp. 199-233.

Knoll, D. and Schreiber, L. (1998) Influence of epiphytic micro-organisms on leaf wettability: wetting of the upper leaf surface of *Juglans regia* and of model surfaces in relation to colonization by micro-organisms. *New Phytologist* 140, 271-282.

Knoll, D. and Schreiber, L. (2000) Plant-microbe interactions: wetting of ivy (*Hedera helix* L.) leaf surfaces in relation to colonization by epiphytic microorganisms. *Microbial Ecology* 40, 33-42.

Leveau, J.H.J. and Lindow, S.E. (2001) Appetite of an epiphyte: Quantitative monitoring of bacterial sugar consumption in the phyllosphere. *Proceedings of the National Academy of Sciences, USA* 98, 3446-3453.

Lindow, S.E. (1982) Population dynamics of epiphytic ice nucleation active bacteria on frost sensitive plants and frost control by means of antagonistic bacteria. In: Li, P.H. and Sakai, A. (eds) *Plant Cold Hardiness*. Academic Press, New York, USA. pp. 395-416.

Lindow, S.E. (1996) Role of immigration and other processes in determining epiphytic bacterial populations: implications for disease management. In: Morris, C.E., Nicot, P.C. and Nguyen-the, C. (eds) *Aerial Plant Surface Microbiology*. Plenum Publishing Co., New York, USA. pp. 155-168.

Lindow, S.E. and Andersen, G.A. (1996) Influence of immigration in establishing epiphytic bacterial populations on navel orange. *Applied and Environmental Microbiology* 62, 2978-2987.

Lindow, S.E. and Brandl, M.T. (2003) Microbiology of the phyllosphere. *Applied and Environmental Microbiology* 69, 1875-1883.

Lindow, S.E. and Suslow, T.V. (2003) Temporal dynamics of the biological control agent *Pseudomonas fluorescens* strain A506 in flowers in inoculated pear trees. *Phytopathology* 93, 727-737.

Lindow, S.E., Andersen, G. and Beattie, G. (1993) Characteristics of insertional mutants of *Pseudomonas syringae* with reduced epiphytic fitness. *Applied and Environmental Microbiology* 59, 1593-1601.

Lindow, S.E., Arny, D.C. and Upper, C.D. (1978) Distribution of ice nucleation active bacteria on plants in nature. *Applied and Environmental Microbiology* 36, 831-838.

Lindow, S.E., Desurmont, C., Elkins, R., McGourty, G., Clark, E. and Brandl, M.T. (1998) Occurrence of indole-3-acetic acid-producing bacteria on pear trees and their association with fruit russet. *Phytopathology* 88, 1149-1157.

Lindow, S.E. and Leveau, J.H.J. (2002) Phyllosphere microbiology. *Current Opinion in Biotechnology* 13, 238-243.

Marcell, L.M. and Beattie, G.A. (2002) Effect of leaf surface waxes on leaf colonization by *Pantoea agglomerans* and *Clavibacter michiganensis*. *Molecular Plant-Microbe Interactions* 12, 1236-1244.

Mercier, J. and Lindow, S.E. (2000) Role of leaf surface sugars in colonization of plants by bacterial epiphytes. *Applied and Environmental Microbiology* 66, 369-374.

Miller, W.G., Brandl, M.T., Quinones, B. and Lindow, S.E. (2001) Biological sensor for sucrose availability: relative sensitivities of various reporter genes. *Applied and Environmental Microbiology* 67, 1308-1317.

Monier, J.-M. and Lindow, S.E. (2003) Differential survival of solitary and aggregated bacterial cells promotes aggregate formation on leaf surfaces. *Proceedings of the National Academy of Sciences, USA* 100, 15977-15982.

Monier, J.-M. and Lindow, S.E. (2004). Frequency, size and localization of bacterial aggregates on bean leaf surfaces. *Applied and Environmental Microbiology* 69, 346-355.

Morris, C. E., Monier, J.-M. and Jacques, M.-A. (1997) Methods for observing microbial biofilms directly on leaf surfaces and recovering them for isolation of culturable microorganisms. *Applied and Environmental Microbiology* 63, 1570-1576.

Mueller, J.A. and Mueller, W.P. (1970) *Colpoda cucullus*: a terrestrial aquatic. *The American Midland Naturalist* 84, 1-12.

Newsham, K.K., Low, M.N.R., McLeod, A.R., Greenslade, P.D. and Emmett, B.A. (1997) Ultraviolet-B radiation influences the abundance and distribution of phylloplane fungi on pedunculate oak (*Quercus robur*). *New Phytologist* 138, 287-297.

Ngugi, H.K and Scherm, H. (2004) Pollen mimicry during infection of blueberry flowers by conidia of *Monilinia vaccinii-corymbosi. Physiological and Molecular Plant Pathology* 64, 113-123.

Ngugi, H.K., Scherm, H. and Lehman, J.S. (2002) Relationships between blueberry flower age, pollinaton, and conidial infection by *Monilinia vaccinii-corymbosi. Phytopathology* 92, 1104-1109.

Nuclo, R.L., Johnson, K.B. Stockwell,V.O. and Sugar, D. (1998) Secondary colonization of pear blossoms by two bacterial antagonists of the fire blight pathogen. *Plant Disease* 82, 661-668.

O'Brien, R.D. and Lindow, S.E. (1989) Effect of plant species and environmental conditions on epiphytic population sizes of *Pseudomonas syringae* and other bacteria. *Phytopathology* 79, 619-627.

Pusey, P.L. and Curry, E.A. (2004) Temperature and pomaceous flower age related to colonization by *Erwinia amylovora* and antagonists. *Phytopathology* 94, 901-911.

Quinones, B., Pujol, C.J. and Lindow, S.E. (2004) Regulation of AHL production and its contribution to epiphytic fitness in *Pseudomonas syringae. Molecular Plant-Microbe Interactions* 17, 521-531.

Quinones, B., Dulla, G. and Lindow, S.E. (2005) Quorum sensing regulates exopolysaccharide production, motility, and virulence in *Pseudomonas syringae. Molecular Plant-Microbe Interactions* 18, 682-693.

Reiderer, M. and Schreiber, L. (2001) Protecting against water loss: analysis of the barrier properties of plant cuticles. *Journal of Experimental Botany* 52, 2023-2032.

Sabaratnam, S. and Beattie, G.A. (2003) Differences between *Pseudomonas syringae* pv. *syringae* B728a and *Pantoea agglomerans* BRT98 in epiphytic and endophytic colonization of leaves. *Applied and Environmental Microbiology* 69, 1220-1228.

Schönherr, J. and Baur, P. (1996) Cuticle permeability studies: a model for estimating leaching of pant metabolites to leaf surfaces. In: Morris, C.E. Nicot, P.C. and Nguyen-The C. (ed.) *Aerial Plant Surface Microbiology*, Plenum Press, New York, USA. pp. 1-23.

Schreiber, L. (1996) Wetting of the upper needle surface of *Abies grandis*: influence of pH, wax chemistry and epiphyllic microflora on contact angles. *Plant, Cell and Environment* 19, 455-463.

Schreiber, L., Krimm, U., Knoll, D., Sayed, M., Auling, G. and Kroppenstedt, R.M. (2005) Plant-microbe interactions: identification of epiphytic bacteria and their ability to alter leaf surface permeability. *New Phytologist* 166, 589-594.

Schreiber, L., Skrabs, M., Hartmann, K.D., Diamantopoulos, P., Simanova, E. and Santrucek, J. (2001) Effect of humidity on cuticular water permeability of isolated cuticular membranes and leaf disks. *Planta* 214, 274-282.

Shulski, M.D., Walter-Shea, E.A., Hubard, K.G., Yuen, G.Y. and Horst, G. (2004) Penetration of photosynthetically active and ultraviolet radiation into alfalfa and tall fescue canopies. *Agronomy Journal* 96, 1562-1571.

Stockwell, V.O., McLaughlin, R.J., Henkels, M.D., Loper, J.E., Sugar, D. and Roberts, R.G. (1999) Epiphytic colonization of pear stigmas and hypanthia by bacteria during primary bloom. *Phytopathology* 89, 1162-1168.

Stout, J.D. (1960) Bacteria of soil and pasture leaves at Claudelands showgrounds. *New Zealand Journal of Agricultural Research* 3, 413-430.

Sundin, G.S. (2002) Ultraviolet radiation on leaves: its influence on microbial communities and their adaptations. In: Lindow, S.E., Hecht-Poinar, E.I. and Elliott, V.J. (eds) *Phyllosphere Microbiology*. APS Press, St. Paul, Minnesota, USA. pp. 27-42.
Sundin, G.W. and Jacobs, J.L. (1999) Ultraviolet radiation (UVR) sensitivity analysis and UVR survival strategies of a bacterial community from the phyllosphere of field-grown peanut (*Arachis hypogeae* L.). *Microbial Ecology* 38, 27-38.
Thomson, S.V. and Gouk, S.C. (2003) Influence of age of apple flowers on growth of *Erwinia amylovora* and biological control agents. *Plant Disease* 87, 502-509.
Thompson, I. P., Bailey, M.J., Ellis, R.J., Lilley, A.K., McCormack, P.J., Purdy, K.J. and Rainey, P.B. (1995) Short-term community dynamics in the phyllosphere microbiology of field-grown sugar beet. *FEMS Microbiology Ecology* 16, 205-211.
Thompson, I. P., Bailey, M.J. Fenlon, J.S., Fermor, T.R., Lilley, A.K., Lynch, J.M., McCormack, P.J., McQuilken, M.P. and Purdy, K.J. (1993) Quantitative and qualitative seasonal changes in the microbial community from the phyllosphere of sugar beet (*Beta vulgaris*). *Plant and Soil* 150, 177-191.
Upper, C.D., Hirano, S.S., Dodd, K.K. and Clayton, M.K. (2003) Factors that affect spread of *Pseudomonas syringae* in the phyllosphere. *Phytopathology* 93, 1082-1092.
Vanderhoff, L.N. and Dute, R.R. (1981) Auxin-regulated wall loosening and sustained growth in elongation. *Plant Physiology* 67, 146-149.
Von Bodman, S.B., Bauer, W.D. and Coplin, D.L. (2003) Quorum sensing in plant-pathogenic bacteria. *Annual Review of Phytopathology* 41, 455-482.
Walcott, R.R., Gitaitis, R.D. and Castro, A.C. (2003) Role of blossoms in watermelon seed infestation by *Acidovorax avenae* subsp. *citrulli*. *Phytopathology* 93, 528-534.
Wichmann, G. and Lindow, S.E. (2004) War between kingdoms: Induction of programmed cell death in filamentous fungi by *Pseudomonas syringae*. *Phytopathology* 94, S110.
Wilson, M. and Lindow, S.E. (1992) Relationship of total and culturable cells in epiphytic populations of *Pseudomonas syringae*. *Applied and Environmental Microbiology* 58, 3908-3913.
Wilson, M. and Lindow, S.E. (1994a) Coexistence among epiphytic bacterial populations mediated through nutritional resource partitioning. *Applied and Environmental Microbiology* 60, 4468-4477.
Wilson, M. and Lindow, S.E. (1994b) Ecological differentiation and coexistence between epiphytic Ice^+ *Pseudomonas syringae* strains and an Ice^- biological control agent. *Applied and Environmental Microbiology* 60, 3128-3137.
Wilson, M., Hirano, S.S. and Lindow, S.E. (1999) Location and survival of leaf-associated bacteria in relation to pathogenicity and potential for growth within the leaf. *Applied and Environmental Microbiology* 65, 1435-1443.
Wilson, M. and Lindow, S.E. (1999) Viable but nonculturable cells in plant-associated bacterial populations. In: Colwell, R.R. and Grimes, D.J. (eds) *Nonculturable microorganisms in the environment*. ASM Press, Washington DC., USA. pp. 229-241.
Yang, C.-H., Crowley, D.E, Borneman, J. and Keen, N.T. (2001) Microbial phyllosphere populations are more complex than previously realized. *Proceedings of the National Academy of Sciences, USA* 98, 3889-3894.
Yuen, G.Y., Jochum, C.C., Giesler, L.J., Shulski, M.D., Walter-Shea, E.A., Hubbard, K.G. and Horst, G.L. (2002) UV-B Biodosimetry in turfgrass canopies. *Crop Science* 42, 859-868.
Zhang, S. and Sundin, G.W. (2004). Long-term effect of mutagenic DNA repair on accumulation of mutations in *Pseudomonas syringae* B86-17. *Journal of Bacteriology* 186, 7807-7810.

Chapter 2

Microbial Diversity in the Phyllosphere and Rhizosphere of Field Grown Crop Plants: Microbial Specialisation at the Plant Surface

T.M. Timms-Wilson[1], K. Smalla[2], T.I. Goodall[1] A. Houlden[1,3], V. Gallego[1,4] and M.J. Bailey[1]
[1]Centre for Ecology and Hydrology, Mansfield Road, Oxford, OX1 3SR, UK; [2]Institute for Plant Virology, Microbiology and Biosafety, Federal Biological Research Centre for Agriculture and Forestry, Messeweg 11/12, D-38104 Braunschweig, Germany; [3]Department of Animal and Plant Sciences, The University of Sheffield, Western Bank, Sheffield, South Yorkshire S10 2TN, UK; [4]Department of Microbiology and Parasitology, Faculty of Pharmacy, University of Seville, 41012, Seville, Spain
tmt@ceh.ac.uk

Introduction

The phyllosphere is broadly defined as the surfaces and internal parts of the aerial structures of plants, including flowers, fruits, stems and leaves. Specialised microbial colonists, phytopathogens, spoilage organisms and periodic immigrants have all been described as residents of this physically diverse habitat. As plants represent one of the most important features of our landscape and our primary food source it is somewhat surprising that only a limited number of detailed investigations have been conducted that describe this above-ground microbiology. This is in contrast to the attention that soil rhizosphere systems receive. Pertinent investigations reveal a diverse and specialised microbial community that is distinct from the rhizosphere. Bacteria are by far the most numerically abundant colonisers; typical community densities up to of 2 x10^7 cells per cm^2 of leaf surface have been recorded by traditional methods (Andrews and Harris, 2000). Although given the varied nature of the environment that aerial portions of plants provide numbers can vary from 10s to in excess of 1 x 10^{12} in arid and senescing leaves respectively.

Densities also vary over the plant surface. In leaf buds or on young emerging leaves densities of bacteria may be two orders of magnitude greater than those established in mature leaves. These differences can in part be explained by the increased availability of water, as these regions are often protected or collect moisture derived from rainfall or dew. In addition nutrient loss from young, developing leaves provides suitable carbon (C) and nitrogen (N) sources for colonising bacteria. Colonising bacteria typically form complex mixed assemblages or biofilms that facilitate survival. Microbes must demonstrate considerable genotypic and phenotypic flexibility in their response to the fluctuating conditions at the leaf surface, and recent studies confirm that they carry a number of novel genes that are specifically induced in response to local conditions in the phyllosphere. Conditions in the phyllosphere vary diurnally and over the longer term, as leaves mature their permeability often changes as the protective waxy cuticle develops and nutrients become scarce. The production of assemblages or biofilms is thought to aid survival by protecting cells from desiccation and they provide suitable hot-spots for cellular activity and the horizontal transfer of mobile genetic elements which carry unique traits that facilitate further local adaptation of cells and populations to this specialised niche.

The source of phyllosphere or leaf surface colonists remains somewhat obscure, and the role of immigrants and founder populations has yet to be clearly demonstrated under field conditions for non-pathogenic bacteria or fungi. But the leaf surface supports a diverse collection of specialist microorganisms that have evolved to exploit this environment. In studies undertaken to describe the microbiology of emerging seedlings the bacteria identified are generally typical of those also found in the spermosphere and rhizosphere. But as plants develop so too does the phyllosphere community, it becomes more distinct, both in terms of relative species abundance and community complexity when compared to the below-ground environment associated with the bulk soil or the root surface (rhizosphere). For example, common rhizosphere and soil associated bacteria such as those of the genera *Rhizobium* or *Azospirillum* do not effectively colonise leaves (Jurkevitch and Shapira 2000, O'Brien and Lindow, 1989) and with few exceptions (Thompson *et al.,* 1995), plant growth promoting bacteria (PGPB), particularly *Pseudomonas* and *Xanthomonas* isolates, establish very limited and transient population densities on undamaged, mature leaves when compared to roots following seed or soil inoculation (Espinosa-Urgel *et al.,* 2002). Even the direct topical application of bacterial suspensions rarely establishes large persisting densities on leaves (Hirano and Upper 2000). These limited successes may be due to the laboratory conditions selected for the growth of the inocula although it is apparent that such strategies have proved useful in the delivery of biological control strains that have proved effective in reducing insect pest damage by *Bacillus thuringiensis*, the use of *Klebsiella* spp. or *Beijerinckia* spp. to augment nitrogen fixing capability, bacteria and bacterial consortia applied as biofertilisers and for combating diseases such as fire-blight or reducing frost damage following the introduction of inocula that out compete the natural pathogen (Lindow 1987).

In the study by Lilley *et al.,* an unusual approach was followed to monitor the colonisation, survival and transfer of phyllosphere bacteria from plant to plant. In this example the direct application of inocula to leaves resulted in the establishment of active, colonising populations of bacteria that persisted on the leaf surface. Lilley *et al.* (1997) followed the point source transfer of an established phyllosphere bacterial population on sugar beet, a natural isolate *Pseudomonas fluorescens* SBW25. Effective bacterial transfer from leaf to leaf was observed from established populations of bacteria to the emerging leaves of adjacent field plants grown in the open under field conditions. Direct leaf-to-leaf physical contact, rain splash (real or simulated) or direct inoculation of suspensions of bacteria grown in microbiological media were not effective and only resulted in a transient "contamination". However, seed inoculation at the time of planting resulted in the successful and reproducible establishment of colonising populations of SBW25 on leaves and roots of sugar beet. When the phytophageous lepidopteran pest *Mamestra brassicae* was introduced at high densities to these plants we observed rapid defoliation and transfer of bacterial inocula to adjacent plants (including those that had not received an original seed inoculation with SBW25). It was observed that the third to fourth instar lepidopteran larvae passively transferred the inocula to adjacent plants, particularly to the heart of the sugar beet where new leaf sets emerged. These were rapidly colonised at an increased density from the mature leaves of the "donor" plant and established populations equivalent to the normal carrying capacity and phyllosphere density observed for the seed inoculated SBW25 (Bailey and Lilley, 2002). This is one example of "natural" conditions of transfer. It was assumed that the bacteria were carried on the surface of the insects as SBW25 does not survive passage though the insect gut. It was also assumed that the insect frass provided additional nutrients that aided bacterial survival, proliferation and hence pre-emptive colonisation that excluded other immigrants that arrived from the soil, air currents or other methods of dissemination. The assessment of the true extent of dispersal and distribution that results in the establishment of "natural populations" at the plant surface remains a difficult area to study. For the most part these studies are limited by the practical restraints associated with the collection of representative samples of sufficient magnitude to allow assessment, and the sensitivity of the analytical procedures required to determine habitat specific phenotype (Jacques and Morris, 1995).

Despite current limitations to the assessment of the genetic and ecological factors that contribute to successful colonisation of the phyllosphere, methods for evaluating the diversity, distribution and relative abundance of microorganisms have advanced significantly over the last decade. Molecular techniques have been employed for the identification of isolates and communities, certain methods produce finger prints or barcodes of the accessed diversity. This approach allows comparisons to be made between plant types, leaf types, growing seasons and assessments of potential impacts such as pollution events or climate change. Methods regularly applied to study microbial diversity and functionality also include community level physiological profiling (CLPP) on BIOLOG plates that contain a range of sole carbon sources. When samples are compared an assessment can be made of the substrate utilisation capacity of the sampled habitat, this is a valuable tool for evaluating change or the physiological state of the community. The total culturable populations can also be determined using traditional microbiological methods such as the assessment of the number of colony forming units on agars, which can be general or selective. The application of molecular methods for strain characterisation or the use of PCR based techniques allows the rapid and detailed analyses of total extracted communities. These samples can be analysed to assess the presence of a particular gene, possibly associated with a known function or as a diagnostic to identify a plant pathogen. However analysis of the sampled species diversity is perhaps the most common application where variation in the sequence of ribosomal RNA or the ribosomal RNA gene (DNA) is used to determine phylogenetic relationships against well-established data bases that include 16S ribosomal RNA gene sequence for bacteria and 18S ribosomal RNA gene sequence for fungi. These methods rely on the use of highly conserved oligonucleotide primers for general description or the use and application of group- or taxon-specific oligonucleotides. The advantage is that total cell extracts or total nucleic extracts can be examined to estimate diversity or relative abundance without the need to isolate bacterial colonies by traditional methods.

Molecular techniques have revolutionised our ability to describe the diversity of communities, predominantly on the basis of 16S ribosomal RNA gene phylogeny. An advantage of molecular techniques is that they can be developed to analyse the total nucleic acid fraction (DNA and/or RNA) extracted directly from the environment. Methods include reverse array hybridisation and clone library screening, as well as the well described PCR-based community profiling techniques such as denaturing gradient gel electrophoresis (DGGE) and temperature gradient gel electrophoresis (TGGE) (Muyzer *et al.,* 1993; Heuer and Smalla, 1999; Yang; *et al.,* 2001). These approaches are now highly reproducible within and between samples. However detailed inventories that allow identification and relative enumeration of individuals in a sampled community require more analysis. This can be achieved by the production of clone libraries of ribosomal RNA genes, or other phylogenetic targets, cDNA libraries of extracted ribosomal RNA or the use of microarrays or phylochips as described by Andersen *et al.,* in Chapter 18 of this book. To identify component taxa can be time consuming and expensive in resources and technically demanding. Sensitivity depends on the analytical facilities available but DGGE (TGGE/tRFLP), cloning and phylochips will readily detect component populations that represent considerably less than 0.1% of the sample. This sensitivity can be increased with the use of target specific probes to reach sensitivities of 10 or less cells per sample. Indeed, the *in situ* analysis of individual cells (and their identification) is also possible for specific groups or even species by the use of fluorescent *in situ* hybridisation (FISH) and variants that determine distribution at the microscale (relative to the body size of the bacteria) and critically to assess activity based on either reporter gene expression (see Lindow and Mornier chapters) or related methods that detect gene expression and cellular activity such as substrate-tracking autoradiographic fluorescent *in situ* hybridisation (STARFISH), (Ouverney and Ruhrman 1999; Gray *et al.,* 2000) or *in situ* RT-PCR (Wieland 2001).

In this chapter we provide working examples of the value of molecular methodologies for assessing microbial population diversity and ecology in three crop plants grown under field conditions. These data reveal the extent of habitat specialisation in these communities, how they distinguish plant type, and provide a demonstration of the dynamic nature of these communities as they respond to changes in local conditions and plant maturation over the growing season. To illustrate the need for detailed assessments we have used DGGE to produced "diversity" assessments and analysed the resultant replicated profiles to demonstrate changes in diversity and relative abundance of individual taxa, or operational taxonomic units (OTUs). These data reveal not only the diversity of phyllosphere bacteria when compared to the rhizosphere but also the sensitivity of the technique in accessing the previously unknown diversity. Many of the sequences obtained from the study are unique and as such are typical of molecular studies of natural habitats where the majority of bacteria cannot be isolated or grown on laboratory media; either because we lack the knowledge to reproduce the growth requirements, because the bacterial cells are in a stage of their life cycle where they are no longer able to respond to and grow on the nutrients available in typical bacteriological agars or perhaps they only grow as part of a consortia dependent on other cells for essential factors.

When total cell counts using DNA stains are made and compared with counts of metabolically active cells with vital stains and compared to the cells that form colonies on agar we find that leaf age and type strongly influence not only the density of the community but also relative activity. Newly emerging leaves support more microorganisms (including bacteria, fungi and protozoa) than older mature leaves. This is due to the increased humidity and nutrient availability in the young leaves where the active component approaches 100% of which as many as 60% of cells can produce colonies in the laboratory. On mature leaves the availability of water and nutrients is less predictable and population densities per unit area decline by three or more orders of magnitude with less than 1% of bacteria able to grow in the laboratory. These values do in fact change depending on local climactic conditions, wind speed, precipitation, and rates of dispersal and immigration of airborne microorganism from other sources and on the concentration of nutrients, especially pollen dispersed on the wind. Finally in some plants the greatest diversity, metabolic activity and cellular abundance (biomass) can be found on senescing or decaying leaves where there is rotting organic matter, in highly favourable conditions.

We investigated the microbiology of three crop plants, grown in the same soil under identical field conditions. Field experiments were undertaken at the University of Oxford field station, Wytham. The field soil is classified as heavy clay with 24.88% sand, 21.60% silt, and 53.52% clay. The soil pH is 7.7 with an organic matter content of 8.57%. The seeds used were untreated commercially available varieties; *Pisium satvium* var. *quincy* (pea), *Triticum aestivum* var. *pena wawa* (wheat), *Beta vulgaris* var. *amythyst* (sugar beet). Three 2 m by 2 m plots were established for each plant type and sown with the seeds in June 2002. Bulk soil samples were collected to assess microbial diversity and each plant type was sampled four times according to their developmental stage, seedling, mature, flowering and harvest (seedling, mid-summer, end of summer and end of autumn for sugar beet). For each plant species, three plants were pooled to form one sample from each of the three plots as independent replicates. Whole individual plants were collected and separated into above- and below-ground portions (phyllosphere and rhizosphere) and all of the material collected used in the sample. All non-tightly adhering soil was removed from the roots prior to the extraction of the plant surface associated microbial community by homogenisation in PBS (5% w/v). Microbial population densities were estimated using 1:10 serial dilutions of the suspension on three microbiological media: aerobic heterotrophic bacteria were isolated on Tryptone Soya Broth Agar (TSBA; (0.1 mg ml^{-1}) (Difco-Oxoid, UK); pseudomonad population densities were estimated on Pseudomonad Selective Agar (PSA) (Difco-Oxoid, UK) supplemented with cyclohexamide (0.1 mg ml^{-1}), centrimide (10 μg ml^{-1}), fucidin (10 μg ml^{-1}), and cephalosporin (50 μg ml^{-1}); fungal population densities were estimated on Potato Dextrose Agar (PDA; 0.32 mg ml^{-1} aureomycin) (Difco-Oxoid, UK). Plates were incubated at 28°C for 2 days and colonies counted.

The potential substrate utilisation pattern for samples was estimated using a variation of the Community level physiological profiling (CLPP) approach (Garland and Mills, 1994; Ellis *et al.,* 1995) using BIOLOG™ GN2 plates. Bacterial suspensions were washed in PBS, diluted, inoculated at ca. $2x10^4$ cells ml^{-1} per well of the 96 well plate (95 sole C/N sources and a blank control) and incubated at 15°C for 7 days. Optical densities of each well were measured on a Rosys Anthos Lucy I plate reader (Switzerland) at a wavelength of 600 nm.

DNA and RNA was extracted from environmental samples using a modification of the bead beating-CTAB method described by (Griffiths *et al.,* 2000), an additional freeze/thaw lysis step was included. Extracted DNA was analysed using standard method for PCR amplification and subsequent Denaturing gradient gel electrophoresis (DGGE) of 16S and 18S ribosomal RNA genes essentially as described by Timms-Wilson *et al.*, (2005)

For the three plant species investigated, pea, wheat and sugar beet, their microbial population dynamics were studied using complementary methods. The data showed that changes in diversity and community succession were affected by the plant species and the growth stage of the plant (Smalla *et al.,* 2001) and that there were fundamental differences between the rhizospheres and phyllospheres across the growing season. Population dynamics on the rhizosphere and phyllosphere of the three plant types was estimated on selective media as described. Total culturable heterotrophs, pseudomonads and fungi were estimated and these data are represented in Fig. 2.1.

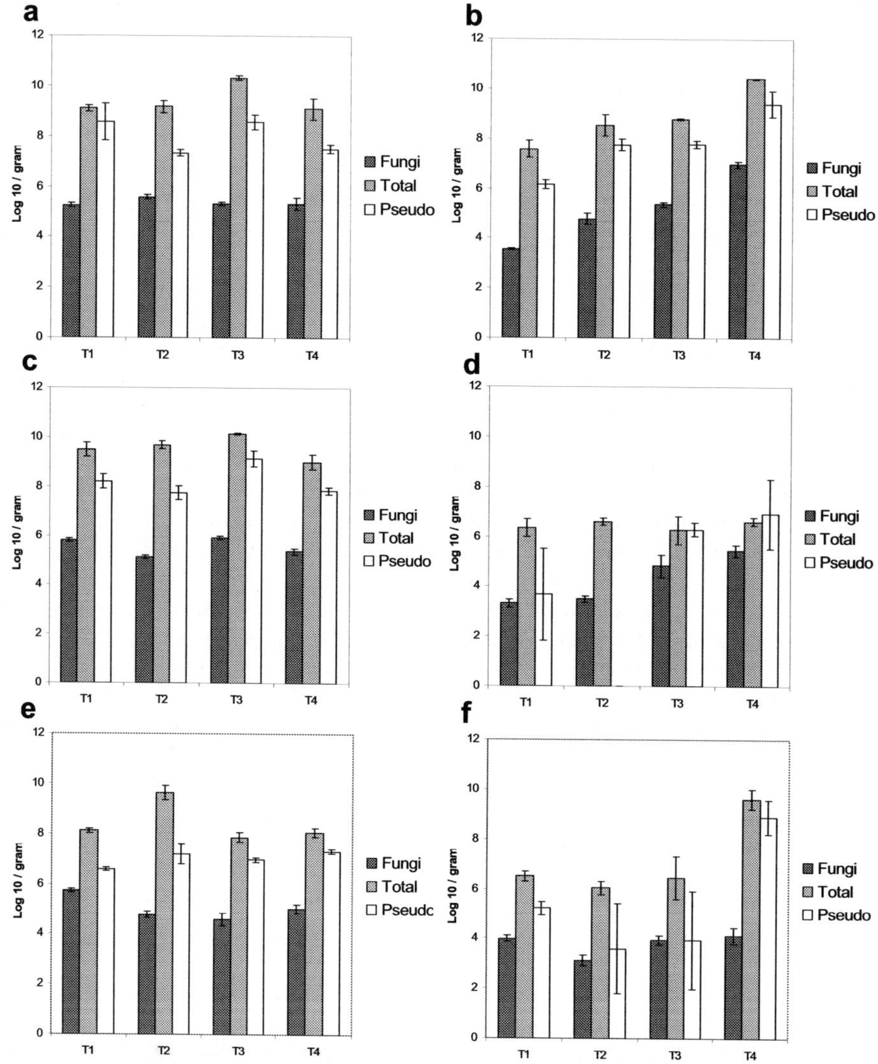

Fig. 2.1. Bar graphs showing population dynamics of total culturable heterotrophs (**Total**) (estimated on Tryptone soya broth agar, TSBA), pseudomonas (**Pseudo**) (estimated on Pseudomonad selective agar, PSA) and fungal communities (**fungi**) (estimated on Potato dextrose agar, PDA) in the rhizosphere and phyllosphere of field grown crops. Selective medium was incubated at 28°C for 2 days before enumeration. The three plants investigated were wheat, pea and sugar beet. **a** Wheat rhizosphere. **b** Pea rhizosphere. **c** Sugar beet rhizosphere. **d** Wheat phyllosphere, **e** Pea phyllosphere **f** Sugar beet phyllosphere.

Fungal populations in both the phyllosphere and rhizosphere of the three plant types remained fairly constant throughout the growing season. This equated to approximately 2.56×10^4 pfu/g (propagule forming units) wet tissue in the phyllosphere and 2.09×10^5 pfu/g wet weight in the rhizosphere tissue. Similarly the average numbers of total culturable heterotrophs isolated in the phyllosphere were 2.97 x 10^7 cfu/g fresh weight tissue versus 1.39 x 10^9 cfu/g fresh weight tissue in the rhizosphere. Total numbers of culturable pseudomonads were 1.95 x 10^6 cfu/g fresh weight tissue in phyllosphere and 5.63 x 10^7 in cfu/g fresh weight tissue versus rhizosphere.

Although counts for each microbial population varied according to plant type (see Fig. 2.1) it is apparent that significant differences can be found in the population densities recorded in rhizosphere and the phyllosphere communities. This may be attributed to the relative stability of the rhizosphere environment where temperature, moisture and pH remain relatively constant on any particular day and change gradually over time. Nutrients too may be more readily available in the form of root exudates and decaying organic matters. By contrast the phyllosphere is subject to considerable diurnal fluctuations in temperature, moisture and nutrient availability, which may change rapidly. The leaf is also subject to other physical extremes, rain, wind and UV. All these factors require considerable resilience in the colonizing microbial population and it is reasonable to speculate that the phyllosphere supports a more specialised population of bacteria and fungi adapted to this environment of extremes.

The analysis of CLPP at different stages of plant development was undertaken to identify the influence of plant type or habitat (phyllosphere or rhizosphere) on the carbon utilisation profiles of the sampled bacteria, a measure of their diversity and physiological potential. The three microbial populations extracted form the phyllosphere were distinct from each other (Fig. 2.2a) providing a strong correlation between resource (plant type) and selection of a specific functional diversity (physiological profiles). This was also seen in the rhizosphere (Fig. 2.2b). Replicates from each time point clustered closely together demonstrating the reproducibility of the method and biological system. All plant types separated at a significant level with even distribution. Distinct differences were seen in the data produced following the comparative analysis of the different environments sampled except for in the sugar beet where polygons produced for data from the rhizosphere and phyllosphere overlapped (Fig. 2.2c.). Population carbon utilisation profiles clearly changed over time throughout the growing season, these were distinct not only for each plant type but between rhizosphere and phyllosphere.

Community Analysis of the Phytosphere of Three Crop Plants by DGGE

Molecular fingerprinting analysis allowed the estimation of changes in bacterial and fungal diversity in the phytosphere of three crop plants to be assessed over a growing season. Diversity changes were estimated by comparing banding pattern changes in the DGGE profiles (Fig. 2.3) produced by PCR. This confirms our previous findings from an acid grassland soil (Griffiths *et al.,* 2000; 2002) where we observed only differences in RNA or DNA in profiles. It is assumed that the DNA and RNA targets both represent the predominant, accessible, active, and presumably viable OTUs present in the sample. It is likely that ribosomal RNA may be long lived in soil bacteria.

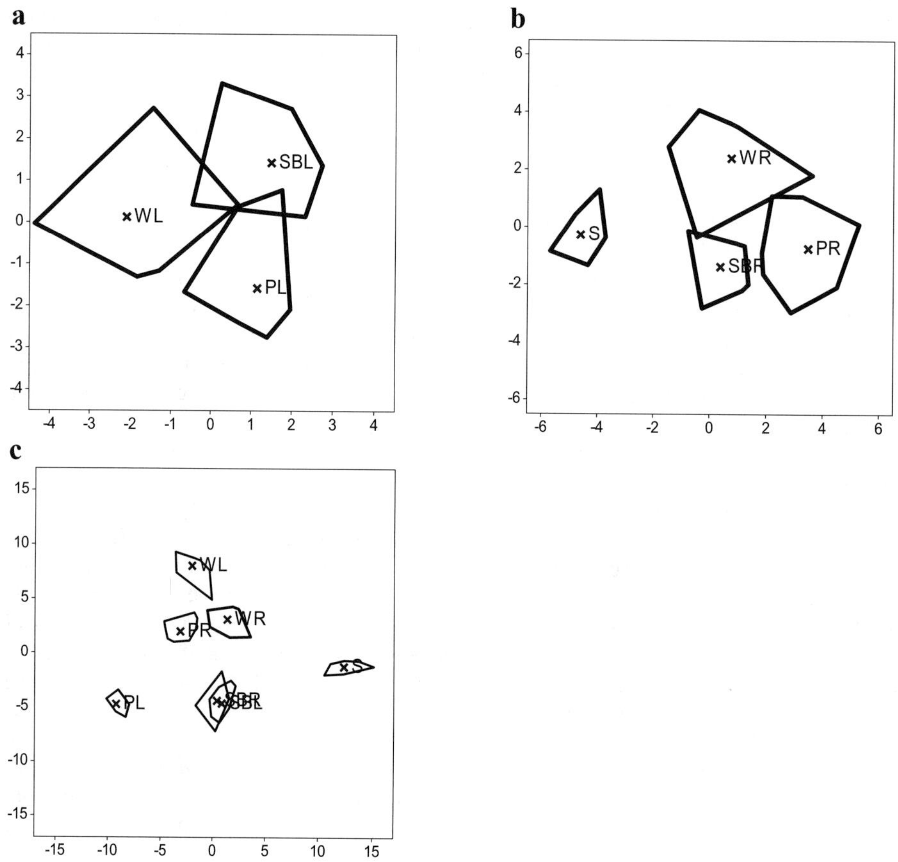

Fig. 2.2a. Plant phyllosphere: discriminatory analysis of community level physiological profiling, CLPP data from Biolog GN2 plates produced from bacterial populations from. WL = Wheat phyllosphere; PL = Pea phyllosphere; SBL = Sugar beet phyllosphere.
Fig. 2.2b. Plant rhizosphere: discriminatory analysis of CLPP from Biolog GN2 plates.produced by bacterial populations from: WR = Wheat rhisosphere; PR = Pea; SBR = Sugarbeet rhizosphere; S = bulk soil.
Fig. 2.2c. Plant phytoplane: discriminatory analysis of Biolog GN2 plates, produced by bacterial populations from: WR = Wheat rhizosphere; WL = Wheat phyllosphere; PR = Pea rhizosphere; PL = Pea phyllosphere; SBR = Sugarbeet rhizosphere; SBL = Sugarbeet phyllosphere; S = bulk soil.
Polygons represent the extent of data spread (n = 12) for each plant region over 4 time points. Plot produced in GenStat v8.1 (VSN International Ltd). Data produced from microtitre plate well Optical Density as a proportion of the total Optical Density (SynegyHT Plate Reader).

The approach is most effective for recording changes in diversity due to natural community succession that results from plant type and plant maturation. Principal component analysis of within gel diversity for each plant type were produced for the root (rhizosphere) (Figs 2.3 and 2.4) and shoot (phyllosphere) (Figs 2.3 and 2.4) for both bacterial and fungal diversity. In all cases these profiles revealed that diversity and community structure and successsion was directly determined by plant type growth stage and the two plant habitats (rhizosphere and phyllosphere). Visual assessments of the banding patterns in the gels illustrated in Fig. 2.3 provide clear distinctions in the relative abunance of diversity in respect of the intensitities of bands (OTUs) and the numbers of bands. For example there are potentially hundreds of fient bands in rhizosphere samples, indicative of extensive diversity. In contrast phyllosphere samples have less over all diversity but more intense banding patterns indicating the presence of fewer but more abundant indicidual populations. The different plant types support distinct populations. In the example provided in Fig. 2.3 each sampling date consists of three independent replicates which confirm their closer similarity, particularly in the rhizosphere, when compared to the different plant type samples. In the phyllosphere sample variation between replicates and diffferents samples is also higher over the growing season. Fig. 2.4 details discriminatory analysis of data derived from banding patterns for each sampled community. This clearly shows that the plant rhizospheres have similar diversity to each other and to the bulk soil (S). The phyllosphere habitats are statistically distinct from the rhizosphere diversity. Of particular interest is the marked difference between the phyllosphere communities of the broad leaf dicot and monocot plants.

The plant type, growth stage and plant tissue (rhizosphere and phyllosphere) influence and presumably direct succession in community composition and function. To illustrate the extent of diversity a number of individual bands common to the DGGE gels produced from rhizosphere samples were removed to identify the key OTUs. Following sequence analysis and BLAST searches of the ca. 200 bp fragment spanning the V3 region revealed typical plant associated bacteria such as *Pseudomonas* spp., *Flavobacterium*, *Enterobacteriacea*, *Sphingomonas* spp., *Cytophagales* and other previously reported "yet to be cultured" soil bacteria in the RDP database. Fungal species identified in the same way using ca. 500 bp (as described in Timms-Wilson *et al.*, 2005) of the beginning of the 18S rDNA gene revealed *Pythium* spp., *Albatrellus skamanius, Glomus versiforme, Glomus* sp., *Pichia mexicana*, zygomycota such as *Umbelopsis isabellina, Mortierella* spp. and Ascomycota such as *Geosmithia putterillii, Chaetomium globosum, Penicillium* spp. and *Verticillum* spp. (data not shown).

Concluding Remarks

The plant phytosphere has significant effects on community structure and function this has been well documented in the rhizosphere and is know as the "rhizosphere effect" (Smalla *et al.*, 2001; Dunfield and Germida, 2003). The culturable populations of microorganisms had high stability in the plant phytosphere except in the sugar beet rhizosphere whose populations demonstrated more seasonal variation. Different plant species had different carrying capacities of indigenous microorganisms. The comparison of CLPP demonstrated distinct separation of profiles between young plant and those of older plants, each plant species having its own unique profile. Molecular community analysis by DGGE by the amplification of the 16S rRNA gene demonstrated strong growth stage dependant shifts in bacterial community profiles with the separation of plant species. Fungal (18S rRNA gene) community analysis demonstrated gradual seasonal shifts of growth stages with very strong distinct separation of plant types.

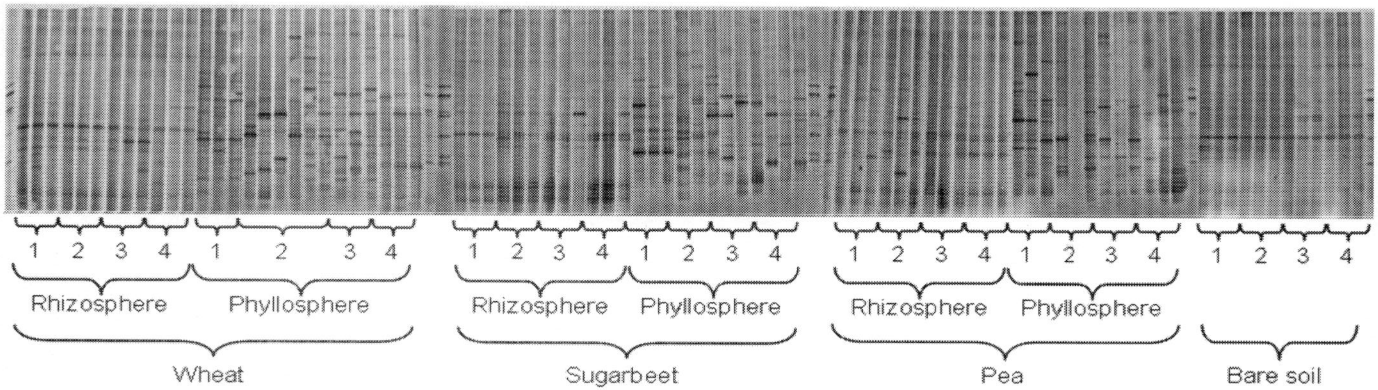

Fig. 2.3. Composite gel of 16S denaturing gradient gels electrophoresis (DGGE) from the plant phytoplane: produced from 16S rDNA populations of: Wheat rhizosphere, Wheat phyllosphere, Pea rhizosphere, Pea phyllosphere, Sugar beet rhizosphere, Sugar beet phyllosphere and bulk soil. Numbers represent the sampling times 1 = seedling, 2 = maturing plant, 3 = flowering plant, 4= harvest. Each growth stage has three independent replicates.

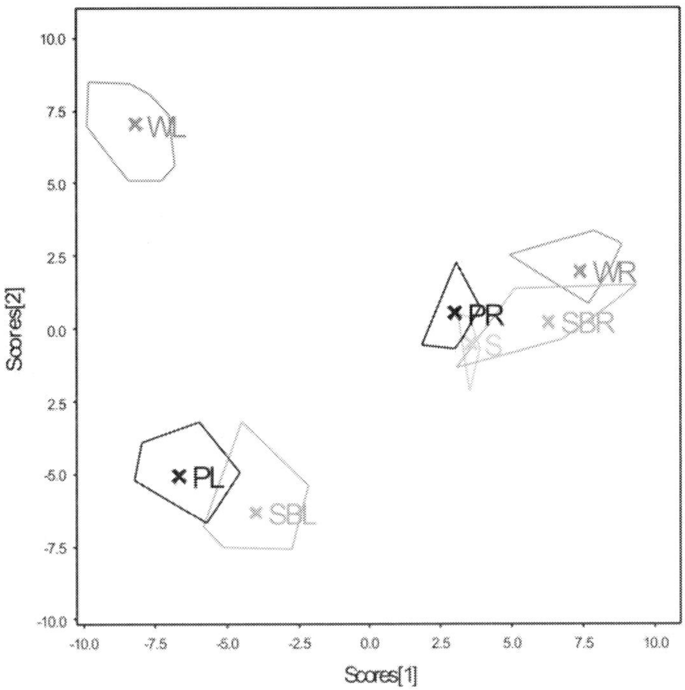

Fig. 2.4. Plant phytoplane: discriminatory analysis of density gradient gel electrophoresis, DGGE band matrix (see Fig. 2.3.) produced from 16S rDNA populations of: WR = Wheat rhizosphere; WL = Wheat phyllosphere; PR = Pea rhizosphere; PL = Pea phyllosphere; SBR = Sugar beet rhizosphere; SBL = Sugar beet phyllosphere; S = bulk soil. Polygons represent the extent of data spread (n = 12) for each plant region over 4 time points representative of plant growth stage. Plot produced in GenStat v8.1 (VSN International Ltd), band matrix produced by Phoretix 1D v5.0 (Nonlinear Dynamics Ltd).

Total aerobic heterotroph bacterial population densities (per unit gram of plant material) in the pea did not differ significantly between growth stages indicating a stable carrying capacity of the pea. When individual population groups were assessed it was observed, for example, that pseudomonad populations gradually increased in density from early seedling to flowering stage, before returning to lower population densities after harvest and at the onset of senescence. Fungal population densities reduced as the pseudomonads increased which may have been a result of the suppressive effect the pseudomonads may have on the indigenous fungi populations (Mazzola *et al.,* 1992; O'Sullivan and O'Gara, 1992). Population densities in the wheat rhizosphere of the culturable fraction also demonstrated high stability. Total heterotroph densities, but not diversity, were stable throughout the entire growing season, with a similar effect recorded in the pseudomonad population, a transient reduction in the population density was recorded at flowering. The fungal populations were stable with no significant difference recorded over the growing season. Densities were lower, however, for aerobic heterotrophs and pseudomonads when compared to the pea rhizosphere. This was the opposite for the phyllosphere where numbers in the wheat were significantly higher.

The sugar beet rhizosphere appeared to be a less stable environment in comparison to the pea and wheat with significant fluctuations in population densities. Total heterotroph densities fluctuated between growth stages. This was also reflected in the pseudomonad densities but not to the same extent. An increase in bacterial carrying capacity may have been as a result of the specific of sugar beet development. As the plant matures periods of rapid growth are recorded as the taproot develops and penetrates deeper into the soil strata (Elliott and Weston 1995). The storage cells in the root then swell with the production and storage of carbohydrates, e.g. sugars. This may increase the availability of nutrients, increase cell sloughing and exudation at the root tip (Dandurand and Knudsen, 2002), there may also be a coincident improvement in water supply to the plant, which again results in a relative increase in exudates levels in the rhizosphere and phyllosphere. The fungal populations appear to have a very gradual decline in density in the maturing sugar beet plant before a slight increase in the fully mature plant as leaves begin to senesce. No direct link could be established between fungal and bacterial population fluctuations.

Population density data indicate that the pea phytosphere consistently supports larger populations of bacteria and fungi throughout the growing season in comparison to the other plant species. Sugar beet, on the other hand, consistently supports fewer microorganisms than the other plant species, with the wheat plants fluctuating between the high and low microorganism population densities depending on the organisms being measured. In comparison to bulk soil, all plant species enrich the populations of culturable bacteria. This effect has been attributed to the high level of nutrients available to the microorganisms inhabiting this environment. From this data it can be inferred that pea plants have the highest level of available niches for colonisation of the three plant species followed by wheat and then sugar beet. The field conditions may result in less favourable conditions for bacterial colonisation or be a result of the high variability identified in the sugar beet rhizosphere.

The comparison between growth stages and plant type identified a strong and significant plant growth stage effect in the CLPP. Seasonal shifts in CLPP patterns have been demonstrated in other systems (Natsch *et al.*, 1997; Thirup *et al.*, 2003), however this variance is not consistent (Heuer *et al.,* 2002) for all plant studies (Duineveld *et al.,* 1998; Miethling *et al.,* 2003). In our study young seedlings of all plant species were significantly different to each other in their CLPP with the greatest separation being between the wheat and the other two plants. All subsequent growth stages were readily distinguished for each plant species. This would demonstrate that the plant species has a significant effect on the selection of specific communities, which is reflected in the carbon utilisation profile of the microbial communities. This impact of plant species has been demonstrated before (Ellis *et al.,* 1995; Fang *et al.,* 2001; Miethling *et al.,* 2003).

Analysis of the 16S rRNA gene DGGE identified plant specific effects when the three plant species were compared. The comparison between growth stages for individual plant species identified that there were strong growth stage effects on the bacterial community structure. The comparison of each plant species community profiles at individual growth stages confirmed the plant effects on the community structure. The separation of plant types by principle component analysis was distinct at each growth stage, demonstrating a higher degree of separation than was indicated by looking at overall communities. The analysis of community structure by 16S rRNA gene DGGE analysis has been wildly undertaken in which there have been many conclusions, which support or contradict this data. Some researchers have found little effect of plant on community structure (Duineveld *et al.,* 1998), whereas others have found significant shifts (Gomes *et al.,* 2001; Marschner *et al.,* 2001; Smalla *et al.,* 2001; Thirup *et al.,* 2003). These conflicting data may be a result of the highly complex and diverse nature of the phytosphere. Different soils which effect the rhizosphere may have higher resilience to perturbations, which include planting, in comparison to others due to their composition. Soil type has been shown to be influential in the effects of rhizosphere colonisation (Latour *et al.,* 1996). It was therefore vital to understand the level of significance that the plant species have in the soil environment.

The 18S rRNA gene DGGE community analysis (data not shown) comparisons between the three plant species identified significant plant effects on the fungal community structure. All known non-fungal bands were not included in analysis, therefore this effect was not a result of 18S rRNA genes amplified from the host plant or other amplified non-fungal 18S rRNA gene sequences. This proves that the effects seen on the community structure are a direct result of plant interactions. The comparison of growth stages for individual plant species identified gradual shifts between growth stages of the fungal community structure. This would suggest a gradual succession of fungal colonisers of the rhizosphere in response to the plant. The study of fungal populations by DGGE in rhizosphere environment is relatively new. Plant enrichment and seasonal shifts have recently been reported in the maize rhizosphere with limited differences in bulk soil (Gomes *et al.,* 2003). Molecular analysis of fungal communities has been undertaken by other techniques (Smit *et al.,* 1999; Glandorf *et al.,* 2001) but no measures of seasonal variation were undertaken. Comparison of each plant species community profile at individual growth stages confirmed the plant effects on the community structure. The separation of plant species by principle component analysis increased from young seedlings with each subsequent growth stage. This indicates a changing but increasingly plant species-specific effect on the fungal community.

The three plant species were sampled at four growth stages throughout their life cycle. Studies of the culturable populations were undertaken, providing good correlative data on the resource capacity of the habitat as reflected in the diversity and population densities achieved. The molecular techniques provide insight of the unculturable fraction, and assessments of the accessible total diversity, but they are at present only semi-quantitative and reflect a selective bias influenced by the relative abundance of the target nucleic acid of individual OTUs in any sample. Nonetheless comparative data are essential and when combined with other methods it is possible to demonstrate that in the phytosphere the microorganisms, particularly bacteria, need to be metabolically active to survive, perhaps this is reflected in recent observations that up to 70% of rhizosphere communities identified by molecular techniques can actually be isolated on laboratory media (Kowalchuk, personal communication).

The phytosphere of plants is a highly variable and selective habitat as reflected in the microbial diversity recorded. Plant development directly influences community activity and succession. Further details can be found in related studies (Timms-Wilson *et al.,* 2005; Griffiths *et al.,* 2003; Houlden 2005). The phyllosphere is a highly selective environment, which is exposed to a huge range of biotic and abiotic factors that influence successful colonisation and survival of key groups of bacteria and fungi. Counts for both bacterial and fungal populations were stable but the communities composition was changing with plant growth and development. Data from both community profiling (DGGE analysis) and physiological profiling (CLPP) confirm the phyllosphere and rhizosphere of the same and different plants to be distinct environments throughout the growing season. The net effect of immigration was not measured in these studies, however it is assumed that the potential of a common source of inocula was provided for each plant type as each were planted in the same soils and grown under identical conditions in the field. Observed differences reflect plant physiology and architecture, which affect the rate of nutrient release at the plant surface. It is apparent that the phyllosphere attracts specialist and opportunist microorganisms. The challenge is to identify not only the genetic mechanisms that facilitate colonisation and survival but also the underlying ecological factors that lead to success of a given population. These are considered in other chapters in this book but the development and application of high througput techniques, particularly environmental genomics and community sequencing projects, will help to identify common gene sets that may identify essential traits. This work is progressing in a number of laboratories but the technical challenge is considerable and will require a more complete systems approach to gather the understanding necessary to increase plant productivity.

Acknowledgements

This work was supported by the NERC and DEFRA, UK (award RG0113).

References

Andrews, J.H. and Harris, R.F. (2000) The ecology and biogeography of microorganisms on plant surfaces. *Annual. Review of Phytopathology* 38, 145-180.

Bailey, M.J. and Lilley, A.K. (2002) Niche colonisation and the dispersal of bacteria and their genes in the natural environment. In: J.M. Bullock, J.M. Kenwood and R.S. Hails (eds) *Dispersal Ecology*, Proceedings of the 42nd British Ecological Society. Blackwell Publishing, Oxford, UK. pp. 219-236.

Dandurand, L.C. and Knudsen, G.R. (2002). Sampling microbes from the rhizosphere and phyllosphere. In: C.J. Hurst, R.L. Crawford, G.R. Knudsen, M.J. McInerney, and L.D. Stetzenbach (eds) *Manual of Environmental Microbiology*, 2nd edition, American Society for Microbiology, Washington, USA. pp. 516-526.

Duineveld, B.M., Rosado, A.S., van Elsas, J.D. and van Veen, J.A. (1998) Analysis of the dynamics of bacterial communities in the rhizosphere of the chrysanthemum via denaturing gradient gel electrophoresis and substrate utilization patterns. *Applied and Environmental Microbiology* 64, 4950-4957.

Dunfield, K.E. and Germida, J.J. (2003) Seasonal changes in the rhizosphere microbial communities associated with field-grown genetically modified canola (*Brassica napus*). *Applied and Environmental Microbiology* 69, 7310-7318.

Elliott, M.C. and Weston, G.D. (1995) Biology and physiology of the sugar-beet plant. In: D.A. Cooke and R.K. Scott (eds) *The Sugar Beet Crop*, 1st edition, Chapman & Hall, London, UK. pp. 37-66.

Ellis, R.J., Thompson, I.P. and Bailey, M.J. (1995) Metabolic profiling as a means of characterizing plant associated microbial communities. *FEMS Microbiology Ecology* 16, 9-17.

Espinosa-Urgel, M., Kolter, R. and Ramos, J-L.(2002) Root colonisation by *Pseudomonas putida*: love at first sight. *Microbiology* 148, 341-343.

Fang, C.W., Radosevich, M. and Fuhrmann, J.J. (2001) Characterization of rhizosphere microbial community structure in five similar grass species using FAME and BIOLOG analyses. *Soil Biology and Biochemistry* 33, 679-682.

Garland, J.L. and Mills. A.L. (1994) A Community-level physiological approach for studying microbial communities. In: K. Ritz, J. Dighton and K.E. Giller (eds) *Beyond the Biomass*, 1st edition. John Wiley & Sons, Chichester, UK. pp. 77-83.

Glandorf, D.C., Verheggen, M.P., Jansen, T., Jorritsma, J.W., Smit, E., Leeflang, P., Wernars, K., Thomashow, L.S., Laureijs, E., Thomas-Oates, J.E., Bakker, P. and Van Loon, L. C. (2001) Effect of genetically modified *Pseudomonas putida* WCS358r on the fungal rhizosphere microflora of field-grown wheat. *Applied and Environmental Microbiology* 67, 3371-3378.

Gomes, N.C.M., Heuer, H., Schonfeld, J., Costa, R., Mendonca-Hagler, L. and Smalla, K. (2001) Bacterial diversity of the rhizosphere of maize (*Zea mays*) grown in tropical soil studied by temperature gradient gel electrophoresis. *Plant and Soil* 232, 167-180.

Gomes, N.C.M., Fagbola, O., Costa, R., Rumjanek, N.G., Buchner, A., Mendonca-Hagler, L. and Smalla, K. (2003) Dynamics of fungal communities in bulk and maize rhizosphere soil in the tropics. *Applied and Environmental Microbiology* 69, 5737-5737.

Gray, N.D., Howarth, R., Pickup, R.W., Gwyn Jones J. and Head, I.M. (2000) Use of combined microautoradiography and fluorescence *in situ* hybridization to determine carbon metabolism in mixed natural communities of uncultured bacteria from the genus Achromatium. *Applied and Environmental Microbiology* 66, 4518-4522.

Griffiths, R.I., Whiteley, A.S., O'Donnell, A.G. and Bailey, M.J. (2000) Rapid method for coextraction of DNA and RNA from natural environments for analysis of ribosomal DNA and rRNA-based microbial community composition. *Applied and Environmental Microbiology* 66, 5488-5491.

Griffiths, R.I., Whiteley, A.S., O'Donnell, A.G. and Bailey, M.J. (2002) Influence of depth and sampling time on bacterial community structure in an upland grassland soil. *FEMS Microbiology Ecology* 43, 135-43.

Griffiths, R.I., Whiteley, A.S., O'Donnell, A.G. and Bailey, M.J. (2003) Physiological and community responses of established grassland bacterial populations to water stress. *Applied and Environmental Microbiology* 69, 6961-6968.

Heuer, H. and Smalla, K. (1999) Bacterial phyllosphere communities of *Solanum tuberosum* L. and T4-lysozyme producing transgenic variants. *FEMS Microbiology Ecology* 28, 357-371.

Heuer, H., Kroppenstedt, R.M., Lottmann, J., Berg, G. and Kornelia Smalla, K. (2002) Effects of T4 lysozyme release from transgenic potato roots on bacterial rhizosphere communities are negligible relative to natural factors. *Applied and Environmental Microbiology* 68, 1325-1335.

Hirano, S. S. and Upper, C.D. (2000) Bacteria in the leaf ecosystem with emphasis on *Pseudomonas syringae* - a pathogen, ice nucleus, and epiphyte. *Microbiology and Molecular Biology Reviews* 64, 624-653.

Houlden A. (2005) Bacterial and fungal diversity effects and the activity of biocontrol agents in the rhizosphere of crop plants. PhD Thesis, Cardiff University, UK.

Jacques, M-A. and Morris, C.E. (1995) A review of issues related to the quantification of bacteria from the phyllosphere. *FEMS Microbiology Ecology* 18, 1-14.

Jurkevitch, E.J. and Shapira, G. (2000) Structure and colonization dynamics of epiphytic bacterial communities and of selected component strains on tomato *Lycopersicon esculentum* leaves. *Microbial Ecology* 40, 300-308.

Latour, X., Corberand, T.S., Laguerre, G., Allard, F. and Lemanceau, P. (1996) The composition of fluorescent pseudomonad populations associated with roots is influenced by plant and soil type. *Applied and Environmental Microbiology* 62, 2449-2456.

Lilley, A.K., Hails, R.S., Cory, J.S. and Bailey, M.J. (1997) The dispersal and establishment of pseudomonad populations in the phyllosphere of sugar beet by phytophagous caterpillars. *FEMS Microbiology Ecology* 24, 151-158.

Lindow, S.E., (1987). Competitive exclusion of epiphytic bacteria by Ice$^-$ *Pseudomonas syringae* mutants. *Applied and Environmental Microbiology* 53, 2520-2527.

Marschner, P., Yang, C.H., Lieberei, R. and Crowley, D.E. (2001) Soil and plant specific effects on bacterial community composition in the rhizosphere. *Soil Biology and Biochemistry* 33, 1437-1445.

Mazzola, M., Cook, R.J., Thomashow, L.S., Weller, D.M. and Pierson, L.S. (1992) Contribution of phenazine antibiotic biosynthesis to the ecological competence of fluorescent pseudomonads in soil habitats. *Applied and Environmental Microbiology* 58, 2616-2624.

Miethling, R., Ahrends, K. and Tebbe, C.C. (2003) Structural differences in the rhizosphere communities of legumes are not equally reflected in community level physiological profiles. *Soil Biology and Biochemistry* 35, 1405-1410.

Muyzer, G., de Waal, E.C. and Uitterlinden, A.G. (1993) Profiling of complex microbial populations by denaturing gradient gel electrophoresis analysis of polymerase chain reaction-amplified genes coding for 16S rRNA. *Applied and Environmental Microbiology* 59, 695-700.

Natsch, A., Keel, C., Hebecker, N., Laasik, E. and Defago, G. (1997) Influence of biocontrol strain *Pseudomonas fluorescens* CHA0 and its antibiotic overproducing derivative on the diversity of resident root colonizing pseudomonads. *FEMS Microbiology Ecology* 23, 341-352.

O'Brien, R.D. and Lindow, S.E. (1989) Effect of plant species and environmental conditions on epiphytic population sizes of *Pseudomonas syringae* and other bacteria. *Phytopathology* 79, 619-627.

O'Sullivan, D.J. and O'Gara, F. (1992) Traits of fluorescent *Pseudomonas* spp. involved in suppression of plant root pathogens. *Microbiological Reviews* 56, 662-676.

Ouverney, C.C. and Fuhrman, J.A. (1999) Combined microautoradiography–16S rRNA probe technique for determination of radioisotope uptake by specific microbial cell types *in situ. Applied and Environmental Microbiology* 65, 1746-1752.

Smalla, K., Wieland, G., Buchner, A., Zock, A., Parzy, J., Kaiser, S., Roskot, N., Heuer, H. and Berg, G. (2001) Bulk and rhizosphere soil bacterial communities studied by denaturing gradient gel electrophoresis: Plant dependent enrichment and seasonal shifts revealed. *Applied and Environmental Microbiology* 67, 4742-4751.

Smit, E., Leeflang, P., Glandorf, B., van Elsas, J.D. and Wernars, K. (1999) Analysis of fungal diversity in the wheat rhizosphere by sequencing of cloned PCR amplified genes encoding 18S rRNA and temperature gradient gel electrophoresis. *Applied and Environmental Microbiology* 65, 2614-2621.

Thirup, L., Johansen, A. and Winding, A. (2003) Microbial succession in the rhizosphere of live and decomposing barley roots as affected by the antagonistic strain *Pseudomonas fluorescens* DR54-BN14 or the fungicide imazalil. *FEMS Microbiology Ecology* 43, 383-392.

Thompson, I.P., Lilley, A.K., Ellis, R.J., Bramwell, P.A. and Bailey, M.J. (1995) Survival, colonisation and dispersal of genetically modified *Pseudomonas fluorescens* SBW25 in the phytosphere of field grown sugar beet. *Nature Biotechnology*. 13, 1493-1497.

Timms-Wilson. T.M., Kilshaw, K. and Bailey, M.J. (2005) Risk assessment for engineered bacteria used in biocontrol of fungal disease in agricultural crops. *Plant and Soil* 266, 57-67.

Wieland, G., Neumann, R. and Backhaus, H. (2001) Variation of microbial communities in soil, rhizosphere, and rhizoplane in response to crop species, soil type, and crop development. Applied and Environmental Microbiology 67, 5849-5854.

Yang, C-H., Crowley, D.E., Bornmeman, J. and Keen, N.T. (2001) Microbial phyllosphere populations are more complex than previously realized. *Proceedings of the National Academy of Science*, USA 98, 3889-3894.

Chapter 3

Diversity, Scale and Variation of Endophytic Fungi in Leaves of Tropical Plants

P. Bayman
Departamento de Biología, Universidad de Puerto Rico - Río Piedras, PO Box 23360, San Juan, PR 00931, USA
pbayman@uprrp.edu

Introduction

Diversity makes life interesting. Diversity of microorganisms in leaves makes leaves interesting, but also makes them complex and unpredictable in ways which are still largely unexplored.

This chapter reviews aspects of diversity of endophytic fungi: how sampling strategies affect results of diversity studies; how communities of endophytic and epiphytic fungi differ in a single leaf; how endophytic fungi vary from leaf to leaf; how endophytes are transmitted; and how endophytes may affect other microorganisms and host plants. (As used here, 'endophytes' are organisms that inhabit internal plant tissues without apparent disease symptoms; Petrini, 1991). The chapter focuses principally on literature published since Phyllosphere 2000 and on tropical plants. Clavicipitaceous endophytes of grasses are excluded (e.g. *Epichlöe* and *Neotyphodium*) because they are well known, mostly temperate, and have been reviewed extensively (Stone *et al.*, 2000; Schardl *et al.*, 2004). The genus *Xylaria* is used as an example of differences between endophytes of tropical and temperate plants, and of unresolved problems of diversity and distribution of endophytes. Although the chapter focuses on endophytic fungi, examples from epiphytes, bacteria and temperate endophytes are also included, because their diversity is better understood in many respects. It concludes that a more integrative approach is needed to understand the diversity and importance of phyllosphere microorganisms.

Why Study Endophytes of Tropical Plants?

Only a small proportion of fungal species have been described, less than 100,000 out of an estimated 1,500,000 (Hawksworth and Rossman, 1997). Where are the rest? There are four reasons to focus on endophytes of tropical plants as reservoirs of undescribed species. First, endophytes are speciose, making an important but cryptic contribution to fungal biodiversity (Frölich and Hyde, 2001). The ratio of endophyte species to plant species, if constant, can provide an estimate of how many endophyte species remain to be discovered (Bills, 1996; Hawksworth and Rossman, 1997). If this is true, then endophyte diversity should be highest in the tropics, where plant diversity is highest. However, it has also been suggested that the ratio is not constant, and that fewer plant-dependent organisms are host-specific in the tropics than in temperate regions (May, 1988). For this reason, diversity and host specificity of tropical endophytes are important unknowns for estimating global fungal biodiversity (Hawksworth and Rossman, 1997; Frölich and Hyde, 2001).

Second, infection frequencies of endophytes are generally higher in leaves of tropical plants than in temperate plants (Arnold *et al.,* 2001, 2005), which may imply higher species richness in the tropics (assuming evenness is similar). Third, despite their higher frequency, much less is known about endophytes of tropical plants than of temperate plants. In general, the mycota of the tropics have been less explored. And fourth, tropical leaves provide a more stable environment than temperate leaves: there are fewer deciduous plants in many areas, leaf turnover is less simultaneous and less drastic, and many tree leaves live for several years. This stability should lead to a more complex microbial community, although this has not been demonstrated.

Xylaria - A Poster Child for Endophytes of Tropical Leaves

The ascomycete genus *Xylaria* and its anamorphs (collectively referred to here as *Xylaria*) are a good example of how leaf endophytes in the tropics differ from those of temperate areas. *Xylaria* spp. are among the most commonly isolated endophytes from tropical plants, including important crop plants such as coffee, banana and the acai palm (Rodrígues, 1994; Petrini *et al.,* 1995; Lodge *et al.,* 1996; Rodrigues and Petrini, 1997; Bayman *et al.,* 1997, 1998; Gamboa *et al.,* 2002; Santamaria and Bayman, 2005). They have also been isolated from non-vascular plants (Davis *et al.,* 2003). However, in temperate areas *Xylaria* species are best known as wood-rotters and pathogens, and are rarely seen as leaf endophytes (Petrini *et al.,* 1995; Rogers, 2000). *Xylaria* is also interesting for its production of diverse classes of secondary metabolites. Some of these metabolites have potential uses as drugs or insecticides (Whalley, 1996).

Identification of endophytic *Xylaria* species is difficult. Many isolates of *Xylaria* do not form stromata and perithecia readily in culture, making identification to species or even to genus difficult (Rogers, 2000). Poor sporulation in culture is a problem with many tropical endophytes and epiphytes, and results in many fungi being grouped in 'morphospecies' based on colony morphology in culture rather than being identified to species (Arnold *et al.,* 2001; Gamboa and Bayman 2001). However, these morphospecies appear to correspond fairly well to terminal clades based on DNA sequence analysis (Arnold *et al.,* 2001, 2003).

A similar argument can be made for *Guignardia* (anamorph *Phyllosticta*), another genus of tropical endophytes that appears to be as widely distributed as *Xylaria* (Pandey *et al.,* 2003; Rodrigues *et al.,* 2004). As with *Xylaria*, a single plant or population can contain a diverse population of *Guignardia.* Unlike *Xylaria*, many *Guignardia* strains are pathogens; endophytic strains of *G. citrifolia* were distinct from pathogenic strains, based on DNA polymorphisms (see Azevedo *et al.,* 2000).

Host Specificity

As mentioned above, endophytes of tropical plants may include many thousands of undescribed species, assuming host:endophyte ratios are similar to those of temperate plants. Because of this assumption, research on tropical endophytes has focused on host specificity. In general, some studies have found evidence for host specificity in endophytes of tropical plants at the community level (e.g. Arnold *et al.,* 2001) while others have not (e.g. Bayman *et al.,* 1997; Cannon and Simmons, 2002). However, studies vary greatly in sampling strategies and levels of taxonomic resolution, so it is difficult to compare results. Many fungi occur in multiple hosts, but at very different frequencies; these distribution patterns are quantitative preferences rather than absolute specificity (Lodge, 1997).

There may also be host specificity or host preference at the intraspecific level (i.e. host races of endophytes), a level rarely addressed in these studies (Pandey *et al.,* 2003; Ganley *et al.,* 2004; Rodrigues *et al.,* 2004). Conversely, intraspecific variation in the plant host may affect its relationship with endophytes. For example, in hybrid cottonwoods in the southwestern USA, plants with a higher percentage of the *Populus fremontii* genotype had higher frequency of twig endophytes (>40% higher; Bailey *et al.,* 2005). These trees also had lower tannin concentration in twigs than other hybrids, so the effect on endophytes may have been mediated by tannins. These data suggest that plant evolution can affect relationships with endophytes (Schweitzer *et al.,* Chapter 7). Since host specificity of endophytes has been amply discussed in the cited articles, it will not be reviewed further here.

Sampling Strategy and Scale

What size leaf pieces should be used to sample endophytic fungi, in order to maximise diversity of fungi and efficiency of sampling? Several studies have established an effect of leaf fragment size on frequency and diversity of fungi isolated. In one study, Douglas fir needles were surface-sterilised, cut into 0.25 x 0.4 mm pieces, and plated on nutrient agar until fungal colonies grew out (Carroll, 1995). Results were compared with expected results from the same leaves, if they had been cut into much larger pieces. It was assumed that the most rapidly-growing fungus present would dominate each piece, inhibiting other fungi present. Under these assumptions, the infection rate for *Rhabdocline parkeri* would decrease from 94% to 13% and for *Nodulosporium* from 6% to 0.01%. Two less frequent fungi, *Cryptocline abietina* and *Phaeocryptopus gaumannii*, would have been overlooked altogether using larger pieces (Carroll, 1995). Based on these data, Carroll proposed a sampling strategy to reveal maximum diversity: reduce size of sampling units until only 10-20% yield endophytes. However, such a low yield is less efficient because it implies waste of materials.

As predicted by Carroll, cutting leaf pieces of the same size into different numbers of fragments affects the number of fungal endophytes isolated. We took 2 x 2 cm pieces from leaves of five diverse tropical plants, surface-sterilised them, and plated them on potato dextrose agar (Gamboa *et al.,* 2002). Replicate pieces were plated whole (2 x 2 cm) cut into four pieces (each 1 x 1 cm), 16 pieces (each 0.5 x 0.5 cm), and 64 pieces (each 0.25 x 0.25 cm). Number and type of endophyte colonies isolated from each piece was recorded; some of the taxa were not identified, but were grouped into morphospecies based on similarity in culture. For all five plant species, the number of fungal morphospecies isolated was correlated significantly with the number of pieces. A ln-ln plot of the regression suggested that cutting the leaf into even smaller fragments would increase the total number of fungi isolated, to 16±3 morphospecies. Smaller is better, when it comes to choosing leaf fragment size for fungal diversity studies.

The reason for this strong relation between leaf fragment size and fungal diversity is probably competition (Carroll, 1995; Gamboa *et al.,* 2002). The first colony to grow out of a leaf piece on agar medium can often inhibit the growth of others, even if viable inoculum of other fungi is present. This implies that the optimal sample size for leaf fragments will contain a single colony - in some cases, perhaps as small as a single leaf cell.

Colony Size

The optimum size of leaf pieces for isolation of endophytes partly depends on the size of endophyte colonies in the leaf, but this is largely unknown. Few of the studies discussed here have attempted to define the size or extent of individual fungal colonies. In distribution maps of endophyte taxa within leaves (e.g. Carroll, 1995; Lodge *et al.,* 1996; Gamboa and Bayman, 2001) adjacent leaf fragments are often occupied by the same fungal taxon. It is unclear if these fragments contain the same physical and genetic individual, or a patchwork of different genetic individuals.

The difference between one large individual and many small individuals is important for several reasons. First, intraspecific fungal antagonism can cause secondary metabolite production, and these metabolites can affect both the fungus and the plant. Second, if a substrate is divided among many small, incompatible colonies, each colony may not obtain enough resources to be able to sporulate, particularly in the case of fungi with large fruiting bodies. Third, although discussion of biodiversity usually focuses on species diversity, intraspecific genetic diversity also influences community structure and function.

Histological studies can determine colony size of endophytes more precisely than culture studies, but relatively few such studies have been done. Some endophyte colonies are limited to single host cells (Stone, 1987; Suske and Acker, 1989), in which case sampling units will be larger than colony size (due to practical limitations). However, these studies were done on host-specific endophytes of conifer needles in temperate areas, and it is not clear if they apply to less specific endophytes such as *Xylaria*.

Variation Among Leaves

How much does microbiota differ from one leaf to the next? Three leaves were compared by intensive sampling for fungal endophytes (Lodge *et al.,* 1996). The three leaves came from different individuals in a single population of *Manilkara bidentata*, a rainforest tree. Two of the three leaves yielded similar arrays of fungi, but the third was different. For instance, *Xylaria* spp. were isolated from 73% and 74% of leaf pieces from the first two leaves, but only from 21% of the third. Various factors may influence diversity of fungal endophytes in leaves of tropical plants: leaf age, location, climate and conservation status (Rodrigues, 1994; Arnold *et al.,* 2001; Gamboa and Bayman, 2001; Suryanarayanan *et al.,* 2002). However, the three *M. bidentata* leaves were chosen to be as similar as possible, with no visible difference between them, and they differed nonetheless. Rather than using their sampling strategy (three leaves x 50 pieces/leaf), Lodge *et al.* (1996) recommend increasing the number of leaves and decreasing the number of pieces sampled from each leaf.

How could three identical leaves differ so much in endophyte biotas? An answer may be found in the theory of island biogeography, which has been used as a model to study distribution of microorganisms on leaves (Andrews and Harris, 2000). Two recent extensions of this theory, metapopulation models and the neutral theory of biodiversity (Hubbell, 2001) can provide further insight for addressing this question. If each leaf is considered an island (disregarding the tree from which it came), then endophyte species richness and composition will be functions of leaf size, isolation (or distance from potential sources of inoculum), and leaf age. If the source community is very large compared to the community of each leaf and each leaf is a random sample of the source community, it is likely that leaves will differ (Curtis and Sloan, 2004). And since dispersal is driven largely by chance, it is not surprising that species richness, composition and relative abundances differ greatly between leaves. Data from cacao in Panamá support the applicability of these models: similarity of endophyte communities among leaves decreased significantly as distance between trees increased (Arnold *et al.,* 2003).

Culturing vs Direct Amplification

The above discussion on sampling strategy and variation among leaves is based on culture studies; for PCR-based studies that use direct amplification of total DNA extractions from leaves, the issue of sampling unit size may not be as critical. However, PCR studies are affected by other kinds of competition: competition for primers and polymerase among different targets in a sample. A target with lower homology to a primer sequence, or whose DNA is more difficult to extract, may be outcompeted and overlooked as easily as a slowly growing fungus in a culture study. So sampling unit size can still affect results.

Fungal endophytes have been cultured and subsequently identified by DNA sequencing, RFLPs or ARDRA (amplified ribosomal DNA restriction analysis), DGGE, RAPDs, and related techniques (e.g. Davis *et al.,* 2003; Pandey *et al.,* 2003). However, these studies do not recover unculturable fungi which would be revealed by direct amplification. Relatively few studies to date have used direct amplification (also called environmental PCR) to study diversity of endophytes of tropical plants (Guo *et al.,* 2001; Arnold *et al.,* 2005).

How many species of fungal endophytes are revealed by direct amplification and sequencing compared to culturing? In loblolly pine, direct amplification revealed more species than culturing per unit of effort (Arnold *et al.,* 2005). However, in some groups culturing revealed more diversity than direct amplification (e.g. Sordariomycetes), and culturing tended to reveal more high-level groups than direct amplification. The two techniques are mutually re-enforcing, because direct amplification may reveal the presence of interesting clades for which new culturing methods can then be developed (Arnold *et al.,* 2005). In contrast, studies of epiphytic bacteria suggest that higher diversity at all levels is revealed by direct amplification (Yang *et al.,* 2001).

It is likely that direct amplification and sequencing from plant tissues will reveal unknown, major clades of fungi. However, endophytes are sufficiently unexplored that much can still be learned through culture studies. For example, a new species of *Penicillum* was recently isolated as an endophyte of coffee (Peterson *et al.,* 2005). *Penicillium* is one of the most common and best known genera of fungi, and is easy to isolate in culture, but it is not well known as an endophyte; coffee is one of the world's most important crops, but it is just beginning to be explored for endophytes (Santamaría and Bayman, 2005). So this study is novel, even though *Penicillium* endophytes in coffee might be expected to be well-catalogued already.

For this reason, large-scale diversity studies should include a culturing component in addition to direct amplification: a small amount of effort diverted to culturing will probably reveal more diversity than the same amount of effort if used for additional sequencing, as shown on the production possibility frontier in Fig. 3.1.

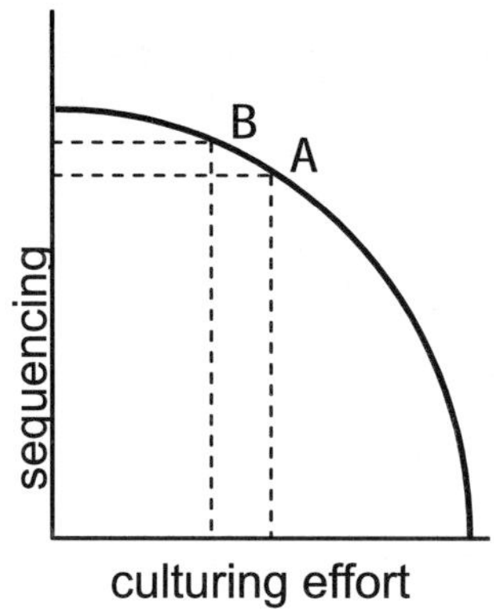

Fig. 3.1. Production possibility frontier showing the relationship between sequencing effort and culturing effort for microbial diversity studies. The area under the curve is the total number of taxa found in the study. From point A to point B, the number of new taxa revealed by sequencing decreases for every additional unit of sampling effort. At this point of the curve, more new taxa would be revealed if the same additional effort were invested in culturing.

Endophytes vs. Epiphytes

Most studies on endophytes have ignored epiphytes, and vice versa. In practice, endophytes are considered to be any fungi isolated from a leaf that are not killed by whatever surface-sterilisation protocol is used (Weber and Anke, Chapter 15). Any epiphyte able to survive surface sterilisation would thus be counted as an endophyte. Surface sterilisation methods have been compared to address this problem (Schulz *et al.,* 1993; Bills, 1996). However, the distinction between endophytes and epiphytes is less clear than one would gather from the literature, because some fungi may be both. More precise distinctions between epiphytes and endophytes would require microscopy, but identification of fungi *in planta* by microscopy can be difficult.

In a recent study, we compared epiphytes and endophytes of coffee leaves (Santamaría and Bayman, 2005). We asked whether epiphyte and endophyte biotas of a single coffee leaf were the same, and equally rich in species. We hypothesised that since plants presumably have more control over colonisation of internal tissues than external tissues, the composition of epiphyte and endophyte communities would be different, and epiphyte communities would be richer in species. We also compared biota of leaves in different sites; since endophytes live in a more protected and constant environment inside the leaf, we predicted that endophyte biotas would vary less among sites than epiphyte biotas.

We cultured pieces of coffee leaves on PDA (Santamaría and Bayman, 2005). Pieces for epiphytes were placed on agar for an hour and then removed; pieces for endophytes were surface sterilised with ethanol and sodium hypochlorite. We sampled five populations, three plants per population, five leaves per plant, and from each leaf we sampled four pieces for epiphytes and four for endophytes. We isolated a total of 821 colonies and grouped them into 131 morphospecies; the most common were identified by morphology or by sequencing the nuclear ribosomal ITS. As predicted, the epiphytic and endophytic biotas were very different: of the five most common genera, two were significantly more common as endophytes (*Botryosphaeria* and *Pestalotia*) and three were significantly more common as epiphytes (*Colletotrichum*, *Xylaria* and *Guignardia*). However, contrary to our hypothesis, the endophyte communities and epiphyte communities were equally diverse (66 vs. 63 morphospecies observed). Species accumulation curves and the jack-knife estimator of richness showed the two communities were about equal, and the Chao1 estimator predicted that the endophyte community was richer - the reverse of what we expected. Furthermore, differences among sites were as pronounced for endophytes as for epiphytes, despite having a more protected environment within the leaf. However, most of the common endophytes were occasionally found as epiphytes, and *vice versa.* There is probably substantial overlap between epiphytes and endophytes, just as there is among endophytes, saprotrophs and pathogens (Schulz *et al.,* 1999).

It is impressive that epiphyte and endophyte biotas, existing millimetres apart, are so distinct, and that the endophyte biota is as rich as the epiphyte biota. And since endophytes arriving on a the leaf presumably have to interact with the epiphytes living on the surface, interactions between epiphytes and endophytes may affect which endophytes are able to penetrate and establish themselves successfully. But little is known about interactions between epiphytes and endophytes.

Tissue Specificity

Just as fungal biotas differ between the surface and the interior of the leaf, they may also differ from one internal tissue to another. For instance, fungal biotas in leaf blades differ from those in petioles (Lodge *et al.*, 1996; see Carroll, 1995 for examples from conifers). In Douglas fir, *Rhabdocline parkeri* inhabits epidermal cells and *Phyllosticta abietis* spongy mesophyll cells (Stone, 1987). These fungi might be detected in the same leaf piece, but they inhabit different tissues and presumably do not interact much. Tissue specificity may reflect environmental differences among tissues: parts of the leaf may differ in relative humidity, oxygen concentration and CO_2 concentration, carbohydrate and protein availability, and temperature. Differences in tissue specificity may also reflect differences among microorganisms in how they enter the leaf, or host responses to the presence of microorganisms.

Sources of Inoculum

The endophytic habit implies a paradox (T.S. Eliot, *Four Quartets*):

> In order to arrive there,
> To arrive where you are, to get from where you are not,
>   You must go by a way wherein there is no ecstasy.
> In order to arrive at what you do not know
>   You must go by a way which is the way of ignorance.
> In order to possess what you do not possess
>   You must go by the way of dispossession.
> In order to arrive at what you are not
>   You must go through the way in which you are not.
> And what you do not know is the only thing you know
> And what you own is what you do not own
> And where you are is where you are not.

Since endophytes are (by definition, if not in fact) asymptomatic and entirely contained within living plant tissues, they must become something other than endophytes in order to sporulate, infect a new host, and become endophytic once more. If an organism were strictly endophytic in living plants, it would have limited opportunities for dispersal, unless it were dispersed by herbivores. However, it is likely that some fungi are accidental endophytes, and that certain populations in leaves are rarely able to reproduce and repeat the life cycle.

Endophytes can sporulate, disseminate and infect a new host in four ways: sporulation as an endophyte or pathogen within host tissues, followed by rupture of the tissues; sporulation as an epiphyte or pathogen on plant surfaces, or sporulation as a saprotroph in or on dead tissues. The spores or other forms of inoculum can be transmitted by wind, water, soil, direct contact, or an animal agent. The remaining possibility for transmission is vertical transmission through the seed.

Unlike some well-known grass endophytes in the Clavicipitaceae, it appears that most endophytes of woody, tropical plants are transmitted horizontally rather than vertically. Vertical transmission has been tested in several tropical plants, by isolations from surface-sterilised seeds. In cacao, neither seeds nor leaves of seedlings grown from surface-sterilised seeds yielded the endophytes found in leaves (Arnold *et al.,* 2003); in citrus and *Manilkara bidentata*, endophytes were not isolated from surface-sterilised seeds (Bayman *et al.*, 1998; Azevedo *et al.,* 2000). However, in *Casuarina equisitifolia*, *Xylaria* was isolated from seeds in some sites (Bayman *et al.,* 1998). Also, *Phomopsis casuarinae* was reported from the seeds of *C. equisitifolia*, and was assumed to be vertically inherited, though this assumption was not tested (see Bayman *et al.,* 1998).

Endophytes vs. Pathogens

Are endophytes latent pathogens? Recent studies have examined relationships between endophytes and pathogens using diverse and creative approaches. A phylogenetic approach was used to compare endophytes and pathogens of western white pine in the Rocky Mountains, USA (Ganley *et al.,* 2004). Ninety per cent of endophytes and three species of common pathogens belonged to the family Rhytismataceae. However, none of the 2,000 endophytes isolated was conspecific with the pathogens, as determined by ITS sequences of strains of each morphospecies. In most cases the closest known sequence was from a pathogen of a different species of pine. It is not clear if these fungi have specialised as endophytes or if they are pathogens of other hosts that do not cause disease on white pine; Ganley *et al.* preferred the former explanation, based on percent sequence divergence and geographic distribution of hosts.

In some cases, endophytes may be derived from pathogens that have lost pathogenicity. For example, a non-pathogenic endophyte was derived in the laboratory from a pathogenic *Colletotrichum* (Redman *et al.,* 1999). Also, phylogenetic trees that include both endophytes and pathogens have been interpreted as evidence of multiple, independent derivations of endophytism from parasitism (Arnold *et al.,* 2005). However, this approach assumes that a given organism is either an endophyte or a pathogen, and does not account for organisms that may switch back and forth depending on host and environment.

It is likely that some endophytes are opportunistic pathogens, or pathogens on certain plants and endophytes of others. Many common endophytes are conspecific with hemibiotrophic pathogens (Rodriguez *et al.,* 2004; Weber and Anke, Chapter 15). Endophytes of barley and larch produced more phytotoxic secondary metabolites than pathogens (Schulz *et al.,* 1999). This suggests that the endophytes had not lost pathogenicity, but that the plants were able to detect them and limit their growth. In this sense endophytism may be viewed as a balanced antagonism, in which each partner tries to attack the other (Schulz *et al.,* 1999). From this perspective, any factor that weakens the plant's ability to limit growth of the endophyte could allow the endophyte to become pathogenic.

Functional Aspects: Interactions of Endophytes with Other Microorganisms

Evidence of protective mutualisms between plants and endophytic fungi comes mostly from clavicipitaceous endophytes of grasses. The protection is most dramatic in cultivated grasses from temperate areas. These results have stimulated similar studies on wild grasses and woody plants, including studies on tropical plants, but protective roles for endophytes have been much more difficult to demonstrate than in temperate grasses (Azevedo *et al.,* 2000).

Most studies on interactions among endophytes, or between endophytes and pathogens, have been motivated by interest in disease control. There are four mechanisms by which an endophyte might reduce pathogen population size or virulence: by competition for resources, by production of toxic or inhibitory metabolites, by parasitism of the pathogen, or by triggering defensive responses in the plant (Redman *et al.,* 1999).

Some studies have established correlations between presence of endophytes and presence of pathogens, and some have tested inhibition of pathogen growth by endophytes *in vitro*. For example, fungi and actinomycetes isolated from rice (mainly *Fusarium* and *Streptomyces*) inhibited growth of rice pathogens *in vitro* (Tian *et al.,* 2004). (The pathogens tested were *Magnaporthe grisea*, *Rhizoctonia solani*, *Fusarium moniliforme* and *Xanthomonas oryzae* pv. *oryzae*.) Culture filtrates of the endophytes also inhibited pathogen growth, suggesting a role for secondary metabolites. Similarly, culture filtrates of endophytes from *Palicourea* and *Strychnos* inhibited pathogen growth *in vitro* (Lima de Souza *et al.,* 2004). In orchid and coffee leaves, presence of *Colletotrichum* in leaf fragments was negatively correlated with presence of *Xylaria* and *Guignardia*, possibly suggesting competition or antagonism (Santamaría and Bayman, 2005).

However, these studies tend to have two inherent limitations: 1) correlations in populations of microorganisms do not imply causality, and 2) interactions or metabolite production *in vitro* does not necessarily reflect the situation *in planta* (Weber and Anke, Chapter 15). Two of the most significant such studies involving tropical plants are discussed in the following paragraphs.

Populations of endophytic bacteria differed in citrus plants with and without *Xylella fastidiosa*, the causal agent of citrus variegated chlorosis (CVC) (Azevedo *et al.,* 2000; Lacava *et al.,* 2004). Branches with *X. fastidiosa* had significantly more total endophytic bacteria than branches without *X. fastidiosa*. For leaves, differences were not significant. *Methylobacterium* spp. were significantly more common in stems with CVC than in stems without. In contrast, *Curtobacterium flaccumfaciens* was significantly less common in leaves with CVC than in leaves without. Culture filtrate from *C. flaccumfaciens* reduced growth of *X. fastidiosa in vitro*, whereas filtrates from two species of *Methylobacterium* did not. The authors suggested that *Methylobacterium* may contribute to symptoms of CVC, while *C. flaccumfaciens* may produce metabolites that inhibit it, although no experimental evidence was provided. Genetically modified endophytic bacteria were inoculated into *Citrus* plants and recovered, suggesting that it is possible to alter the endophyte biota.

These results are promising for the Brazilian citrus industry, which has suffered hundreds of millions of dollars in losses because of CVC (Azevedo *et al.,* 2000).

Fungal endophytes were able to protect cacao leaves from damage by the pathogen *Phytophthora infestans* (Arnold *et al.,* 2003). Endophyte-free plants were grown from surface-sterilised seeds in a sterile greenhouse. Leaves were inoculated with a mixed suspension of spores from seven common endophytes, including *Xylaria*, *Colletotrichum*, and *Fusarium*; these plants (and control plants) were later challenged with the pathogen. Leaves inoculated with endophytes had significantly lower mortality and significantly smaller lesions than control plants without endophytes. The protective effect was most apparent in older leaves, which tend to have lower levels of chemical defences against fungi than younger leaves.

Functional Aspects: Effects of Endophytes on Plants

Effects of endophytes on photosynthesis, nitrogen fixation and resistance to stress have been studied. As with interactions between endophytes and microorganisms, the best-documented cases are clavicipitaceous endophytes of temperate grasses (Schardl *et al.*, 2004). Tropical plants often face different environmental challenges than temperate plants, so it is not clear if the results of these studies can be extrapolated.

Endophytes can affect photosynthesis in tropical plants, but the effects have not always been significant. Banana plants inoculated with *Colletotrichum musae* and maize plants inoculated with *Fusarium moniliforme* were compared to endophyte-free plants to determine effects of endophytes on photosynthesis (Rodrigues Costa Pinto *et al.,* 2000). Results varied between plants: in maize, *Fusarium* decreased total chlorophyll content, whereas in banana, *Colletotrichum* did not. In both plants, endophyte-infected plants had lower maximum photochemical capacity than endophyte-free plants, but differences were more pronounced for banana. Similarly, endophyte-infected seedlings of *Guarea guidonea* had lower photosynthetic rates than endophyte-free seedlings, but results were not consistent (Gamboa *et al.,* 2005).

The evidence that endophytes can help plants survive environmental stress has been recently reviewed; such symbioses might protect plants from some effects of global climate change (Rodriguez *et al.,* 2004). For example, *Dichanthelium lanuginosum* grows in geothermal areas of the western USA, where soil temperatures can reach 57° C. Plants with a *Curvularia* endophyte resisted high soil temperatures and drought better than plants without *Curvularia* (Redman *et al.,* 2002).

Nitrogen-fixing diazotrophic bacteria are common endophytes of tropical plants, especially grasses, and may contribute to N budgets of plants. The best-studied is *Acetobacter diazotrophicus* in sugarcane. N fertilisation decreased populations of *A. diazotrophicus* in sugarcane, suggesting that fertilisation may inhibit symbiotic nitrogen fixation (Fuentes-Ramires *et al.,* 1999). In some grasses diazotrophic bacteria are vertically transmitted. These bacteria are of interest as natural fertilisers.

Most studies on effects of endophytes on plant function have observed high levels of variation in the parameters measured. It is not clear how much of this variation is due to variation in endophytes among plants.

Conclusions

So how many fungal species are in a single, healthy leaf of a tropical tree? There are no direct answers in the literature, but an estimate can be extrapolated from various studies. Let's assume there are 25 culturable endophytes, based on estimates of 25-28 for *Manilkara bidentata* (Lodge *et al.,* 1996) and 26-33 for *Guarea guidonia* (Gamboa and Bayman, 2001). Comparison of epiphytes and endophytes on coffee suggest there are about equal numbers of each (Santamaría and Bayman, 2005). However, some of the common epiphytes were also encountered as endophytes, so we might estimate 20 additional species of epiphytes. However, although some of these fungi were identified by sequencing, all were isolated as colonies; how many more species were present but unculturable (at least on the culture media used)? This is difficult to estimate. For bacteria, the majority are unculturable (Yang *et al.,* 2001) but for fungi in general, it appears a lower percentage are unculturable. In a study on soil fungi, there was little overlap between the fungi identified by direct PCR amplification and those isolated as cultures, but no attempt was made to determine the size of either group (Viaud *et al.,* 2000). So assuming there are an equal number of unculturable fungi (a conservative estimate), another 45 species may be present. That gives a total of about 90 fungi in (and on) a single leaf. This question could be better answered using species abundance curves, as has been done for bacteria (Curtis and Sloan, 2004), but so far no species abundance curves have been constructed using both epiphytic and endophytic fungi, and both culturable and non-culturable fungi. A more robust answer will be possible when more extensive sequence data sets are available.

However, lack of sequence data is not the only reason that microbial diversity in leaves of tropical plants is still unknown. There are other reasons, many of which reflect the limitations of the studies that have been done in this field. First, most studies consider a single guild of a single group of organisms (e.g. epiphytic bacteria or endophytic fungi). Endophytic fungi are as likely to interact with bacteria as with other fungi, and endophytes must interact with epiphytes to enter the leaf, but each group is usually considered separately. Studies of endophytes of leaves are reminiscient of the parable of the blind men and the elephant, in which each man describes the elephant differently, depending on the part of the animal he has in his hand. Second, many groups are difficult to identify and taxonomically unresolved. Third, tissue specificity of endophytes is poorly understood. Fourth, sources of inoculum are often unknown. And fifth, sampling strategy affects results of all these studies, but is rarely considered carefully.

The variability of the microbiota from leaf to leaf influences many aspects of plant biology. However, many biologists who work with leaves do not take the phyllosphere into account, much less its variability. A more integrative approach to the biodiversity of the phyllosphere may lead to an increased awareness of its influence on other fields of research.

Acknowledgements

I thank Marcela Zalamea Bustillo for advice on the manuscript, and NOAA-SeaGrant, NIH-SCORE and NASA-IRA for financial support.

References

Andrews, J.H. and Harris, R.F. (2000) The ecology and biogeography of microorganisms on plant surfaces. *Annual Review of Phytopathology* 38,145-180.

Arnold, A.E., Maynard, Z. and Gilbert, G. (2001) Fungal endophytes in dicotyledonous neotropical trees: patterns of abundance and diversity. *Mycological Research* 105, 1502-1507.

Arnold, A.E., Mejía, L., Kyllo, D., Rojas, E., Maynard, Z. and Herre, E.A. (2003) Fungal endophytes limit pathogen damage in a tropical tree. *Proceedings of the National Academy of Sciences, USA* 100, 15649-15654.

Arnold, A.E., Miadlikowska, J., Higgins, K.L., Dalling, J.W., Gallery R.E., Henk, D.A., Eells, R.L.,Vilgalys, R.J. and Lutzoni, F. (2005) What can environmental PCR tell us about foliar fungal endophyte communities? *Inoculum* 56(4), 7 [abstract].

Azevedo, J.L., Maccheroni, W. Jr, Pereira, J.O. and Araújo, W.L. (2000) Endophytic micro-organisms: a review on insect control and recent advances on tropical plants. Electronic Journal of Biotechnology. http://www.ejbiotechnology.info/content/vol3/issue1/full/4/index.html.

Bailey, J.K., Deckert, R., Schweitzer, J.A., Rehill, B.J., Lindroth, R.L., Gehring, C. and Whitham, T.G. (2005) Host plant genetics affect hidden ecological players: links among *Populus*, condensed tannins, and fungal endophyte infection. *Canadian Journal of Botany* 83, 356-361.

Bayman, P., Angulo-Sandoval, P., Báez-Ortiz, Z. and Lodge, D.J. (1998) Distribution and dispersal of *Xylaria* endophytes in two tree species in Puerto Rico. *Mycological Research* 102, 944-948.

Bayman, P., Lebrón, L.L., Tremblay, R.L. and Lodge, D.J. (1997) Fungal endophytes in roots and leaves of *Lepanthes* (Orchidaceae). *New Phytologist* 135, 143-149.

Bills, G. (1996) Isolation and analysis of endophytic fungal communities from woody plants. In: Redlin, S.C. and Carris, L.M. (eds) *Endophytic Fungi in Grasses and Woody Plants*. APS Press, St. Paul, Minnesota, USA. pp. 31-65.

Cannon, P.F. and Simmons, C.M. (2002) Diversity and host preference of leaf endophytic fungi in the Iwokrama Forest Reserve, Guyana. *Mycologia* 94, 210-220.

Carroll, G. (1995) Forest endophytes: pattern and process. *Canadian Journal of Botany* 73, S1316-S1324.

Curtis, T.P. and Sloan, W.T. (2004) Prokaryotic diversity and its limits: microbial community structure in nature and implications for microbial ecology. *Current Opinion in Microbiology* 7, 221-226.

Davis, E.C., Franklin, J.B., Shaw, A. J. and Vilgalys, R. (2003) Endophytic *Xylaria* (Xylariaceae) among liverworts and angiosperms: phylogenetics, distribution, and symbiosis. *American Journal of Botany* 90, 1661-1667.

Frölich, J. and Hyde, K.D. (1999) Biodiversity of palm fungi in the tropics: are global fungal diversity estimates realistic? *Biodiversity and Conservation* 8, 977-1004.

Fuentes-Ramires, L.E., Caballero-Mellado, J., Sepulveda, J. and Martinez-Romero, E. (1999) Colonization of sugarcane by *Acetobacter diazotrophicus* is inhibited by high N-fertilization. *FEMS Microbiology and Ecology* 29, 117-128.

Gamboa, M.A. and Bayman, P. (2001) Communities of endophytic fungi in leaves of a tropical timber tree (*Guarea guidonia*: Meliaceae). *Biotropica* 33, 352-360.

Gamboa, M.A., Laureano, S. and Bayman. P. (2002) Does size matter? Estimating endophytic fungal diversity in leaf fragments. *Mycopathologia* 156, 41-45.

Gamboa, M.A., Wen, S., Fetcher, N. and Bayman, P. (2005) Effects of fungicides on endophytic fungi and photosynthesis in seedlings of a tropical tree, *Guarea guidonia* (Meliaceae). *Acta Biológica Colombiana* 10: 45-51.

Ganley, R.J., Brunsfeld, S.J. and Newcombe, G. (2004) A community of unknown, endophytic fungi in western white pine. *Proceedings of the National Academy of Sciences, USA* 101, 10107-10112.

Guo, L.D., Hyde, K.D. and Liew, E.C.Y. (2001) Detection and taxonomic placement of endophytic fungi within frond tissues of *Livistona chinensis* based on rDNA sequences. *Molecular Phylogenetics and Evolution* 20, 1-13.

Hawksworth, D.L. and Rossman, A.Y. (1997) Where are all the undescribed fungi? *Phytopathology* 87: 888-891.

Lacava, P.T., Araújo, W.L., Marcon, J., Maccheroni, Jr., W. and Azevedo, J.L. (2004) Interaction between endophytic bacteria from citrus plants and the phytopathogenic bacteria *Xylella fastidiosa*, causal agent of citrus-variegated chlorosis. *Letters in Applied Microbiology* 39, 55–59.

Lima de Souza, A.Q., Leão de Souza, A.D., Astolfi Filho, S., Belém Pinheiro, M.L., de Moura Sarquis, M.I. and Pereira, J.O. (2004) Atividade antimicrobiana de fungos endofíticos isolados de plantas tóxicas da amazônia: *Palicourea longiflora* (Aubl.) Rich e *Strychnos cogens* Bentham. *Acta Amazonica* 34, 185-195.

Lodge, D.J., Fisher, P.J. and Sutton, B.C. (1996) Endophytic fungi of *Manilkara bidentata* leaves in Puerto Rico. *Mycologia* 88, 733-738.

May, R.M. (1988) How many species are there on Earth? *Science* 241, 14419.

Pandey, A.K., Reddy, M.S. and Suryanarayanan, T.S. (2003) ITS RFLP and ITS sequence analysis of a foliar endophytic *Phyllosticta* from different tropical trees. *Mycological Research* 107, 439–444.

Peterson, S.W., Vega, F.E., Posada, F. and Nagai, C. (2005) *Penicillium coffeae,* a new endophytic species isolated from a coffee plant and its phylogenetic relationship to *P. fellutanum, P. thiersii* and *P. brocae* based on parsimony analysis of multilocus DNA sequences. *Mycologia* 97, 659-666.

Petrini, O. (1991) Fungal endophytes of tree leaves. In: Andrews, J. and Hirano, S.S. (eds) *Microbial Ecology of Leaves*. Springer-Verlag, NY, USA. pp. 179-197.

Petrini, O., Petrini, L.E. and Rodrigues, K. (1995) Xylariaceaous endophytes: an exercise in biodiversity. *Fitopatologica Brasiliensis* 20, 531-539.

Redman, R.S., Ranson J.C. and Rodriguez, R.J. (1999) Conversion of the pathogenic fungus *Colletotrichum magna* to a nonpathogenic, endophytic mutualist by gene disruption. *Molecular Plant–Microbe Interactions* 12, 969–975.

Redman, R.S., Sheehan, K.B., Stout, R.G., Rodriguez, R.J. and Henson, J.M. (2002) Thermotolerance conferred to plant host and fungal endophyte during mutualistic symbiosis. *Science* 298, 1581.

Rodrigues, K.F. (1994) The foliar fungal endophytes of the Amazonian palm *Euterpe oleracea*. *Mycologia* 86, 376-385.

Rodrigues, K.F. and Petrini, O. (1997) Biodiversity of endophytic fungi in tropical regions. In: Hyde, K.D. (ed.) *Biodiversity of Tropical Microfungi*. Hong Kong University Press, Hong Kong. pp. 57-69.

Rodrigues, K.F., Sieber, T.N., Grünig, C.R., and Holdenrieder, O. (2004) Characterization of *Guignardia mangiferae* isolated from tropical plants based on morphology, ISSR PCR amplifications and ITS1-5.8S-ITS2 sequences. *Mycological Research* 108, 45–52.

Rodrigues Costa Pinto, L.S., Azevedo, J.L., Pereira, J.O., Carneiro Vieira, M.L. and Labate, C.A. (2000) Symptomless infection of banana and maize by endophytic fungi impairs photosynthetic efficiency. *New Phytologist* 147, 609-615.

Rodriguez, R.J., Redman, R.S. and Henson, J.M. (2004) The role of fungal symbioses in the adaptation of plants to high stress environments. *Mitigation and Adaptation Strategies for Global Change* 9, 261-272.

Rogers, J.D. (2000) Thoughts and musings on tropical Xylariaceae. *Mycological Research* 104, 1412-1420.

Santamaría, J. and Bayman, P. (2005) Fungal epiphytes and endophytes of coffee leaves (*Coffea arabica*). *Microbial Ecology* 50, 1-8.

Schardl, C.L., Leuchtmann, A. and Spiering, M.J. (2004) Symbioses of grasses with seedborne fungal endophytes. *Annual Review of Plant Biology* 55, 315-40.

Schulz, B., Rommert, A.K., Dammann, U., Aust, H.J. and Strack, D. (1999) The endophyte-host interaction: a balanced antagonism? *Mycological Research* 103, 1275-1283.

Schulz, B., Wanke, U., Draeger, S. and Aust, H.J. (1993) Endophytes from herbaceous plants and shrubs: effectiveness of surface sterilisation methods. *Mycological Research* 97, 1447-1450.

Stone, J.K. (1987) Initiation and development of latent infections by *Rhabdocline parkeri* on Douglas-fir. *Canadian Journal of Botany* 65, 2614-2621.

Stone, J.K., Bacon, C.W. and White, J.F. (2000) An overview of endophytic microbes: endophytism defined. In: Bacon, C.W. and White, J.F. (eds) *Microbial Endophytes*. Marcel Dekker, New York, USA. pp. 3-29.

Suryanarayanan, T.S., Murali, T.S. and Venkatesan, G. (2002) Occurrence and distribution of fungal endophytes in tropical forests across a rainfall gradient. *Canadian Journal of Botany* 80, 818-826.

Suske, J. and Acker, G. (1989) Identification of endophytic hyphae of *Lophodermium piceae* in tissues of green, symptomless Norway spruce needles by immunoelectron microscopy. *Canadian Journal of Botany* 67, 1768-1774.

Tian, X.L., Cao, L.X., Tan, H.M., Zeng, Q.G., Jia, Y.Y., Han, W.Q. and Zhou. S.N. (2004) Study on the communities of endophytic fungi and endophytic actinomycetes from rice and their antipathogenic activities *in vitro*. *World Journal of Microbiology and Biotechnology* 20, 303–309.

Viaud, M., Pasquier, A. and Brygoo Y. (2000) Diversity of soil fungi studied by PCR-RFLP of ITS. *Mycological Research* 104, 1027-1032.

Whalley, A.J.S. (1996) The xylariaceous way of life. *Mycological Research* 100, 897-922.

Yang, C.H., Crowley, D.E., Borneman, J., Keen, N.T. (2001) Microbial phyllosphere populations are more complex than previously realized. *Proceedings of the National Academy of Sciences, USA* 98, 3889-3894.

Chapter 4

Microorganisms in the Phyllosphere of Temperate Forest Ecosystems in a Changing Environment

T. Müller[1], K. Strobel[2] and A. Ulrich[1]
[1]Leibniz-Centre for Agricultural Landscape Research (ZALF) e.V. Müncheberg, Institute of Landscape Matter Dynamics, Eberswalder Str. 84, D-15374 Müncheberg, Germany; [2]Bayreuth Institute for Terrestrial Ecosystem Research, University of Bayreuth, D-95440, Germany
tmueller@zalf.de

Introduction

Temperate forests are summer-green deciduous, evergreen coniferous or mixed forests in climatic zones with an approximate six-month growth season. They occur mainly in the northern hemisphere in three major regions, namely western and central Europe, eastern Asia, and eastern North America. The area of these forests covers 2 million km^2, which is about 8% of the total vegetated continental area (Whittaker and Likens, 1975). The total leaf surface area of all temperate forests on the earth is 95 million km^2, with 60 million km^2 evergreen, and 35 million km^2 deciduous, i.e. about 15% of the total leaf area on earth (Whittaker and Likens, 1975).

Forests play a major role in climate stabilisation and water management, and may provide an important source of renewable energy for the future. However, they are subject to progressive global change processes that are characterised by climatic stress, rising atmospheric CO_2 concentration (Anon., 2001), increase in UV radiation resulting from the erosion of the stratospheric ozone layer, eutrophication due to increasing atmospheric N deposition, and a large-scale acidification of soils and water, due to acid deposition that is linked to SO_2 and N emissions (Bouwman *et al.*, 2002). Therefore, the responses of temperate forest ecosystems to environmental changes have received considerable attention during the last few decades. With the area of these forests increasing in many industrial countries, and with forests becoming a major focus in current endeavours for engineering the biology of the global carbon cycle, these aspects will also be of considerable interest in the future.

The changing environmental conditions are likely to affect nutrient cycling and energy flows in ecosystems. The identification and elucidation of interactions involved in these processes is crucial for predicting the inorganic and organic matter dynamics and it is suggested that in understanding these interactions, one needs to consider the relationships between biological processes in different ecosystem strata, above and below ground (Schulze, 2000).

Ecosystem studies on organic matter cycling in temperate forests have suggested the forest floor as the primary source of dissolved organic matter (DOM) (Andersen and Gundersen, 2000). However, the exchange of matter between trees and their environment takes place not only in soil through the roots, but also via the canopies (Harrison *et al.*, 2000). It is very likely that microorganisms colonising the surface of leaves and needles take up parts of this matter for assimilation and metabolic processes and release it again in particulate form, either as cell mass or as DOM after cell death and lysis.

In light of the large leaf surface area of trees, the response of forest ecosystems to atmospheric pollutants, and the potential change in the transformation processes due to microorganisms in the phyllosphere, forest canopies can be expected to be as important for DOM processes as soil. In particular, through the colonisation of the leaf surface, active transformation of organic matter directly links microorganisms with large-scale biogeochemical cycles. In this chapter, we will explore this special role of phyllosphere microorganisms which have so far received little attention in the literature on ecosystem function, we also present some of our findings on their responses to air pollutants and insect infestation, and point out possible consequences for the flow of matter in temperate forest ecosystems.

Exchange of Organic Matter via the Canopy of Trees in Temperate Forests

The average daily net primary production, the net carbon gain, of temperate forests is 6.2 g carbon per m^2 of ground area during the growth season. This is slightly less than that of tropical forests (6.8 g carbon per m^2 and day), but twice as much as that of crops (Gower, 2002).

The leaves take up atmospheric CO_2, as well as pollutant N, from wet, dry and gaseous deposition in exchange for O_2, CO_2 and water, as well as carbohydrates and proteins that are produced in their primary metabolism, plus many secondary compounds, such as isoprene and other volatile compounds, resins, tannins and terpenoids (Roshchina and Roshchina, 1993; Harrison *et al.*, 2000). Additionally, all above-ground parts of trees are leached by aqueous solutions, such as rain, dew and fog (Tukey, 1970). The leached compounds reach the forest floor with the throughfall precipitation (canopy drip plus stemflow). The chemical composition of the throughfall is the result of canopy exchange processes, which are characterised by the leaching and uptake of materials derived from foliage, woody parts and epiphytic microorganisms, and the solution flowing over them. The throughfall is rich in dissolved organic C, N and P (Qualls *et al.,* 1991). When quantifying the role of dissolved organic carbon (DOC) and nitrogen (DON) in the C and N cycles of Central European deciduous forest ecosystems, Solinger *et al.* (2001) showed that the total annual DON fluxes in the throughfall of beech and oak are nearly as high as those in forest floor leachates, and represent a significant portion of the total N turnover in forests (Fig. 4.1). Only the DOC flux in the Oa layer (forest floor, litter, humus) exceeded that in the throughfall by about four-fold. The chemical composition of DON in the throughfall and the leachates of the humus layer in a Norway spruce forest were similar: 60-80% of the total soluble N was amino N, and 6-13% originated from amino sugar N (Michalzik and Matzner, 1999). According to Yu *et al.* (2002), DON accounted for 77-99% of the total dissolved N in forest floor leachates. The DON fraction consists of 48-74% combined amino acids, suggesting that proteins and peptides were the main components of DON in

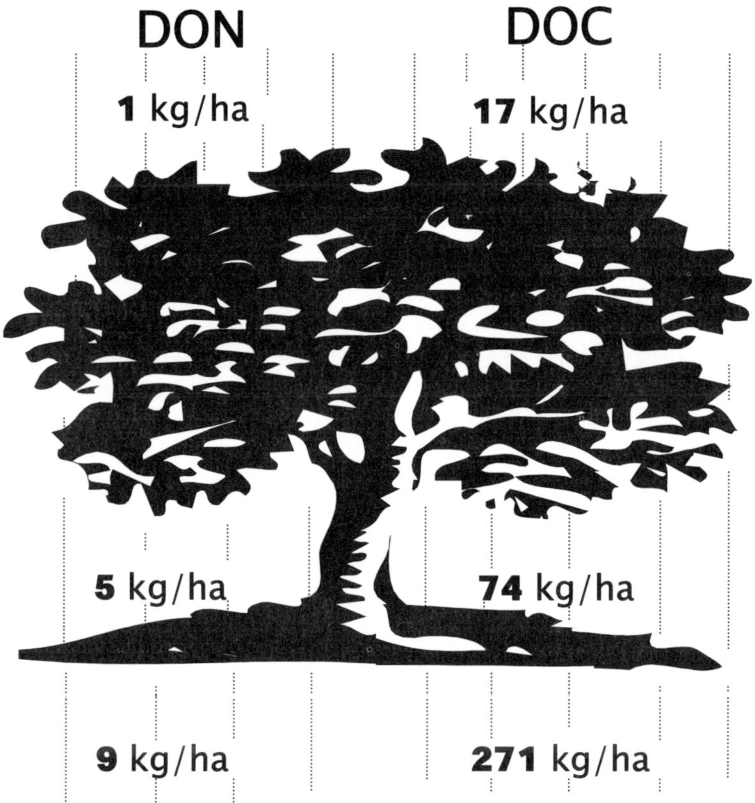

Fig. 4.1. Annual fluxes of dissolved organic nitrogen (DON) and carbon (DOC) in precipitation, throughfall and humus layer of a Central European deciduous forest according to Solinger *et al.*, (2001). Means of data measured in 1997 and 1998.

these leachates. Nitrogen in free amino acids and alkyl amines ranged from 1.5-10.6% of the total DON content. Low molecular weight DON compounds (e.g. urea, amino acids, polyamines and small polypeptides) are taken up by plant roots and microorganisms, whereas most DON in the soil solution is suggested to be of a high molecular weight recalcitrant nature (Jones *et al.,* 2005).

Microorganisms of the Phyllosphere in Temperate Forests Involved in C and N Transformation Processes

Additional to the high net primary production, temperate forests are characterised by a high DOM production, with canopy leaching and litter decomposition as the primary processes (McDowell and Likens, 1988, Table 4.1). Microbial processes play a central role in the transformation and degradation of these compounds in all compartments of the ecosystem.

Table 4.1. Fluxes of dissolved organic matter in bulk precipitation and throughfall of coniferous and deciduous forest trees (kg per hectare per year).

DON		DOC		Forest type	Reference
Bulk precipitation	Throughfall	Bulk precipitation	Throughfall		
		4.0	23.3	deciduous	McDowell and Likens (1988)
		15.3-18.4	35.3-51.6	coniferous	Guggenberger and Zech (1994)
1.0	3.6	18.0	83.0	coniferous	Michalzik and Matzner (1999)
1.2-1.5	4.7-5.3	11.9-22.0	53.4-94.6	deciduous	Solinger *et al.* (2001)

The processes of C and N cycling in the soil and litter of temperate forests have been well investigated. For example, Persson *et al.* (2000) had determined the C mineralisation rate, which was dependent on field N-fertilisation, as a function of soil temperature and moisture in conifer and broadleaf forest stands, and had calculated the turnover time of carbon for different soil layers and sites. The influence of N inputs through biological fixation, as well as wet and dry deposition on forest structure and function that include plant uptake and soil processes was the subject of a survey given by Nadelhoffer (2001). However, the canopy as a habitat for microorganisms colonising leaves has so far hardly been considered in this context. One study by Guggenberger and Zech (1994) found higher proportions of hexoses and deoxy-hexoses as compared with pentoses in the canopy precipitation throughfall of conifers. The ratio of (man + gal)/(ara + xyl) was about 3, which indicates a predominance of microbial metabolites in the fraction of carbohydrates. They suggested that microbial C transformation takes place on the needles. If this is correct, the influence of the canopy on nutrient cycling reaches far beyond its role as a source of leaf litter, since in this case, the canopy is a habitat of active microbial transformation processes that affect the vertical nutrient fluxes. This suggestion is also supported by investigations by Papen *et al.* (2002) and Müller *et al.* (2004), who found that microorganisms are involved in N transformations in the phyllosphere of conifers. Papen *et al.* (2002) presented evidence of chemolithoautotrophic ammonia and nitrite oxidizers colonising, in appreciable cell numbers, the phyllosphere of spruce trees in a forest ecosystem exposed for decades to high levels of atmospheric nitrogen. These bacteria are mainly located inside the spruce needles, most likely within the stomatal cavities. They are characterised by high substrate turnover rates, increased mineral atmospheric nitrogen uptake, and hence may affect the above-ground N supply of trees. Müller *et al.* (2004) identified high activities of leucine arylamidase (EC 3.4.11.2) in the washings of needles from the canopy of Scots pine. This indicates the activity of an exopeptidase in the final part of the proteolytic cascade, in which complex proteins are broken into smaller fragments by proteinase enzymes (endo-peptidases). Furthermore, these peptides serve as substrates for exo-peptidases that release terminal amino acids. It was shown that this exopeptidase activity was due to epiphytic microorganisms on the needles, and elimination of these microorganisms by surface sterilisation resulted in a loss of about 75% of the enzyme activity in the washings of untreated needles. The additional disintegration of the needle tissue with pestle and mortar led to similar results, indicating that the leucine arylamidase largely originated from microbial epiphytes and not from the plant tissue. Strains of bacteria and yeasts isolated from the leaves of forest trees were tested for their potential leucine arylamidase activity. About 80% of all bacterial strains (n = 126) and more than half of the yeast strains (n = 60) were arylamidase positive. Most of this enzyme activity was shown to be associated with living cells of the cultures. In the washings of Scots pine, the activity of the enzyme correlated positively with the number of culturable bacteria on the needles.

Responses of Phyllosphere Microorganisms to Environmental Changes

General Responses of Microorganisms

Microorganisms generally are highly responsive to environmental changes because their large surface in relation to their small volume facilitates closer contact with their environment. Emissions of air pollutants and increased temperature or atmospheric CO_2 concentrations have been shown to alter microbial activities and/or population densities and structures in terrestrial ecosystems. For example, elevated atmospheric N depositions in northern hardwood forests affected the microbial community composition in soil, as assessed using phospholipid fatty acid (PLFA) analysis (Waldrop *et al.*, 2004). Bacteria and endomycorrhizal fungi decreased in relative abundance, but two Gram-negative PLFAs and a fungal PLFA increased. In contrast, the ratio of fungal to bacterial activity measured by isotope incorporation into fungal and bacterial lipids revealed a significant decrease in fungal:bacterial activity with N deposition. Microbial soil biomass increased by 29% and microbial respiration by 63% under elevated N supply in a black oak/white oak forest.

Ultraviolet radiation, especially in the shortest and most energetic UV-B wavelengths, is particularly damaging to microbial cells. Smits *et al.* (1996) demonstrated that a given UV-B diffuse radiation was up to twice as detrimental as the same amount of total solar radiation for the conidial persistence of the hyphomycete *Paecilomyces fumosoroseus*.

Otherwise, microbial communities have an amazing capacity to adapt to new situations and to evolve under selection pressure. The great diversity of heterotrophic microorganisms, for example in soil, is a major reason of a functional redundancy in the community. Therefore, microbial processes of C and N mineralisation are affected less by, for example, acid rain than processes that rely on a more restricted community, such as autotrophic nitrifiers (Myrold and Nason, 1992).

Responses of Phyllosphere Microorganisms to Gaseous Pollutants and UV-B Radiation

Studies in the 1980s showed that phyllosphere communities or parts of them are affected by pollutants from anthropogenic sources such as SO_2, NO_x, SO_4^{2-} , and other emissions from fossil fuel combustion, fertiliser production, intensive animal husbandry etc. (for review see Magan and McLeod, 1991a). For example, the effects of SO_2 on the mycobiota populations on ripening barley were determined with an open-air fumigation system (Magan and McLeod, 1991b). In general, the total epiphytic populations were reduced by a SO_2 concentration of 0.05 ppm. *Sporobolomyces roseus* and *Cryptococcus* spp. were sensitive to this concentration. However, populations of *Aureobasidium pullulans* were significantly greater on flag leaves treated with up to 0.038 ppm SO_2, while those of *Cladosporium* spp. were relatively unaffected by these treatments.

While many of these studies focused on epiphytic yeast and filamentous fungal populations, little attention has been given to bacteria. Additionally, Magan and McLeod (1991a) pointed out the lack of studies on short-term and long-term effects of pollutant depositions on epiphytic microorganisms and regarding interactions between several environmental factors and pollutants.

Under ambient UV-B level, *Aureobasidium pullulans* and *Sporobolomyces roseus* colonised preferentially the abaxial (lower) surface of oak saplings' leaves, but a 30% elevation above the ambient level on the adaxial (upper) surface of the leaves did reduce both species significantly (Newsham *et al.*, 1997). These results suggest that current levels of short wavelength radiation already influence the distribution of fungi on leaf surfaces, and that future increases in UV-B radiation will directly affect the abundances of specific phylloplane fungi. Jacobs and Sundin (2001) found that solar UV-B radiation did not affect the total population size of culturable bacteria on the leaves of peanut plants. However, it did alter the composition of the bacterial community. It is apparent that solar UV-B does exert some selection pressure that impacts the survival and relative fitness of microorganisms in the phyllosphere (Sundin, 2002). Adaptations to UV radiation are: the colonisation of protected sites of the leaves or preferential growth lower in the plant canopy; pigmentation in bacteria and fungi; and enhanced DNA repair capabilities (for review see Sundin, 2002).

Response of Phyllosphere Microorganisms to Insect Infestation

Environmental change from a microbial point of view is not only a consequence of the deposition of pollutants, but can also result from the presence of leaf-feeding or sap-sucking insects. Particularly instructive examples come from accidentally or deliberately introduced insect species, which sometimes develop into pests. For example, the hemlock woolly adelgid (*Adelges tsugae*) was introduced in the 1950s to eastern North America, where it fed destructively on hemlock trees (*Tsuga canadensis* and *T. carolineana*) resulting in accelerated rates of needle loss and ultimately in the demise of those trees within a time span of 10-15 years (McClure, 1989). Insects are also likely to show pronounced responses to pollutants, climate change and atmospheric CO_2 increase in forest ecosystems. Their marked phenotypic plasticity, high generation and offspring numbers are beneficial traits to adjust quickly to environmental perturbations. For example, under elevated sulphur, nitrogen or ozone deposition in many parts of Europe and North America, aphids often showed increased growth rates on their respective host plants (Whittaker, 2001). The expectation is that an increased abundance of insects also affects the abundance and community structure of phyllosphere microorganisms. For example, Stadler *et al.* (2005) showed that hemlock woolly adelgids do affect the abundance of phyllosphere microorganisms co-occurring on the same trees: culturable bacteria, yeasts and filamentous fungi were 2-3 log-units more abundant on medium and heavily infested trees than uninfested trees.

Effects of Aphids on Microbial Growth, Enzyme Activity and Organic Matter Flows

The abundance of phytophagous insects, such as aphids, is affected by plant quality, natural enemies and weather conditions. Changing climatic and environmental conditions are suggested to affect plant susceptibility to herbivores and may lead to higher frequencies of outbreaks of insect pests (Ayres, 1993). Temperature is the main abiotic factor controlling aphid development. Their prodigious rates of population increase contribute to the pest status of many species. Aphids feed on the phloem sap of their host plants and excrete copious amounts of honeydew; for example, between 400 and 700 kg fresh mass per hectare per year were estimated for different *Cinara* species on conifers (Eckloff, 1972). Honeydew is low in nitrogen, but rich in carbon and contains mainly glucose, fructose and melezitose (Maurizio, 1985). Much of this substrate is consumed by other insects, but considerable amounts might be available for microorganisms in the phyllosphere of the trees. In contrast to these energy-rich carbon compounds in honeydew, nitrogen does not seem to be a limited resource to microorganisms in the forests of Central Europe. This is because ammonia, as well as nitric oxide and nitrogen dioxide (NO_x), are the primary compounds of the atmospheric N deposition in these forest ecosystems (Harrison *et al.*, 2000) and are thus abundantly available in the canopies of trees. The potential supply with energy from honeydew suggests that a microbial phyllosphere community might benefit from an infestation of their host plants by phytophagous insects. Dik and van Pelt (1992) showed that honeydew on wheat leaves can stimulate the growth of pathogenic filamentous fungi, while bacteria remained unaffected. This is in contrast to our studies on microbial groups colonising the needles and leaves of forest trees (Stadler and Müller, 1996; 2000), where the growth of moulds, yeasts, and particularly bacteria was found to be stimulated by aphid honeydew. When compared to uninfested trees, significantly larger numbers of colony forming units (CFU) of bacteria and yeasts were found on the needles and leaves of Norway spruce (*Picea abies*), beech (*Fagus sylvatica*) and oak (*Quercus petraea*) that were infested with aphids in the summer, shortly after aphid populations have peaked. The most pronounced differences were recorded on spruce infested by *Cinara pilicornis*, with almost 3 log-units for bacteria and 2 log-units for yeasts. The honeydew dry mass excreted by this species was significantly larger when compared to aphids on beech and oak (Stadler and Müller, 2000). *Cinara pilicornis* produced large droplets of honeydew which were available readily to microorganisms under moist conditions. The lowest amount of honeydew was excreted by *Phyllaphis phagi*. This species feeds on beech and produces wax wool, which is less available to microorganisms. Consequently, the differences in the numbers of CFUs of bacteria and yeasts on infested and uninfested leaves (0.5 and 0.6 log-units, respectively) were less pronounced on beech leaves (Stadler and Müller, 2000).

The infestation of forest trees by aphids affects not only the densities of phyllosphere microorganisms, but also several ecosystem processes. For example, fluxes of inorganic nitrogen (NH_4-N, NO_3-N) are often significantly reduced and those of dissolved organic C and N increased in the throughfall beneath infested trees (Stadler *et al.*, 2001a; Mühlenberg and Stadler, 2005). It is very likely that microbial metabolism is involved in these transformation processes, since it is known that phyllosphere microorganisms assimilate NH_4-N, as well as NO_3-N. Indeed, 74% (n = 83) of bacterial isolates from oak leaves and 61% (n = 168) of those from beech leaves were able to use both N-sources (Müller *et al.*, 2003), as it increased fluxes of amino-sugar N within the DON fraction of the throughfall beneath infested trees recorded (Mühlenberg and Stadler, 2005). Amino-sugar N only originate only from the cell walls of microorganisms and not from plants.

Where aphid populations were small or absent but residues of honeydew and wax wool were still visible on the leaves, investigations on the densities of the bacterial and fungal populations, microbial enzyme activities and concentrations of organic matter in the washings of both beech leaves that were infested or uninfested with *Phyllaphis phagi,* were carried out recently (Fig. 4.2). The number of bacteria and filamentous fungi increased significantly under the influence of aphids, showing an extended positive effect of the phytophagous insects beyond their physical presence. Analysing the bacterial community of infested and uninfested leaves by T-RFLP of 16S rDNA revealed strong differences in the community structure (Fig. 4.3). Whereas the analysis of the replicates resulted in similar profiles, differences in both the presence and absence of T-RFs, as well as the relative abundance of certain peaks, were obvious in the profiles derived from infested and uninfested leaves. This suggests a shift in the dominant groups of the bacterial community due to the infestation by *P. phagi*.

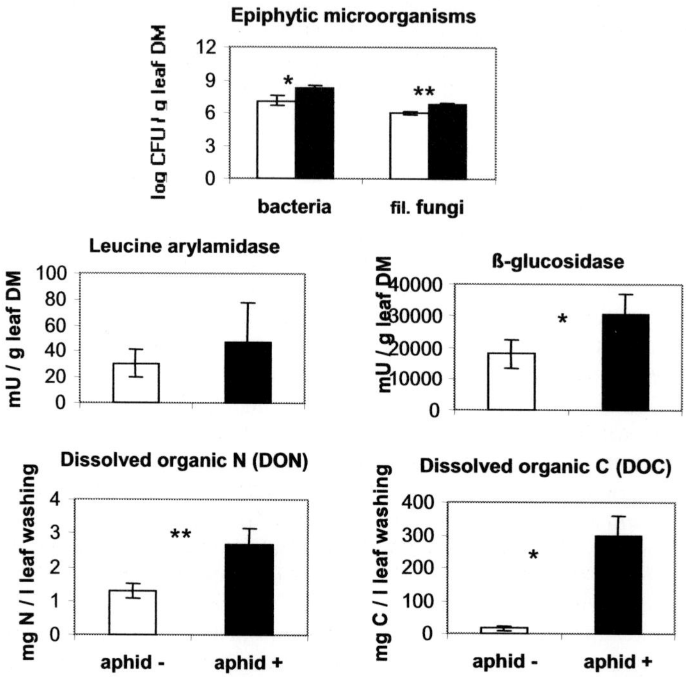

Fig. 4.2. Densities of phyllosphere microorganisms, enzyme activities, and concentrations of organic matter in the washings of beech leaves infested (aphid +) or uninfested (aphid -) with *Phyllaphis phagi*. Each point is the mean of three samples, error bars indicate the standard deviations. Means are significantly different (*t*-test:, * $P < 0.05$).

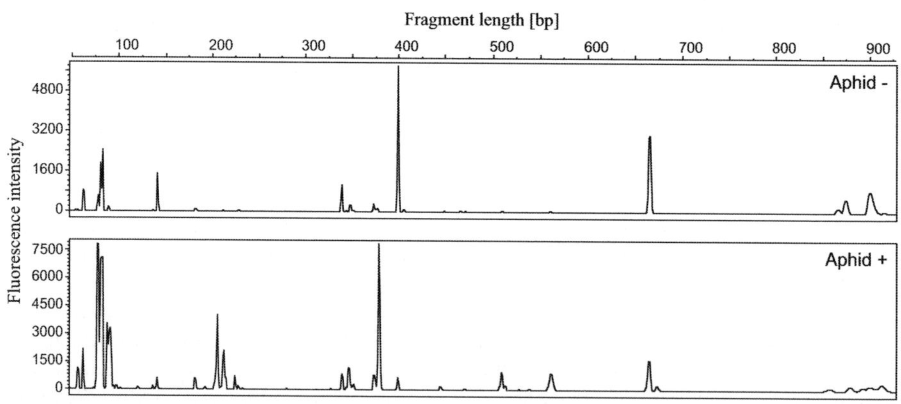

Fig. 4.3. T-RFLP profiles of bacterial communities derived from leaves infested (Aphid+) or uninfested (Aphid-) with *Phyllaphis phagi*. The profiles were based on 16S rDNA amplified from total community DNA with the primers 799f (labelled with 6-FAM) (Chelius and Triplett, 2001) and 1525r. Terminal fragments were generated by a *Hha*I digestion.

The leucine arylamidase activity was not significantly different on leaves from infested and uninfested trees (Fig. 4.2). However, the activity of β-glucosidase (EC 3.2.1.21) showed a significant response to the presence of aphids. This enzyme catalyses the hydrolysis of carbohydrates with β-D-glycosidic bonds and plays an important role in the complete degradation of cellulose to glucose. Therefore, it can be used as an indicator of C-cycling in the phyllosphere of forest trees. β-glucosidase has been detected in bacteria, yeasts, filamentous fungi, plants and animals. We have also determined it in the tissue of beech leaves after surface sterilisation of intact leaves, but not on their surface. Cultures of bacteria and yeasts isolated from leaves of beech and oak were tested for their potential β-glucosidase activity under standardised conditions in the laboratory. About 79% of all bacterial (n = 141) and 82% of the yeast isolates (n = 49) were positive (Müller, unpublished). The method we used to measure β-glucosidase activity on leaves was a modified procedure that originated from soil microbiology (Alef, 1991). The activity of this enzyme was significantly higher on beech leaves infested with *P. phagi* as compared to uninfested ones (Fig. 4.2). The increased enzyme activity is very likely to be an indicator of elevated activity of epiphytic microbial populations, which thrive on the availability of the energy-rich carbohydrates in honeydew. This input was reflected in an 18-fold increase of the DOC concentration in the washings of aphid infested leaves. There was also a significant two-fold increase of DON in the washings of infested leaves in comparison to uninfested. While DOC originates undoubtedly from the excreta of aphids, DON might reflect increased enzyme activities, metabolism and cell lysis of phyllosphere microorganisms. Ultimately, high amounts of organic matter reach the forest floor as a result of interactions between aphids and microorganisms in the canopy of forest trees. This might have major impacts on forest floor processes such as decomposition, soil respiration and nutrient cycling (Reynolds and Hunter, 2001).

Effects of Insect Invasion and Elevated Nitrogen and Sulphur Deposition

A manipulation experiment with Sitka spruce (*Picea sitchensis*) in Scotland was undertaken in order to investigate the relative effects of increasing numbers of invasive insects under stressful conditions on phyllosphere microorganisms. The trees were either infested or uninfested with the green spruce aphid (*Elatobium abietinum*) and were sprayed with rain containing elevated concentrations of nitrogen and sulphur (Sheppard *et al.*, 1999; Stadler *et al.*, 2001b).

The green spruce aphid is an example of a pest species, widely distributed throughout Europe, that causes serious defoliation and growth loss of *Picea sitchensis* but not of its native host *Picea abies*. They grow to large numbers and cause heavy needle losses especially in trees under stress. The frequency and intensity of *E. abietinum* infestations are expected to increase in response to climate change and accidental introduction into new environments.

The experimental design was described in Sheppard *et al.* (1999) and is summarised here. Treatments were made up in rainwater collected at the site and applied to the upper canopy as mist droplets from full cone sprayers. The N treatment provided additional N at 48 kg per hectare per year, the S treatment 50 kg S per hectare per year, and the NS acid treatment at pH 2.5 provided the same amounts of S and N. The untreated Control received no spray. Treatments were supplied as NH_4NO_3 (N), Na_2SO_4 (S), and NH_4NO_3 + H_2SO_4 (NS Acid). Background N deposition at the site was approximately 8 kg N per hectare (wet + dry) and the treatment solution provided the sprayed trees with an additional 10% rainfall. The mean pH of the rainwater was about 4.5. One-year-old shoots were sampled from the periphery of the canopies after aphid population peak.

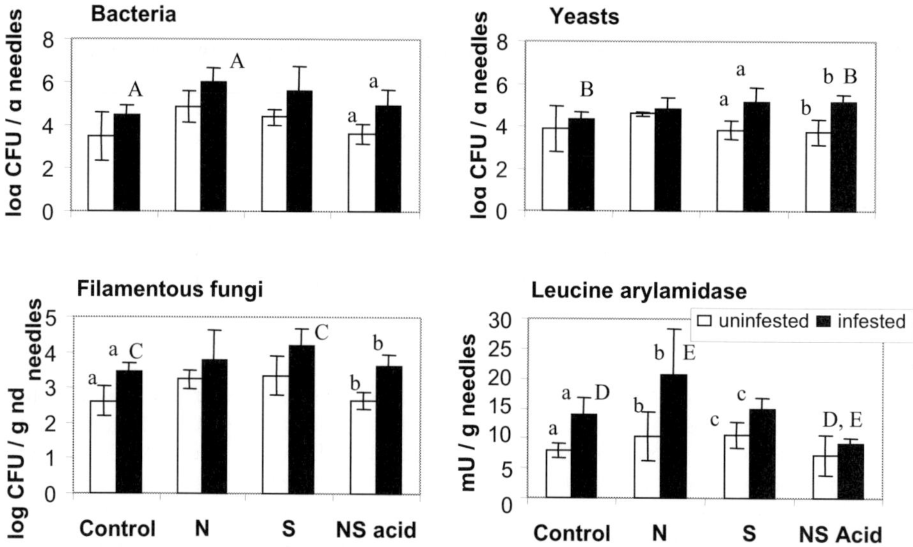

Fig. 4.4. Densities of epiphytic microorganisms and activities of the leucine arylamidase on needles of Sitka spruce infested or uninfested with *Elatobium abietinum* and exposed or unexposed (Control) to elevated concentrations of NH_4NO_3 (N), Na_2SO_4 (S), and NH_4NO_3 + H_2SO_4 (NS Acid). Each point is the mean of five samples, error bars indicate the standard deviations. Equal small letters indicate significant differences (*t*-test, $P < 0.05$) between means of infested and uninfested trees of the same treatment. Capital letters refer to differences between treatment groups.

The effects of the infestation of Sitka spruce needles by *E. abietinum*, as well as those of elevated nitrogen and sulphur depositions, on the growth of epiphytic microorganisms and the activity of the leucine arylamidase are shown in Fig. 4.4. Cell densities of all microorganism groups, as well as enzyme activity, were increased by *E. abietinum* infestation, and the arylamidase activity correlated positively with the density of bacteria on the spruce needles ($r^2 = 0.506$, $P = 0.01$). In the NS acid treatment, *E. abietinum* showed the highest infestation rates (Stadler *et al.*, 2001b) and the responses of all microbial groups to the presence of aphids were most pronounced. However, arylamidase activities did not increase under these conditions. One reason for the lower enzyme activities seems to be related to the low pH of the mist droplets applied. The pH optimum of this exopeptidase is in the neutral range.

Irrespective of treatment, DOC fluxes in the throughfall generally were higher beneath aphid infested shoots. Multivariate analyses showed that DON fluxes were affected primarily by the application of the pollutants, whereas the DOC fluxes were affected significantly by the aphids (Stadler *et al.*, 2001a). It was strikingly clear that the various pollutants only caused an increase in the enzyme activities or microbial densities when the trees were infested simultaneously by *E. abietinum*. This is likely to be because honeydew provides energy for epiphytic microorganisms, as was discussed above. Aphid infestation alone had significant effects on most parameters in various treatments, indicating a stronger influence of this biotic environmental factor on microorganisms than that of pollutants. This suggests that environmental perturbation is more complex than previously anticipated, and that the impacts of atmospheric pollutants on forest ecosystems need to be considered in more detail, especially in combination with insect-microbial interactions in the phyllosphere.

Effects of CO_2 Enrichment in the Atmosphere on Phyllosphere Microorganisms

Although recent studies have indicated the positive effect of the predicted elevated atmospheric CO_2 concentration on the physiology and growth performance of woody plants, little is known how a microbial community colonising the leaf surface is affected in a high CO_2 environment. Changes in the communities' compositions are very likely, because of differences in CO_2 tolerance between microbial groups, genera and species. For example, it was found that the total fungal population on winter wheat leaves had thrived in an open-air CO_2 enriched environment, whereas yeasts had not been affected markedly (Magan and Baxter, 1996). The fungal colony-forming cells of the genus *Cladosporium* had increased significantly, but common leaf colonisers such as *Aureobasidium pullulans*, *Epicoccum nigrum* and fusaria were seldom isolated.

Large-scale free-air CO_2 enrichment (FACE) experiments enable canopy research to determine the community response to rising atmospheric CO_2 concentrations under natural growth conditions and a complexity of biotic interactions. The Swiss Canopy Crane Project (SCC) at Hofstetten near Basel, Switzerland aims to elucidate the response of different tree species to elevated atmospheric CO_2 concentrations, with respect to canopy architecture, functional leaf attributes, reproductive dynamics, changes in storage reserves, sapflow and the dynamics of stable isotopes. The site represents a typically high altitude, mature mixed forest consisting of about 120-year-old conifers and deciduous trees (http://www.unibas.ch/botschoen/07/06/e.shtml). The CO_2 exposure system consists of a CO_2-release system that is installed within the crown of adult trees with the use of a 45 m tower crane, a CO_2 monitoring system and an automated regulation system (Pepin and Körner, 2002). Pure CO_2 gas is released from a network of small tubes that have been woven into the canopy. The seasonal daytime mean CO_2 concentration is 520 µmol per mol, i.e. about 40% above the current atmospheric level.

The unique opportunity to obtain samples was seized, in order to investigate microbiological parameters in the phyllosphere of hornbeam (*Carpinus betulus*) trees in June 2004. To collect the throughfall, samplers were installed within the canopies that were either exposed to ambient or elevated atmospheric CO_2 concentrations.

Preliminary analyses of the growth of the microbial epiphytes revealed different responses to enriched CO_2 in different microbial groups. Whereas the population densities of bacteria and yeasts hardly seemed to be influenced, fungi appear to be suppressed in an elevated CO_2 atmosphere (Fig. 4.5). Similarly, the structure of the bacterial community as analysed by T-RFLP displayed only a comparably low, negligible effect of the CO_2 application (Fig. 4.6).

Because of high variations within the sparse data obtained (only tow replicates), it was not possible to indicate any differences with respect to CO_2 application in the enzyme activities in leaf washings (Fig. 4.5). The same was true for DON and DOC fluxes in the throughfall, although the number of replicates was considerably larger ($n = 12$) in this case. Possibly, the amounts of organic matter in throughfall were affected by aphids of the species *Myzocallis carpini*, which infested parts of the canopies to a moderate extent.

Conclusion

Climate warming, elevation in atmospheric CO_2 concentration, UV-radiation and air pollutants affect all compartments of an ecosystem. This includes the epiphytic microbial populations on the leaf surfaces, which respond to the direct and indirect impacts of

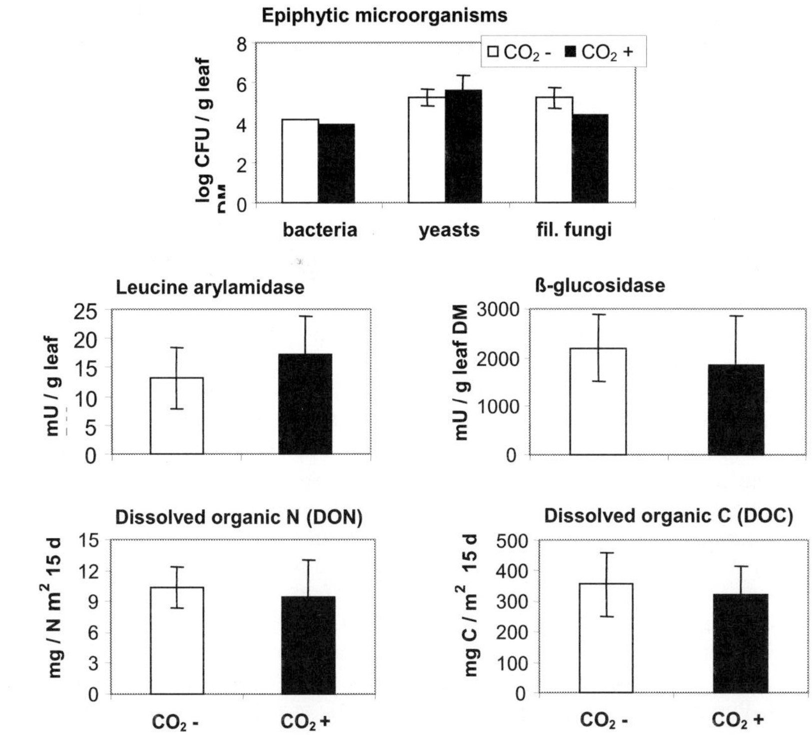

Fig. 4.5. Free-air CO_2 enrichment (FACE) experiment. Densities of epiphytic microorganisms and enzyme activities on leaves of hornbeam exposed to an atmosphere of ambient (370 µmol per mol, CO_2 -) and elevated (520 µmol per mol, CO_2 +) CO_2 content. Means of twofold repetitions. Fluxes of dissolved organic nitrogen and carbon in the canopies. Means of 12 samples, error bars indicate the standard deviations.

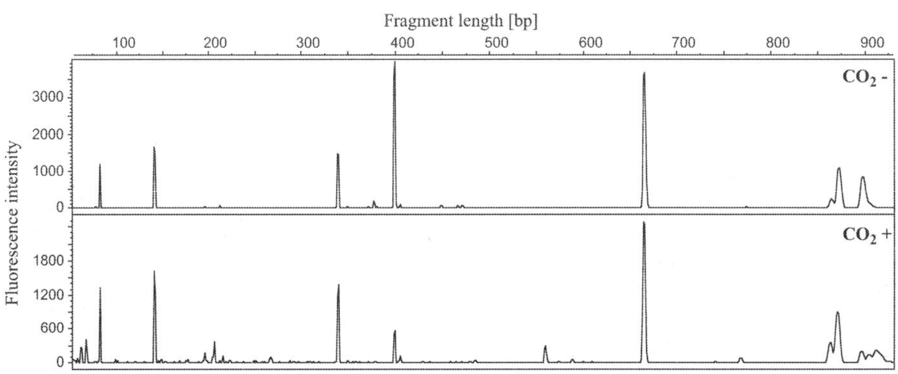

Fig. 4.6. T-RFLP profiles of bacterial communities derived from hornbeam leaves exposed to an atmosphere of ambient (370 µmol per mol, CO_2 -) and elevated (520 µmol per mol, CO_2 +) CO_2 content. The profiles were generated as described in Fig. 4.3.

environmental change by altering their growth and metabolic activity. The studies reviewed in this chapter suggest that environmental perturbations influence the life on leaf surfaces in more complex ways than previously anticipated. It became especially clear that studies on the impacts of atmospheric pollutants on forest ecosystems need to be integrated with the associated biotic changes affecting the epiphytic microorganisms.

The canopy of temperate forest trees is one of the most abundant above-ground habitat types on earth, and a site of active transformation processes that affect large-scale nutrient cycling. There is little doubt that phyllosphere microorganisms play a key role in these processes. However, their contribution to ecosystem function needs further elucidation, both qualitatively and quantitatively, to enable a better prediction of the influence of environmental changes on nutrient dynamics in forest ecosystems.

As indicated in this chapter, the microbial communities of the phyllosphere in temperate forests are responsive to biotic and abiotic environmental changes, but it is unclear how microbial species diversity will be affected by the changing environments. Environmental changes will most likely not only have effects on microbial population densities, but might also lead to shifts in species composition and/or metabolic activity of functional groups and individual species of microorganisms. Therefore, future research should aim at linking genetic and functional diversity with biogeochemical processes in order to elucidate fully the consequences of environmental changes on ecosystem function.

Acknowledgement

This work was undertaken following support by the Bundesministerium für Verbraucherschutz, Ernährung und Landwirtschaft, the Ministerium für Ländliche Entwicklung, Umwelt und Verbraucherschutz des Landes Brandenburg, and the Deutsche Forschungsgemeinschaft (Projekt 521/3-1). We thank Bernhard Stadler for his helpful discussions and comments on the manuscript.

References

Alef, K. (1991) *Methodenhandbuch Bodenmikrobiologie*. Ecomed Verlagsgesellschaft, Landsberg/Lech.

Andersen, D.R. and Gundersen, P. (2000) Nitrogen and carbon interactions of forest soil water. In: Schulze, E.-D. (ed.) *Carbon and Nitrogen Cycling in European Forest Ecosystems*. Springer, Berlin, Germany. pp. 332-340.

Anon. (2001) Contribution of Working Group I to the Third Assessment Report of the Intergovernmental Panel on Climate Change. In: Houghton, J.T., Ding, Y., Griggs, D.J., Noguer, M., van der Linden, P.J., Dai, X., Maskell, K. and Johnson, C.A. (Eds) Climate Change 2001: The Scientific Basis. Cambridge University Press, Cambridge, UK. pp.

Ayres, M.P. (1993) Plant defence, herbivory and climate change. In: Kareiva, P., Kingsolverand, J.G. and Huey, R.B. (eds) *Biotic Interactions and Global Change*. Sinauer Associates Inc., Sunderland, UK. pp. 75-94.

Bouwman, A.F., van Vuuren, D.P., Derwent, R.G. and Posch, M. (2002) A global analysis of acidification and eutrophication of terrestrial ecosystems. *Water, Air and Soil Pollution* 141, 349-382.

Chelius, M.K. and Triplett, E.W. (2001) The diversity of archaea and bacteria in association with the roots of *Zea mays* L. *Microbial Ecology* 41, 252-263.

Dik, A.J. and van Pelt, J.A. (1992) Interaction between phyllosphere yeasts, aphid honeydew and fungicide effectiveness in wheat under field conditions. *Plant Pathology* 41, 661-675.

Eckloff, W. (1972) Beitrag zur Ökologie und forstlichen Bedeutung bienenwirtschaftlich wichtiger Rindenläuse. *Zeitschrift für Angewandte Entomologie* 70, 134-157.

Gower, S.T. (2002) Productivity of terrestrial ecosystems. In: Mooney, H.A. and Canadell, J. (eds) *Encyclopedia of Global Change. Vol. 2.* Blackwell Scientific, Oxford, UK. pp. 516-521.

Guggenberger, G. and Zech, W. (1994) Composition and dynamics of dissolved carbohydrates and lignin-degradation products in two coniferous forests, N.E. Bavaria, Germany. *Soil Biology and Biochemistry* 26, 19-27.

Harrison, A.F., Schulze, E.-D., Gebauer, G. and Bruckner, G. (2000) Canopy uptake and utilisation of atmospheric pollutant nitrogen. In: Schulze, E.-D. (ed.) *Carbon and Nitrogen Cycling in European Forest Ecosystems.* Springer, Berlin, Germany. pp. 171-188.

Jacobs, J.L. and Sundin, G.W. (2001) Effect of solar UV-B radiation on a phyllosphere bacterial community. *Applied and Environmental Microbiology* 67, 5488-5496.

Jones, D.L., Healey, J.R., Willett, V.B., Farrar, J.F. and Hodge, A. (2005) Dissolved organic nitrogen uptake by plants – an important N uptake pathway? *Soil Biology and Biochemistry* 37, 413-423.

Magan, N. and Baxter, E.S. (1996) Effect of increased CO_2 concentration and temperature on the phyllosphere mycoflora of winter wheat flag leaves during ripening. *Annals of Applied Biology* 129, 189-195.

Magan, N. and McLeod, A.R. (1991a) Effects of atmospheric pollutants on phyllosphere microbial communities. In: Andrews, J.H. and Hirano, S.S. (eds) *Microbial Ecology of Leaves.* Springer, New York, USA. pp. 379-400.

Magan, N. and McLeod, A.R. (1991b) Effect of open-air fumigation with sulphur dioxide on the occurrence of phylloplane fungi on winter barley. *Agriculture, Ecosystems and Environment* 33, 245-261.

Maurizio, A. (1985) Honigtau-Honigtauhonig. In: Kloft, W.J. and Kunkel, H. (eds) *Waldtracht und Waldhonig in der Imkerei.* Ehrenwirth, Munich, Germany. pp. 268-295.

McClure, M.S. (1989) Evidence of a polymorphic life-cycle in the hemlock woolly adelgid, *Adelges tsugae* (Homoptera, Adelgidae). *Annals of the Entomological Society of America* 82, 50-54.

McDowell, W.H. and Likens, G.E. (1988) Origin, composition and flux of dissolved organic carbon in the Hubbard Brook Valley. *Ecological Monographs* 58, 177-195.

Michalzik, B. and Matzner, E. (1999) Dynamics of dissolved organic nitrogen and carbon in a Central European Norway spruce ecosystem. *European Journal of Soil Sciences* 50, 579-590.

Mühlenberg, E. and Stadler, B. (2005) Effects of altitude on aphid-mediated processes in the canopy of Norway spruce. *Agricultural and Forest Entomology* 7, 133-143.

Müller, T., Müller, M., Behrendt, U. and Stadler, B. (2003) Diversity of culturable phyllosphere bacteria on beech and oak: the effects of lepidopterous larvae. *Microbiological Research* 158, 291-297.

Müller, T., Müller, M. and Behrendt, U. (2004) Leucine arylamidase activity in the phyllosphere and the litter layer of a Scots pine forest. *FEMS Microbiology Ecology* 47, 153-159.

Myrold, D.D. and Nason, G.E. (1992) Effect of acid rain on soil microbial processes. In: Mitchell, R. (ed.) *Environmental Microbiology.* Wiley-Liss, New York, USA. pp. 59-81.

Nadelhoffer, K.J. (2001) The impacts of nitrogen deposition on forest ecosystems. In: Follett, R.F. and Hatfield, J.L. (eds) *Nitrogen in the Environment: Sources, Problems, and Management.* Elsevier, Amsterdam, NL. pp. 311-331.

Newsham, K.K., Low, M.N.R., McLeod, A.R., Greenslade, P.D. and Emmett, B.A. (1997) Ultraviolet-B radiation influences the abundance and distribution of phylloplane fungi on pedunculate oak (*Quercus robur*). *New Phytologist* 136, 287-297.

Papen, H., Geßler, A., Zumbusch, E. and Rennenberg, H. (2002) Chemolithoautotrophic nitrifiers in the phyllosphere of a spruce ecosystem receiving high atmospheric nitrogen input. *Current Microbiology* 44, 56-60.

Pepin, S. and Körner, C. (2002) Web-FACE: a new canopy free-air CO_2 enrichment system for tall trees in mature forests. *Oecologia* 133, 1-9.

Persson, T., Karlsson, P.S., Seyferth, U., Sjöberg, R.M. and Rudebeck, A. (2000) Carbon mineralisation in European forest soils. (2000) The carbon and nitrogen cycle of forest ecosystems. In: Schulze, E.-D. (ed.) *Carbon and Nitrogen Cycling in European Forest Ecosystems.* Springer, Berlin, Germany. pp. 257-275.

Qualls, R.G., Haines, B.L. and Swank, W.T. (1991) Fluxes of dissolved organic nutrients and humic substances in a deciduous forest. *Ecology* 72, 254-266.

Reynolds, B.C. and Hunter, M.D. (2001) Responses of soil respiration, soil nutrients and litter decomposition to inputs from canopy herbivores. *Soil Biology and Biochemistry* 33, 1641-1652.

Roshchina, V.V. and Roshchina, V.D. (1993) *The Excretory Function of Higher Plants*. Springer, Berlin, Germany.

Schulze, E.-D. (2000) The carbon and nitrogen cycle of forest ecosystems. In: Schulze, E.-D. (ed.) *Carbon and Nitrogen Cycling in European Forest Ecosystems.* Springer, Berlin, Germany. pp. 4-13.

Sheppard, L.J., Crossley, A., Cape, J.N., Harvey, F., Parrington, J. and White, C. (1999) Early effects of acid mist on Sitka spruce planted on acid peat. *Phyton* 39, 1-25.

Smits, N., Rougier, M., Fargues, J., Goujet, R. and Bonhomme, R. (1996) Inactivation of *Paecilomyces fumosoroseus* conidia by diffuse and total solar radiation. *FEMS Microbiology Ecology* 21, 167-173.

Solinger, S., Kalbitz, K. and Matzner, E. (2001) Controls on the dynamics of dissolved organic carbon and nitrogen in a Central European deciduous forest. *Biogeochemistry* 55, 327-349.

Stadler, B. and Müller, T. (1996) Aphid honeydew and its effect on the phyllosphere microflora of *Picea abies* (L.) Karst. *Oecologia* 108, 771-776.

Stadler, B. and Müller, T. (2000) Effects of aphids and moth caterpillars on epiphytic microorganisms in canopies of forest trees. *Canadian Journal of Forest Research* 30, 631- 638.

Stadler, B., Solinger, S. and Michalzik, B. (2001a). Insect herbivores and the nutrient flow from the canopy to the soil. *Oecologia* 126, 104-113.

Stadler, B., Müller, T., Sheppard, L. and Crossley, A. (2001b) Effects of *Elatobium abietinum* on nutrient fluxes in Sitka spruce canopies receiving elevated nitrogen and sulphur deposition. *Agricultural and Forest Entomology* 3, 253-261.

Stadler, B., Müller, T., Orwig, D. and Cobb, R. (2005) Hemlock woolly adelgid in New England forests: canopy impacts transforming ecosystem processes and landscapes. *Ecosystems* 8, 233-247.

Sundin, G.W. (2002) Ultraviolet radiation on leaves: its influence on microbial communities and their adaptions. In: Lindow, S.E., Hecht-Poinar, E.I. and Elliot, V.J. (eds) *Phyllosphere Microbiology*. The American Phytopathological Society, St. Paul, Minnesota, USA. pp. 27-41.

Tukey, H.B. (1970) The leaching of substances from plants. *Annual Review of Plant Physiology* 21, 305-324.

Waldrop, M.P., Zak, D.R. and Sinsabaugh, R.L. (2004) Microbial community response to nitrogen deposition in northern forest ecosystems. *Soil Biology and Biochemistry* 36, 1443-1451.

Whittaker, J.B. (2001) Insects and plants in a changing atmosphere. *Journal of Ecology* 89, 507-518.

Whittaker, R.H. and Likens, C.E. (1975) The biosphere and man. In: Lieth, H. and Whittaker, R.H. (eds) *Primary Productivity of the Biosphere. Ecological Studies, vol. 14*. Springer, Berlin, Germany. pp. 305-328.

Yu, Z., Zhang, Q., Kraus, T.E.C., Dahlgren, R.A., Anastasio, C. and Zasoski, R.J. (2002) Contribution of amino compounds to dissolved organic nitrogen in forest soils. *Biogeochemistry* 61, 173-198.

Chapter 5

Bacterial Biofilm Formation, Adaptation and Fitness

S.A. Rice[1,2] and S. Kjelleberg[1,2]
The Centre for Marine Biofouling and Bio-Innovation[1], and The School of Biotechnology and Biomolecular Sciences[2], The University of New South Wales, Sydney NSW 2052, Australia
scott.rice@unsw.edu.au

Introduction

Much of the foundational knowledge in microbiology is based on laboratory studies of planktonic bacteria. Those laboratory studies have been critically important in developing our fundamental understanding of the genetics, physiology and behaviour of microbes. However, it is now known that many, if not all bacteria stick to surfaces. It has been recognised that bacteria form biofilms on almost all surfaces, whether they are industrial surfaces such as water pipes, environmental surfaces such as riverine rocks, or on host tissues such as the surfaces of plants and dental plaque.

The biofilm provides community members with a number of opportunities and challenges and hence is quite different from its planktonic counterparts (for reviews see Stoodley *et al.*, 2002 and Sauer *et al.*, 2004). This is reflected in a number of features of the biofilms. For example, proteome analyses have shown that the expression of 40-100 proteins is unique to biofilm cells of *Pseudomonas putida* and *Pseudomonas aeruginosa* compared to their planktonic counterparts (for a review see Stoodley *et al.,* 2002). It is particularly interesting that microarray analysis of biofilm cells shows that many of the genes that are up or down regulated in a biofilm specific fashion are genes with unknown functions. For example, Whitely *et al.* (reviewed in Stoodley *et al.,* 2002) determined that 37% of biofilm specific genes in *P. aeruginosa* were of unknown function. The implication is that these functions are unknown because they do not have ramifications for planktonic growth and thus have been overlooked in planktonic-based studies, further reflecting past bias towards planktonic cells as model systems. Clearly, it is important to begin to assess the role of genes not only during planktonic growth, but also in relation to biofilm phenotypes in order to understand the functions of many of these genes, which presumably give biofilms their unique characteristics. At the phenotypic level, biofilms have been shown to exhibit increased resistance to a range of stresses, including oxidative stress, heavy metal exposure and antibiotics (reviewed in Hall-Stoodley *et al.,* 2004). In fact, biofilm cells may be up to 1000 times more resistant to antibiotics than their planktonic counterparts (Høiby, 2002). Recent estimates suggest that 60-70% of all infections are biofilm related, including infections such as otitis media, kidney stones, urinary tract infections and endocarditis (Parsek and Sing, 2003).

The same biofilm structures that have been observed in the laboratory have also been observed in the environment. For example, using fluorescent *in situ* hybridisation, micro-colony and filamentous biofilm structures have been observed on the surfaces of marine alga (Tujula *et al.*, and Case *et al.*, unpublished data) and Monier and Lindow (2004) have observed bacterial biofilms on leaf surfaces where more than 50% of the bacterial numbers could be attributed to clusters or biofilms. This suggests that biofilm formation on living surfaces follows similar processes as on inanimate surfaces. Thus, it is highly probable that the processes of biofilm formation (apart from the mechanisms of attachment which may be specific to one surface) that drive host-pathogen interactions in mammals are likely to apply across all levels of interaction of microbes with living surfaces more generally, and hence information on biofilm formation on one surface is likely to be applicable to biofilm formation on other surfaces.

Technology has Improved the Ability to Study Biofilms

This new appreciation for the important role biofilms play, both in the environment and in various hosts, can be largely attributed to new technologies that allow for the study of biofilms as living, hydrated surfaces. These include the development of techniques such as confocal laser microscopy and differential fluorescent stains (e.g. BacLite Live-Dead® staining-Molecular Probes and constitutive reporters such as fluorescent proteins) which allow investigators to generate three-dimensional images of living bacterial biofilms and give them the ability to monitor changes and development of living biofilms over time, without the need for destructive sampling. These techniques provide significant advantages over more traditional techniques such as electron microscopy, which requires significant processing of samples that often results in alteration of structural features, and light microscopy which is not able to resolve individual cells in thick three-dimensional structures due to background interference.

Model Bacteria Used in the Study of Biofilms

While biofilm formation by many different bacteria has been reported, *Escherichia coli* and particularly *P. aeruginosa*, have been established as model organisms for biofilm work. Reasons for this are the ease of genetic manipulation and culturing, predominance in both the environment and scientific literature, and relevance to human health. Studies on these organisms have been used to build the current model for biofilm formation which is comprised of: attachment, reversible followed by irreversible; development, aggregate formation followed by micro-colony formation; maturation, including cell death within the micro-colonies and dispersal of cells for subsequent recolonisation of new habitats. Similar processes and structure formation have now been observed for a broad range of bacteria including: *Vibrio cholerae*, *Vibrio vulnificus* (McDougald *et al.*, unpublished), *Pseudoaltermonas tunicata* (Mai-Prochnow *et al.,* 2004), *Burkholderia cepacia* (Huber *et al.,* 2001) and *Aeromonas hydrophilia* (Lynch *et al.,* 2002) to name only a few examples.

Co-operative Behaviour in the Biofilm

It has also been suggested that the high density environment of a biofilm provides an opportunity for sharing of resources or for co-operative metabolic functions. For example, Molin *et al.* (see Christensen *et al.*, 2002) have shown that co-metabolism of substrates can occur within a biofilm and that under such conditions, the two co-operating strains within the biofilm are closely integrated within the biofilm matrix. The biofilm also provides a clear example of a high density, diffusion limited environment where cell-cell communication between bacteria could occur. Indeed, it has been shown for several bacteria that cell-cell communication, or quorum sensing, is important for the biofilm process. For example, reporter fusions have been used to show that both the *las* and *rhl* quorum sensing systems of *P. aeruginosa* are expressed within the biofilm, and that their expression is stage specific; the *las* quorum sensing circuit was first active during irreversible attachment, while the *rhl* system was not induced until the early maturation stage (aggregate formation) (for a review see Stoodley *et al.*, 2002). Thus, the cells within a biofilm are capable of interacting, either through sharing of nutrient resources as well as co-ordinately regulating the expression of group level phenotypes.

Regulation of Biofilm Formation

There have been many suggestions for how biofilm morphology develops, ranging from the strictly physical-chemical nature of the biofilm and liquid flow, to a complex network of gene regulation resulting in a predetermined, differentiated biofilm structure, akin to tissue differentiation of higher eukaryotes. Modelling of biofilm formation suggests that the key parameters are flow rates and nutrient diffusion rates, which affect the ability of cells to colonise void areas between large groups of cells which might otherwise cause a local exhaustion of nutrients. Other studies have concluded that biofilm formation is dependent on the expression of a functional motility system or that the expression of surface structures such as EPS or fimbrae play significant roles in micro-colony formation and biofilm development (see Parsek and Singh, 2003 and Hall-Stoodley *et al.*, 2004). More recently, Klausen *et al.* (2003) have shown that the typical mushroom-like structures observed for *P. aeruginosa* are the result of an active process of cell migration and differential activity. This is achieved by a combination of non-motile cells that form the sessile stalks and twitching competent cells that then migrate to the top of the stalk to form the cap of the mushroom shaped micro-colony (Klausen *et al.*, 2003). Finally, it has now been established that quorum sensing (QS), plays a role in biofilm formation *in vitro*. Collectively, and perhaps not surprisingly, these studies suggest that biofilm formation is a complex process that involves all of these aspects to culminate in the formation of the micro-colony structures that are associated with biofilm formation.

Quorum Sensing (QS) and Bacteria

In addition to regulating surface colonisation *in vitro*, it is now appreciated that quorum sensing (QS) is an important phenomenon in nature. Numerous strains of bacteria isolated from the rhizosphere regulate the expression of phenotypes using quorum sensing. These organisms range from *Rhizobia* spp. which produce a wide range of QS compounds for an as yet unidentified phenotype, to *Agrobacterium tumefaciens* which regulates the transfer of plasmids that are required for infection of host plants, to the plant pathogen *Erwinia carotovora* which regulates the expression of virulence factors using the QS response (Zhang and Kerr, 1991; Lithgow *et al.*, 2000; Welch *et al.*, 2000). In some instances, bacteria use the QS system to ward off competitors, such as in *Chromobacterium violaceum* where QS regulates the expression of the antibiotic compound violacein (McClean *et al.*, 1997). The presence of QS competent bacteria appears to be widespread on plants. For example, Steidle *et al.* (2001) added mutant strains of *P. putida* and *S. marcescens* (note, the *Serratia liquefaciens* strain used, MG44, has been reclassified as *S. marcescens*; Rice *et al.*, 2005a) that do not make their own signal, carrying QS reporter constructions to the roots of tomato plants and were able to demonstrate that the strains colonised the roots and that they responded to signals being produced by the rhizosphere bacteria present (Steidle *et al.*, 2001). Quorum sensing clearly is also important in the phyllosphere as well as in the rhizosphere. Data indicate that just under 10% of epiphytic bacteria produce compounds that stimulate QS (Dulla *et al.*, presented at Phyllosphere 2005). Furthermore, they demonstrated that approximately 18% of leaf isolates produced compounds that interfere with QS in *Pseudomonas syringae*. In addition, Quninones *et al.* (2004) have shown that QS mutants of *P. syringae* had reduced fitness under desiccation stress when inoculated onto the surface of bean leaves, demonstrating that QS matters for these organisms (Quninones *et al.*, 2004).

QS Blocking by Eukaryotes

QS is not restricted to bacteria-bacteria communication. For example, it has been shown that the marine red alga, *Delisea pulchra*, produces compounds called furanones that act as QS blockers (Rice *et al.*, 2005b). The plant appears to produce these compounds as a chemical defence to reduce or prevent bacteria from colonising the plant's surface, a factor that is ordinarily detrimental to the plant in terms of reduced photosynthetic capacity, increased drag due to fouling, or disease mediated by the colonists. Terrestrial plants also appear to participate in this game of communication and molecular trickery (Teplitski *et al.*, 2000; Mathesius *et al.*, 2003,). For example, Tepliski *et al.* (2000) demonstrated that pea and crown vetch make substances that can induce or inhibit QS responses in bacteria and that the legume *Medicago truncatula* responded to the presence of QS molecules (Mathesius *et al.*, 2003), as shown through both proteomic studies and through the use of specific plant reporter gene fusions. Genes that were affected included both auxin and chalcone synthase responsive genes, indicating that the plant response was specific rather than a general hypersensitive type of response (Mathesius *et al.*, 2003).

QS and the Immune Response

There have now been several reports on the effect of QS molecules on mammalian immune responses (Telford *et al.*, 1998; Ritchie *et al.*, 2003; Tateda *et al.*, 2003). The effects of the QS signals include shifts in the Th1-Th2 response of the host (Telford *et al.*, 1998) as well as effects on apoptosis of macrophages (Tateda *et al.*, 2003). Further work in this area has supported evidence that QS signals can modulate the host immune response, but that the specific response is also dependent on the underlying genetic background of the host (Ritchie *et al.*, 2003). Taken together, these data suggest that bacterial biofilms, which are high density communities of bacteria, are actively producing, secreting and responding to the production of QS molecules and moreover, that this process is not one sided, as eukaryotes also participate in the process, and can either produce compounds to inhibit or encourage bacterial processes. In some instances, the release of such compounds from the biofilm may be a strategy on the part of the pathogen to increase its virulence (e.g. by modifying the immune response) or the host organism may recognise those signals to mount a specific response (e.g. defence response of plants).

Biofilm Formation by *Serratia marcescens*

In the next part of this chapter, recent data on biofilm formation by *Serratia marcescens* are presented. *S. marcescens* has been selected because it has many features that are found in a broad range of other bacteria. For example, *S. marcescens* is an opportunistic pathogen, but it is also commonly found in the soil and can be found associated with plants; indeed, the strain used in the present study was isolated from a rotting cucumber. Moreover, there is a strong understanding of its growth, physiology and regulatory systems. Surface colonisation by this bacterium is well understood, for swarming motility, surface attachment and biofilm formation. Regarding the latter, it has been shown that *S. marcescens* makes two types of biofilm depending on prevailing growth conditions and, as described below, forms stable variants the differ from the parental strain in colonisation phenotypes. *S. marcescens* has also been shown to use a cell-cell communication system to regulate many of these processes. Thus, based on these many attributes, this bacterium serves well as a model organism for a range of processes. Themes addressed in this chapter include the formation of a filamentous biofilm compared to the micro-colony type of biofilm, the role of quorum sensing, quorum sensing controlled genes and nutrient conditions in biofilm development, the generation of variants from the biofilm and the potential role of these biofilms as mechanisms for protection from common environmental stresses, such as grazing by protozoa. Furthermore, attempts are made to relate the types of biofilms observed and factors controlling biofilm formation to what is currently known about such phenotypes and morphotypes in the environment.

Quorum Sensing Regulation of Surface Colonisation by *Serratia marcescens*

S. marcescens has been shown to make a biofilm that is distinct from the micro-colony biofilm type previously described for *P. aeruginosa* and other bacteria (Labbate *et al.*, 2004), and that quorum sensing regulates this process. Using continuous flow biofilm set-ups, it has been observed that the biofilm formed by *S. marcescens* consisted of a base layer of cells attached to the substratum, interspersed with polarly attached, elongated cells (Fig. 1, A-C). At the tips of the elongated cells, which may extend from 12 to 200 μm up from the substratum, were aggregates of cells that resembled tree-tops. From the aggregates, long filaments and cell chains seem to radiate out and intertwine between aggregates, forming a filamentous lattice that is the top of the biofilm. Thus, these structures are considerably more diffuse than the micro-colonies observed for *P. aeruginosa* and the cell chains could be observed moving with the flow of the medium. The observation of filamentous biofilms on the surface of marine algae by an environmental *Aeromonas* isolate demonstrates that the filamentous biofilm is relevant outside of the laboratory conditions (Case *et al.*, unpublished data). Given that this biofilm type is distinct from what has been observed previously in many laboratory studies, it was of interest to revişit the potential genes and conditions required for the formation of this filamentous biofilm. In light of earlier observations that the acylated homoserine lactone-regulated QS system regulates swarming motility in *S. marcescens,* and that QS regulates biofilm development in other bacteria such as *P. aeruginosa*, *B. cepacia* and *A. hydrophilia*, it was of interest to survey the various stages of biofilm development for QS control in *S. marcescens*. Previously, Givskov *et al.* (1998) had demonstrated, by 2D gel electrophoresis, that approximately 28 proteins were QS regulated. These included genes such as *lipB* and *swrA*, which encodes a non-ribosomal peptide synthetase that produces the surfactant serrawetin which is required for swarming motility (Givskov *et al.,* 1998). Using a transposon-based reporter system, it was determined that approximately 170 Tn insertions were QS regulated (Queck *et al.*, unpublished). The pattern of regulation was complex, with some genes being up regulated in either logarithmic or stationary phase while other genes were down regulated during the different growth phases. Whitely *et al.* (1999) also observed similar patterns for regulation of QS controlled genes, and divided those genes into four classes, some of which were induced early in response to signal addition and others that were induced much later, possibly when they were in or entering stationary phase. This differential regulation of QS genes suggests that they are influenced by at least a second layer of regulation that is growth phase or physiology related. This is supported by the work of Yarwood *et al.* (2005), who proposed that QS expression is conditional on the nutrient status of the cells.

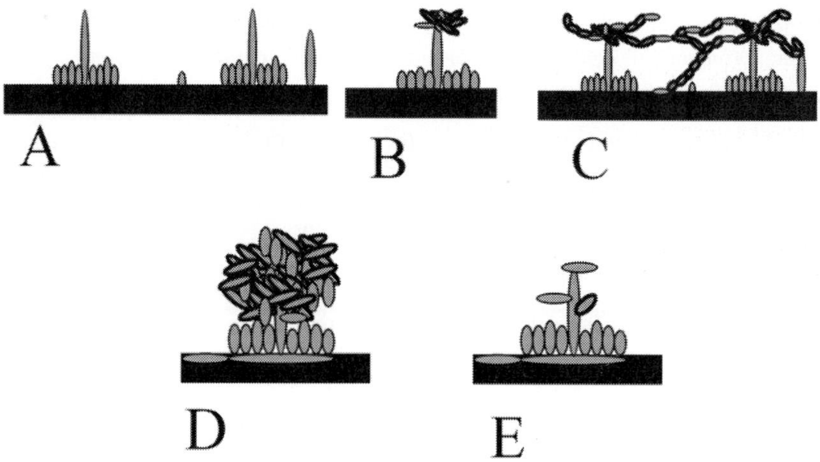

Fig. 5.1. Filamentous biofilm formation by *Serratia marcescens* MG1. A) Filamentous biofilm formation starts with the attachment of cells to the substratum and the vertical alignment of elongated cells; B) cell clusters form at the tips of the elongated cells; C) cell chains and filaments extend from the clusters to form an intertwined network of filaments; D) biofilm formation by the *bsmA* mutant results in large cluster formation at the cell tips, but no filaments or cell chains; E) the *bsmB* mutant accumulates very few cells at the cell tips and also fails to form filaments.

Attachment and Swarming

Because QS clearly plays an important role in regulating bacterial behaviour and has a role in biofilm formation, it was interesting to explore the role of QS in surface colonisation by *S. marcescens*. The QS regulated transposon insertion mutants obtained from the study above and 19 transposon mutants previously generated (M. Givskov) were tested in screens for defects in attachment, swarming and biofilm formation. In the screen for attachment deficient mutants, it was demonstrated that ten of the mutants were deficient in initial attachment. Sequence analysis of four mutants indicated that the transposon had inserted into genes with significant homology to *lipB* and *rssA* and two genes with no homologues in sequence databases. These were designated *bsmA* and *bsmB* (Labbate *et al.,* 2004). LipB is part of the secretory apparatus involved in the secretion of lipase, S layer proteins and protease, while RssA is a regulatory protein of a two component signal transduction system. While BsmA and BsmB did not match any known proteins, they appear to have domains that are indicative of regulatory functions, and hence may regulate other genes that make up structural components of the biofilm as opposed to encoding those structural components directly (Labbate *et al.,* 2004). In addition to the ten attachment deficient mutants identified above, the signal synthase deficient mutant *S. marcescens* MG44 (*swrI*) was also defective in attachment to polystyrene 96-well microtitre plates. Interestingly, no insertions in the attachment deficient mutants were found in flagella or putative EPS regions as have been identified for other organisms such as *E. coli* and *P. aeruginosa,* although two of the attachment mutants were in putative fimbriae related genes (Queck *et al.,* unpublished). Further testing of these strains for swarming defects, revealed that two were defective in swarming motility. One of the mutants carried insertions in non-ribosomal peptide synthases which is potentially involved in the production of surfactant. It was shown previously that the gene *swrA*, encoding a non-ribosomal peptide synthase, was required for swarming and that SwrA was responsible for the production of a surfactant that reduced the surface tension thereby allowing swarming (Givskov *et al.,* 1998). The second mutant carried an insertion in a putative iron sulphur binding protein. While, the role of this protein in swarming motility is unclear, this family of proteins is generally involved in post-translational modifications and thus may play a similar role in *S. marcescens*. Moreover, this mutant was not completely defective in swarming, but rather swarming was delayed for 48 h compared to 24 h for the wild-type.

Biofilm Formation

The role of these QS mutants in biofilm development was also tested using a combination of a microtitre-based screens and flow cell experiments, similar to approaches taken for other organisms. The *swrI* mutant was again found to be defective in biofilm development, in that this strain could colonise the surface of the flow cell and could form the polarly attached, elongated cells, but failed to progress any further in biofilm development (Labbate *et al.,* 2004). Similarly, the *bsmA* and *bsmB* mutants did not form a mature biofilm. The *bsmA* mutant forms large clusters of cells at the tip of the elongated cells while the *bsmB* mutant has reduced numbers of cells in the clusters compared to the wild-type (Fig. 1, D-E) (Labbate *et al.,* 2004). Given that these two proteins may play regulatory roles in biofilm formation, it is interesting that mutants in these genes have slightly different effects on the same phenotype (cluster formation at the tips of the elongated cells), and that both fail to generate the typical cell chains and filaments. The microtitre screen identified a further eight mutants that were defective in biofilm formation. In contrast to the *bsmA* and *bsmB* mutants, these eight showed biofilm morphologies that were largely similar to the wild-type and differed in thickness of the biofilm, or the timing of biofilm formation. This may indicate that filamentous biofilm formation in flow cells is dependent on multiple genes and hence, to obtain mutants with obvious structural defects requires either multiple mutations in structural genes, or mutations in regulatory pathways such as *bsmA* and *bsmB*.

Dispersal

The final stage of biofilm development is detachment or dispersal and it is suggested that this stage is important for bacteria to escape the biofilm in order to colonise new sites. The *S. marcescens* biofilm consistently detaches after approximately 75 h of development and generally does so in large masses. However, when the *swrI* mutant was examined, little or no detachment of the biofilm was observed (Rice *et al.,* 2005a). While it appears that QS controls detachment of the biofilm in *S. marcescens*, it is not clear which other QS genes or products are involved in detachment of *S. marcescens*. It has also been shown that QS is involved in dispersal phenotypes in other bacteria, such as *Rhodobacter sphaeroides* (see Puskas *et al.,* 1997 in Stoodley *et al.,* 2002), *Xanthomonas campestris* (Dow *et al.,* 2003) and for *P. aeruginosa* (Schooling *et al.,* 2004). The QS system of *P. aeruginosa* appears to regulate the expression of rhamnolipid production, which is involved in biofilm detachment. However, when the surfactant mutant *swrA* was tested in *S. marcescens*, there was no difference in detachment, suggesting that other factors are involved in detachment in this organism (Rice *et al.,* 2005a). QS is not the only factor that controls biofilm detachment and dispersal. For example, Sauer *et al.* (2004) have recently shown that increases in nutrient content can lead to biofilm dispersal, and Webb *et al.* (2003) have shown that increases in reactive nitrogen species and phage excision within biofilm clusters is correlated with dispersal from the biofilm. As for the systems described above, QS clearly regulates multiple genes that are involved in surface colonisation in *S. marcescens* and that some of the genes, e.g. *bsmA* and *bsmB*, are involved in multiple stages of biofilm formation. One of the exciting challenges for future research will therefore be to understand how such differential regulation is achieved in conjunction with the QS system.

Physiological Regulation of Biofilm Formation and Variant Formation

In addition to QS, it has also been noted that nutrient conditions can affect biofilm formation, and indeed, in some cases the role of the QS system can be subordinate to nutrient cues in biofilm development process (Heydorn *et al.,* 2002; Purevdorj *et al.,* 2002). This also appears to be the case in *S. marcescens*, where the QS mutant forms the filamentous biofilm under conditions of increased nutrients (Rice *et al.,* 2005a). In contrast, when grown under nutrient limiting conditions, *S. marcescens* forms a micro-colony type of biofilm (Rice *et al.,* 2005a). Monitoring of the biofilm by confocal microscopy (Fig. 2) indicated that the biofilm developed a limited number of micro-colonies by day 2, then hollow or empty colonies are typically observed around days 3-4 and, interestingly, by days 5-7, there is a huge radiation of micro-colony types and the flow chambers become virtually clogged with the biofilm at this stage. Subsequently, cell death is observed and the biofilm begins to detach over the next few days (days 8-10), after which time the biofilm begins to re-establish itself (Koh *et al.,* unpublished data). This process, with the exception of the rapid expansion of the biofilm over a 24 h period, is similar to that reported for other organisms such as *P. aeruginosa*, *P. tunicata* and *Vibrio cholerae*. Interestingly, from the time when the biofilm begins its expansion on day 4, morphological colony variants begin to appear in the effluent, consisting of approximately 6 distinct morphotypes. Of these, three were investigated in detail, a small colony variant (NSCV), a smooth variant which has a sticky colony type (SSV) and a mucoid variant (SUMV) (Fig. 5.3) (Koh *et al.,* unpublished data). The timing and pattern of variant formation is consistent, with some variants, such as the smooth variant always appearing first and the mucoid variant always appearing at the latter stages of the biofilm, when dispersal or sloughing is occurring. Currently, the genes involved in the formation of these variants in *S. marcescens* have not been identified. However, the colony morphotypes are stable when passaged in the laboratory, suggesting that there is a genetic change involved as opposed to transient gene expression. In addition, there may be multiple genes involved in the formation of these variants as it has been observed that the effluent from a biofilm made from the smooth variant eventually gives rise to the mucoid variant, and that the timing of appearance of the mucoid variant is consistent with its appearance relative to the smooth variant when observing biofilms made by the wild-type strain (Koh *et al.*, unpublished data). Phenotypic characterisation of these variants indicated that they are altered from the parental strain in a number of properties including biofilm formation, where some variants make more and some make less biofilm, as well as alterations in protease production and some virulence factors. These variants only appear when a micro-colony biofilm type is formed and have not to date been observed during filamentous biofilm formation (Koh *et al.*, unpublished data). This would suggest that the internal conditions for the two biofilm types may differ dramatically, where the micro-colony biofilm may provide the appropriate conditions (e.g. stress) required to induce the formation of genetic variants.

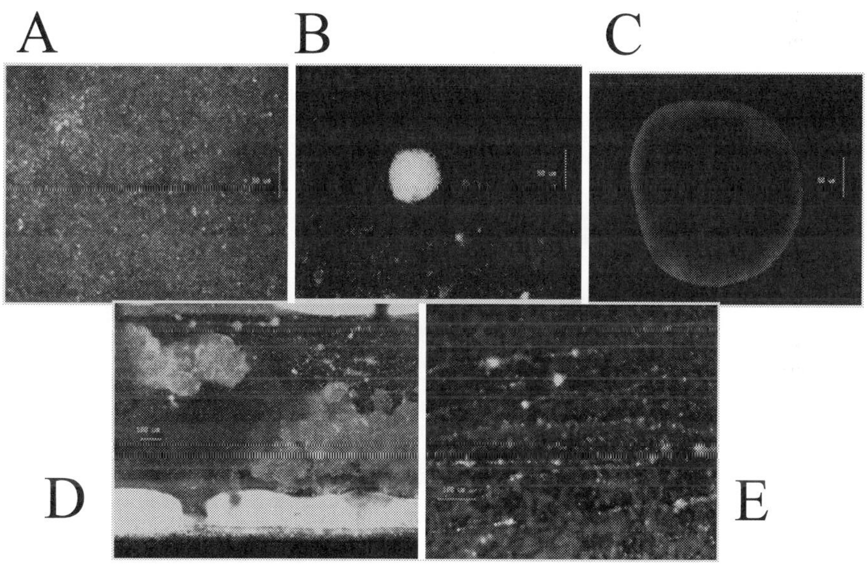

Fig. 5.2. Micro-colony development of *Serratia marcescens* MG1. Bacteria were cultivated in flow cells, stained with a fluorescent stain and imaged using confocal microscopy on different days. Representative images are presented for: A) day 1, initial attachment; B) day 2, formation of small clusters; C) days 3-4, hollow colony formation; D) days 5-7, biofilm expansion; E) days 8-10, detachment and biofilm reformation. Scale bars: 50 μm in A, B and C; 100 μm in D and E.

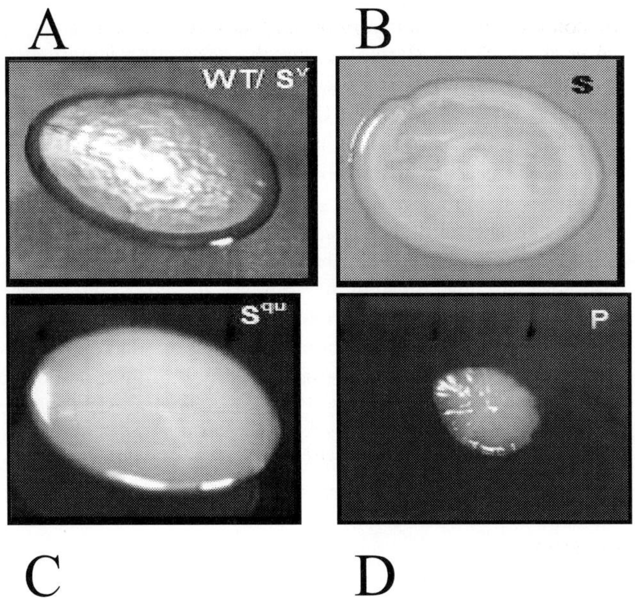

Fig. 5.3. Colony variants from biofilms of *Serratia marcescens*. A) Colony morphology of wild-type *S. marcescens*; B) the Smooth "SSV" variant; C) the mucoid "SUMV" variant, and D) the small "NSCV" variant.

Variant Formation and Consequences

This process of variant formation from biofilms is a common theme that is emerging from a range of other bacteria including *P. aeruginosa* (Webb *et al.,* 2004), *P. tunicata* (Mai-Prochnow *et al.,* 2004), *V. vulnificus* (McDougald *et al*., unpublished data) and *V. cholerae* (McDougald *et al*., unpublished data). Small colony variants have been observed in *P. aeruginosa* and this small colony variant has been associated with increased biofilm formation (Webb *et al.,* 2004) and antibiotic resistance (Drenkard and Ausubel, 2002). In *P. aeruginosa*, the formation of mucoid variants has been associated with poor outcomes for cystic fibrosis patients (Høiby, 2002), presumably because the mucoid variant forms a thicker, more resistant biofilm than the smooth parental strain. The formation of the mucoid variant is commonly associated with point mutations in the anti-sigma factor, mucA and this mutation is enhanced under conditions of oxidative stress (e.g. H_2O_2 production by polymorphonuclear cells). Variant formation has also been well documented for static biofilms of *Pseudomonas fluorescens* which generates multiple variants with different biofilm properties. Under static conditions, a radiation of growth types can be observed, with some cells growing at the bottom of the tube, some in the middle and some on the surface (see Spiers *et al.,* 2003). It was demonstrated that those growing on the liquid surface, deemed 'wrinkly spreader', carried a point mutation in the *wss* gene, the result of which is the over expression of a cellulose polymer (Spiers *et al.,* 2003). Spiers *et al.* (this volume) have also shown that *in vitro* biofilm formation at the liquid-air interfaces was a common property of environmental strains of *P. fluorescens* (70% of isolates), and that cellulose was a major component of those biofilm matrices. Interestingly, *wss* mutants were less competitive when grown in competition on roots and leaf surfaces, indicating that variation is both common in various strains and has consequences for survival in the environment (Gal *et al.*, 2003). In *P. fluorescens*, the gene *sss*, a homolgoue of the *E. coli* gene *xerD* (a lambda type recombinase), has been identified as playing a role in the regulation of variant formation (Martinez-Granero *et al.*, 2005). In *P. putida*, a *xerD* mutation is not different from the wild-type in planktonic cultures, but when cultured with root tips, shows the formation of significantly fewer variants than the wild-type suggesting *XerD* plays a role in root colonisation; indeed, these *xerD* mutants were out-competed by the wild-type cells in a root model (Martinez-Granero *et al.*, 2005). Sanchez-Contrares (see Martinez-Granero *et al.*, 2005 for reference) noted for *P. fluorescens* that the wild-type cells attached to the entire root system, while the morphological variants attached in a more specialised fashion and were mostly recovered from either the central part of the root or from the last centimeter. Given that *S. marcescens* also exists in the rhizosphere, the formation of such variants may have similar implications for survival on roots. For all of these systems, the formation of variants has been suggested to be a response to optimise fitness either to exclude other, less fit strains from the biofilm (enhanced biofilm formers), or to increase the options for colonisation of different surfaces (dispersal cells).

Grazing Protection

One of the main factors contributing to the mortality of bacteria in the environment is grazing by protozoans (Matz *et al.*, 2004a, b, 2005). Protozoa make up an important part of marine, soil, and plant associated communities and hence are relevant to pressures on bacteria in a variety of habitats. The intense selection put upon the bacteria by protozoans has been suggested to have selected for size and shape variants, which may be more grazing resistant. Furthermore, recent studies have demonstrated that protozoans influence biofilm formation behaviour by bacteria, where grazing pressure can select for the formation of micro-colonies or increased size of micro-colonies (Matz *et al.*, 2004a). Because *S. marcescens* makes a filamentous biofilm morphology, it was decided to challenge this biofilm type with grazers to determine if it provides protection from grazing. These studies investigated whether there was a role for QS in mediating grazing resistance, whether feeding type of the grazer was important for the mechanism of resistance and whether the biofilm morphology was an important factor in grazing protection. Results demonstrated that feeding type was important for QS mediated protection. In the presence of a suspension feeder, *Bodo saltans*, the QS system was not important for grazing resistance, where both the wild-type and QS deficient strain were equally protected in a batch (static) biofilm system. However, the study did show that the presence of the grazer led to the formation of micro-colonies, which were presumably resistant to the effects of the grazer. Neither the wild-type nor the QS mutant were protected in the batch system in the presence of the surface feeder, *Acanthamoeba polyp aga*. In contrast, the QS system was important in mediating resistance in the presence of a surface feeder, *A. polyphaga* when grown in a flowing system, where the wild-type forms a filamentous biofilm. In this system, *S. marcescens* wild-type was protected from the amoeba, however the QS mutant, which does not form a filamentous biofilm, was completely grazed by *A. polyphaga*. There are two potential mechanisms which explain these results. First, many bacteria produce factors that are toxic to grazers, as has been shown for *P. aeruginosa*, *V. cholerae* and *C. violaceum* (Matz *et al.*, 2004a, b, 2005). In *V. cholerae*, a *hapR* mutant, a transcriptional activator in the quorum sensing pathway of *V. cholerae* was non-toxic to grazers, while the wild-type was toxic, suggesting that the QS system regulates production of a toxic factor (Matz *et al.*, 2005). Another potential mechanism is a mechanical one, where the shear size of the bacterium is too great to be accommodated within the food vacuole of *A. polyphaga*. Indeed as mentioned above, intense grazing pressure has been shown to lead to changes in cell size and morphology, selecting for the formation of filamentous cells that are too long to be consumed by the protozoa. Morphologically, those filamentous cells resemble the cell chains and filaments formed in the *S. marcescens* biofilm under flowing conditions and hence may be the mechanism of grazing defence. A secondary role of the filamentous biofilm may be that it does not present a solid, stable surface that the amoeba can use to gain a foothold, and hence it cannot associate stably with the filamentous biofilm.

One question that remains to be answered is the role of the variants in mediating grazing protection from protozoa. For example, in a previous study, it was observed that the mucoid form of *P. aeruginosa* was protected from grazing and this may also be the case for the mucoid *S. marcescens* variant described above. Moreover, it would be particularly exciting to investigate whether grazing pressure contributes to the process of variant formation in *S. marcescens*. In *V. cholerae*, grazing pressure clearly selects for the rugose (mucoid) morphotype, presumably because the excess capsule prevents effective feeding, and thus excess capsule production may represent a selective advantage for the rugose variants (Matz *et al.,* 2005). These selective pressures are also in play in the phyllosphere as Lindow *et al.* reported (Phyllosphere 2005) the presence of grazers on leaf surfaces. Moreover, microscopic studies of the leaf surface indicate that bacteria may in fact form aggregates rather than exist solely as individual cells, which may in part be a protective mechanism to avoid grazing pressure.

Conclusions

It is clear that bacteria in the environment grow as biofilms, and it has been shown that the biofilm offers a protective niche for the bacteria to survive and persist. Laboratory-based studies have been key in helping to unravel some of the regulatory processes and the physiological status of bacteria within these biofilms. Such work has shown that bacteria rely on a complex regulatory network involving cell-cell communication as well as internal measures of physiological status to control the life-style, planktonic or attached, as well as biofilm morphology. Furthermore, laboratory work has indicated that grazing pressure from protozoa has the ability to select for biofilm formation, a condition that protects the bacteria from feeding pressure. Not only does the biofilm offer a protective niche, but it also appears to be a cauldron for the formation of genetic variants that are prepared to move on and colonise new sites. The next step is to investigate how these processes interact in the real world, and indeed numerous studies have begun to explore and identify biofilms on living surfaces. Plants, both aquatic and terrestrial, have proven to be ideal for studying such processes. Clearly bacteria do rely on biofilm formation and signalling in the environment to enhance their fitness. Because bacteria have evolved in parallel with the living surfaces on which they grow, some plants have developed their own chemistries that can antagonise biofilm formation. Fig. 5.4 presents a marine model for the central role that biofilms play in the interaction of bacteria with the environment, where they may interact with their substratum, living surfaces, as well as interacting with eukaryotes in the planktonic phase to either protect themselves from predation or to influence the settlement of fouling organisms. Further work will no doubt help to elucidate in increasing detail the complexity of such interactions that appear to centre around bacterial biofilms, which may indeed be not only the glue that holds a bacterial community together, but may also play integral roles in the biology of higher eukaryotes and the bio-geochemical processes that are essential for continued survival.

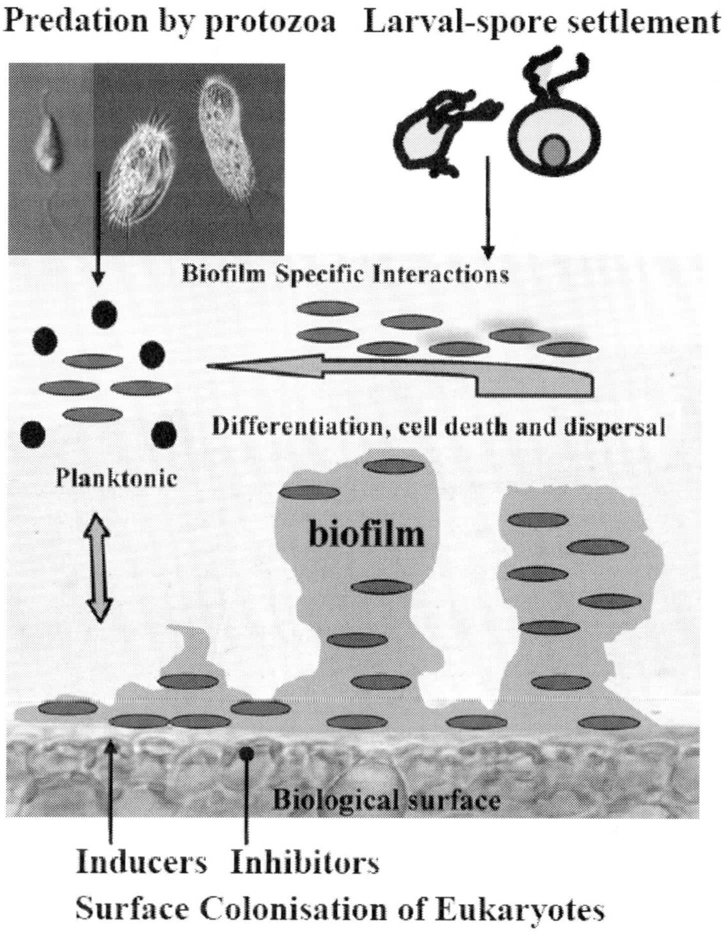

Fig. 5.4. Interaction of biofilms in the marine environment. Planktonic cells (ellipses) migrate to the surface, form a biofilm surrounded by extracellular matrix on a living surface, and can disperse from the biofilm back into the planktonic phase. The living surface may present attractants (= inducers) that induce biofilm formation (e.g. secretion of metabolites) or inhibitors that prevent biofilm formation. The biofilm in turn may produce compounds that induce or inhibit the settlement of eukaryotes. One of the key mortality factors faced by bacteria in the environment is predation by grazing protozoa, from which the biofilm offers protection.

Acknowledgements

We would like to thank M. Givskov for transposon strains used in some of this work. This work was supported by funding from the Centre for Marine Biofouling and Bio-Innovation (UNSW), the Australian Research Council (S. Kjelleberg), and the National Health and Medical Research Council (S. Rice and S. Kjelleberg). Thanks to Dr D. McDougald for critical reading of this manuscript.

References

Christensen, B.B., Haagensen, J.A.J., Heydorn, A. and Molin, S. (2002) Metabolic commensalism and competition in a two-species microbial consortium. *Applied Environmental Microbiology* 68, 2495-2502.

Dow, J.M., Crossman, L., Findlay, K., He, Y.Q., Feng, J.X. and Tang, J. L. (2003) Biofilm dispersal in *Xanthomonas campestris* is controlled by cell-cell signaling and is required for full virulence to plants. *Proceedings of the National Academy of Sciences, USA* 100, 10995-11000.

Drenkard, E. and Ausubel, F.M. (2002) *Pseudomonas* biofilm formation and antibiotic resistance are linked to phenotypic variation. *Nature* 416, 740-743.

Gal, M., Preston, G.M., Massey, R.C., Spiers, A.J. and Rainey, P.B. (2003) Genes encoding a cellulosic polymer contribute toward the ecological success of *Pseudomonas fluorescens* sbw25 on plant surfaces. *Molecular Ecology* 12, 3109-3121.

Givskov, M., Ostling, J., Eberl, L., Lindum, P.W., Christensen, A.B., Christiansen, G., Molin, S. and Kjelleberg, S. (1998) Two separate regulatory systems participate in control of swarming motility of *Serratia liquefaciens* MG1. *Journal of Bacteriology* 180, 742-745.

Hall-Stoodley, L., Costerton, J.W. and Stoodley, P. (2004) Bacterial biofilms: From the natural environment to infectious diseases. *Nature Microbiology Reviews* 2, 95-108.

Heydorn, A., Ersbøll, B.K., Kato, J., Hentzer, M., Parsek, M.R., Nielsen, T.A., Givskov, M. and Molin, S. (2002) Statistical analysis of *Pseudomonas aeruginosa* biofilm development: Impact of mutations in genes involved in twitching motility, cell-to-cell signaling, and stationary-phase sigma factor expression. *Applied Environmental Microbiology* 68, 2008–2017.

Huber, B., Riedel, K., Hentzer, M., Heydorn, A., Gotschlich, A., Givskov, M., Molin, S. and Eberl, L. (2001) The *cep* quorum-sensing system of *Burkholderia cepacia* H111 controls biofilm formation and swarming motility. *Microbiology* 147, 2517-2528.

Høiby, N. (2002) New antimicrobials in the management of cystic fibrosis. *Journal of Antimicrobial Chemotherapy* 49, 235-238.

Klausen, M., Aaes-Jorgensen, A., Molin, S. and Tolker-Nielsen, T. (2003) Involvement of bacterial migration in the development of complex multicellular structures in *Pseudomonas aeruginosa* biofilms. *Molecular Microbiology* 50, 61-68.

Labbate, M., Queck, S.Y., Koh, K. S., Rice, S.A., Givskov, M. and Kjelleberg, S. (2004) Quorum sensing controlled biofilm development in *Serratia liquefaciens* MG1. *Journal of Bacteriology* 186, 692-698.

Lithgow, J.K., Wilkinson, A., Hardman, A., Rodelas, B., Wisniewski-Dye, F., Williams, P. and Downie, J. A. (2000) The regulatory locus *cinRI* in *Rhizobium leguminosarum* controls a network of quorum-sensing loci. *Molecular Microbiology* 37, 81-97.

Lynch, M.J., Swift, S., Kirke, D.F., Keevil, C.W., Dodd, C E. and Williams, P. (2002) The regulation of biofilm development by quorum sensing in *Aeromonas hydrophila. Environmental Microbiology* 4, 18-28.

Mai-Prochnow, A., Evans, F., Dalisay-Saludes, D., Stelzer, S., Egan, S., James, S., Webb, J.S. and Kjelleberg, S. (2004) Biofilm development and cell death in the marine bacterium *Pseudoalteromonas tunicata. Applied Environmental Microbiology* 70, 3232-3238.

Martinez-Granero, F., Capdevila, S., Sanchez-Contreras, M., Martin, M. and Rivilla, R. (2005) Two site-specific recombinases are implicated in phenotypic variation and competitive rhizosphere colonization in *Pseudomonas fluorescens. Microbiology* 151, 975-983.

Mathesius, U., Mulders, S., Gao, M., Teplitski, M., Caetano-Anolle, G., Rolfe, B.G. and Bauer, W.D. (2003) Extensive and specific responses of a eukaryote to bacterial quorum-sensing signals. *Proceedings of the National Academy of Sciences, USA* 100, 1444-1449.

Matz, C., Bergfeld, T., Rice, S.A. and Kjelleberg, S. (2004a) Microcolonies, quorum sensing and cytotoxicity determine the survival of *Pseudomonas aeruginosa* biofilms exposed to protozoan grazing. *Environmental Microbiology* 6, 218-226.

Matz, C., Deines, P., Boenigk, J., Arndt, H., Eberl, L., Kjelleberg, S. and Jürgens, K. (2004b) Impact of violacein-producing bacteria on survival and feeding of bacteriovorous nanoflagellates. *Applied Environmental Microbiology* 70, 1593-1599.

Matz, C., Mcdougald, D., Moreno, A.M., Young, P.Y., Yildiz, F. H. and Kjelleberg, S. (2005) Biofilm formation and phenotypic variation enhance predation-driven persistence of *Vibrio cholerae*. *Proceedings of the National Academy of Sciences, USA* 102, 16819-16824.

McClean, K.H., Winson, M.K., Fish, L., Taylor, A., Chhabra, S.R., Camara, M., Daykin, M., Lamb, J.H., Swift, S., Bycroft, B.W., Stewart, G. and Williams, P. (1997) Quorum sensing and *Chromobacterium violaceum*: Exploitation of violacein production and inhibition for the detection of N acylhomoserine lactones. *Microbiology* 143, 3703-3711.

Monier, J. and Lindow, S.E. (2004) Frequency, size, and localization of bacterial aggregates on bean leaf surfaces. *Applied Environmental Microbiology* 70, 346-355.

Parsek, M.R. and Singh, P. K. (2003) Bacterial biofilms: An emerging link to disease pathogenesis. *Annual Review of Microbiology* 57, 677-701.

Purevdorj, B., Costerton, J.W. and Stoodley, P. (2002) Influence of hydrodynamics and cell signaling on the structure and behavior of *Pseudomonas aeruginosa* biofilms. *Applied Environmental Microbiology* 68, 4455-4464.

Quninones, B., Pujol, C.J. and Lindow, S.E. (2004) Regulation of AHL production and its contribution to epiphytic fitness in *Pseudomonas syringae*. *Molecular Plant-Microbe Interactions* 17, 521-531.

Rice, S.A., Koh, K. S. K., Queck, S.Y., Labbate, M., Lam, K.W. and Kjelleberg, S. (2005a) Biofilm formation and sloughing in *Serratia marcescens* is controlled by quorum sensing and nutrient cues. *Journal of Bacteriology* 187, 3477-3485.

Rice, S.A., Mcdougald, D., Kumar, N. and Kjelleberg, S. (2005b) The use of quorum sensing blockers as therapeutic agents for the control of biofilm associated infections. *Current Opinion in Investigational Drugs* 6, 178-184.

Ritchie, A.J., Yam, A. O. W., Tanabe, K. M., Rice, S.A. and Cooley, M.A. (2003) Modification of *in vivo* and *in vitro* T- and B-cell-mediated immune responses by the *Pseudomonas aeruginosa* quorum sensing molecule n-(3-oxododecanoyl)-L homoserine lactone. *Infection and Immunity* 71, 4421-4431.

Sauer, K., Cullen, M.C., Rickard, A.H., Zeef, L.A.H., Davies, D.G. and Gilbert, P. (2004) Characterization of nutrient-induced dispersion in *Pseudomonas aeruginosa* PAO1 biofilm. *Journal of Bacteriology* 186, 7312-7326.

Schooling, S.R., Charaf, U. K., Allison, D.G. and Gilbert, P. (2004) A role for rhamnolipid in biofilm dispersion. *Biofilms* 1, 91-99.

Spiers, A.J., Bohannon, J., Gehrig, S.M. and Rainey, P.B. (2003) Biofilm formation at the air-liquid interface by the *Pseudomonas fluorescens* SBW25 wrinkly spreader requires an acetylated form of cellulose. *Molecular Microbiology* 50, 15-27.

Steidle, A., Sigl, K., Schuhegger, R., Ihring, A., Schmid, M., Gantner, S., Stoffels, M., Riedel, K., Givskov, M., Hartmann, A., Langebartels, C. and Eberl, L. (2001) Visualization of N-acylhomoserine lactone-mediated cell-cell communication between bacteria colonizing the tomato rhizosphere. *Applied Environmental Microbiology* 67, 5761-5770.

Stoodley, P., Sauer, K. and Costerton, J.W. (2002) Biofilms as complex differentiated communities. *Annual Review of Microbiology* 56, 187-209.

Tateda, K., Ishii, Y., Horikawa, M., Matsumoto, T., Miyairi, S., Pechere, J.C., Standiford, T. J., Ishiguro, M. and Yamaguchi, K. (2003) The *Pseudomonas aeruginosa* autoinducer N-3-oxododecanoyl homoserine lactone accelerates apoptosis in macrophages and neutrophils. *Infection and Immunity* 71, 5785-5793.

Telford, G., Wheeler, D., Williams, P., Tomkins, P.T., Appleby, P., Sewell, H., Stewart, G.S.A.B., Bycroft, B.W. and Pritchard, D. (1998) The *Pseudomonas aeruginosa* quorum sensing signal molecule N-(3-oxydodecanoyl)-L-homoserine lactone has immunomodulatory activity. *Infection and Immunity* 66, 36-42.

Teplitski, M., Robinson, J.B. and Bauer, W.D. (2000) Plants secrete substances that mimic bacterial N-acyl homoserine lactone signal activities and affect population density-dependent behaviors in associated bacteria. *Molecular Plant-Microbe Interactions* 13, 637-648.

Webb, J.S., Lau, M. and Kjelleberg, S. (2004) Bacteriophage and phenotypic variation in *Pseudomonas aeruginosa* biofilm development. *Journal of Bacteriology* 186, 8066-8073.

Webb, J. S., Thompson, L. S., James, S., Charlton, T., Tolker-Nielsen, T., Koch, B., Givskov, M. and Kjelleberg, S. (2003) Cell death in *Pseudomonas aeruginosa* biofilm development. *Journal of Bacteriology* 185, 4585–4592.

Welch, M., Todd, D.E., Whitehead, N. A., Mcgowan, S.J., Bycroft, B.W. and Salmond, G.P.C. (2000) N-acyl homoserine lactone binding to the CarR receptor determines quorum-sensing specificity in *Erwinia*. *EMBO Journal* 19, 631-641.

Whiteley, M., Lee, K.M. and Greenberg, E.P. (1999) Identification of genes controlled by quorum sensing in *Pseudomonas aeruginosa*. *Proceedings of the National Academy of Sciences, USA* 96, 13904-13909.

Zhang, L. and Kerr, A. (1991) A diffusible compound can enhance conjugal transfer of the Ti plasmid in *Agrobacterium tumefaciens*. *Journal of Bacteriology* 173, 1867-1872.

Chapter 6

Bacterial Assemblages on Plant Surfaces

J.-M. Monier
Laboratoire d'Ecologie Microbienne, UMR 5557, Université Claude Bernard, Lyon, F-69622, France
jm.monier@free.fr

Introduction

Over the last decade, an increasing amount of research has reported the presence of bacterial assemblages on plant surfaces, revealing the segregated spatial distribution of epiphytic bacteria. While some of the first reports of their presence on plant surfaces were published over 40 years ago (Ruinen, 1961; Rovira and Campbell, 1974), it is only recently that the ubiquity and biological significance of these assemblages have been shown and investigated in more details (Fig. 6.1). This finding that a majority of epiphytic bacteria are found aggregated on leaf surfaces has raised many questions as to how aggregates form, the benefit they confer to individual bacteria or how they influence interactions between bacteria on plants. As a consequence, research on bacterial assemblages in the phyllosphere has gained much interest these past few years (Fig. 6.2), with a growing number of research projects addressing their impact on the biology and ecology of epiphytic bacteria.

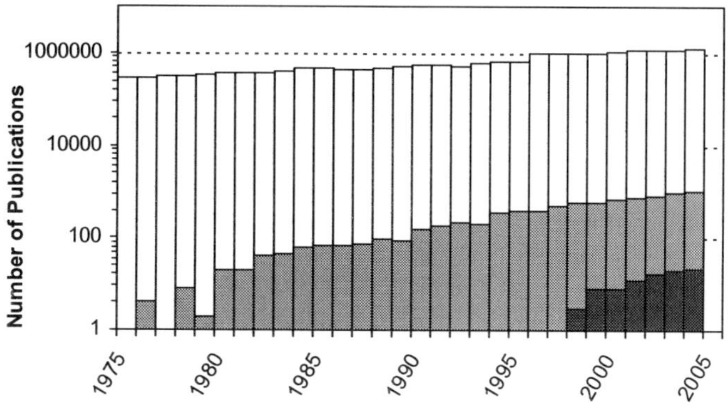

Fig. 6.1. Bar chart showing the number of research articles published related to bacterial assemblages or biofilms in the phyllosphere (black) and to biofilms regardless of the environment (grey) as a function of the year of publication. This search was performed using Scopus™ (www.scopus.com, Scopus™ Elsevier B.V.). White bars correspond to the total number of publications referenced in the database for the given year.

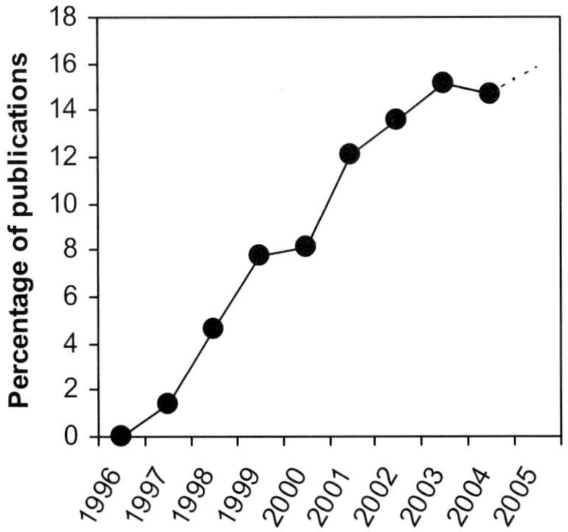

Fig. 6.2. Graph showing percentage of publications on phyllosphere microbiology related to or addressing bacterial assemblages (source: www.scopus.com, Scopus™ Elsevier B.V.).

Bacterial assemblages have been defined in various ways and are often referred to as biofilms, aggregates or microcolonies. While the latest clearly portrays a cluster of cells growing at a given site, it is somehow misleading and often associated with a notion of clonal monospecific colonies resulting from the growth of a single cell. It is however symptomatic of the lack of harmony on the definition and term used to describe epiphytic bacterial assemblages. Such discrepancy probably reflects our still limited knowledge of these assemblages (often descriptive) but also the fact that a consensual definition would have to encompass their extraordinary potential diversity in size, shape, taxonomic composition, and therefore function. A definition broad enough to include such diversity has been proposed by Costerton *et al.* (1995) to describe biofilms formed in aqueous environments. The authors defined these structures as "matrix-enclosed bacterial populations adherent to each other and/or to surfaces or interfaces". While some structural traits of biofilms can be considered universal, it seems reasonable to presume that bacterial assemblages formed on a transient unsaturated environment such as plant surfaces are markedly different, at least in their structure, from biofilms formed in aqueous environments. Some authors consider that epiphytic bacterial assemblages with an extent and structure similar to those of biofilms that develop in aquatic habitats are probably found only on long-lived leaves in moist climates and that bacterial assemblages on most other plants might best be called aggregates (Andrews and Harris, 2000; Lindow and Brandl, 2003). Although the term "biofilm" is now often associated with the mushroom shaped objects through which channels allowing movement of water are formed (stromatolites), historically, this term has encompassed a huge diversity of different structures ranging from patchy monolayers of cells such as those observed on plant surfaces (Morris *et al.*, 1997; Fett and Cooke, 2005) to macroscopic communities (Wimpenny *et al.*, 2000). Morris *et al.* (2002) stated that although the analogy between biofilms formed in aquatic environment and those formed on leaf surfaces might be naive, research advance on biofilms in aquatic environment (Fig. 6.1) provides a stimulating working model for the field of phyllosphere microbiology.

Regardless of the semantic debate around a still structure-based definition of epiphytic bacterial assemblages, it is important to keep in mind that the aggregated community on a leaf surface almost certainly comprises a huge diversity of bacterial assemblages ranging from monospecific clusters constituted of a few cells to biofilm-like structures of several hundred microns long (Morris *et al.*, 1997; Monier and Lindow, 2002). If we now also consider that every bacterial assemblage encountered on a leaf surface is unique (differing in its localisation, size, taxonomic composition, or spatial organisation), it may be easiest to define such structures in terms of what they are not, i.e. solitary cells. But then again, how many cells are needed to start referring to an aggregate, would two cells constitute an aggregate? Is the presence of an exopolymeric matrix required to start referring to a biofilm? While much discussion could take place in the search of a "consensual" definition of bacterial assemblages, the existence of such definition would be ultimately simplifying the enormous potential diversity of these structures. Although tempting, it seems difficult to characterize all possible epiphytic assemblages with a single conceptual definition and we may ask ourselves how useful would such definition of epiphytic bacterial assemblages be? Taking into account the heterogeneity of the surface habitat (topography, resource distribution etc.), I would be tempted to say that a small cluster of cells confined to the groove between epidermal plant cells may exhibit density dependent traits just as a biofilm-like structure of hundreds of thousands of cells located in an open space such as the base of a trichome where nutrients are more copious. This raises critical questions about these assemblages; what do they do, and how does their occurrence impact on the biology and ecology of epiphytic communities?

In order to try to answer part of these key issues we will be considering in this chapter some of the common characteristics of bacterial assemblages and their biological significance to epiphytic bacterial communities. First, we will try to understand how frequent and ubiquitous bacterial assemblages are and how they formed on leaf surfaces. The increasing amount of data addressing the spatial distribution of epiphytic bacteria at small scales will help us to propose a model that highlights critical steps in the colonisation process leading to the establishment of bacterial assemblages on leaf surfaces. Finally, we will try to comprehend how bacterial assemblage formation may benefit epiphytic bacteria and its possible consequence on the management of plant pathogenic bacteria. Recent review articles have addressed features of bacterial assemblages or biofilms associated with the rhizosphere and endosphere and the author would like to recommend the following. Morris *et al.*, 2002; Molina *et al.*, 2003; Morris and Monier, 2003. Essentially we will discuss bacterial assemblages formed on the aerial parts of plant surfaces.

Bacterial Assemblage Preview

Culture-based Approaches

Since the pioneering work of Zobell and others reporting that marine bacteria were more numerous on solid surfaces (Zobell and Allen, 1935; Zobell, 1937; Heukelekian and Heller, 1940; Zobell, 1943), bacterial assemblage formation has been described in almost any environment as long as it is exposed to water for at least some time (Characklis, 1973; Costerton *et al.*, 1995; Davey and O'Toole, 2000; Wimpenny *et al.*, 2000; Hall-Stoodley *et al.*, 2004). While plant surfaces do not escape this "rule" and the presence of complex microbial assemblages being reported over 40 years ago (Ruinen, 1961), it is only recently that the ubiquity and abundance of bacterial assemblages were described in the phyllosphere of several plant species (Morris *et al.*, 1997, 1998).

Indirect culture-based methods of studying spatial distribution of bacteria on leaf surfaces such as leaf imprints (Leben, 1988) or quantification of epiphytic population sizes on individual leaf segments (Kinkel *et al.*, 1995) strongly suggested that bacterial populations are highly variable in size even at small spatial scales and provided some of the first evidence that bacteria on leaf surfaces do not occur in a uniform pattern across the leaf (Kinkel *et al.*, 1995; Monier and Lindow, 2004). The first attempt to quantify bacterial assemblage abundance on leaves was undertaken by Morris *et al.* in 1998 using a method that combined leaf washing and differential filtration of the aggregated and solitary populations. In this work, the authors reported that aggregated populations on field-grown broad-leaved endive and parsley leaves constitute between 10% and 40% of the total bacterial populations. Further studies by this group revealed that aggregated populations could constitute up to 80% of the total bacterial populations naturally occurring on the leaf surfaces of several plant species (Morris *et al.*, 2002; Boureau *et al.*, 2004). When naturally occurring epiphytic assemblages were isolated and characterised, individual assemblages were found to harbour a wide range of microorganisms including numerous species of bacteria as well as yeasts and filamentous fungi (Morris *et al.*, 1997).

It is however the development of new approaches that combine molecular technologies with light microscopy techniques that has enabled us to broaden our scale of investigation of epiphytic communities from the whole plant or leaf that had characterised most phyllosphere microbiology research to date, to single cell studies (Morris *et al.*, 1997; Monier and Lindow, 2003a). While the different approaches are complementary and some conclusions coherent, the development of microscopy techniques adapted to the observation of epiphytic bacteria *in situ* has revealed unanticipated features of leaf surface microbial communities and has brought new insights to the processes leading to the structured organisation of bacterial populations on leaves.

Visualisation of Bacteria on Plant Surfaces

Since the description of "small animalcules" by Antonie van Leeuwenhoek, the development of new technologies has allowed the following generations of microscopists to perform detailed analysis of bacterial communities *in situ*. Up until the mid-90s, epiphytic microbial communities had been extensively yet almost exclusively, scrutinised under scanning electron microscopes (SEM) (Bashan *et al.*, 1981; Mew *et al.*, 1984; Timmer *et al.*, 1987; Surico, 1993). Although SEM has proven useful for studying microbial colonisation of plant surfaces, it can only provide limited information due to several drawbacks linked to the fixation and drying steps (e.g. dislodging and removal of bacterial cells, dehydration of the exopolymeric matrix embedding bacterial cells) and giving no, or very limited, information about bacterial cell identity, viability or metabolic activity. It is the development of epifluorescence (EM) and confocal laser scanning (CSLM) microscopy combined with fluorescent molecules (proteins, dyes or conjugates) that brought new light to the study of bacteria in their natural habitat. This advent of new technologies adapted to *in situ* detailed analysis of microorganisms associated with plant tissues had spurred a micro-revolution among phyllosphere microbiologists.

The initial development of new approaches to visualise bacteria directly on leaf surfaces was reported almost 10 years ago (Morris *et al.*, 1997) and has been instrumental in our understanding of bacterial assemblages and their biological significance to phyllosphere microbiology. Since then, a plethora of conjugated and non-conjugated fluorochromes including fluorescent dyes (e.g. acridine orange, propidium iodide, DAPI, ethydium bromide) already used for applications in biological microscopy (Kasten, 1999; Haugland 2002) has been applied to the study of plant associated bacterial communities (Brandl and Monier, 2006). The discovery of the green fluorescent protein (GFP) (Shimomura *et al.*, 1962; Prasher *et al.*, 1992) followed by a cortege of colour variants (Heim *et al.*, 1995) and their applications as intrinsic marker or reporter systems (Chalfie *et al.*, 1994; Errampalli *et al.*, 1999; Larrainzar *et al.*, 2005) was however the major breakthrough in the field of phyllosphere microbiology.

The first reports of plant-associated bacteria labelled with the GFP as an intrinsic marker were published in the mid-90s. Gage *et al.* (1996) labelled *Rhizobium meliloti* with plasmid-born GFP in order to visualize *R. meliloti* cells during infection of the root and nodules formation on alfalfa roots (Gage *et al.*, 1996). The first study using GFP-labelled bacteria in the phyllosphere was published in 1998 by Normander *et al.* where the authors reported the visualisation of transconjugants of *Pseudomonas putida* in relation to the spatial distribution of bacterial cells on the phylloplane of bush beans (Normander *et al.*, 1998). Since then numerous studies have reported the use of fluorescent proteins to study plant-associated bacterial strains. While in this chapter we will only address applications relevant to bacterial assemblages on plant surfaces, the author would like to recommend the following reviews providing an exhaustive description of the different fluorescent proteins and dyes (Tsien, 1998; Haugland, 2005) and their applications in biology (Kasten, 1999; Larrainzar *et al.*, 2005) and plant microbiology (Brandl and Monier, 2006).

Descriptive Analysis of Bacterial Assemblages

Structure and Architecture

Over the last decade, the popularity of fluorescent proteins was such that most reported studies have focused on labelled bacterial strains inoculated on leaves, and paradoxically, native communities have often been neglected. While direct observations of native communities proved to be challenging, several studies, using SEM, EM or CSLM, have revealed the presence of large complexes of aggregated microbial cells (Gras *et al.*, 1994; Morris *et al.*, 1997; Fett, 2000; Fett and Cooke, 2003). Such biofilm-like structures occur naturally on a wide range of plant species including vegetable crops, herbs and trees (Morris *et al.*, 1997; Fett and Cooke, 2003, 2005). Quantification of the dimension of these assemblages revealed their significant sizes. Bacterial assemblages formed on broad-leaved endives were characterised by a high density of cells surrounded by an exopolymeric matrix (Fig. 6.3A) and could reach several hundred microns in length and over 20 microns in depth (Fig. 6.3B) (Morris *et al.*, 1998).

A recent survey of native microbial aggregates naturally occurring on alfalfa, clover and mung bean (Fett and Cooke, 2005) confirmed the architecture observed for bacterial

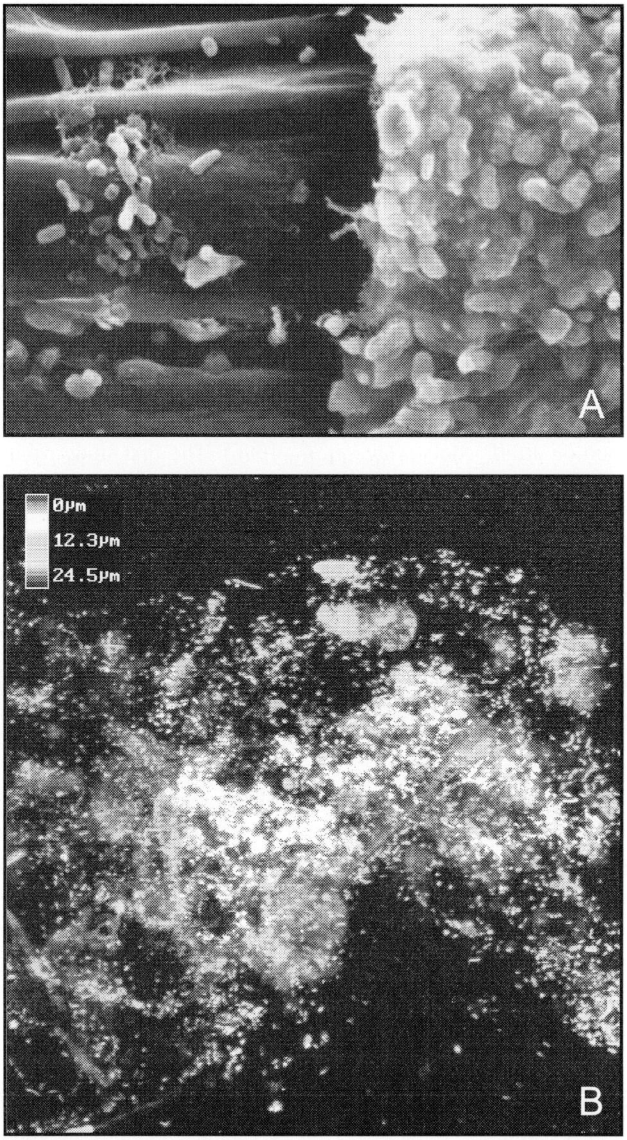

Fig. 6.?. Bacterial assemblages naturally occurring on broad-leaved endives.
(A) Image obtained by scanning electron microscopy and showing epiphytic bacteria attached to the leaf surface (left) and bacteria embedded in an exopolymeric matrix (right).
(B) Confocal laser scanning micrograph of a bacterial assemblage stained with propidium iodide; bar scale indicates the depth of the biofilm that has a central zone about 25 µm thick.

assemblages on broad-leaved endives. These authors, using confocal microscopy in combination with the Live/Dead® BacLight™ Viability Kit (Molecular Probes Inc., Eugene, USA) (Haugland, 2005), reported the presence of assemblages with an estimated area over 7000 μm^2 and a maximum thickness of 10.6 µm, 12.6 µm and 5.4 µm on mung bean sprout, alfalfa and clover cotyledons, respectively (Fett and Cooke, 2005). While only a relatively small number of plant species have been investigated so far, most naturally occurring assemblages imaged on leaf surfaces appeared to be flat structures constituted of a few layers of cells and very few seemed to exhibit more complex architectures (Morris *et al.*, 1997; Fett and Cooke, 2003), contrasting with that of the actual paradigm of stromatolite formation in biofilms in aquatic environments.

There is still an ongoing debate concerning the respective contribution of genetic traits and environmental factors on the architecture of biofilms in aquatic environments (Wimpenny *et al.*, 2000; Hall-Stoodley *et al.*, 2004). While individual epiphytic assemblages appear to be unique and constitute a wide range of diverse microorganisms (Morris *et al.*, 1998), their architecture appears to be relatively homogenous. Without being too Manichean, this would suggest that, although genetic traits may play a significant role in the spatial organisation of cells within individual assemblages, the general architecture of epiphytic assemblages is predominantly driven by the physico-chemical conditions encountered on leaf surfaces. It is most likely that among the factors affecting epiphytic assemblage structure, are factors that are limiting for growth on leaf surfaces (i.e. water and nutrient availability). While providing only indirect evidence of such hypothesis, larger assemblages are preferentially observed under moist environmental conditions or, in the case of broad-leaved endives, localised at the base of the leaf where residual water is often present. Since nutrient composition and concentration play a key role in biofilm structures and taxonomic composition (van Loosdrecht *et al.*, 1995; Olapade and Leff, 2005), slight differences reported in the average thickness, size and complexity of epiphytic assemblages may only reflect local differences in nutrients available for microbial growth.

While CSLM allowed for *in situ* non-destructive analysis of bacterial assemblages in their hydrated state, the presence of a distinct exopolymeric matrix enrobing the cells was revealed by SEM (Fig. 3B) (Gras *et al.*, 1994; Morris *et al.*, 1997; Fett and Cooke, 2003). Often referred to as EPS (somehow misleading since it initially stood for exopolysaccharides and was later replaced by extracellular polymeric substance), the matrix of aquatic biofilms is constituted of polymers of varying chemical composition and is critical to their structural integrity (Sutherland, 2001; Branda *et al.*, 2005). Although generally assumed to be preferentially composed of polysaccharides, it can also contain proteins and nucleic acids either excreted by the cells or resulting from cell lyses (Sutherland, 2001; Hall-Stoodley *et al.*, 2004; Branda *et al.*, 2005). While numerous plant-associated bacteria are also known to produce polysaccharides (Leigh and Coplin, 1992; Denny, 1995), our knowledge of the composition and properties of the matrix, or matrices, of epiphytic assemblages remains speculative. However, characterisation of common features of matrices of epiphytic assemblages might prove to be a difficult task, since such matrices might be as diverse as assemblages, themselves as diverse and heterogeneous as their constituent bacteria and local environmental conditions.

The still limited amount of data on the structure and composition of naturally occurring bacterial assemblages on leaf surfaces makes it difficult to propose a consensual view of their general architecture. I would however hypothesise that the flat architecture observed is probably representative of epiphytic bacterial assemblages formed on other plant species and results from the nature of the leaf surface habitat itself. Indeed, formation of flat biofilms constituted of a few layers of cells would: 1) provide bacteria with an optimised access to already limited nutrient resources provided by the surface on which they grow, and 2) limit their exposure to water stress by restricting their growth to the thin laminar layer on the leaf surface in which moisture is thought to be sequestered.

Size and Spatial Distribution of Bacterial Assemblages

As noted previously, the patchy recovery of bacteria from leaves in leaf imprints (Leben, 1988) or the large variation in population sizes of bacteria on individual leaf segments (Kinkel *et al.,* 1995) strongly suggested that bacteria do not occur in uniform patterns across leaf surfaces but tend to be localised in particular sites. The nature of these sites has been qualitatively reported using SEM and revealed that bacterial colonisation preferentially occurs at the base of trichomes, along the veins, at stomates or in the grooves between epidermal cells (Roos and Hattingh, 1983; Mew and Vera Cruz, 1986; Mansvelt and Hattingh, 1987; Timmer *et al*., 1987; Mansvelt and Hattingh, 1989; Surico, 1993; Bjorklof *et al.*, 2000). The first quantitative analysis of bacterial assemblages on leaf surfaces was reported only recently (Monier and Lindow, 2004). Using epifluorescence microscopy and image analysis, the frequency and size of bacterial assemblages on leaf surfaces of bean plants inoculated with the plant-pathogenic bacterium *Pseudomonas syringae* pv. *syringae* strain B728a was quantitatively described. Bacterial cells were observed to be not randomly distributed on the leaf surfaces and occurred in a wide range of cluster sizes, ranging from single cells to large assemblages of over 104 cells (Fig. 6.3). Indeed, the spatial distribution of bacteria on leaf surfaces could be compared, to some extent, to the spatial distribution of houses in a country, with only a few large assemblages accounting for the majority of the population (Fig. 6.4).

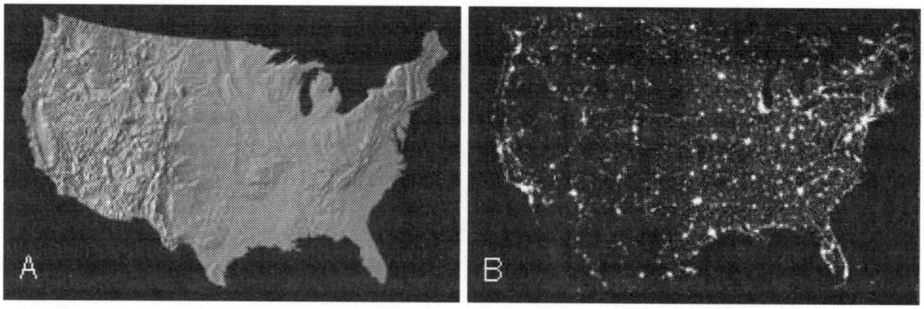

Fig. 6.4. The spatial distribution of bacteria on leaf surfaces could be compared, to some extent, to the spatial distribution of houses in a country, with only a few large assemblages accounting for the majority of the population.
(A) Digital shaded-relief image of the USA created by Thelin and Pike (1991) that shows the shape of the land surface by variations in brightness (US Geological Survey, www.usgs.gov).
(B) USA night lights cloud-free composite image taken by the DMSP from October 1994 to March 1995 (National Oceanic and Atmospheric Administration, www.noaa.gov).

While, the large majority of aggregates observed were small (less than 100 cells) and aggregate sizes exhibited a strongly right-hand-skewed frequency distribution, the few large aggregates observed accounted for the majority of cells present on leaves. The authors estimated that approximately 50% of the cells present on a leaf were located in assemblages constituted of 10^3 cells or more, corroborating previous studies, revealing that aggregated populations could constitute the vast majority of the epiphytic communities naturally occurring on the leaf surfaces of several plant species (Morris *et al.*, 1998, 2002; Boureau *et al.*, 2004).

Descriptions of the spatial distribution of epiphytic bacteria at a scale corresponding to their size using microscopy techniques has provided further information on the localisation of bacterial assemblages on leaf surfaces relevant to our understanding of the processes that lead to successful colonisation of leaf surfaces. Mapping of bacterial assemblages on a leaf surface clearly revealed that bacterial colonisation preferentially occurs at certain features on a leaf. Bacterial assemblages are preferentially formed at the base of glandular trichomes or along the veins (Morris *et al.*, 1997; Cao *et al.*, 2001; Monier and Lindow, 2004) and are significantly larger than those formed at other locations on the leaf surface (Monier and Lindow, 2004). Although glandular trichomes represented on average *ca* 0.6% of the total leaf surface area, over 70% of the aggregated cells of *P. syringae* were observed at the base of glandular trichomes. A study conducted with bean plants inoculated with different bacterial strains constitutively expressing fluorescent proteins as intrinsic markers, revealed that cells of both *Pantoea agglomerans* and *Pseudomonas fluorescens*, like *P. syringae* formed large aggregates at the base of glandular trichomes (Monier and Lindow, 2005a), suggesting that these sites are not species-selective and are conducive for growth of a diversity of epiphytic bacterial species. This also suggests that while the variability in population sizes observed at the plant or leaf levels could result from several environmental and biological factors (Beattie and Lindow, 1999; Hirano and Upper, 2000), the variability in population sizes and the spatial distribution of epiphytic bacteria, observed at a scale corresponding to their size, seems to be driven by the presence or absence of microsites conducive to bacterial growth (Kinkel *et al.*, 2002; Monier and Lindow, 2004). Although leaf surface features may be significant in the way they influence the initial impact of immigrating bacteria (e.g. cells might be expected to collect preferentially near trichomes or veins where water may be last to evaporate), further growth and development of large bacterial assemblages at such sites most likely reflect a localised copiotrophic environment and physico-chemical conditions conducive for growth. Although the majority of the sites on a leaf may not provide conducive conditions for bacterial growth, glandular trichomes are probably well suited for large assemblage formation due to their ability to retain water droplets (Brewer *et al.*, 1991) but also to secrete a wide array of chemical compounds potentially suitable for bacterial growth (Ascencao and Pais, 1998; Valkama *et al.*, 2003). While no direct correlation between bacterial assemblage formation and the presence of growth substrates at such sites has been established yet, indirect evidence would tend to corroborate such an hypothesis. In a recent study, Yadav *et al.* (2005) reported that the total population size of culturable epiphytic bacteria is positively correlated with trichome densities of Mediterranean perennial plant species. Surprisingly, while leaf water content was reported as the primary explanatory attribute for leaves with higher population densities, no correlation was found with soluble sugar contents. However, as stated by the authors, it could not be established whether the factors found to influence population sizes of epiphytic bacteria were related to growth rather than bacterial survival (Yadav *et al.*, 2005). Monier and Lindow (2005a) have recently reported that the fate of immigrant bacteria will depend on the nature of the leaf feature at which they land, and shown the preferential survival of immigrating bacteria at the base of glandular trichomes. Another piece of indirect evidence of glandular trichomes as hot spots for bacterial growth was provided by *in situ* measurement of *P. syringae* cells in relation to their localisation on bean leaf surfaces (Monier and Lindow, 2003b). It is now accepted that bacteria can adapt to nutrient limitation by reducing their cell size, and inversely respond to nutrient addition by cell enlargement, suggesting that cell size can provide information on the nature of microbial habitat and be used as an indicator of the trophic status of the bacteria (Moriarty and Bell, 1993; Givskov *et al.*, 1994). Using such a concept, Monier and Lindow (2003b) reported that cells grown in culture exhibited a normal size distribution, whereas cells recovered from bean plants were on average smaller in size and exhibited a right-hand skewed distribution, thus suggesting that only a small fraction of the cells present on a leaf are exposed to copiotrophic conditions. While cell size measurements of bacteria recovered from leaves provided an indirect estimation of the heterogeneity of nutrient availability on leaf surfaces, *in situ* analysis revealed that bacterial cells associated with glandular trichomes, often found aggregated, were significantly larger than solitary cells located at sites further away from trichomes.

While observations of large bacterial assemblages reported in the literature were preferentially associated with glandular trichomes, distinct bacterial strains appear to differ in the sites at which growth and bacterial assemblage formation on the leaf surface is preferred. *P. agglomerans* 299R, like *P. syringae* B728a cells are able to form large aggregates at the base of glandular trichomes. However, their respective spatial aggregation pattern on bean leaves was significantly different (Monier and Lindow, 2005a). For example, the majority of aggregates formed by *P. agglomerans* cells were located on top of epidermal cells, and while smaller in size than aggregates formed at the base of glandular trichomes, they accounted for *ca* 40% of the total number of aggregated cells. More importantly, the presence of large numbers of cells of this strain, unlike that of other bacteria examined, could somehow alter the physiology of epidermal plant cells. In addition, bacteria located on such epidermal cells appeared significantly larger than neighbouring, more solitary cells, suggesting that such cells were more metabolically active. The authors hypothesised that altered epidermal cells may be a source of increased metabolic activity of bacteria, possibly due to higher nutrient abundance. These observations tend to corroborate the hypothesis that *P. agglomerans* 299R could modify the microhabitat of epiphytic bacteria by secreting the plant growth regulator indole-3-acetic acid (IAA), thereby increasing nutrient leakage from plant cells (Brandl and Lindow, 1996; 1997). The percentage of the total population of *P. agglomerans* found in aggregates on altered epidermal cells, and thus likely to produce and benefit from IAA production, was estimated to be *ca* 20% of the total population. While apparently significant, such observations have proven to be more difficult to detect using conventional culture-based approaches assessing total population sizes. Indeed, an IAA-deficient strain of *P. agglomerans* 299R was reported to be only slightly deficient in epiphytic fitness compared to the wild type (Brandl and Lindow, 1998). Although this report is still anecdotal, producers of the plant growth regulator IAA, or any other compounds susceptible to alter the plant surface properties are commonly found among bacterial epiphytes (Brandl and Lindow, 2003; Schreiber *et al.*, 2005). Further studies conducted with bacterial strains deficient in their ability to produce, or overproducing, exogenous plant hormones should help us improve the understanding of the role and impact of such traits, that potentially enable bacteria to modify their habitat in order to increase their access to nutrient availability, or their ability to colonise leaf surfaces.

Direct evidence of a potential correlation between localised nutrients or water availability and bacterial assemblage formation could come from the development of bacterial whole-cell biosensors fused to luciferase or fluorescent protein genes. Over the past few years, an increasing number of bacterial strains designed to sense their immediate environment have been developed (Janson, 2003). Several biosensors have been designed to sense a variety of plant exudates (Espinosa-Urgel and Ramos, 2001; Casavant *et al.*, 2002; Koch *et al.*, 2002; Dunn *et al.*, 2003). Several have already been developed for and applied to the leaf surface habitat, including sucrose and fructose (Leveau and Lindow, 2001; Miller *et al.*, 2001), iron (Joyner and Lindow, 2000), or water availability (Axtell and Beattie, 2002). Biosensor development has already provided invaluable information about the physico-chemical environment of the leaf surface and further applications should prove to be extremely valuable to our understanding of colonisation patterns and bacterial assemblage formation on leaf surfaces.

Taxonomic Composition and Spatial Organisation of Cells within Assemblages

Bacterial assemblages naturally occurring on leaf surfaces harbour a wide range of microorganisms. Morris *et al.* (1997) have illustrated that diverse microorganisms, including both gram-positive and gram-negative bacteria as well as yeasts and filamentous fungi, may inhabit the same assemblages in the phyllosphere. Individual assemblages isolated from the same leaves or plant species revealed that each bacterial assemblage is unique in its taxonomic microbial composition (Morris *et al.*, 1997). If we also take into consideration that only culturable microorganisms were isolated and analysed in this study, it is tempting to speculate that the microbial diversity observed within each individual assemblage is only the tip of the iceberg and only provides a partial image of the actual diversity of microorganisms present in each bacterial assemblage. In an attempt to determine whether differences existed in bacterial composition within assemblages (aggregated population) compared to that of outside assemblages (solitary population), this research group illustrated that, even though gram-positive bacteria tended to occur more frequently within bacterial assemblages, there were no systematic differences in the composition and structure of the two populations (Boureau *et al.*, 2003). Based on their observations, these authors hypothesised that leaf surface colonisation involves a flux of cells between the aggregated and solitary states that would allow bacteria to exploit the advantages of these two types of existence as environmental conditions fluctuate. It is also important to mention that the topography and the spatial heterogeneity of the physico-chemical environment characterising the leaf surface are most likely to influence the spatial organisation of cells within assemblages. Therefore, needless to say that in light of the apparent extraordinary complexity of these dynamic structures (Fig. 6.5), little is known about the factors that lead to the establishment and the spatial organisation of the different microbial species within bacterial assemblages encountered on leaf surfaces. The development of culture-independent techniques such as fluorescent *in situ* hybridisation (FISH) and applications of high-throughput techniques such as 16s rDNA microarray (Phylochips) to epiphytic communities should soon provide valuable information on the taxonomic composition and spatial organisation of bacterial cells species within complex epiphytic assemblages. Taxonomic composition analysis of individual assemblages, but also, the assessment of their metabolic potential, their ability to produce quorum-sensing molecules and a survey of mobile genetic elements present are needed in order to decipher the potential interactions and functions of the aggregated community on leaf surfaces. This should prove to be a stimulating challenge for phyllosphere microbiologists in the coming years.

Despite the apparent complexity of these structures, a few studies have attempted to decipher the factors that shape the composition and spatial organisation of bacterial assemblages on leaf surfaces. In order to determine to which extend the spatial organisation of bacterial cells within assemblages was influenced by the bacterial strains constituting the assemblage, Monier and Lindow (2005b) established different pair-wise mixtures of three different bacterial species (*P. syringae*, *P. agglomerans* and *P. fluorescens*) constitutively expressing the GFP or CFP (cyan fluorescent protein) on bean leaves and examined the resulting assemblages by epifluorescence microscopy. The authors reported that approximately half of the total number of assemblages observed on leaves were composed of the two strains (mixed) and that the spatial organisation of bacterial cells within such mixed assemblages was influenced by the identity of their bacterial constituents.

Quantification of the degree of segregation revealed that the highest segregation occurred between bacterial species exhibiting negative interactions. In such association, *ca* 0.2% of the cells of the two populations were in direct contact, whereas cells of two isogenic strains were more randomly distributed within the assemblage and on average *ca* 8.0% of the cells of the two populations were in direct contact (Monier and Lindow, 2005b).

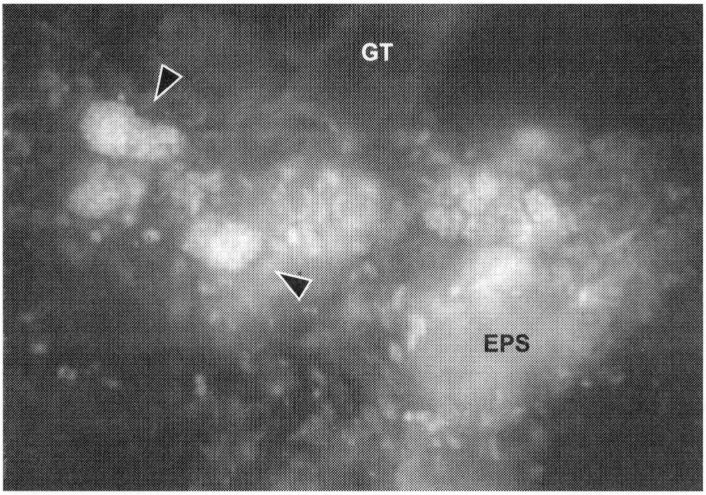

Fig. 6.5. Detail of a bacterial assemblage naturally occurring on a leaf of field-grown broad-leaved endive illustrating the complex spatial arrangement of bacterial cells within the assemblage. The image was obtained by epifluorescence microscopy after staining with acridine orange. Arrows point out cell clusters occurring in discrete locations within the structure. GT, glandular trichome; EPS, putative exopolymeric substances enrobing the cells. (Courtesy of J.-M. Monier and C.E. Morris)

Interestingly, the authors reported that the segregation of the strains exhibiting negative interactions (i.e. *P. syringae* and *P. agglomerans*) were significantly higher in larger aggregates reaching a quorum of *ca* 500 cells (Fig. 6.6), and hypothesised that such segregation may result from traits regulated in a cell-density dependent manner. However, the variability in spatial organisations of mixed assemblages observed was such that it did not permit to identify traits dictating their structures. As suspected, bacterial interactions *per se* are not the only mechanisms shaping the structure of epiphytic bacterial assemblages. Environmental factors such as the leaf topography and leaf features at which bacteria assemble also influences their spatial organisation. Regardless of the apparent complexity of the factors governing the spatial structures of epiphytic communities and bacterial assemblage formation, direct bacterial interactions between distinct populations appeared to be limited to a few sites and involved only of small fraction of the bacterial populations, corroborating the model developed by Kinkel *et al.* (2002) on resources aggregation in the phyllosphere.

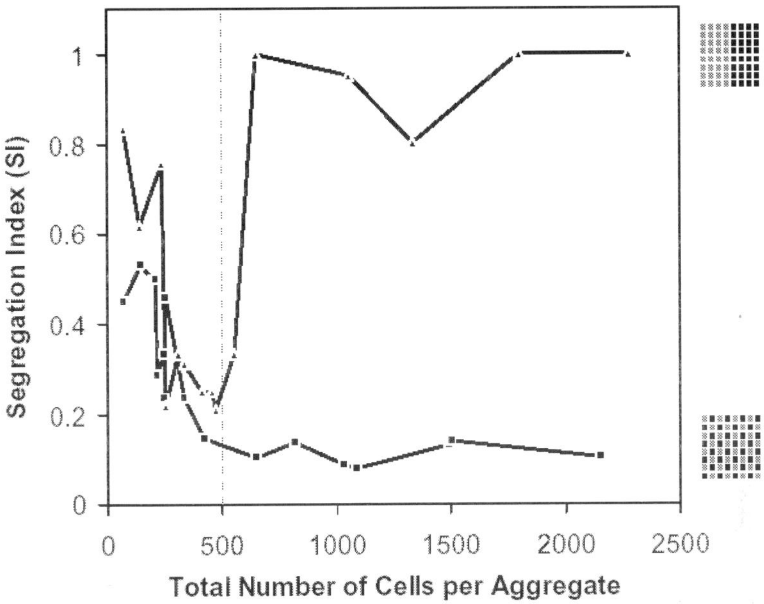

Fig. 6.6. Segregation index (SI) values in relation to the total number of cells per aggregate obtained for mixed aggregates formed by *Pseudomonas agglomerans* 299R and *P. agglomerans* 299R (squares) and by *P. agglomerans* 299R and *Pseudomonas syringae* B728a (triangles). Segregation index was defined as SI = (R1/C1) + (R2/C2), where C1 and C2 correspond to the numbers of clusters formed by strains 1 and 2, respectively, and R1 and R2 the proportions of cells of strain 1 and strain 2, respectively, in an aggregate. SI values range from 0 (cells randomly distributed) to 1 (cells highly clustered) as illustrated on the right-hand side of the figure.

Bacterial Assemblage Formation: a Unifying Hypothesis

The increasing amount of data addressing the spatial distribution of epiphytic bacteria at small scales corresponding to their size has provided us with important biological and ecological information leading to improved understanding of the processes that lead to successful colonisation of leaf surfaces. The following model is proposed to highlight critical steps in the colonisation process leading to the establishment of bacterial assemblages on leaf surfaces:

Initial Immigration to Leaf Surfaces

Relatively few and rather dispersed initial colonisation events on leaf surfaces occur. The initial invasion of leaves by pioneer bacteria results most likely in an initial random spatial distribution of cells on the leaf surface. The number of initial immigrants is almost certainly far lower than the number of bacteria that are eventually colonising a leaf. While during the drying process following waterborne deposition (i.e. rain drops) cells might be expected to collect preferentially where water is last to evaporate (i.e. trichomes and veins), bacterial cells are initially found across the entire leaf surface.

Heterogeneous Bacterial Growth across the Leaf Surface

Only a few sites on a leaf surface offer conditions suitable for the survival and subsequent growth of immigrating bacteria. Those sites, such as glandular trichomes, are conducive for growth of a diversity of epiphytic bacterial species and do not exhibit any bacterial species-specificity. However, and regardless of environmental conditions, growth is highly heterogeneous across the leaf and among given leaf features, and is highest at sites where nutrients are most abundant. Nutrient availability varies dramatically across the leaf and is abundant only in a few sites. Presumably, most bacterial growth is at such sites, and they are the location where bacterial assemblages can eventually form. While epiphytic bacteria are able to adapt to nutrient-limiting conditions by reducing their cell size, the majority of immigrant solitary cells landing on non-conducive sites will eventually die, even if plants are temporarily exposed to moist conditions that are optimal for bacterial growth if other resources are not limiting. That is, most immigrants will arrive at locations where there are insufficient nutrients for growth and their fate is to slowly die. Such cells pass through a reduced cell size stage in this journey. Those cells fortunate enough to arrive at a site of more abundant resources can be distinguished by their larger size and the presence of progeny. The death of starved cells results in a shift, from a random to a clustered spatial distribution of bacteria on leaves, with a majority of cells located in a few large assemblages.

Periodic Stressful Conditions Promote a Higher Level of Spatial Segregation

While moist conditions favour bacterial growth and large assemblages formation, the constantly changing environmental conditions encountered by bacteria on the leaf surface result in the differential survival of epiphytic bacteria in a density-dependent manner (Fig. 6.6). While solitary cells and cells located in small clusters will die after repeated exposure to stressful conditions (e.g. desiccation, UV radiation, bactericides), cells located in large aggregates or biofilm-like structures will survive by mutually protecting each other as well as by expressing resistance phenotypes. Further subsequent growth will occur if environmental conditions again are conducive. This will result in an even higher spatial segregation of bacteria on leaf surfaces with the entire bacterial population located within a few large aggregates. The preferential survival of aggregated cells reflects a physiological condition, presumably enabled by high nutrient availability that allows cells to express traits that are protective against stress.

Microhabitat Modification and further Growth of Bacterial Cells

Microhabitat modification can either result from, or lead to, bacterial assemblage formation depending on the nature of the site where immigrant bacteria of a given genotype are deposited. Conducive sites, such as glandular trichomes, offer optimal conditions for survival and growth of immigrant bacteria and lead to the formation of aggregates. Such aggregated cells can escape stressful conditions encountered on leaves by physically protecting each other by the production of the cell aggregate, and/or by expressing protective traits regulated in a cell density-dependent manner (e.g. exopolymeric matrix, phytotoxins leading to an increased availability of nutrients), allowing further growth of cells within the aggregate (this seems to be the strategy of bacteria such as *P. syringae*).

Alternatively, bacteria such as *P. agglomerans* landing on non-conducive sites can modify their local environment, and therefore increase their probability of survival, by stimulating plant exudation via production of exogenous plant hormones leading to an increased local nutrient availability. Such habitat modification will allow rapid bacterial growth and aggregate formation on sites that were initially non-conducive (e.g. epidermal plant cells). Other such methods of habitat modification may exist and their identification should prove of great interest in the study of plant-microbes interactions.

Composition and Spatial Organisation of Cells within Epiphytic Assemblages

While small aggregates formed on leaf surfaces may either be monospecific (i.e. the progeny of a single cell of a given species) or mixed (i.e. composed of several bacterial species), larger assemblages (biofilm-like structures) are most likely composed of a wide diversity of microbial species. Mixed assemblages can result from growth and convergence of monospecific assemblages formed by adjacent microcolonies of different bacterial species, by incorporation and growth of immigrant bacteria landing in an established assemblage, or motile bacteria migrating to established assemblages. However, the incorporation of bacteria in an existing assemblage depends strongly on the nature of the interactions between bacterial species. Incorporation may be enhanced when established aggregated bacteria may benefit from multicellular cooperation (e.g. synergism) and therefore allow access to resources that are not accessible by another population. In contrast, aggregated cells will collectively defend against the incorporation of species that are antagonistic or non-beneficial to the existing assemblage. This results in a highly segregated spatial organisation of bacterial populations on leaf surfaces as well as within bacterial assemblages and only a small fraction of each bacterial population is directly interacting with the others.

Why do Epiphytic Bacteria Form Assemblages?

It is now accepted that bacterial assemblage formation is a widespread phenomenon in nature and that the majority of bacteria on the planet have adopted such aggregated modes of life (Costerton *et al.*, 1995; Davey and O'Toole, 2000). Recent data demonstrate that bacterial assemblages are also ubiquitous on plant surfaces. Such ubiquity strongly suggests that assemblage formation must offer some advantages to their constituents. To date several hypotheses have been proposed including protection from a wide range of environmental challenges, metabolic cooperativity or acquisition of new genetic traits (Davey and O'Tolle, 2000; Morris *et al.*, 2002; Morris and Monier, 2003; Hall-Stoodley *et al.*, 2004). While some of them remain speculative, several studies have addressed the biological significance of bacterial assemblage formation by bacterial communities. In this section, we will present some emerging evidence of the ecological advantages that bacterial assemblage formation may confer epiphytic populations.

Although the fate of a bacterium landing on a leaf surface will be governed by a set of environmental conditions and biological interactions, the initial colonisation of leaf surfaces can be considered a stochastic phenomenon with some sites constituting hot spots for the survival of immigrant bacteria and the subsequent formation of bacterial assemblages. In regard to the question "Why do epiphytic bacteria form assemblages?" I would initially be tempted to reply, "Because they can". There is however growing evidence that epiphytic bacteria may benefit from the formation of bacterial assemblages. Several traits important in plant-microbe interactions have been shown to be regulated in a cell density-dependent manner (von Bodman and Farrand, 1995; Dumenyo *et al.*, 1998, Koiv and Mae, 2001; Loh *et al.*, 2002; von Bodman *et al.*, 2003; Crossman and Maxwell Dow, 2004; Quiñones *et al.*, 2004, 2005) by a process known as quorum sensing. Bacteria that use such mechanisms can assess their local population density by detecting the production of a diffusible signal molecule that increases with the local population density, resulting in a coordinated expression of new genetic traits. A majority of epiphytic bacteria found in large assemblages would suggest that epiphytic fitness may be controlled in a density-dependent manner. While direct evidence of the production of such signal molecules in epiphytic bacterial assemblages has yet to be reported, it seems reasonable to presume that bacterial cells in assemblages, in which such compounds would most likely accumulate, may exhibit a markedly different epiphytic fitness than more solitary cells.

Large assemblages might provide a protected site in which bacteria could escape harsh conditions encountered on the leaf surface. The presence of an EPS matrix might provide a different ecological niche for the majority of the cells present on a leaf than the more solitary cells which had previously been considered typical of epiphytes. Two different studies have reported that bacterial assemblages offer protection against desiccation stress (Monier and Lindow, 2003; Jacques *et al.*, 2005). Monier and Lindow have reported the differential survival of solitary and aggregated cells of *P. syringae* strain B728a on bean plants exposed to desiccation stress. These authors showed that the fraction of cells in large assemblages that were alive on plants exposed to desiccation stress was much higher than more solitary cells and that the preferential survival of cells in aggregates promotes a highly clustered spatial distribution of bacteria on leaf surfaces (Fig. 6.7). Indeed, while following inoculation cells were randomly distributed over the leaf surface, a highly aggregated pattern of colonisation was apparent within 7 days, and over 90% of the living cells were located in assemblages of 100 cells or more (Monier and Lindow, 2003). The differential survival of aggregated and solitary cells on bean leaves was also reported for *Xanthomonas axonopodis* pv. *phaseoli* (Jacques *et al.*, 2005). These authors reported that while aggregated population sizes remained stable throughout the growing season, solitary population sizes varied with climate. In controlled conditions, application of a desiccation stress resulted in a decrease of the solitary populations while the aggregated fraction remained stable and suppression of the hydric stress allowed solitary bacterial populations to increase again. While such studies are still anecdotal as applied to desiccation stress, they clearly suggest that bacterial assemblages may provide protection to epiphytic bacteria against environmental stresses.

The presence of large clusters of bacteria in close contact on leaves may also facilitate metabolic and genetic exchange. Several studies have reported that remarkably high rates of transfer of conjugative plasmids can occur on plants (Normander *et al.*, 1998; Bailey *et al.*, 2002; Van Elsas *et al.*, 2003; Sorensen *et al.*, 2005). In an attempt to demonstrate the role of aggregates in plasmid exchange among epiphytes Jacques *et al.* (1996) have characterised plasmid distribution among different strains of fluorescent pseudomonads in an individual aggregate isolated from broad-leaved endive and reported that half of the strains contained the same mercury-resistant plasmid. Bjorklof *et al.* (1995) have reported that plasmid transfer was highly efficient when plants were incubated under high relative humidity but could also occur under low relative humidity and when bacterial growth had ceased (Bjorklof *et al.*, 2000). These authors suggested that plasmid transfers occurred in small aggregates observed on the leaf surface. They have also reported that when bacteria were inoculated separately in time, plasmid transfer rates decreased and have suggested that the formation of a mucoid matrix may have acted as a barrier for gene transfer by preventing cell-to-cell contact (Bjorklof *et al.*, 2000). Thus, patchy distribution of nutrients leading to bacterial assemblages on leaves may be important breeding grounds for microbial diversity in plant-associated bacteria (Davison, 1999; Van Elsas *et al.*, 2003; De Vries and Wackernagel, 2004) and could also enhance stabilisation of the bacterial assemblage structure (Molin and Tolker-Nielsen, 2003).

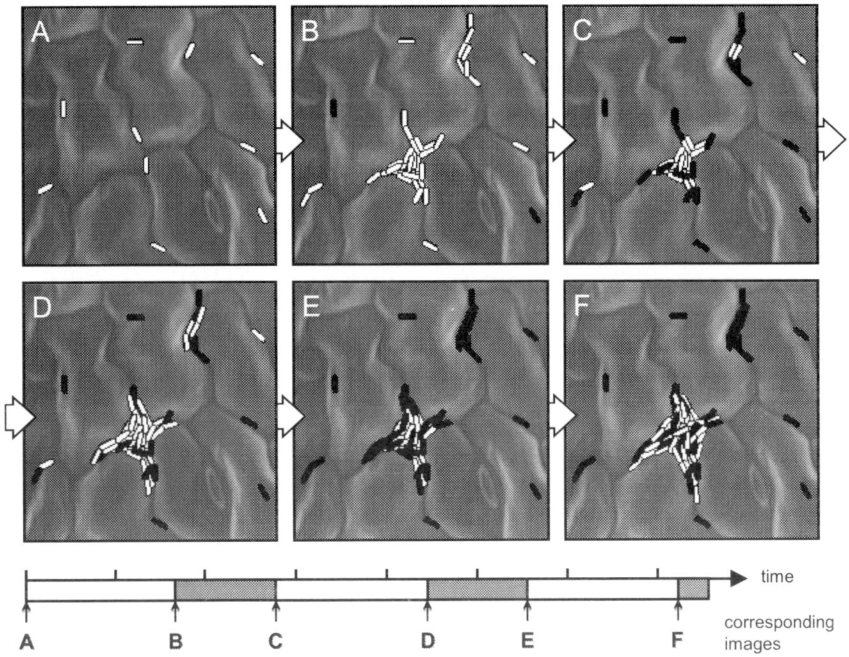

Fig. 6.7. Hypothetical model describing the shift in the spatial distribution of epiphytic bacteria, resulting from exposure to periodic stress, from solitary cells randomly distributed on the leaf surface (A), to a highly clustered distribution with the majority of viable cells found in aggregates (F). Viable cells are represented in white and non-viable cells in black. Periods of exposure to stress and periods conducive for bacterial growth are represented in grey and in white on the time scale, respectively.

Assemblage formation among epiphytes has significant ecological implications that must be considered when designing strategies to control plant pathogenic bacteria. As suggested above, at least some cells in large aggregates may be protected from bactericides applied to plants especially if the cells are embedded in a mucoid matrix. Since most antibacterial compounds have been developed based on tests of their activity against planktonic cells in synthetic media, they probably differ in efficacy against aggregated epiphytic populations. In addition, a clustered colonisation pattern of the leaf surface, with a majority of the cells located in a few large aggregates, may significantly limit microbial interactions to a few sites. On the other hand, indirect evidence that assemblage formation by plant pathogenic bacteria and cell-cell signalling might be required for disease development. However, determining if assemblage formation by plant pathogenic bacteria is required before disease occurrence and thus, if aggregates are sites where leaf invasion preferentially occurs and symptoms are initiated might proved to be a challenging task. While observation of individual necrotic plant cells shows that these were often heavily colonised by pathogenic bacteria (Fig. 6.8), it is difficult to determine whether the wounded plant cell resulted from the presence of an aggregate formed by this pathogen or whether an aggregate resulted from the presence of a wounded plant cell. However, such hypothesis could be more easily tested indirectly by establishing correlation between either, the frequency of occurrence of aggregates on a given leaf and the number of infection sites subsequently seen, or between the localisation of aggregates and the localisation of infection sites (e.g. determine if infection sites are preferentially found at the base of glandular trichomes).

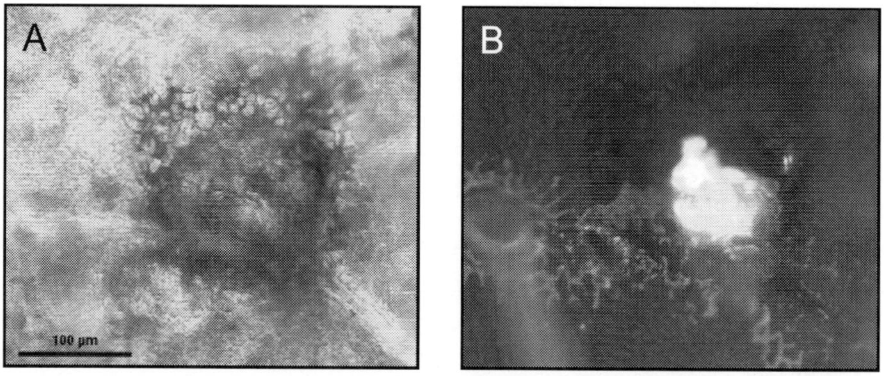

Fig. 6.8. Micrographs of a bacterial brown spot lesion caused by the plant pathogenic bacterium *P. syringae* strain B728a on a bean leaf. (A) Bright field image of the lesion formed along a vein at the base of a trichome. (B) Micrograph of the same lesion observed by epifluorescence microscopy revealing the presence of an assemblage formed by cells of *P. syringae* harbouring a green fluorescent protein marker. (bar=100 µm)

Identification of genes expressed in a density-dependent manner within bacterial assemblages on leaf surfaces and the determination of their spatial and temporal expression during epiphytic colonisation and/or disease development should prove to be very informative to our understanding of the biological significance of bacterial assemblage formation on plant surfaces. We could expect that the difference in behaviour of solitary and aggregated cells is due to a differential expression of genes in a density-dependent manner. Quantification the significance of intercellular communication (cross-talk) between plant pathogenic bacteria and saprophytic bacterial species commonly found on leaf surfaces. The potential occurrence of interfering organisms may suggest that chemical cross-talk could account for some of the segregation of communities on leaves. Further investigation should allow to us to understand how intercellular communication and concerted density-dependent activities may influence utilisation of resources and survival of populations of bacteria on leaf surfaces.

References

Andrews, J.H. and Harris, R.F. (2000) The ecology and biogeography of microorganisms on plant surfaces. *Annual Review of Phytopathology* 38, 145-180.

Ascensao, L. and Pais, M.S. (1998) The leaf capitate trichomes of *Leonotis leonurus*: histochemistry, ultrastructure and secretion. *Annals of Botany* 81, 263-271.

Axtell, C.A. and Beattie, G.A. (2002) Construction and characterization of a *proU-gfp* transcriptional fusion that measures water availability in a microbial habitat. *Applied and Environmental Microbiology* 68, 4604-4612.

Bailey, M.J., Rainey, P.B., Zhang, X.-X. and Lilley, A.K. (2002) Population dynamics, gene transfer and gene expression in plasmids, the role of the horizontal gene pool in local adaptation at the plant surface, In: S.E. Lindow, E.I. Hecht-Poinar and V.J. Elliott (eds), *Phyllosphere Microbiology*. American Phytopathological Society, St. Paul, Minnesota, USA. pp. 173-191.

Bashan, Y., Sharon, E., Okon, Y. and Henis, Y. (1981) Scanning electron and light microscopy of infection and symptom development in tomato leaves infected with *Pseudomonas tomato*. *Physiological Plant Patholology* 19, 139-144.

Beattie, G.A. and Lindow, S.E. (1999) Bacterial colonization of leaves: A spectrum of strategies. *Phytopathology* 89, 353-359.

Bjorklof, K., Suoniemi, A., Haahtela, K. and Romantschuk, M. (1995) High frequency of conjugation versus plasmid segregation of RP1 in epiphytic *Pseudomonas syringae* populations. *Microbiology* 141, 2719-2727.

Bjorklof, K., Nurmiaho-Lassila, E.L., Klinger, N., Haahtela, K. and Romantschuk, M. (2000) Colonization strategies and conjugal gene transfer of inoculated *Pseudomonas syringae* on the leaf surface. *Journal of Applied Microbiology* 89, 423-432.

von Bodman, S.B. and Farrand, S.K. (1995) Capsular polysaccharide biosynthesis and pathogenicity in *Erwinia stewartii* require induction by an N-Acylhomoserine lactone autoinducer. *Journal of Bacteriology* 177, 5000-5008.

von Bodman, S. B., Bauer, W. D., and Coplin, D. L. (2003) Quorum sensing in plant-pathogenic bacteria. *Annual Review of Phytopathology* 41, 455-482.

Boureau, T., Jacques, M.-A., Berruyer, R., Dessaux, Y., Dominguez, H. and Morris, C.E. (2003) Comparison of the structures of biofilm and solitary components of epiphytic bacterial populations on broad-leaved endive. *Microbial Ecology* 47, 87-95.

Branda, S.S., Vik, S., Friedman, L. and Kolter, R. (2005) Biofilms: the matrix revisited. *Trends in Microbiology* 13, 20-26.

Brandl, M.T. and Lindow, S.E. (1996) Cloning and characterization of a locus encoding an indolepyruvate decarboxylase involved in indole-3-acetic acid synthesis in *Erwinia herbicola*. *Applied and Environmental Microbiology* 62, 4121-4128.

Brandl, M.T. and Lindow, S.E. (1997) Environmental signals modulate the expression of an indole-3-acetic acid biosynthetic gene in *Erwinia herbicola*. *Molecular Plant-Microbe Interactions* 10, 499-505.

Brandl, M.T. and Lindow, S.E. (1998) Contribution of indole-3-acetic acid production to the epiphytic fitness of *Erwinia herbicola*. *Applied and Environmental Microbiology* 64, 3256-3263.

Brandl, M.T. and Lindow, S.E. (2003) Microbiology of the phyllosphere. *Applied and Environmental Microbiology* 69, 1875-1883.

Brandl, M.T. and Monier, J.-M. (2006) Methods in microscopy for the visualization of bacteria and their behavior on plants, In: G.M. Sapers, J.R. Gorny and A.E. Youssef (eds), *Microbiology of Fruits and Vegetables*. CRC Press, Taylor & Francis Group, Boca Raton, Florida, USA. pp. 595-619.

Brewer, C.A., Smith, W.K. and Vogelmann, T.C. (1991) Functional interaction between leaf trichomes, leaf wettability and the optical properties of water droplets. *Plant Cell and Environment* 14, 955-962.

Cao, H., Baldini, R.L. and Rahme, L.G. (2001) Common mechanisms for pathogens of plants and animals. *Annual Review of Phytopathology* 39, 259-284.

Casavant, N.C., Beattie, G.A., Phillips, G.J. and Halverson, L.J. (2002) Site-specific recombination-based genetic system for reporting transient or low-level gene expression. *Applied and Environmental Microbiology* 68, 3588-3597.

Chalfie, M., Tu, Y., Euskirchen, G., Ward, W.W. and Prasher, D.C. (1994) Green fluorescent protein as a marker for gene expression. *Science* 263, 802-805.

Characklis, W.G. (1973) Attached microbial growths II. Frictional resistance due to microbial slimes. *Water Research* 7, 1249-1258.

Costerton, J.W., Lewandowski, Z., Caldwell, D.E., Korber, D.R. and Lappin-Scott, H.M. (1995) Microbial biofilms. *Annual Review of Microbiology* 49, 711-745.

Crossman, L. and Maxwell Dow, J. (2004) Biofilm formation and dispersal in *Xanthomonas campestris*. *Microbes and Infection* 6, 623-629.

Davey, M.E. and O'Toole, G.A. (2000) Microbial biofilms: from ecology to molecular genetics. *Microbiology and Molecular Biology Reviews* 64, 847-867.

Davison, J. (1999) Genetic exchange between bacteria in the environment. *Plasmid* 42, 73-91.

Denny, T.P. (1995) Involvement of bacterial polysaccharides in plant pathogenesis. *Annual Review of Phytopathology* 33, 173-197.

De Vries, J. and Wackernagel, W. (2004) Microbial horizontal gene transfer and the DNA release from transgenic crop plants. *Plant and Soil* 266, 91-104.

Dumenyo, C. K., Mukherjee, A., Chun, W. and Chatterjee, A. K. (1998) Genetic and physiological evidence for the production of N-acyl homoserine lactones by *Pseudomonas syringae* pv. *syringae* and other fluorescent plant pathogenic *Pseudomonas* species. *European Journal of Plant Pathology* 104, 569-582.

Dunn, A.K., Klimowicz, A.K. and Handelsman, J. (2003) Use of a promoter trap to identify *Bacillus cereus* genes regulated by tomato seed exudate and a rhizosphere resident *Pseudomonas aureofaciens*. *Applied and Environmental Microbiology* 69, 1197-1205.

Van Elsas, J.D., Turner, S. and Bailey, M.J. (2003) Horizontal gene transfer in the phytosphere. *New Phytologist* 157, 525-537.

Errampalli, D., Leung, K., Cassidy, M.B., Kostrzynska, M., Blears, M., Lee, H. and Trevors, J.T. (1999) Applications of the green fluorescent protein as a molecular marker in environmental microorganisms. *Journal of Microbiological Methods* 35, 187-199.

Espinosa-Urgel, M. and Ramos, J.-L. (2001) Expression of a *Pseudomonas putida* aminotransferase involved in lysine catabolism is induced in the rhizosphere. *Applied and Environmental Microbiology* 67, 5219-5224.

Fett, W.F. (2000) Naturally occurring biofilms on alfalfa and other types of sprouts. *Journal of Food Protection* 63, 625-632.

Fett,W.F. and Cooke, P.H. (2003) Scanning electron microscopy of native biofilms on mung bean sprouts. *Canadian Journal of Microbiology* 49, 45-50.

Fett,W.F. and Cooke, P.H. (2005) A survey of native microbial aggregates on alfalfa, clover and mung bean sprout cotyledons for thickness as determined by confocal scanning laser microscopy. *Food Microbiology* 22, 253-259.

Gage, D.J., Bobo, T. and Long, S.R., (1996) Use of green fluorescent protein to visualize early events of symbiosis between *Rhizobium meliloti* and alfalfa (*Medicago sativa*). *Journal of Bacteriology* 178, 7159-7166.

Givskov, M., Eberl, L., Moller, S., Poulsen, L.K. and Molin, S. (1994) Responses to nutrient starvation in *Pseudomonas putida* KT2442: Analysis of general cross-protection, cell shape, and macromolecular content. *Journal of Bacteriology* 176, 7-14.

Gras, M.-H., Druet-Michaud, C. and Cerf, O. (1994) La flore bactérienne des feuilles de salade fraîche. *Science des Aliments* 14, 173–188.

Hall-Stoodley, H., Costerton, J.W. and Stoodley, P. (2004) Bacterial biofilms: from the natural environment to infectious disease. *Nature Reviews* 2, 95-108.

Haugland, P.H. (2005) *A Guide to Fluorescent Probes and Labeling Technologies*, 10th edn., Molecular Probes, Inc., Eugene, USA.

Heim, R., Cubitt, A.B. and Tsien, R.Y. (1995) Improved green fluorescence. *Nature* 373, 663-664.

Heukelekian, H. and Heller, A. (1940) Relation between food concentration and surface for bacterial growth. *Journal of Bacteriology* 40, 547-558.

Hirano, S.S. and Upper, C.D. (2000) Bacteria in the leaf ecosystem with emphasis on *Pseudomonas syringae*: a pathogen, ice nucleus, and epiphyte. *Microbiology and Molecular Biology Reviews* 64, 624–653.

Jacques, M.-A., Morris, C.E. and Bailey, M.J. (1996) The microbial diversity of a phyllosphere micro-aggregate, a site for enhanced gene transfer? Proceedings, BAGECO 5, 25-29 May 1996, Nafplion, Greece. S1. 1:135.

Jacques, M.-A., Josi, K., Darrasse, A. and Samson, R. (2005) *Xanthomonas axonopodis* pv. *phaseoli* var. *fuscans* is aggregated in stable biofilm population sizes in the phyllosphere of field-grown beans. *Applied and Environmental Microbiology* 71, 2008-2015.

Jansson, J.K. (2003) Marker and reporter genes: illuminating tools for environmental microbiologists. *Current Opinion in Microbiology* 6, 310-316.

Joyner, D.C. and Lindow, S.E. (2000) Heterogeneity of iron bioavailability on plants assessed with a whole-cell GFP-based bacterial biosensor. *Microbiology Reading* 146, 2435–2445.

Kasten, F.H. (1999) Introduction to fluorescent probes: properties, history and applications. In: Mason, W.T. (ed.), *Biological Techniques Series; Fluorescent and Luminescent Probes for Biological Activity:* A *Practical Guide to Technology for Quantitative Real-Time Analysis*, 2nd edn., Academic Press, New York, USA. pp. 17-39.

Kinkel, L.L., Newton, M.R. and Leonard, K.J. (2002) Resource aggregation in the phyllosphere: implications for microbial dynamics across spatial scales. In: S.E. Lindow, E.I.Hecht-Poinar and V.J. Elliot (eds), *Phyllosphere Microbiology*. APS Press, St. Paul, Minnesota, USA. pp. 317-340.

Kinkel, L.L., Wilson, M. and Lindow, S.E. (1995) Effects of sampling scale on the assessment of epiphytic bacterial populations. *Microbial Ecology* 29, 283-297.

Koch, B., Nielsen, T.H., Sørensen, D., Andersen, J.B., Christophersen, C., Molin, S., Givskov, M., Sørensen, J. and Nybroe, O. (2002) Lipopeptide production in *Pseudomonas* sp. strain DSS73 is regulated by components of sugar beet seed exudate *via* the Gac two-component regulatory system. *Applied and Environmental Microbiology* 68, 4509-4516.

Koiv, V., and Mae, A. (2001) Quorum sensing controls the synthesis of virulence factors by modulating *rsmA* gene expression in *Erwinia carotovora* sp. *carotovora*. *Molecular Genetics and Genomics* 265, 287-292.

Larrainzar, E., O'Gara, F. and Morrissey, J.P. (2005) Applications of autofluorescent proteins for *in situ* studies in microbial ecology. *Annual Review of Microbiology* 59, 257-277.

Leben, C. (1988) Relative humidity and the survival of epiphytic bacteria with buds and leaves of cucumber plants. *Phytopathology* 78, 179-185.

Leigh, J.A. and Coplin, D.L. (1992) Exopolysaccharides in plant-bacterial interactions. *Annual Review of Microbiology* 46, 307-346.

Leveau, J.H.J. and Lindow, S.E. (2001) Appetite of an epiphyte: quantitative monitoring of bacterial sugar consumption in the phyllosphere. *Proceedings of the National Academy of Sciences*, USA 98, 3446-3453.

Lindow, S.E. and Brandl, M.T. (2003) Microbiology of the Phyllosphere. *Applied and Environmental Microbiology* 69, 1875-1883.

Loh, J., Pierson, E.A., Pierson, L.S., 3rd, Stacey, G., and Chatterjee, A. (2002) Quorum sensing in plant-associated bacteria. *Current Opinion in Plant Biology* 5, 285-290.

van Loosdrecht, M.C.M., Eikelboom, D., Gjaltema, A., Mulder, A., Tijhuis, L. and Heijnen, J.J. (1995) Biofilm structures. *Water Science and Technology* 32, 35-43.

Mansvelt, E.L. and Hattingh, M.J. (1987) Scanning electron microscopy of colonization of pear leaves by *Pseudomonas syringae* pv. *syringae*. *Canadian Journal of Botany* 65, 2517-2522.

Mansvelt, E.L. and Hattingh, M.J. (1989) Scanning electron microscopy of invasion of apple leaves and blossoms by *Pseudomonas syringae* pv. *syringae*. *Applied and Environmental Microbiology* 55, 533-538.

Mew, T.W., Mew, I.C. and Huang, J.S. (1984) Scanning electron microscopy of virulent and avirulent strains of *Xanthomonas campestris* pv. *oryzae* on rice leaves. *Phytopathology* 74, 635-641.

Mew, T.W. and Vera Cruz, C.M. (1986) Epiphytic colonization of host and non-host plants by phytopathogenic bacteria. In: N.J. Fokkema and J. van den Heuvel (eds), *Microbiology of the Phyllosphere*. Cambridge University Press, New York, USA. pp. 269-282.

Miller, W.G., Brandl, M.T., Quiñones, B. and Lindow, S.E. (2001) Biological sensor for sucrose availability: relative sensitivities of various reporter genes. *Applied and Environmental Microbiology* 67, 1308-1317.

Molin, S. and Tolker-Nielsen, T. (2003) Gene transfer occurs with enhanced efficiency in biofilms and induces enhanced stabilisation of the biofilm structure. *Current Opinion in Biotechnology* 14, 255-261.

Molina, M.A., Ramos, J.-L. and Espinosa-Urgel, M. (2003) Plant-associated biofilms. *Reviews in Environmental Science and Biotechnology* 2, 99-108.

Monier, J.-M. and Lindow, S.E. (2003a) *Pseudomonas syringae* responds to the environment on leaves by cell size reduction. *Phytopathology* 93, 1209-1216.

Monier, J.-M. and Lindow, S.E. (2003b) Differential survival of solitary and aggregated bacterial cells promotes aggregate formation on leaf surfaces. *Proceedings of the National Academy of Sciences,* USA 100, 15977-15982

Monier, J.-M. and Lindow, S.E. (2004) Frequency, size and localization of bacterial aggregates on bean leaf surfaces. *Applied and Environmental Microbiology* 70, 346-355.

Monier, J.-M. and Lindow, S.E. (2005a) Aggregates of resident bacteria facilitate survival of immigrant bacteria on leaf surfaces. *Microbial Ecology* 49, 343-352.

Monier, J.-M. and Lindow, S.E. (2005b) Spatial organization of dual-species bacterial aggregates on leaf surfaces. *Applied and Environmental Microbiology* 71, 5484-5493

Moriarty, D.J.W. and Bell, R.T. (1993) Bacterial growth and starvation in aquatic environments. In: S. Kjelleberg (ed.), *Starvation in Bacteria.* Plenum Press, New York, USA. pp. 25-53.

Morris, C.E., Barnes, M.B. and McLean, R.J.C. (2002) Biofilms on leaf surfaces: Implications for the biology, ecology and management of populations of epiphytic bacteria, In: S.E. Lindow, E.I. Hecht-Poinar and V.J. Elliott (eds), *Phyllosphere Microbiology*. APS Press, St. Paul, MN, USA. pp. 139-155.

Morris, C.E. and Monier, J.-M. (2003) The ecological significance of biofilm formation by plant-associated bacteria. *Annual Review of Phytopathology* 41, 429-453.

Morris, C.E., Monier, J.-M. and Jacques, M.-A. (1997) Methods for observing microbial biofilms directly on leaf surfaces and recovering them for isolation of culturable microorganisms. *Applied and Environmental Microbiology* 63, 1570-1576.

Morris, C.E., Monier, J.-M. and Jacques, M.-A. (1998) A technique to quantify the population size and composition of the biofilm component in communities of bacteria in the phyllosphere. *Applied and Environmental Microbiology* 64, 4789-4795.

Normander, B., Christensen, B.B., Molin, S. and Kroer, N. (1998) Effect of bacterial distribution and activity on conjugal gene transfer on the phylloplane of the bush bean (*Phaseolus vulgaris*). *Applied and Environmental Microbiology* 64, 1902-1909.

Olapade, O.A. and Leff, L.G. (2005) Seasonal response of stream biofilm communities to dissolved organic matter and nutrient enrichments. *Applied and Environmental Microbiology* 71, 2278-2287.

Prasher, D.C., Eckenrode, V.K., Ward, W.W., Prendergast, F.G. and Cormier, M.J. (1992) Primary structure of the *Aequorea victoria* green-fluorescent protein. *Gene* 111, 229-233.

Quiñones, B., Pujol, C.J. and Lindow, S.E. (2004) Regulation of AHL production and its contribution to epiphytic fitness in *Pseudomonas syringae*. *Molecular Plant-Microbe Interactions* 5, 521-531.

Quiñones, B., Dulla, G. and Lindow, S.E. (2005) Quorum sensing regulates exopolysaccharide production, motility, and virulence in *Pseudomonas syringae*. *Molecular Plant-Microbe Interactions* 7, 682-693.

Roos, I.M.M. and Hattingh, M.J. (1983) Scanning electron microscopy of *Pseudomonas syringae* pv. *morsprunorum* on sweet cherry leaves. *Phytopathology Zeitschrift* 180, 18-25.

Rovira, A.D. and Campbell, R. (1974) Scanning electron microscopy of micro-organisms on the roots of wheat. *Microbiology and Ecology* 1, 15–23.

Ruinen, J. (1961) The phyllosphere. I. An ecologically neglected milieu. *Plant Soil* 15, 81-109.

Schreiber, L., Krimm, U., Knoll, D., Sayed, M., Auling G. and Kroppenstedt, R.M. (2005) Plant–microbe interactions: identification of epiphytic bacteria and their ability to alter leaf surface permeability. *New Phytologist* 166, 589-594.

Shimomura, O., Johnson, F.H. and Saiga, Y. (1962) Extraction, purification and properties of aequorin, a bioluminescent protein from the luminous hydromedusan, *Aequorea*. *Journal of Cellular and Comparative Physiology* 59, 223-239.

Sørensen, S.J., Bailey, M., Hansen, L.H., Kroer, N. and Wuertz, S. (2005) Studying plasmid horizontal transfer *in situ*: a critical review. *Nature* 3, 700-710.

Stoodley, P., Boyle, J.D., Dodds, I. and Lappin-Scott, H.M. (1997) Consensus model of biofilm structure. In: J.W.T. Wimpenny, P.S. Handley, P. Gilbert, H.M. Lappin-Scott and M. Jones (eds), *Biofilms: Community Interactions and Controls*. BioLine, Cardiff, UK. pp. 1-9.

Surico, G. (1993) Scanning electron microscopy of olive and oleander leaves colonized by *Pseudomonas syringae* subsp. *savastanoi*. *Journal of Phytopathology* 138, 31-40.

Sutherland, I.W. (2001) The biofilm matrix: an immobilized but dynamic microbial environment. *Trends in Microbiology* 9, 222-227.

Thelin, G.P. and Pike, R.J. (1991) Landforms of the conterminous United States; A digital shaded-relief portrayal. US *Geological Survey Miscellaneous Investigations* Map I-2206, scale 1:3,500,000.

Timmer, L.W., Marois, J.J. and Achor, D. (1987) Growth and survival of xanthomonads under conditions nonconducive to disease development. *Phytopathology* 77, 1341-1345.

Tsien, R.Y. (1998) The green fluorescent protein. *Annual Review of Biochemistry* 67, 509-544.

Valkama, E., Salminen, J.P., Koricheva, J. and Pihlaja, K. (2003) Comparative analysis of leaf trichome structure and composition of epicuticular flavonoids in Finnish birch species. *Annals of Botany* 91, 643-655.

Wimpenny, J., Manz, W. and Szewzyk, U. (2000) Heterogeneity in biofilms. FEMS *Microbiology Reviews* 24, 661-671.

Yadav, R.K.P., Karamanoli, K. and Vokou, D. (2005) Bacterial colonization of the phyllosphere of Mediterranean perennial species as influenced by leaf structural and chemical features. *Microbial Ecology* 50, 185-196.

Zobell, C.E. (1937) The influence of solid surfaces upon the physiological activities of bacteria in seawater. *Journal of Bacteriology* 33-86.

Zobell, C.E. (1943) The effects of solid surfaces upon bacterial activity. *Journal of Bacteriology* 46, 39-56.

Zobell, C.E. and Allen, E.C. (1935) The significance of marine bacteria in the fouling of submerged surfaces. *Journal of Bacteriology* 29, 239-251.

Chapter 7

The Role of Plant Genetics in Determining Above- and Below-ground Microbial Communities

J.A. Schweitzer[1], J.K. Bailey[2], R.K. Bangert[2], S.C. Hart[1] and T.G. Whitham[2]

[1]School of Forestry, Northern Arizona University, Box 15018, Flagstaff AZ-86011, USA; [2]Department of Biological Sciences, Northern Arizona University, Box 5640, Flagstaff AZ-86011

jennifer.schweitzer@nau.edu

Introduction

In ecological research, hypotheses that have broad general applicability are rare. Recent studies in community genetics, however, have shown that specific genetically based plant traits have direct and indirect effects that extend beyond the individual possessing the trait to affect associated communities and ecosystem processes. Such genetically based traits may be general organising factors of systems that are largely structured by bottom-up forces. Unfortunately, little is known about the specific role of plant genes in structuring associated communities, microbial or otherwise. This is likely to be due to the fact that understanding the genetic basis to community structure and ecosystem processes is an emerging field requiring extensive interdisciplinary integration (Whitham *et al.,* 2003). To date, none of the prevailing theories on the determinants of community structure have addressed the potential genetic basis of community structure (e.g. Lawton, 1999; Hubbell, 2002).

The majority of work on the genetic basis of communities has been done with terrestrial arthropod communities, which may represent a parallel and analogous area of study to microbial communities (Horner-Devine *et al.,* 2003). Specific plant genes and quantitative trait loci (QTL) have been linked to particular arthropods and their resistance or susceptibility (Via and Hawthorne, 2002; Tan *et al.,* 2004). However, we are aware of no studies linking specific genes to arthropod community structure. Previous studies have shown a link between arthropod communities and plant genetic diversity (Aguilar and Boecklen, 1992; Wimp *et al.,* 2005) as well as genetic-based traits such as plant architecture, phenology (Johnson and Agrawal, 2004) and leaf chemistry (Dungey *et al.,* 2000; Abrahamson *et al.,* 2003; Bangert *et al.,* 2005b). The data linking plant genetics and arthropod communities have mostly focused on genetic-based plant traits that directly or indirectly structure a community that is dependent upon these plants for their survival. For example, variation in traits such as plant architecture or phenology among evening primrose

(*Oenothera* spp.) genotypes has been shown to strongly affect associated community variables (i.e. arthropod diversity, richness, abundance and biomass; Johnson and Agrawal, 2004; Johnson *et al.,* 2006). Similarly, in both field and common garden (i.e. garden in which trees of known genotype were randomly planted in a common environment) studies, Bangert *et al.,* (2005b) developed the concept of a "genetic similarity rule" that predicts community structure based on the genetic similarity of individual plants. For example, in cottonwoods, they found that individual trees that had similar genetic compositions also had similar arthropod compositions, whereas trees that were genetically more dissimilar supported arthropod communities that also were dissimilar (Fig. 7.1). Importantly, they also found that genetic similarity was correlated with the phytochemical make-up of individual trees, which provides a potential mechanistic basis for this genetic similarity rule. Studies such as these suggest that there is potential to predict community structure through plant genes (which in turn influence traits such as architecture or plant chemistry) that may provide a mechanistic basis for understanding factors that determine communities.

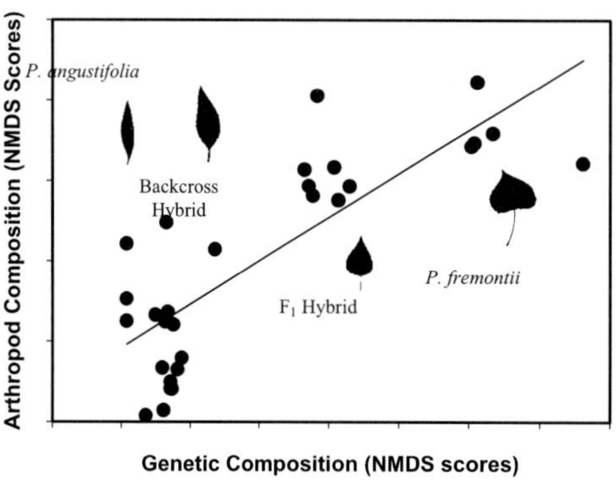

Fig. 7.1. Arthropod communities change with poplar genetic composition. Ordination scores from a non-metric multi-dimensional scaling (NMDS) procedure represent poplar genetic and arthropod compositions of the common garden trees (for use of this procedure see Clarke and Warwick, 2001; Abrahamson *et al.,* 2003; Shuster *et al.,* 2006). Genetic composition is correlated with arthropod composition ($F = 51.89$, $r^2 = 0.66$, $p<0.001$). Figure adapted from Bangert *et al.* (2005b).

Few studies have demonstrated the role of plant genetics on their associated microbial communities in either the phyllosphere or rhizosphere. Due to their intimate associations with plants (i.e. inside or immediately adjacent to the plant tissues) and their potential for rapid evolution (Bohannan and Lenski, 2000), we predict that the potential for such plant-microbial community relationships is high. If we are to understand how microorganisms might influence the evolution of communities and ecosystems, it is crucial to understand the genetic basis of these interactions (Preston *et al.,* 1998; Smith and Goodman, 1999). The importance of plant genetics or genetic-based traits to whole microbial communities is understudied, despite a preponderance of literature on gene-for-gene and multigenic control of resistance or susceptibility to specific microbe taxa at the individual plant genotype or gene level. For example, several studies of foliar endophytes have documented the specificity of host-plant genotypes to endophyte compatibility, as well as genotype specific effects on bacterial population and composition of functional groups (Christianson *et al.,* 1997; Adams and Kloepper, 2002; Cheplick and Cho, 2002; Zhan *et al.,* 2002). Detailed molecular studies have documented genetic relationships with host-plant and endophyte genotypes (Ahlholm *et al.*, 2002) and specific plant genes that are expressed differentially with endophyte infection (Johnson *et al.*, 2003). Perhaps the best examples of genetic control of specific microbe taxa are with *Rhizobia*-plant interactions whereby single plant genes (and *Rhizobium* genes) can determine compatibility and nodule formation (Caetano-Anolles and Gresshoff, 1991; Lieven-Antoniou and Whittam, 1997). Furthermore, studies with arbuscular mycorrhizas, bacterial biocontrol species (such as *Bacillus* spp.), and a range of both plant pathogens and beneficial microorganisms have demonstrated that variation in host plants can determine microbial compatibilities (Lackie *et al.*, 1988; Clement and Habte, 1995; Newcome and Bradshaw, 1996; Smith *et al.*, 1999). Many of these interactions have been linked to specific genes or QTL and show heritable variation in their colonisation patterns (Kaeppler *et al.*, 2000; Onozaki *et al.*, 2004). However, only a handful of studies have demonstrated variation in whole microbial communities in which plant cultivars were important in modification of microbial community size or composition (Diab el Arab *et al.*, 2001; Mazzola and Gu, 2002). For example, Diab el Arab *et al.* (2001) found distinct microbial communities associated with two cultivars of wheat.

In this chapter, we will describe examples of microbial communities, in both the phyllosphere and soil, in which the community composition was shaped by associated plant genetics (i.e. genetically based plant traits), with two case studies in *Populus*. *Populus* is a model system in which to address these questions as it is the first tree genome to be sequenced (Brunner *et al.*, 2004) and is a system in which genetically based traits have been linked to diverse community members and ecosystem processes (Whitham *et al.*, 2003). Despite the difficulties of assessing diversity or communities in microorganisms, we will illustrate that, in general, community similarity is related to the associated tree genotypes in which these communities interact. Data such as this suggest that ecological factors that structure microbial communities follow similar ecological patterns as arthropods indicating qualitative patterns in ecological processes. Furthermore, these data demonstrate the strength of plant genetic based trait(s) to influence community composition and support recent claims of the importance of plant genetics to community and ecosystem processes (Neuhauser *et al.*, 2003; Whitham *et al.*, 2003).

Phyllosphere Microbial Communities

An important phyllosphere community that could be affected by plant genetics is fungal endophytes. Fungal endophytes inhabit the aerial parts of plants without producing symptoms of disease and are present across the taxonomic range of vascular plants and range from parasitic to neutral to mutualistic in their interactions with plants (Saikkonen *et al.*, 1998; Bayman, Chapter 3, this volume). Endophytes have been shown to serve many functions in plants including deterrence of herbivory (Clay, 1996), increased drought tolerance (Elmi and West, 1995), and improved nutrient uptake (Malinowski *et al.*, 2000); endophytes have also been shown to alter ecosystem processes (Rudgers *et al.*, 2004). Documented variation in host-plant genotype to endophyte compatibility, population dynamics and composition indicates that host plant genotype has a strong influence on the dynamics of endophyte populations (Christiansen *et al.*, 1997; Adams and Kloepper, 2002; Ahlholm *et al.*, 2002; Zhan *et al.*, 2002). For example, Adams and Kloepper (2002) found that there were significant effects of cotton (*Gossypium hirsutum*) cultivars on leaf bacterial populations as well as the composition of bacterial functional groups. Variation at the host level for endophyte fitness indicates that traits of the cultivar, such as host-plant physiology, morphology or phytochemistry, may be important to the regulation of endophyte population and community dynamics.

Populus Case Study

In 2002, we examined the genetic basis of endophyte communities across pure and hybrid *Populus* trees both in the field and in a common garden (Bailey *et al.*, 2005). These studies took place in a natural *Populus* hybrid zone (*Populus fremontii* x *P. angustifolia*) along the Weber River, Utah, USA, and a nearby 14-year-old common garden with trees of known genotype planted randomly within the site (for specific site and garden details see Schweitzer *et al.*, 2004; Bangert *et al.*, 2005b). Natural hybridisation and introgression creates a genetic continuum between the two hybridising species that is ideal for examining the evolutionary implications of gene flow on community processes. Our observations within the Weber River hybrid zone demonstrated that *Populus fremontii* had significantly greater twig endophytes than F1 and backcross *P. angustifolia* genotypes, based on an equal expectation among tree types. In the field, fungal endophytes infected 43% more *P. fremontii* than expected, assuming equal colonisation potential. In contrast, F1 hybrids and backcross/*P. angustifolia* (pooled as they are morphologically indistinguishable in the field) had 95% and 74% lower infection than expected, respectively (Fig. 7.2). Rather than any single species of fungal endophyte being affected, these observational data suggested that plant traits related to cottonwood hybridisation might have a generalised effect on fungal endophytes as a whole. These data supported the hypothesis that genetic factors related to plant hybridisation are in part responsible for the genetic differences among cottonwoods that affect fungal endophytes.

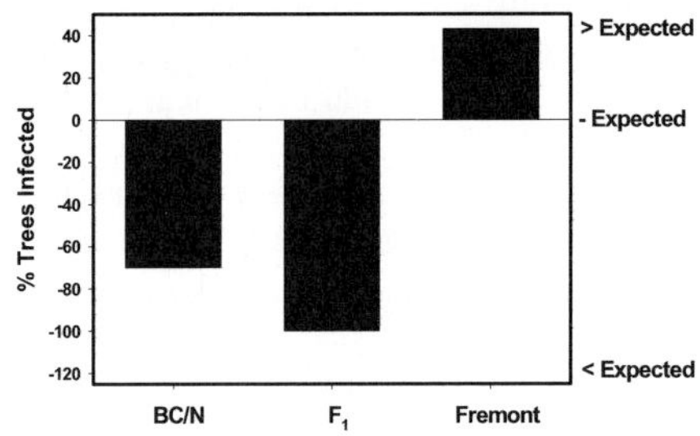

Fig. 7.2. Fungal endophytes vary in their infection of the pure and hybrid poplars. In the field, endophyte infection is greater than expected on *P. fremontii* than on *P. angustifolia* and their naturally occurring hybrids. These observational data suggest a genetic basis to fungal endophyte infection. Adapted from Bailey *et al.* (2005).

To test our findings experimentally in the wild we repeated our studies in a common garden where we could largely eliminate environmental effects to study the potential genetic impacts of plant hybridisation on endophytes. We confirmed the observational patterns in the common garden; *P. fremontii* genotypes continued to have significantly greater endophyte infection than F1 hybrids and backcross/*P. angustifolia*. With restriction fragment length polymorphisms (RFLP) molecular marker data (Martinsen *et al.*, 2001), we were able to trace the introgression of *P. fremontii* molecular markers where we found a significant positive correlation with fungal endophyte infection (Fig. 7.3). We found that fungal endophyte infection was negatively correlated with the concentration of condensed tannins, which is known to have anti-fungal properties (Scalbert, 1991; Kraus *et al.*, 2003).

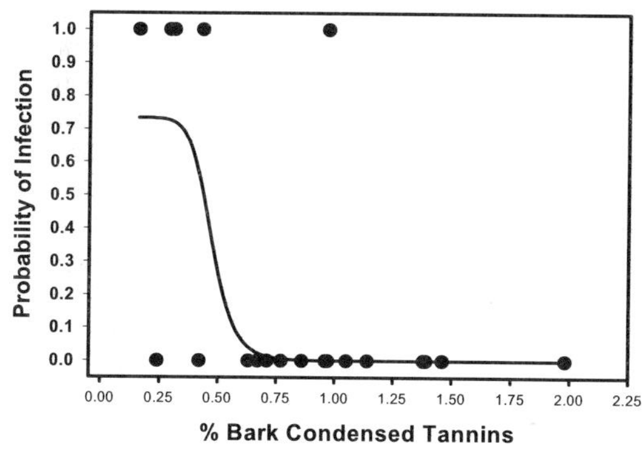

Fig. 7.3. Fungal endophyte infection is related to poplar genetic composition. In a common environment planted with trees of known genotype, the probability of fungal endophyte infection was negatively related to inner bark tannin concentration. Adapted from Bailey *et al.* (2005).

No significant correlations were found with salicortin, HCH-salicortin, and total phenolic glycosides (chemical methods described in Bailey *et al.*, 2005). We were able to address unequivocally the linkages among plant genotype, plant secondary chemistry, and a generalised response of fungal endophytes because this study was conducted in a common garden where all pure and hybrid cross types are randomised throughout the garden, and the age of the tissue sampled was controlled. Overall, our results showed that fungal endophyte infection is related to condensed tannin concentration in the bark tissue of poplars (Bailey *et al.*, 2005), which in turn is under strong genetic control (Whitham *et al.*, 2003). Although there is much that we do not understand about fungal endophytes and their interactions with plant genetics (Saikkonen *et al.*, 1998) or genetic-based traits, we believe this case study provides an example of the potential impacts of genetic introgression on clandestine organisms (i.e. hidden players) such as fungal endophytes that can impact other community members and ecosystem processes.

Soil Microbial Communities

The documented couplings between above- and below-ground components in terrestrial ecosystems suggest that the effects of plant genes, and the traits they control, should interact with soil microbial community composition (Bardgett and Wardle, 2003; Reynolds *et al.*, 2003; Wardle *et al.*, 2004). However, little is known about the importance of plant genetics for structuring soil microbial communities (Horner-Devine *et al.*, 2003). Soil microorganisms are considered hyper-diverse (Finlay *et al.*, 1999) and ubiquitous, however, shifts in community response have been documented under variable biotic or abiotic conditions, which suggest that feedbacks between plants and their associated microbial communities do occur (Grayston *et al.*, 1998; Wardle *et al.*, 2004). Feedbacks between plants and microorganisms have been shown to lead to microbial community compositional shifts with changes to the above-ground plant community at the plant species, functional and perhaps genotype scales (Bever *et al.*, 1997; Madritch and Hunter, 2002; Zak *et al.*, 2003). For example, Zak *et al.* (2003) found microbial community composition shifts after seven years of plant species treatments (1, 2, 4, 8, 16 species mixtures), as well as

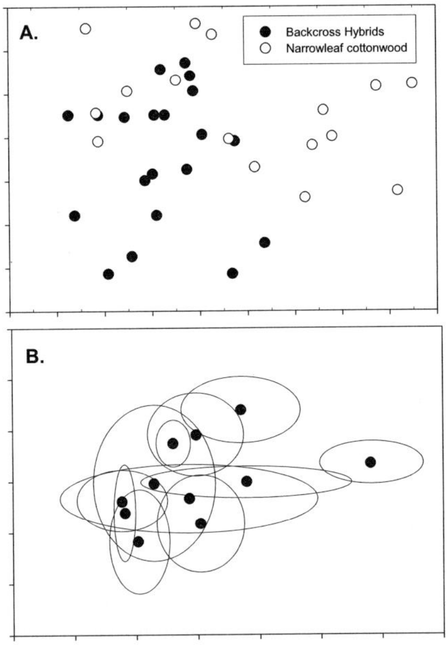

Fig. 7.4. Microbial community compositions differ underneath tree types and individual tree genotypes in a common garden. (A) There are distinct microbial community compositional differences in soils beneath two tree types (*P. angustifolia* and backcross hybrid tree types). (B) Distinct microbial community compositions also exist under individual *P. angustifolia* and backcross hybrid genotypes. Each symbol in the NMDS ordinations represents the mean microbial community dissimilarity on a single tree, surrounded by the standard error of the mean; axes are unitless. Adapted from Schweitzer *et al.* (*in review*).

differences in microbial biomass, respiration and net nitrogen (N) mineralisation. Similarly, Bartelt indicating-Ryser *et al.* (2005) found legacy effects of microbial community composition in soils after plants of different species were grown and then removed from plots, indicating that plant species can have long-term effects on microbial community composition. At the level of plant genotype or cultivar, recent studies have found that plants can play a dominant role in controlling the ecology of associated soil microbial communities. For example, several plant species (mostly agricultural species such as wheat, canola and rice) have been shown to influence the soil microbial community associated with individual cultivars (Azad *et al.,* 1987; Diab el Arab *et al.,* 2001; Mazzola and Gu, 2002; Tan *et al.,* 2003). Non-cultivated plants that differ in their resistance or susceptibility to arthropod herbivores have also been shown to support distinct microbial or mycorrhiza communities (Gehring and Whitham, 2003; Kuske *et al.,* 2003). These results at the genotype or cultivar level strongly suggest the potential for plant genes to direct these microbial-plant interactions.

Populus Case Study

In 2004 as part of a larger study, we quantified the soil microbial community under three to five replicates of five different genotypes each of *P. angustifolia* and six genotypes of backcross hybrids growing in a common garden (same garden as described above; Schweitzer *et al., in review*). The genotypes and replicates were completely randomised in this garden design, with no spatial auto-correlations detected. We analysed the microbial community with phospholipid fatty acid analysis (PLFA) in which the microbial community composition is identified by signature fatty acids that are unique to major taxonomic groups (White and Ringleberg, 1998; Zelles 1999; Kirk *et al.,* 2004). The community data were analysed using non-metric multidimensional scaling ordinations (NMDS; Minchin, 1987, Shuster *et al.,* 2006). We found distinct differences in microbial community composition between the replicate genotypes of *P. angustifolia* and backcross hybrids (trees that differed by seven RFLP molecular markers; Martinsen *et al.,* 2001) at two levels of genetic complexity (the crosstype and genotype level; Fig. 7.4). Soils beneath pure *P. angustifolia* and backcross hybrid genotypes supported significantly different microbial community compositions (Fig. 7.4A; ANOSIM, r=0.087, p=0.043). At the genotype level, the microbial community compositions became more distinct (Fig. 7.4B: ANOSIM, r=0.247, p=0.001). These results suggest that microbial communities can distinguish between individual tree genotypes, presumably due to the specific physiological or chemical traits that the cottonwood genotypes express.

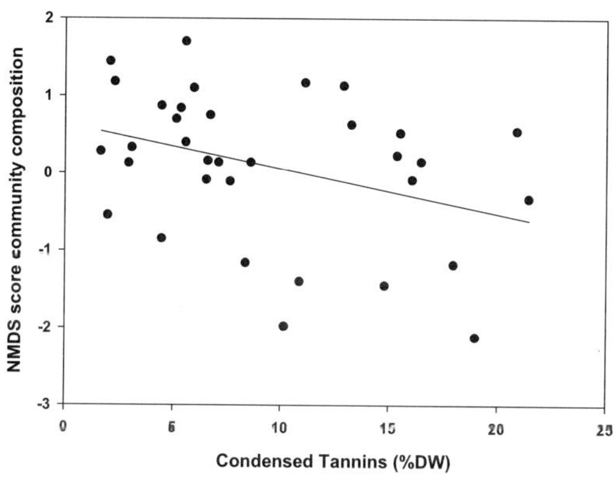

Fig. 7.5. Microbial community composition in the soil underneath the crown of a poplar tree is related to the condensed tannin concentration of the tree foliage. Ordination scores (i.e. community similarity) are correlated to foliar condensed tannins ($F = 4.59$, $r^2 = 0.126$, $p = 0.039$), suggesting a potentially important mechanistic linkage between above-ground plant chemistry (a genetic-based trait, Whitham *et al.,* 2003) and below-ground community processes.

Similar to the fungal foliar endophytic community described above, we found that soil microbial community composition was correlated to the concentration of condensed tannin in the foliage of the associated tree genotype (chemical methods described in Schweitzer *et al.,* 2004; Fig.7.5). The NMDS score of microbial community composition was related positively to foliar condensed tannin in the associated tree genotype ($r^2 = 0.126$, $p = 0.039$), indicating that the soil microbial communities under a given tree genotype co-varied with the foliar condensed tannin concentration of that tree. No other measured plant phytochemical trait (i.e. foliar nitrogen, carbon, C:N ratio) was related to microbial community composition in these genotypes. Thus, as was found with foliar endophytes, soil microbial communities are also influenced by this genetic-based plant phytochemical that is expressed differentially among cottonwood genotypes. Because the soil microbial community mediates litter decomposition, N mineralisation, nitrification, denitrification, and other soil transformations (Nannipieri *et al.,* 2002), genetic factors that influence microbial community composition (i.e. plant chemical, physiological and morphological traits) are also likely to alter these ecosystem processes.

Microbial Communities and Chemical Traits in *Populus*

Why would these disparate organisms (endophytes in the phyllosphere and microorganisms in the soil) be related to the same genetic-based trait of phytochemistry? Plant polyphenols, which include condensed tannins, are a class of secondary compounds that are involved in plant pigmentation, ultraviolet protection and allelopathy, as well as chemical deterrents to herbivores and pathogens (Palo, 1984; Hemingway and Karchesy, 1989). They have been shown to retard leaf litter decomposition by reducing palatability to arthropods and microorganisms and by forming protein-binding complexes (Heal *et al.*, 1997, Krauss *et al.*, 2003). The effects of polyphenols on decomposition processes and their putative ability to inhibit nitrification have led researchers to hypothesize that polyphenols can regulate plant available nutrient pools (Northup *et al.,* 1998; Hättenschwiler and Vitousek, 2000; Kraus *et al.,* 2003). The multiple functions of condensed tannins indicate that their production may be adaptive under many circumstances (Kraus *et al.,* 2003) and have impacts on disparate communities, as demonstrated by the two specific studies reported here. For example, numerous studies in *Populus* have documented the effects of condensed tannins on a suite of different ecological processes including beaver herbivory (Bailey *et al.,* 2004), fine root production (Fischer *et al., in press*), rates of leaf litter decomposition (Schweitzer *et al.,* 2004; Leroy *et al.,* 2006), and the availability of soil nutrients (Schweitzer *et al.,* 2004). These studies demonstrate the general importance of condensed tannins in these riparian ecosystems.

The genetic link to endophytes and soil communities, mediated by plant chemistry, indicates the importance of host plant chemical traits to both phyllosphere and rhizosphere communities. We are aware of only a handful of studies that specifically examined the role of condensed tannins on microorganisms. In reviews of the anti-microbial properties of condensed tannins, both Scalbert (1991) and Field and Lettinga (1992) found that multiple genera of microorganisms, including fungi, yeasts and bacteria, were influenced by condensed tannins, and some microbial groups were completely inhibited by their presence. Similarly, high molecular weight condensed tannins from poplars have been shown to inhibit microbial respiration and N_2 fixation (by co-occurring alder; Schimel *et al.,* 1998), as well as bind extracellular enzyme substrates resulting in carbon (C) and N limitation of soil microorganisms (Fierer *et al.,* 2001). Bending and Read (1996) found that while ericoid fungi are able to degrade polyphenols, ectomycorrhizal fungi are largely inhibited by condensed tannins as a result of the inactivation of cell wall enzymes. The patterns we observed with a negative association between bark chemistry and fungal endophytes, and foliar chemistry and soil microbial communities are additional examples of microorganisms that are inhibited or influenced by condensed tannins. Our data suggest that the variation in the chemical environment of the plant genotype is an important mechanism for controlling the degree of microbial activity within the host (Espinosa-Garcia *et al.,* 1996), as well as the soil microbial community associated with specific trees.

Genetic Effects

The relationship between plant genetic traits and both phyllosphere and rhizosphere microbial communities (as described in the literature and with our case studies), indicate the strength of a genetic approach to community dynamics at fine scales and across disparate micro-environments (i.e. plant tissues and soils). Similar to what has been documented to date for arthropod communities, plant genetics (as mediated by plant traits; Bangert *et al.,* 2005b) may be an important factor for the determination of microbial community structure. Plant genetics have been shown to affect selective foraging patterns of ecosystem engineers (Bailey *et al.,* 2004), arthropod community composition and biodiversity (Dungey *et al.,* 2000; Hochwender and Fritz, 2004; Wimp *et al.,* 2005; Bangert *et al.,* 2005a), and nitrogen, water, and carbon cycles (Treseder and Vitousek, 2001; Madritch and Hunter, 2002; Fischer *et al.,* 2004; Schweitzer *et al.,* 2004).

Recent studies have shown a heritable basis to community structure and ecosystem processes (Johnson and Agrawal, 2004; Bailey *et al.,* 2006; Shuster *et al.,* 2006). Although community phenotypes are the result of genetic, environment and genetic by environment interactions, these studies clearly show that plant genes have a general effect on species interactions across multiple levels of organisation (i.e. population, community, and ecosystem). Furthermore, these studies in disparate habitats (i.e. herbaceous fields, oak, aspen and eucalyptus woodlands, riparian forests) suggest that community composition and ecosystem processes can change as a result of evolution in plants (Whitham *et al.,* 2003). For example, Bailey *et al.,* (2006) show that species interactions are an important component to community heritability and demonstrate that gene flow between plant species is related to predictable changes in community composition. Plant genotypes that have similar molecular markers (amplified fragment length polymorphisms, AFLP) supported similar arthropod communities, and those that vary more in their markers support more dissimilar arthropod communities. These studies suggest that differences in the expression of ecologically important genes or genetically based traits are an important mechanism related to differences in ecological interactions that can structure whole community processes.

While characterisation of community composition does not in and of itself reveal community function (Waldrop *et al.,* 2000), it is a first step to understanding the ecological genetics of multi-species interactions and provides the means to develop testable hypotheses on the evolutionary determinants of community structure and subsequent ecosystem processes. These data further our understanding of ecological genetics and indicate that further investigations into the genetic basis for species interactions and evolution are warranted. Moreover, incorporating a genetics perspective into theories of community structure will be important as new molecular tools permit a better understanding of the molecular details of species interactions.

Acknowledgements

We thank the Ogden Nature Centre and the Utah Department of Natural Resources for supporting our restoration and common garden studies. Funding was provided by NSF grants DEB-0078280, and DEB-0425908.

References

Abrahamson, W.G., Hunter M.D., Melika G. and Price P.W. (2003) Cynipid gall-wasp communities correlate with oak chemistry. *Journal of Chemical Ecology* 29, 209-223.

Adams, P.D. and Kloepper, J.W. (2002) Effect of host genotype on indigenous bacterial endophytes of cotton (*Gossypium hirsutun* L.). *Plant and Soil* 240, 181-189.

Aguilar J.M. and Boecklen W.J. (1992) Patterns of herbivory in the *Quercus grisea* x *Quercus gambelii* species complex. *Oikos* 64, 498-504.

Ahlholm, J.U., Helander, M., Henriksson, J., Metzler, M. and Saikkonen, K. (2002) Environmental conditions and host genotype direct genetic diversity of *Venturia ditricha*, a fungal endophyte of birch trees. *Evolution* 56, 1566-1573.

Azad, H.R., Davis, J.R., Schnathhorst, W.C. and Kado, C.L. (1987) Influence of verticillium wilt resistance and susceptible potato genotypes on populations of antagonistic rhizosphere and rhizoplane bacteria and free nitrogen fixers. *Applied Microbiology and Biotechnology* 26, 99-104.

Bailey, J.K., Schweitzer, J.A., Rehill, B.J., Lindroth, R.L. and Whitham, T.G. (2004) Beavers as molecular geneticist: a genetic basis to the foraging of an ecosystem engineer. *Ecology* 85, 603-608.

Bailey, J.K., Deckert, R., Schweitzer, J.A., Rehill, B.J., Lindroth, R.L., Gehring, C. and Whitham, T.G. (2005) Host-plant genetics affect hidden ecological players: Links among *Populus*, condensed tannins and fungal endophyte infection. *Canadian Journal of Botany* 83, 356-362.

Bailey, J.K., Wooley, S.C., Lindroth, R.L. and Whitham, T.G. (2006) Importance of species interactions to community heritability: A genetic basis to trophic-level interactions. *Ecology Letters* 9, 78-85.

Bangert, R.K., Turek, R.J., Martinsen, G.D., Wimp, G.M., Bailey, J.K. and Whitham, T.G. (2005a) Benefits of conservation of plant genetic diversity to arthropod diversity. *Conservation Biology* 19, 379-390.

Bangert, R.K., Turek, R.J., Rehill, B.J., Wimp, G.M., Schweitzer, J.A., Allan, G.J., Bailey, J.K., Martinsen, G.D., Keim, P., Lindroth, R.L. and Whitham, T.G. (2005b) A genetic similarity rule determines arthropod community structure. *Molecular Ecology* 13, 1379-1391.

Bartelt-Ryser, J., Joshi, J., Schmid, B., Brandl, H. and Balser, T. (2005) Soil feedbacks of plant diversity on soil microbial communities and subsequent plant growth. *Perspectives in Plant Ecology, Evolution and Systematics* 7, 27-49.

Bardgett, R.D. and Wardle, D.A. (2003) Herbivore-mediated linkages between above-ground and below-ground communities. *Ecology* 84, 2258-2268.

Bending, G.D. and Read, D.J. (1996) Effects of soluble polyphenol tannic acid on the activities of ericoid and ectomycorrhizal fungi. *Soil Biology and Biochemistry* 28, 1595-1602.

Bever, J.D., Westover, K.M. and Antonovics, J. (1997) Incorporating the soil community into plant population dynamics:The utility of the feedback approach. *Journal of Ecology* 85, 561-573.

Bohannan, B.J.M. and Lenski, R.E. (2000) Linking genetic change to community evolution: insights from studies of bacteria and bacteriophage. *Ecology Letters* 3, 362-377.

Brunner, A.M., Busov, V.B. and Strauss, S.H. (2004) Poplar genome sequence: functional genomics in an ecologically dominant plant species. *Trends in Plant Sci*ence 9, 49-56.

Caetano-Anolles, G. and Gresshoff, P.M. (1991) Plant genetic control of nodulation. *Annual Review of Microbiology* 45, 345-382.

Cheplick, G.P. and R. Cho (2002) Interactive effects of host endophyte infection and host genotype on growth and storage in *Lolium perenne*. *New Phytologist* 158, 183-191.

Christianson, M.J., Ball, O.J.-P., Bennett, R.J. and Schardl, C.L. (1997) Fungal and host genotype effects on compatibility and vascular colonization by *Epichloë festucae*. *Mycological Research* 101, 493-501.

Clay, K. (1996) Interactions among fungal endophytes, grasses, and herbivores. *Research in Population Ecololgy* 38, 191-201.

Clement, C.R. and Habte, M. (1995) Genotypic variation in vesicular-arbuscular mycorrhizae dependence of the pejibaye palm. *Journal Plant Nutrition* 18, 1907-1916.

Diab el Arab, H. G., Vilich, V. and Sikora, R.A. (2001) The use of phospholipid fatty acid analysis (PL-FA) in the determination of rhizosphere specific microbial communities (RSMC) of two wheat cultivars. *Plant and Soil* 228, 291-297.

Dungey H.S., Potts B.M., Whitham T.G. and Li H.-F. (2000) Plant genetics affects arthropod community richness and composition: evidence from a synthetic eucalypt hybrid population. *Evolution* 54, 1938-1946.

Elmi, A.A. and West, C. P. (1995) Endophyte infection effects on stomatal conductance, osmotic adjustment, and drought recovery of tall fescue. *New Phytologist* 131, 61-67.

Espinosa-Garcia, F.J., Rollinger, J.L. and Langenheim, J.H. (1996) Coastal redwood leaf endophytes: their occurrence, interactions and response to host volatile terpenoids. In: Redlin, S.C. and Carris, L.M. (eds) *Endophytic Fungi in Grasses and Woody Plants: Systematics, Ecology, and Evolution.* APS Press, St. Paul, Minnesota, USA. pp. 101-120.

Field, J.A. and Lettinga, G. (1992) Toxicity of tannic compounds to microorganisms. In: Hemingway, R.W. and Laks, P.E. (eds) *Plant Polyphenols, Synthesis, Properties, Significance.* Plenum Press, New York, USA. pp. 673-692.

Fierer, N., Schimel, J.P., Cates, R.G. and Zou, J. (2001) Influence of balsam poplar tannin fractions on carbon and nitrogen dynamics in Alaskan tiaga floodplain soils. *Soil Biology and Biochemistry* 33, 1827-1839.

Finlay, B.J., G.F. Esteban, J.L. Olmo and P.A. Tyler. (1999) Global distribution of free-living microbial species. *Ecography* 22, 138-144.

Fischer, D.G., Hart, S.C., Rehill, B.J., Lindroth, R.L., Keim, P. and Whitham, T.G. (2006) Do high tannin leaves require more roots? *Oecologia* (*in press*).

Fischer, D.G., Hart, S.C. and Whitham, T.G. (2004) Genetic variation in water-use parameters in cottonwoods: ecosystem implications. *Oecologia* 139, 288-297.

Gehring, C.A. and Whitham, T.G. (2002) Mycorrhiza-herbivore interactions: population and community consequences. In: van der Heijden, M.G.A and Sanders, I.R. (eds) *Mycorrhizal Ecology*. pp. 295-320. Springer Verlag, Berlin, Germany.

Grayston, S.J., Wang, S., Campbell, C.D. and Edwards, A.C. (1998) Selective influence of plant species on microbial diversity in the rhizosphere. *Soil Biology and Biochemistry* 30, 369-378.

Hättenschwiler, S. and Vitousek, P.M. (2000) The role of polyphenols in terrestrial ecosystem nutrient cycling. *Trends in Ecology and Evolution* 15, 238-243.

Heal, O.W., Anderson, J.M. and Swift, M.J. (1997) Plant litter quality and decomposition: an historical overview. In: Cadisch, G. and Giller, K.E. (eds) *Driven by Nature. Plant Litter Quality and Decomposition.* CAB International. Wallingford, UK. pp. 3-32.

Hemingway, R.W. and Karchesy, J.J. (1989) *Chemistry and Significance of Condensed Tannins.* Plenum Press, New York, USA.

Hochwender C.G. and Fritz R.S. (2004) Plant genetic differences influence herbivore community structure: evidence from a hybrid willow system. *Oecologia* 138, 547-557.

Horner-Devine, M.C., Carney, K.M. and Bohannan, B.J.M. (2003) An ecological perspective on bacterial biodiversity. *Proceedings of the Royal Society of London* 271, 113-122.

Horner-Devine, M.C., Lage, M., Hughes, J. and Bohannan, B.J.M. (2004) A taxa-area relationship for bacteria. *Nature* 432, 750-753.

Hubbell, S.P. (2002) *The Unified Neutral Theory of Biodiversity and Biogeography.* Princeton University Press, Princeton, NJ, USA.

Johnson, M.T.J. and Agrawal, A.A. (2004) Plant genotype and the environment interact to shape a diverse arthropod community on evening primrose (*Oenothera biennsis*). *Ecology* 18, 549-551.

Johnson, M.T.J., Lajeunesse, M.J. and Agrawal, A.A. (2006) Additive and interactive effects of plant genotypic diversity on arthropod communities and plant fitness. *Ecology Letters* 9, 24-34.

Johnson, L.J., Johnson, R.D., Schardl, C.L. and Panaccione, D.G. (2003) Identification of differentially expressed genes in the mutualistic association of tall fescue with *Neotyphodium ceonophyialum. Physiological and Molecular Plant Pathology* 63, 305-317.

Kaeppler, S.M., Parke, J.L., Mueller, S.M., Senior, L., Stuber, C. and Tracy, W.F. (2000) Variation among maize inbred lines and detection of quantitative trait loci for growth at low phosphorus and responsiveness to arbuscular mycorrhizal fungi. *Crop Science* 40, 358-364.

Kirk, J.L., Beaudette, L.A., Hart, M., Moutoglis, P., Klironomos, J.N., Lee, H. and Trevors, J.T. (2004) Methods of studying soil microbial diversity. *Journal of Microbiological Methods* 58, 169-188.

Kraus, T.E.C., Dahlgren, R.A. and Zasoski, R.J. (2003) Tannins and nutrient dynamics of forest ecosystems – a review. *Plant and Soil* 256, 41-66.

Kuske, C.R., Busch, J.D., Ticknor, L.O., Gehring, C.A. and Whitham, T.G.. (2003) The pinyon rhizosphere, plant stress, and herbivory affect the abundance of microbial decomposers in soils. *Microbial Ecology* 45, 340-352.

Lackie, S.M., Bowley, S.R. and Peterson, R.L. (1988) Comparison of colonization among half-sib families of *Medicago sativa* L. by *Glomus versiforme* (Daniels and Trappe) Berch. *New Phytologist* 108, 477-482.

Lawton, J.H. (1999) Are there general laws in ecology? *Oikos* 84, 177-192.

LeRoy C.J., Whitham T.G., Keim P. and Marks, J.C. (2006) Plant genes link forests and streams. *Ecology* 87(1), 255-261.

Lieven-Antoniou, C.A. and Whittam, T.S. (1997) Specificity of the symbiotic relationship between *Lotus corniculatus* and *Rhizobium luti* from natural populations. *Molecular Ecology* 6, 629-639.

Madritch M.D. and Hunter M.D. (2002) Plant phenotypic diversity influences ecosystem functioning in an oak sandhills community. *Ecology* 83, 2084-2090.

Malinowski D.P., Alloush, G.A. and Belesky, D.P. (2000) Leaf endophyte *Neotyphodium coenophialum* modifies mineral uptake in tall fescue. *Plant and Soil* 227, 115-126.

Martinsen, G.D., Whitham, T.G., Turek, R.J. and Keim, P. (2001) Hybrid populations selectively filter gene introgression between species. *Evolution* 55, 1325-1335.

Mazzola, M. and Gu, Y.-H. (2002) Wheat genotype-specific induction of microbial communities supressive to disease incited by *Rhizoctonia solani* anastomostis group (AG)-5 and AG 8. *Phytopathology* 19, 1300-1307.

Minchin P.R. (1987) Simulation of multidimensional community patterns: towards a comprehensive model. *Vegetatio* 71, 145-156.

Moreira, L.A., Mollema, C. and van Heusden, S. (1999) Search for molecular markers linked to *Liriomyza trifolii* resistance in tomato. *Euphytica* 109, 149-156.

Nannipieri, P., Kandeler, E. and Ruggiero, P. (2002) Enzyme activities and microbiological and biochemical processes in soil. In: Burns, R.G. and Dick, R.P. (eds) *Enzymes in the Environment: Activity, Ecology and Applications.* Marcel Dekker, Inc. New York, USA. pp. 1-33.

Neuhauser, C., Andow, D., Heimpel, G.E., May, G., Shaw, R.G. and Wagenius, S. (2003) Community genetics: expanding the synthesis of ecology and genetics. *Ecology* 84, 545-558.

Newcombe, G., and Bradshaw, H.D. Jr. (1996) Quantitative trait loci conferring resistance in hybrid poplar to *Septoria populicola*, the cause of leaf spot. *Canadian Journal of Forest Research* 26, 1943-1950.

Northup, R.R., Dahlgren, R.A. and McColl, J.G. (1998) Polyphenols as regulators of plant-litter-soil interactions in Northern California's pygmy forest: a positive feedback? *Biogeochemistry* 42, 189-220.

Onozaki, T., Tanikawa, N., Taneya, M., Kudo, K., Funayama, T., Ikeda, H. and Shibata, M. (2004) A RAPD-derived STS marker is liked to bacterial wilt (*Burkholderia caryophylli*) resistance gene in carnation. *Euphytica* 138, 255-262.

Palo, R.T. (1984) Distribution of birch (*Betula* spp.), willow (*Salix* spp.) and poplar (*Populus* spp.) secondary metabolites and their potential role as chemical defence against herbivores. *Journal of Chemical Ecology* 10, 499-520.

Preston, G.M., Haubold, B. and Rainey, P.B. (1998) Bacterial genomics and adaptation to life on plants: implications for the evolution of pathogenicity and symbiosis. *Current Opinion in Microbiology* 1, 589-597.

Reynolds, H.L., Packer, A. Bever, J.D. and Clay, K. (2003) Grassroots ecology: plant-microbe-soil interactions as drivers of plant community structure and dynamics. *Ecology* 84, 2281-2291.

Rudgers, J.A., Koslow, J.M. and Clay, K. (2004) Endophytic fungi alter relationships between diversity and ecosystem properties. *Ecology Letters* **7**, 42-51.

Saikkonen, K., Faeth, S.H., Helander, M. and Sullivan, T.J. (1998) Fungal endophytes. A continuum of interactions with host plants. *Annual Review of Ecology and Systematics* 29, 319-343.

Scalbert, A. (1991) Anti-microbial properties of tannins. *Phytochemistry* 30, 3875-3883.

Schimel, J.P., Cates, R.G. and Ruess, R. (1998) The role of balsam poplar secondary chemicals in controlling soil nutrient dynamics through succession in the Alaskan tiaga. *Biogeochemistry* 42, 221-234.

Schweitzer, J.A., Bailey, J.K., Hart, S.C., Woolbright, S.A. and Whitham, T.G. (2004) A genetic component to ecosystem processes: the role of genetic variation, and condensed tannins on soil nitrogen dynamics. *Ecology Letters* 7, 127-134.

Schweitzer, J.A., Bailey, J.K., Fischer, D.G., LeRoy, C.J., Hart, S.C. and Whitham, T.G. (2006) Microbial communities as heritable plant traits. (*in review*).

Shuster, S.M., Lonsdorf, E.V., Wimp, G.M., Bailey, J.K. and Whitham, T.G. (2006) Community Heritability: The genetic basis of multilevel selection and community Structure. *Evolution* (*in press*).

Smith, K.P. and Goodman, R.M. (1999) Host variation for interactions with beneficial plant-associated microbes. *Annual Review of Phytopathology* 37, 473-491.

Smith, K.P., Handelsman, J. and Goodman, R.M. (1999) Genetic basis in plants for interactions with disease-resistant bacteria. *Proceedings of the National Academy of Sciences, USA* 96, 4786-4790.

Tan, Z., Hurek, T. and Reinhard-Turek, B. (2003) Influence of N-fertilization, host genotype and environmental conditions on *nif*H gene pools of rice. *Environmental Microbiology* 5, 1009-1015.

Tan, G.X.,Weng, Q.M., Ren, X., Huang, Z., Zhu, L.L. and He, G.C. (2004) Two white-backed planthopper resistance genes in rice share the same loci with those for brown leafhopper resistance. *Heredity* 92, 212-217.

Treseder, K.K. and Vitousek, P.M. (2001) Potential ecosystem-level effects of genetic variation among populations of *Metrosideros polymorpha* from a soil fertility gradient in Hawaii. *Oecologia* 126, 266-275.

Via, S. and Hawthorne, D.J. (2002) The genetic architecture of ecological specialization: correlated gene effects on host use and habitat choice in pea aphids. *American Naturalist* 159, S76-S88.

Waldrop, M.P., Balser, T.C. and Firestone, M.K. (2000) Linking microbial community composition to function in a tropical soil. *Soil Biology and Biochemistry* 32, 1837-1846.

Wardle, D.A., Bardgett, R.D., Klironomos, J.N., Setala, H., van der Putten, W.H. and Wall, D.H. (2004) Ecological linkages between aboveground and belowground biota (review). *Science* 304, 1629-1634.

White, D.C. and Ringleberg, D.B. (1998) Signature lipid biomarker analysis. In: Burlage, R.S., Atlas, R., Stahl, D., Gessey, G. and Sayler, G. (eds) *Techniques in Microbial Ecology.* Oxford, NY, USA. pp. 255-272.

Whitham, T.G., Young, W.P., Martinsen, G.D., Gehring, C.A., Schweitzer, J.A., Shuster, S.M., Wimp, G.M., Fischer, D.G., Bailey, J.K., Lindroth, R.L., Woolbright, S. and Kuske, C.R. (2003) Community and ecosystem genetics: a consequence of the extended phenotype. *Ecology* 84, 559-573.

Wimp G.M., Martinsen G.D., Floate K.D., Bangert R.K. and Whitham T.G. (2005) Plant genetic determinants of arthropod community structure and diversity. *Evolution* 59, 61-69.

Zak, D.R., Holmes, W.E., White, D.C., Peacock, A.D. and Tilman, D. (2003) Plant diversity, soil microbial communities and ecosystem function: are there any links? *Ecology* 84, 2042-2050.

Zhan, J., Mundt, C.C., Hoffer, M.E. and McDonald, B.A. (2002) Local adaptation and effect of host genotype on the rate of pathogen evolution: an experimental test in a plant pathosystem. *Journal of Evolutionary Biology* 15, 634-647.

Zelles, L. (1999) Fatty acid patterns of phospholipids and lipopolysaccharides in the characterization of microbial communities in soil: a review. *Biology and Fertility of Soils* 29, 111-129.

Chapter 8

A Survey of A-L Biofilm Formation and Cellulose Expression Amongst Soil and Plant-Associated *Pseudomonas* Isolates

A.J. Spiers[1], D.L. Arnold[3], C.D. Moon[1] and T.M. Timms-Wilson[2]

[1]Department of Plant Sciences, University of Oxford, South Parks Road, Oxford, OX1 3RB, UK; [2]Centre for Ecology and Hydrology, Mansfield Road, Oxford, OX1 3SR, UK; [3] School of Biosciences, Faculty of Applied Sciences, University of the West of England, Coldharbour Lane, Bristol, BS16 1QY, UK

andrew.spiers@plants.ox.ac.uk

Bacterial Aggregations in the Phytosphere

A critical first step in the colonisation of a new environment is the attachment of individual bacteria to an appropriate surface, such as that provided by a leaf or root, detritus or soil particle. Surface attachment is crucial, allowing growth without requiring further energy expenditure to maintain position in an ideal environment (Costerton *et al.,* 1995). Such growth can result in a variety of assemblages ranging from micro-colonies to larger aggregates, biofilms, slimes and flocs of suspended material. These assemblages are evolved strategies enabling individual bacteria to overcome environmental stress, physical disturbance, predation and competition by acting as a cohesive population. Although often initiated and dominated by one species, they are undoubtedly complex communities where composition changes as the local environment is modified and the assemblage develops. Large communities can dominate complex environments where they act as principle foci for further colonisation, or reservoirs to replenish populations in environments with a high degree of physical disturbance or mortality. (For a selection of biofilm reviews, see Costerton *et al.,* 1995; Davey and O'Toole, 2000; Morris and Monier, 2003; Hall-Stoodley et al., 2004; Ramey *et al.,* 2004; Branda *et al.,* 2005).

The pseudomonads are a particularly versatile and successful group of bacteria found in soil and associated with plants as either epiphytes or pathogens. A number of *Pseudomonas* species, such as *P. aeruginosa, P. fluorescens, P. putida* and *P. syringae* are known to be able to synthesise exopolysaccharide (EPS) and readily form biofilms. Although EPS-deficient mutants are attachment-proficient, they often fail to successfully colonise new environments, demonstrating the real importance of being able to transit from a single bacterial cell into a larger, more resistant and versatile assemblage. The ability to form large aggregations on leaves provides resistance to desiccation stress for *P. syringae*, and the enhanced survival of these promotes the highly clustered spatial distribution of bacteria on leaf surfaces (Monier and Lindow, 2003). Alginate contributes to the virulence of *P. syringae* by facilitating colonisation (Yu *et al.,* 1999), and the primary regulator of alginate expression, the alternative sigma factor AlgT, is stimulated by osmotic and chemical stressors, such as those found during the colonisation of plant tissue (Keith and Bender, 1999). In the case of *P. aeruginosa*, the ability to form a biofilm on the roots of sweet basil confers resistance to the plant antimicrobial compound rosmarinic acid, while mutants unable to form a biofilm are killed on the root surfaces (Walker *et al.,* 2004). Epiphytic *P. fluorescens* patchily colonises root surfaces by forming biofilms along epidermal fissures, whereas pathogenic *P. aeruginosa* and *P. syringae* produce dense, confluent biofilms on root surfaces (Ramey et *al.,* 2004).

In general, it has been hard to establish the chemical identity of many EPS species, even when the biosynthesis genes have been identified and shown to be involved using appropriate mutants. Among the pseudomonads, alginate is involved in biofilms and micro-colonies, though alginate is not essential for the archetypical *P. aeruginosa* PA01 biofilm (Branda *et al.,* 2005) (alginate expression is closely correlated with the mucoidal colony phenotype, rather than with biofilm formation *per se*). As well as being classified as an environmental bacterium, *P. aeruginosa* is also a significant opportunistic human pathogen. In a survey of *P. aeruginosa* isolates from cystic fibrosis patents, 77% showed greater biofilm formation than PA01 (Head and Yu, 2004), indicating that biofilm-formation is just as important in human pathogenicity as it appears to be in the phytosphere. Although alginate-expression amongst the pseudomonads is well recognised, other EPS such as the glucose-rich PEL and mannose-rich PSL polymers are known to be expressed (Branda *et al.,* 2005). Recently, the soil and plant associated bacterium *P. fluorescens* SBW25 has been shown to utilise partially acetylated cellulose as the biofilm matrix under experimental conditions (Spiers *et al.,* 2002, 2003).

Cellulose Expression and Biofilm Formation

Cellulose expression in SBW25 was originally observed in the biofilm-forming mutant known as the Wrinkly Spreader (WS) (Plate 1, see colour plate section). This mutant produces a robust biofilm at the air-liquid (A-L) interface of static microcosms, maintained by attachment to the sides of the vial and a generally hydrophobic nature (Spiers et al., 2003) (static microcosms are 6 ml rich KB liquid media in 30 ml universal glass vials which are incubated under stationary conditions). Mutagenesis of the WS enabled the identification of the cellulose biosynthetic operon (*wss*), as well as a key GGDEF domain-containing response regulator, WspR (Bantinaki, 2002; Spiers *et al.,* 2002, 2003). The strength of the WS biofilm depends on the complex interaction between attachment factor, partially acetylated cellulose fibres and lipopolysaccharide (Spiers *et al.,* 2003; Spiers and Rainey, 2005).

Both the WS and WS-like mutants were originally isolated from experimental static microcosms (Rainey and Travisano, 1998) in which random mutations result in the activation of a pathway or pathways culminating in A-L biofilms. Analysis of 26 independently isolated WS-like mutants found that all expressed cellulose (Spiers *et al.,* 2002), suggesting that this is the primary means of biofilm formation in SBW25. Interestingly, *Escherichia coli* A-L biofilm-forming mutants expressing cellulose can also be isolated in the same manner (A.S., unpublished observations). A-L and submerged (liquid-solid surface, L-S) biofilms appear to be quite distinctive, yet both require a solid surface for attachment and subsequent growth. In L-S biofilms, growth is upwards into the liquid column, whereas in the WS A-L biofilm, growth is out from the region of attachment at the meniscus and across the A-L interface, with further growth leading to secondary development in depth.

The meniscus region of WS biofilms is reminiscent of the 'mushroom and valleys' structure of L-S biofilms, in which small bacterial colonies develop on the glass vial walls and overlap to form a continuous biofilm, and there is no obvious differentiation between the attached region at the meniscus and the biofilm proper at the A-L interface (A.S., unpublished observations). SBW25 is also able to form L-S biofilms similar to those seen with *P. aeruginosa* PA01, *P. aureofaciens* ATCC 13985 and *P. putida* KT2440 (Heydorn *et al.*, 2000), though in the SBW25 L-S biofilm, cellulose has been identified as the matrix component (Villavicencio, 2000).

Cellulose expression in SBW25 is not limited to *in vitro* biofilm formation, as the ability to express cellulose is known to confer an advantage in the sugar beet phytosphere from which SBW25 was originally isolated. Analysis of a cellulose-deficient SBW25 mutant has shown a significant reduction of fitness in the rhizosphere and phyllosphere, but not in bulk soil, indicating that the ability to express cellulose has some functional role in this environment (Gal *et al.*, 2003).

Cellulose Expression Amongst the Proteobacteria

The expression of cellulose in the phytosphere by pseudomonads is probably not limited to SBW25, as both *P. syringae* pv. *tomato* DC3000 and *P. putida* KT2440 contain *wss* operon homologues (Nelson *et al.*, 2002; Römling, 2002; Spiers *et al.*, 2002). Cellulose expression has also been reported for a number of enteric bacteria within the gamma-group of proteobacteria, including *E. coli, Salmonella enterica*, and other human gastrointestinal tract isolates (Römling, 2002; Zogaj *et al.*, 2003). The core cellulose synthase subunits are highly conserved amongst the gamma-group, but the cellulose-acetylation subunits have only been identified in SBW25 and DC3000 (Römling, 2002; Spiers *et al.*, 2002). WspR appears to be a primary regulator in SBW25, but in *E. coli* and *S. enterica*, the unrelated AgfD lux-family response regulator and the GGDEF-domain containing AdrA are both involved (Römling, 2002).

Despite the prevalence of cellulose-expression amongst the gamma-group proteobacteria, the functional role of cellulose is unknown in either the phytosphere or gastrointestinal tract. In contrast, cellulose expression by the alpha-group *Gluconacetobacter xylinus* (formerly known as *Acetobacter x.*) provides a means of excluding competitors during the colonisation of the surface of apples, whilst in both *Agrobacterium tumefaciens* and *Rhizobium* spp., cellulose fibres are involved in the attachment of bacteria to plant surfaces and other cells (Ross *et al.*, 1991). Cellulose over-expression in *G. xylinus* results in a large gelatinous mass of material at the bottom of cultures (in the production of vinegar from red wine, this is known as the 'mother of vinegar'), whereas in *A. tumefaciens*, cellulose expression results in small flocs of material which remain in suspension (Ross *et al.*, 1991).

Survey of A-L Biofilm Formation Amongst the Pseudomonads

We are currently surveying two collections of *Pseudomonas* spp. for their ability to produce A-L biofilms and to express cellulose. These collections include a range of soil, plant-associated and plant-pathogenic strains for which phylogenetic relationships are known (Ellis *et al.*, 2000; Yamamoto *et al.*, 2000), as well as a sub-set of fluorescent isolates recovered from the sugar beet phytosphere from which SBW25 had originally been isolated (Thompson *et al.*, 1995). These strains were screened for A-L biofilm formation by repeated culturing in static microcosms, selecting for mutants in which biofilm-formation had been induced by random mutation, as well as for strains which naturally produced biofilms under these laboratory conditions.

Our interim findings are that the ability to produce A-L biofilms after selection in static microcosms is common-place, with 70% of the *pseudomonas* strains able to form biofilms within 2-15 days incubation. The biofilms varied considerably in morphology, apparent strength and ability to remain at the A-L interface. However, there appeared to be four general categories: (i) waxy aggregations; (ii) large viscous masses; (iii) surface floccular material; and (iv) physically cohesive WS-like biofilms, which could be differentiated by visual observation of the biofilms *in situ*, and after transfer from the microcosm into another container (to assess physical integrity) (Plates 2 and 3, see colour plate section).

Strains such as *P. fluorescens* M114 (Fenton *et al.*, 1992) and *P. putida* PH6 (Fuhrmann and Wollum, 1989) formed very thin, waxy-looking rafts of material ('waxy aggregation' or WA-type biofilms). These biofilms had very little distinguishing surface topology and were hard to see visually. They readily broke to give smaller pieces which remained at the A-L interface. Optical microscopy revealed closely packed cells in these fragments (Fig. 8.1), giving the impression that the biofilm was largely the result of cell-cell contact and a general hydrophobic nature. In contrast, *P. fluorescens* R12T (from J.D. van Elsas) and *P. marginalis* CR30 (Fukui *et al.*, 1994) produced visually impressive lumps of material at the A-L interface ('viscous mass' or VM-type biofilms). These lumps were very glutinous and rapidly sank if the microcosm was disturbed. When emptied from the microcosms, the liquid was very viscous and similar in nature to alginate-expressing *P. aeruginosa* PA01 cultures. Optical microscopy did not identify any aggregation of cells from the biofilm material recovered from these microcosms. The biofilm produced by *P. fluorescens* 54/96 (Ellis *et al.*, 1999) was also lumpy in appearance, but rather than large and viscous, the lumps were smaller and floccular in nature ('floccular material' or FM-type biofilms). When disturbed, these lumps slowly settled to the bottom of the microcosm, yet were robust enough to survive transfer into another container.

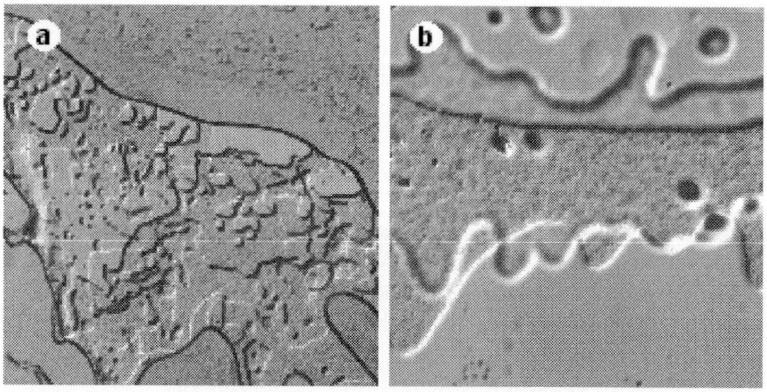

Fig. 8.1. The WA-type biofilm produces thin sheets of material. Shown are biofilm fragments from *P. putida* PH6 at **(a)** 10x magnification (the image is centred on an air-bubble trapped under the microscope cover-piece); and **(b)** 40x magnification of another region showing two layers of material on the slide with a horizontal channel of freely moving bacteria.

Finally, strains such as *P. aureofaciens* PGS12 (Georgakopoulos *et al.*, 1994) and *P. putida* A1 (Fukui *et al.*, 1994) produced A-L biofilms very similar in morphology to the

WS ('physically cohesive' or PC-type biofilms). These biofilms covered the entire A-L interface, often with a wrinkled surface, and clearly attached to the vial walls at the meniscus region. Compared to the WA and VM-type biofilms, these were physically cohesive and resilient, and able to withstand up to 9x greater mass before breaking (Fig. 8.2). Optical microscopy of these biofilms did not reveal large aggregations of bacteria similar to that seen with the WA-type biofilms, and like the VM-type biofilms, it was obvious through the limited motion of cells and the viscous nature of the samples, that some type of extracellular material was present.

The ability to produce strong, cellulose-based, PC-type biofilms was common-place amongst the environmental *Pseudomonas* isolates recovered from Oxfordshire sugar beets (e.g. TW500, 829 and 895, from A.K. Lilley). Interestingly, two other *P. fluorescens* isolates from sugar beet in Belgium (54/96) and Ireland (F113, Fenton *et al.,* 1992) were also found to express cellulose in the A-L biofilms (Fig. 8.3). In contrast to WS and F113, the biofilm produced by 54/96 was particularly weak and did not produce a PC-type biofilm but rather a FM-type biofilm. Static microcosm-selected *P. syringae* DC3000 (Davis *et al.,* 1991) produced a similarly weak biofilm, and was also found to express cellulose when stained with Calcofluor and observed by fluorescent microscopy. However, in both 54/96 and DC3000, small lumps of cellulose fibres were observed, rather than the extensive networks of interconnected fibres and lumps seen in the case of the WS.

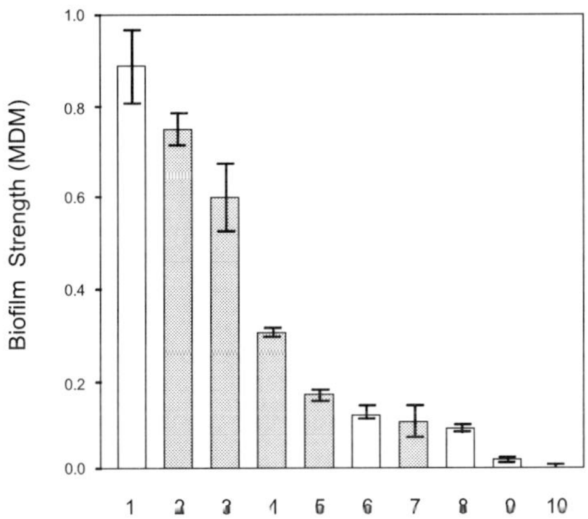

Fig. 8.2. A-L biofilm strengths varied considerably between static microcosm-selected strains. Shown are the MDM determined for **1.** *P. chlororaphis* PGS12 (a PC-type biofilm); **2.** environmental isolate TW643 (PC-type); **3.** *P. putida* A1 (PC-type); **4.** *P. fluorescens* SBW25 (PC-type); **5,** *P. fluorescens* F113 (PC-type); **6.** P. putida PH6 (WA-type); **7.** environmental isolate TW895 (PC-type); **8.** *P. fluorescens* M114 (WA-type); **9.** *P. fluorescens* R12T (VM-type); and **10.** *P. fluorescens* 54/96 (FM-type). Bars in grey (including 10) indicate those in which cellulose was expressed. Mean ± standard errors (8 replicates) are shown.

In contrast, *P. putida* KT2440 (Christensen *et al.,* 1998) which is known to contain a wss homologue, was not found to produce a biofilm after repeated selection in static microcosms (even when cultured at 28°C). Furthermore, no evidence of cellulose expression was found in either microcosm culture or colony material.

Significant amounts of non-cellulose matrix material was also observed through the negative-staining of A-L biofilm material by Calcofluor, such as in the case of *P. fluorescens* Pf-5 (Howell and Stipanovic, 1979) (Fig. 3). However, the identity of the EPS in this and other negatively-stained biofilms was not established. (Very few EPS-specific dyes are available for this type of diagnostic use. Dyes such as Alcian blue have been used previously to stain alginate, but it is not specific for this EPS, and produces a very weak signal compared with Calcofluor bound to cellulose. Most EPS-identification therefore relies on chemical and structural analysis of purified material).

In SBW25, the WspR mutant WspR19 (R129C) converts the wild-type strain into a WS-like phenotype, by activating cellulose and attachment factor expression leading to the formation of a WS-like A-L biofilm when expressed *in trans* (Goymer, 2002). Two *Pseudomonas* strains found to be capable of expressing cellulose were also found to be sensitive to WspR19-activation when transformed with a WspR19-expression plasmid (Goymer, 2002). In both *P. fluorescens* 54/96 and *P. syringae* DC3000, biofilm-formation was significantly enhanced, and in each case, the transformed strains produced strong PC-type biofilms rather than the weaker FM-type biofilms produced by the static microcosm-selected mutants of each strain. This suggests that, just as WspR is involved in the regulation of both cellulose, attachment factor expression, and the production of biofilms in SBW25, the same genetic-physiological regulatory circuits also exist in the closely-related 54/96 and more distantly related DC3000.

GGDEF-containing response regulators, such as WspR, and the associated cyclic-di-GMP signalling molecule, are increasingly implicated in surface colonisation, biofilm and colony development amongst a wide range of bacteria (D'Argenio and Miller, 2004; Jenal, 2004). In *P. aeruginosa* PA01, WspR19 expressed *in trans* affects colony morphology and cellular autoaggregation, but although PA01 has a wsp operon homologue, it does not have cellulose biosynthesis genes and consequently does not express cellulose (D'Argenio *et al.,* 2002).

Although not yet complete, our survey of soil, plant-associated and pathogenic *Pseudomonas* strains suggests that the ability to form biofilms is a common attribute amongst this collection of bacteria. Biofilm formation also appears to be common amongst the members of the *P. fluorescens* and *P. syringae* complexes (Yamamoto *et al.,* 2000), but close phylogenetic-grouping did not necessarily predict the ability to form biofilms. Similarly, the ability to express cellulose was largely restricted to members of these two complexes, whilst the expression of viscous material appeared to be a more general feature of this collection of pseudomonads.

In general, there appeared to be four types of A-L biofilms formed: waxy aggregations (WA), viscous mass (VM), floccular material (FM) and physically cohesive (PC) biofilms. In the first, the hydrophobic nature of rafts of cells seemed to be the means to biofilm formation, whereas in the other three, the expression of extracellular polymers appeared important in maintaining the population of cells at the A-L interface.

It is not possible to comment on whether each of the biofilms observed were the result of a normal physiological response to growth in a static liquid volume with high nutritional value, the result of quorum signalling, or whether the biofilms were the result of random mutation and selection in the growing bacterial population. We favour this latter explanation, as a single point mutation has the ability to derepress a normally well-regulated surface colonisation response (e.g. EPS expression), resulting in the expression in the static microcosm of genes and enzyme activity which subsequently leads to the production of a biofilm (as is known to be the case for the WS). In this sense, *in vitro* biofilm production is an abstracted expression of an environmental function such as surface colonisation. However, there is no *a priori* reason to suppose that the same systems used to produce *in vitro* biofilms serve to produce biofilms under normal environmental conditions (i.e. the environmental phenotype produced by these systems may not in fact be a biofilm *per se*).

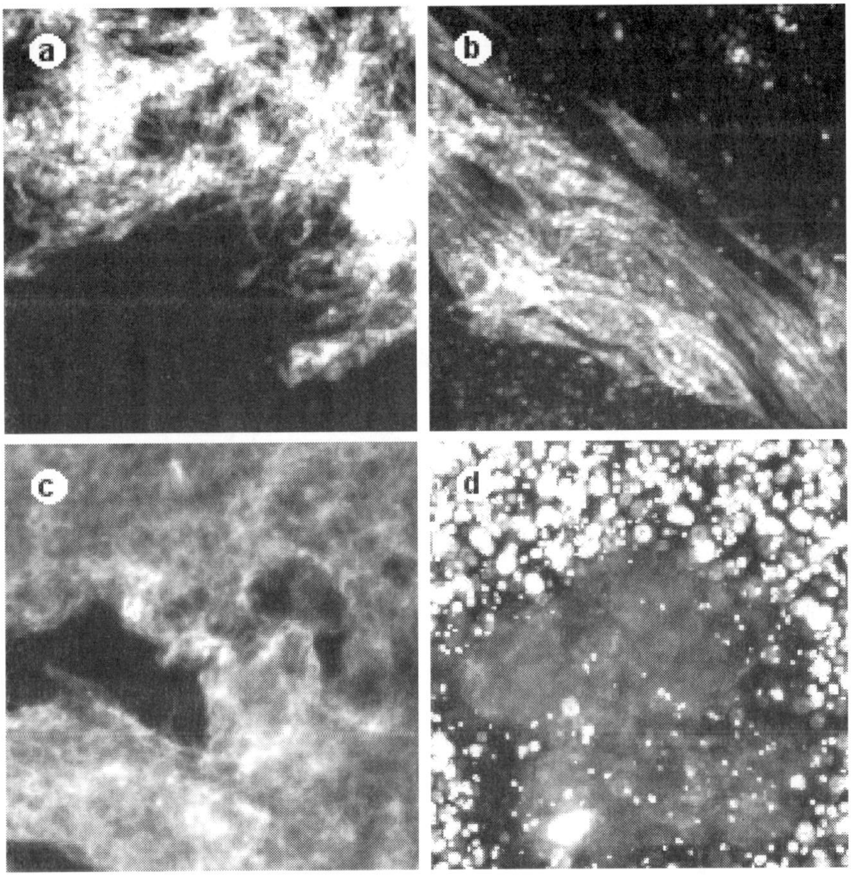

Fig. 8.3. Many of the FM and PC type biofilms contained a cellulose-based matrix which could be identified by staining with the fluorescent dye Calcofluor. Shown are: **(a)** *P. syringae* DC3000; **(b)** *P. putida* A1; and **(c)** environmental isolate TW829 in which cellulose fibres are clearly evident. In contrast, the presence of an unknown matrix component was indicated by Calcofluor-negative staining and image over-exposure in **(d)** *P. fluorescens* Pf-5.

Speculations on Phytosphere Biofilms

It is tempting to speculate whether the four biofilm types can be interpreted in terms of colonisation strategies employed by bacteria in the phyllosphere. Small volumes of rain water trapped on plant surfaces, for example between leaf base and stem, represent an ideal environment for bacterial growth in which photosynthates would be washed down from the leaf or leached from the submerged tissues. Biological growth in such volumes would rapidly establish an oxygen gradient, providing a selective pressure for bacteria to colonise the A-L interface. This could be readily achieved by those bacteria able to modify the cell surface to become more hydrophobic to produce aggregates 'floating' at the surface. Given the protected nature of these small volumes of trapped water, there is no real reason for strong, physically robust biofilms, merely the need to remain at the A-L interface. It is possible that the WA-type biofilms observed after selection in static microcosms is a simple reflection of a common liquid-surface colonisation strategy employed by phyllosphere bacteria.

Bacterial growth is also possible in enclosed volumes such as that found in damaged or infected leaf or stem tissue. In these environments the selective pressure might be to dominate the entire volume rather than just the surface layer. In such cases, the expression of a viscous gel, such as alginate, might be more suitable allowing the colonising bacteria to remain in place as well as denying access to subsequent colonisers. In the case of the vascular pathogens *Xylella fastidiosa* and *Xanthomonas campestris*, viscous gels produced by fastidum and xanthan gum allow colonisation and retention of the pathogen in the xylem vessels of the infected plants (Ramey *et al.,* 2004). This type of volume colonisation strategy might also be expressed in the static microcosms, where large masses of viscous material are sufficient to maintain bacterial populations near the A-L interface.

However, the majority of habitats available in the phyllosphere are probably less than ideal, but are nevertheless colonised by bacteria. Growth on surfaces subjected to more significant physical disturbances, such as leaves, may require a greater emphasis on attachment and resilience, and the resulting bacterial aggregation may look more like scattered micro-colonies rather than biofilms or slimes. If such colonisation mechanisms were to be activated by mutation in the static microcosm, the result might be the formation of a very strong PC-type biofilm well attached to the vial walls. Indeed, although most biofilms required very little strength to maintain position at the A-L interface, some were considerably stronger. This suggests that in these cases, static microcosm biofilm formation was not through the activation of a normal environmental biofilm or volume-colonisation strategy, but through the utilisation of a physically resilient surface colonisation strategy in a new role.

The FM-type biofilm may represent a weaker version of the PC-type biofilm in which insufficient matrix material is produced to form a single, cohesive biofilm. Alternatively, it might represent an alternative to the WA-type biofilm, where the inherent hydrophobicity of the matrix material is sufficient to maintain bacteria at the A-L interface of small volumes of water. In the case of the WS, a single PC-type biofilm is produced through the combined expression of both cellulose and attachment factor (Spiers *et al.,* 2002; 2003). In contrast, *P. fluorescens* 54/96 and *P. syringae* DC3000 each produced the weaker FM-type biofilm after selection in static microcosms. Interestingly, when expressing WspR19 in trans, both produce the stronger PC-type biofilm in which more cellulose was apparent by fluorescent microscopy, suggesting that the amount of cellulose expressed, along with possible attachment factor, is key to forming stronger, physically cohesive biofilms amongst these pseudomonads.

Biofilm matrix EPS show a wide range of physico-chemical properties, in terms of relative solubility and the ability to produce viscous aqueous solutions or gels, ion-interactions influencing the degree of fibre aggregation, and the rigidity or deformability of structures (Sutherland, 2001). Furthermore, modification of the basic polymer is also important. In the case of gellan produced by *Sphingomonas elodea*, a weak gel is produced from the native acylated polymer, whereas the deacylated material produces a rigid but brittle gel (Sutherland, 2001). In the WS, partially acetylated and non-acetylated cellulose fibres and aggregates are similar in nature, yet the strength of biofilms produced by the WS and acetylation-deficient WS mutant were quite different (Spiers *et al.,* 2003). Most PC-type biofilms, but none of the WA or VM-type biofilms identified in this survey, expressed cellulose. Whereas alginate tends to produce viscous gels, β-1,4-linked hexose EPS such as cellulose are more rigid and tend to exclude water to form larger structures (Sutherland, 2001). The relationship between PC and FM-type cellulose biofilms may be very simple: large amounts of cellulose expression might result in PC-type biofilms, whereas lower levels of expression would result in FM-type biofilms. The enhanced expression of attachment factors or secondary matrix components might also lead to the transition of a weaker FM-type biofilm into the stronger PC-type biofilm, as there is increasing evidence to suggest that biofilms are the result of complex interactions between multiple extracellular components and the bacterial cells themselves.

Final comment

With the exception of dental biofilms, most biofilm research is performed *in vitro* and is quite abstracted from the normal environments inhabited by bacteria. Attention has focused on how biofilms are produced under particular conditions, but it is quite probable that biofilm formation is a process that can be applied in a number of different environments and conditions, and may well result in phenotypically distinct types of bacterial aggregations. Although at times contentious, the pelliceous nature of submerged biofilms growing on solid surfaces, biofilms at the air-liquid interface and matrix-containing bacterial colonies suggest that all are in essence biofilms (Branda *et al.*, 2005), and part of a larger continuum of bacterial assemblages, including both slimes and flocs (biofilms were originally defined as matrix-enclosed bacterial populations adherent to each other and/or surfaces or interfaces, including aggregates and floccules, as well as those adherent populations within pore spaces; Costerton *et al.*, 1995).

A particular challenge to our understanding of the role of environmental bacterial aggregations, in terms of bacterial colonisation and survival strategies, is to move from *in vitro* to phytosphere-based experimentation. In doing so, we may discover that bacterial aggregation, community growth and adaptation, are both more complex and more universal than once thought.

Acknowledgements

We thank Julie Stansfield and Zuzana Spiers for their technical assistance in this survey, and for those who provided helpful comments on the manuscript. This work was partially supported by the Department of Plant Sciences, Oxford, and an EU project grant.

References

Bantinaki, E. (2002) Characterisation of a novel chemosensory pathway underlying adaptive evolution in experimental populations of *Pseudomonas fluorescens* SBW25. DPhil Thesis, University of Oxford, UK.

Branda, S.S., Vik, Å., Friedman, L. and Kolter, R. (2005) Biofilms: the matrix revisited. *Trends in Microbiology* 13, 20-26.

Christensen, B.B., Sternberg, C., Andersen, J.B., Eberl, L., Moller, S., Givskov, M. and Molin, S. (1998) Establishment of new genetic traits in a microbial biofilm community. *Applied and Environmental Microbiology* 64, 2247-2255.

Costerton, J.W., Lewandowski, Z., Caldwell, D.E., Korber, D.R. and Lappin-Scott, H.M. (1995) Microbial biofilms. *Annual Reviews in Microbiology* 49, 711-745.

Davey, M.E. and O'Toole, G.A. (2000) Microbial biofilms: from ecology to molecular genetics. *Microbiology and Molecular Biology Reviews* 64, 847-867.

Davis, K.R., Schott, E. and Ausubel, F.M. (1991) Virulence of selected phytopathogenic pseudomonads in *Arabidopsis thaliana. Molecular Plant-Microbe Interactions* 4, 477-488.

D'Argenio, D.A., Calfee, M.W., Rainey, P.B. and Pesci, E.C. (2002) Autolysis and autoaggregation in *Pseudomonas aeruginosa* colony morphology mutants. *Journal of Bacteriology* 184, 6481-6489.

D'Argenio, D.A. and Miller, S.I. (2004) Cyclic di-GMP as a bacterial second messenger. *Microbiology* 150, 2497-2502.

Ellis, R.J., Timms-Wilson, T.M. and Bailey, M.J. (2000) Identification of conserved traits in fluorescent *pseudomonas* with antifungal activity. *Environmental Microbiology* 2, 274-284.

Ellis, R.J., Timms-Wilson, T.M., Beringer, J.E., Rhodes, D., Renwick, A., Stevenson, L. and Bailey, M.J. (1999) Ecological basis for biocontrol of damping-off disease by *Pseudomonas fluorescens* 54/96. *Journal of Applied Microbiology* 87, 454-463.

Fenton, A.M., Stephens, P.M., Crowley, J., O'Callaghan, M. and O'Gara, F. (1992) Exploitation of gene(s) involved in 2,4-diacetylphloroglucinol biosynthesis to confer a new biocontrol capability to a *Pseudomonas* strain. *Applied and Environmental Microbiology* 58, 3873-3878.

Fuhrmann, J. and Wollum, II, A.G. (1989) *In vitro* growth responses of *Bradyrhizobium japonicum* to soybean rhizosphere bacteria. *Soil Biology and Biochemistry* 21, 131-135.

Fukui, R., Schroth, M.N., Hendson, M. and Hancock, J.G. (1994) Interaction between strains of *Pseudomonads* in sugar beet spermospheres and their relationship to pericarp colonisation by *Pythium ultimum* in soil. *Phytopathology* 84, 1322-1330.

Gal., M., Preston, G.M., Massey, R.C., Spiers, A.J. and Rainey, P.B. (2003) Genes encoding a cellulosic polymer contribute toward the ecological success of *Pseudomonas fluorescens* SBW25 on plant surfaces. *Molecular Ecology* 12, 3109-3121.

Georgakopoulos, D.G., Hendson, M., Panopoulos, N.J. and Schroth, M.N. (1994) Cloning of a phenazine biosynthetic locus of *Pseudomonas aureofaciens* PGS12 and analysis of its expression *in vitro* with the ice nucleation reporter gene. *Applied and Environmental Microbiology* 60, 2931-2938.

Goymer, P.J. (2002) The role of the WspR response regulator in the adaptive evolution of experimental populations of *Pseudomonas fluorescens* SBW25. DPhil Thesis, University of Oxford, UK.

Hall-Stoodley, L., Costerton, J.W. and Stoodley, P. (2004) Bacterial biofilms: from the natural environment to infectious diseases. *Nature Reviews Microbiology* 2, 95-108.

Head, N.E. and Yu, H. (2004) Cross-sectional analysis of clinical and environmental isolates of *Pseudomonas aeruginosa*: biofilm formation, virulence, and genome diversity. *Infection and Immunity* 72, 133-144.

Heydorn, A., Nielsen, A.T., Hentzer, M., Sternberg, C., Givskov, M., Ersbøll, B.K. and Molin, S. (2000) Quantification of biofilm structures by the novel computer program COMSTAT. *Microbiology* 146, 2395-2407.

Howell, C.R. and Stipanovic, R.D. (1979) Control of Rhizoctonia solani on cotton seedlings with *Pseudomonas fluorescens* and with an antibiotic produced by the bacterium. *Phytopathology* 69, 480-482.

Jenal, U. (2004) Cyclic di-guanosine-monophosphate comes of age: a novel secondary messenger involved in modulating cell surface structures in bacteria? *Current Opinions in Microbiology* 7, 185-191.

Keith, L.M. and Bender, C.L. (1999) AlgT (sigma22) controls alginate production and tolerance to environmental stress in *Pseudomonas syringae. Journal of Bacteriology* 181, 7176-7184.

Monier, J.M. and Lindow, S.E. (2003) Differential survival of solitary and aggregated bacterial cells promotes aggregate formation on leaf surfaces. *Procedures of the National Academy of Sciences, USA* 100, 15977-15982.

Morris, C.E. and Monier, J.M. (2003) The ecological significance of biofilm formation by plant-associated bacteria. *Annual Reviews in Phytopathology* 41, 429-453.

Nelson, K.E., Weinel, C., Paulsen, I.T., Dodson, R.J., Hilbert, H., Martins dos Santos, V.A., Fouts, D.E., Gill, S.R., Pop, M., Holmes, M., Brinkac, L., Beanan, M., DeBoy, R.T., Daugherty, S., Kolonay, J., Madupu, R., Nelson, W., White, O., Peterson, J., Khouri, H., Hance, I., Chris Lee, P., Holtzapple, E., Scanlan, D., Tran, K., Moazzez, A., Utterback, T., Rizzo, M., Lee, K., Kosack, D., Moestl, D., Wedler, H., Lauber, J., Stjepandic, D., Hoheisel, J., Straetz, M., Heim, S., Kiewitz, C., Eisen, J.A., Timmis, K.N., Dusterhoft, A., Tummler, B. and Fraser, C.M. (2002) Complete genome sequence and comparative analysis of the metabolically versatile *Pseudomonas putida* KT2440. *Environmental Microbiology* 4, 799-808.

Rainey, P.B. and Travisano, M. (1998) Adaptive radiation in a heterogeneous environment. *Nature* 394, 69-72.

Ramey, B.E., Koutsoudis, M., von Bodman, S.B. and Fuqua, C. (2004) Biofilm formation in plant-microbe associations. *Current Opinions in Microbiology* 7, 602-609.

Römling, U. (2002) Molecular biology of cellulose production in bacteria. *Research in Microbiology* 153, 205-212.

Ross, P., Mayer, R. and Benziman, M. (1991) Cellulose biosynthesis and function in bacteria. *Microbiological Reviews* 55, 35-58.

Spiers, A.J., Bohannon, J., Gehrig, S. and Rainey, P.B. (2003) Colonisation of the air-liquid interface by the *Pseudomonas fluorescens* SBW25 wrinkly spreader requires an acetylated form of cellulose. *Molecular Microbiology* 50, 15-27.

Spiers, A.J., Kahn, S.G., Travisano, M., Bohannon, J. and Rainey, P.B. (2002) Adaptive divergence in experimental populations of *Pseudomonas fluorescens*. 1. Genetic and phenotypic bases of wrinkly spreader fitness. *Genetics* 161, 33-46.

Spiers, A.J. and Rainey, P.B. (2005) The *Pseudomonas fluorescens* SBW25 wrinkly spreader biofilm requires attachment factor, cellulose fibre and LPS interactions to maintain strength and integrity. *Microbiology* 151, 2829-2839.

Sutherland, I.W. (2001) Biofilm exopolysaccharides: a strong and sticky framework. *Microbiology* 147, 3-9.

Thompson, I.P., Bailey, M.J., Ellis, R.J., Lilley, A.K., McCormack, P.J., Purdy, K.J. and Rainey, P.B. (1995) Short-term community dynamics in the phyllosphere microbiology of field-grown sugar beet. *FEMS Microbiology Ecology* 16, 205-212.

Villavicencio, E. (2000) Biofilm analysis of *Pseudomonas fluorescens* SBW25. The role of cellulose in biofilm development. MSc Thesis, Technical University of Denmark, Denmark.

Walker, T.S., Bais, H.P., Déziel, E., Schweizer, H.P., Rahme, L.G., Fall, R. and Vivanco, J.M. (2004) *Pseudomonas aeruginosa* - plant root interactions. Pathogenicity, biofilm formation, and root exudation. *Plant Physiology* 134, 320-331.

Yamamoto, S., Kasai, H., Arnold, D.L., Jackson, R.W., Vivan, A. and Harayama, S. (2000) Phylogeny of the genus *Pseudomonas*: intragenic structure reconstructed from the nucleotide sequences of gyrB and rpoD genes. *Microbiology* 146, 2385-2394.

Yu, J., Peñaloza-Vázquez, A., Chakrabarty, A.M. and Bender, C.L. (1999) Involvement of the exopolysaccharide alginate in the virulence and epiphytic fitness of *Pseudomonas syringae* pv. *syringae*. *Molecular Microbiology* 33, 712-720.

Zogaj, X., Bokranz, W., Nimtz, M. and Römling, U. (2003) Production of cellulose and *curli fimbriae* by members of the family *Enterobacteriaceae* isolated from the human gastrointestinal tract. *Infection and Immunity* 71, 4151-4158.

Chapter 9

Biological Control of Plant Diseases by Phyllosphere Applied Biological Control Agents

B.J. Jacobsen
Department of Plant Sciences and Plant Pathology, Montana State University, 119 Plant Biosciences Bldg Bozeman, MT 59717-3150 USA
uplbj@montana.edu

Introduction

Research on the control of plant diseases by phyllosphere applied biological control agents (BCAs) has produced a wealth of information on a wide range of fungal and bacterial BCAs and their applications for controlling diseases affecting leaves, flowers and fruits. Research into their mode of action and ecological adaptation has provided critical insights that have increased commercial utilisation of phyllosphere-applied BCAs worldwide. However, biological control of plant diseases using phyllosphere applied BCAs has proven to be challenging because the leaf surface presents a relatively hostile environment for introduced microbes with a relative paucity of available nutrients (Beattie and Lindow, 1999; Mercier and Lindow, 2000; Beattie, 2002), wide water availability fluxes (Beattie and Lindow, 1995), direct exposure to ultraviolet radiation (Sundin, 2002) and infrared radiation and competition with other phylloplane colonists. It is important to understand that each of these factors will vary with leaf age, fluxes in the physical environment and host plant genetics. Foliar applied BCAs do not have the avoidance strategy of endophytic growth available to pathogens (Beattie and Lindow, 1995) and therefore must primarily utilise a tolerance strategy whereby they colonise and survive on the leaf surface or survive in protected sites that may or may not be the same as the pathogen they are intended to control. Like protective fungicides BCAs must be reapplied regularly to both re-protect old growth as their populations decline and to protect new growth unless they provide disease control by induction of systemic host resistance. Each of these factors and development of stable formulated products must be addressed to achieve high levels of consistent control required for integration into commercial plant protection programmes.

Because of the variable effects of the biological and physical environment, the vast majority of phyllosphere applied BCA research has focused on glasshouse or storage environments where the physical environment is more predictable and stable (Paulitz and Belanger, 2000; Janisiewicz and Korston, 2002). In addition, the majority of research has focused on *Botrytis* grey mould, powdery mildew and fruit storage moulds such as *Penicillium, Monilinia* and *Botrytis*. Reasons for this focus include the relative high value of vegetables, ornamentals and fruit, the lower cost to register a BCA compared to synthetic pesticides, shorter or absent re-entry periods for BCAs compared to synthetic pesticides, an increasingly large organic produce market, and perhaps most important, the greater acceptance of consumers and regulatory agencies for the presence of BCAs on produce. Finally, recent research has provided valuable insights as to how to integrate phyllosphere applied BCAs with host plant resistance, other BCAs, systemic resistance activators, disease prediction programmes and reduced rates of fungicides to achieve the high levels of consistent control needed in the commercial marketplace.

This chapter will review phyllosphere applied BCAs that have provided disease control by their mode of action and provide examples of how integrated control approaches have provided better disease control and reduced fungicide or bactericide use. The mode of action of BCAs can be classified as niche occupation, competition for nutrients, antibiosis, parasitism, and induced systemic resistance. Most BCAs utilise more than one of these modes of action to affect pathogen control. *Trichoderma* species are perhaps the best example since all of the modes of action mentioned above have been implicated for this BCA.

Niche Occupation and Competition for Nutrients

These modes of action for BCAs are not mutually exclusive and will be treated together in this chapter since the BCA commonly successfully competes for an ecological niche on the leaf, flower or fruit by both competing for nutrients and infection niches. Examples involving bacteria include control of fireblight with *Pseudomonas fluorescens* A 506 (Wilson and Lindow, 1993; Temple, *et al.*, 2004) and post harvest control of fruit and potato decay by *P. syringae* L-59-66(ESC-11) and *P. syringae* ESC-10 (Kiewnick and Jacobsen, 1998; Janisiewicz and Korsten, 2002). *Pseudomonas fluorescens* A 506 is marketed as BlightBan A506 (Nufarm Agriculture Division, Burr Ridge, IL) and *P. syringae* L-59-66(ESC-11) and *P.syringae* ESC-10 are sold as Bio-Save 11 LP and 10LP respectively (Jet Harvest Solutions, Longwood, FL.). Wilson and Lindow (1993) demonstrated that *P. fluorescens* A 506 pre-emptively colonised pear pistils by competing for a limiting nutrient, thus making this nutrient unavailable to *Erwinia amylovora*. In 2004, Temple *et al.* demonstrated that apple and pear stigma was an iron-limited environment, and that Fe^{+++} is sequestered by *P. fluorescens* A 506 siderophores such that it is less available to *E. amylovora,* and that the additional of Fe^{+++} is required for *P. fluorescens* A 506 to produce an antibiotic toxic to *E. amylovora*. The case for implication of niche occupation as the mode of action for *P. syringae* L-59-66 is not as clear as the *P. fluorescens* A 506 example. Survival of *P. syringae* L-59-66 or ESC-10 in wounds on both citrus (Bull *et al.,* 1997) and on potato (Kiewnick and Jacobsen, 1998) is required for biocontrol. Rapid colonisation of wounds with more than e 5 cfu was required for control in both the citrus and potato systems. While Bull *et al.* (1998) showed that *P. syringae* L-59-66 produced the antibiotic syringomycin E when grown on some media but were unable to detect this antibiotic in citrus wounds colonised by this BCA (Janisiewicz and Korsten, 2002). Competition for nutrients is also involved with yeasts used for control of apple storage pathogens. Filonow (1998) showed that the yeasts *Cryptococcus laurentii* BSR-Y22 and *Sporobolomyces roseus* FS-43-238 provided control by out competing *Botrytis cinerea* for fructose, glucose and sucrose in apple wounds. Competition for nitrates was identified by Sherm *et al.* (2003) as a factor in the mode of action for the yeast, *Candida guilliermondii*, in the biological control of *Penicillium expansum* rot of apple. In a review paper, Elad and Steward (2004), indicate that competition for space or nutrients was responsible for biological control of apple fruit rots caused by *B. cinerea* by the yeasts *Candida* sp., *Cryptococcus laurentii*, *Sporobolomyces roseus* and *C. oleophila*. Competition for nutrients often results in reduced *B. cinerea* spore germination and slower germ tube growth, thereby reducing pathogen infection potential.

Biological control of *Botrytis cinerea* on a wide range of plants with the BCA *Ulocladium atrum* is another form of competition whereby the BCA competes for colonisation of necrotic tissue with *B. cinerea. Botrytis cinerea* is an unusual pathogen in that it colonises necrotic tissue prior to infection of healthy tissue. *Ulocladium atrum* is an excellent saprotroph and as a result of pre-emptive colonisation of necrotic tissues, it effectively reduces *Botrytis* colonisation resulting in reduced sporulation by *B. cinerea*. Because *U. atrum* colonises necrotic tissues at approximately one-half the rate of *B. cinerea* over a temperature range of 5-25 ^{0}C, it is critical that *U. atrum* conidia be sufficiently dense and even in distribution if optimal biological control is to be achieved (Kessel *et al.,* 2005). The interaction between this BCA and *Botrytis* only takes place in necrotic tissue. This is illustrated by the report of Kessel, *et al.* (2001) where they demonstrate that *U. atrum* is not effective in controlling *B. elliptica* on lily due to the fact that this *Botrytis* sp. can infect healthy leaf tissue.

Antibiosis

Antibiosis is the most common mode of action explored and this most likely reflects a chemical control equivalent mindset of many researchers. It is well known amongst biological control researchers that *in vitro* antibiosis activity does not often predict antibiosis *in planta* (Andrews, 1985) and many reports of antibiosis being responsible for disease control are based only on *in vitro* observations. Antibiotic production *in vitro* is highly dependent on the nutrients available in media that may or may not be available *in planta*. In addition, implication of antibiosis from *in vitro* experiments does not address the spatial constraints that may exist *in planta* between the BCA and the pathogen. Factors to consider when antibiosis is the reported mechanism of action of a BCA are: Were washed cells or cells with the fermentation spent growth media used in control experiments? Were antibiotic non-producing mutants used in comparison experiments? Have the antibiotics been identified *in situ*? The use of washed cells precludes the application of antibiotic substances from fermentation spent growth media. The use of wild type and antibiotic deficient mutants in experiments is perhaps the best way to test the hypothesis that antibiotics are involved in the activity of the BCA. Finally, identification of the antibiotic *in situ* is perhaps the ultimate proof of the involvement of antibiosis as a mode of action.

An excellent example of research to elucidate the role of antibiosis is the work by Wodzinski *et al.* (1994) and Stockwell, *et al.* (2002). These authors studied the control of fireblight on pear using *Pantoea agglomerans* (*Erwinia herbicola*). Wodzinski, *et al.* demonstrated that a crude preparation of antibiotic from strain Eh 318 provided better control of fireblight caused by an *E. amylovora* strain sensitive to the antibiotic than an *E. amylovora* strain insensitive to the antibiotic. In the Stockwell *et al.* paper, they produced an antibiotic deficient mutant of the fireblight control BCA, *Pantoea agglomerans* Eh 252, and compared it in field studies to the wild type BCA. Both stains provided control but the antibiotic producing wild-type strain provided substantially better control than the antibiotic deficient mutant, thus showing that *in planta*, that antibiosis is an important mechanism with this BCA but not the only mode of action employed by this BCA. Other evidence for the implication of antibiosis as a mode of action for Eh 252 is the observation that mixtures of *Pseudomonas fluorescens* A 506 and Eh 252 are not more effective than either BCA used individually, even though both BCAs colonise blossoms at the site of infection for *E. amylovora*. Control achieved by each BCA alone is approximately 50%, the same as when used in combinations (Vanneste *et al.,* 1996) even though these BCAs have different modes of action. A506 produces a produces a protease *in vitro* that inactivates the antibiotics of Eh

252, mccEh 252 and herbicolin O (Anderson *et al.*, 2004). Finally, Temple *et al.* (2004) demonstrate that *P. fluorescens* A506 produces an antibiotic in the presence of Fe^{+++} that may enhance fireblight control based on increased populations of A 506 on apple and pear blossoms.

A diverse array of antibiotics has been implicated in biological control. The antibiotic gramicidin S has been shown to be an important factor in biological control of *Botrytis cinerea* by *Brevibacillus brevis* both *in vitro* and *in vivo*. This was confirmed in experiments utilising a mutant that did not produce gramicidin S and by showing that the amount of gramicidin S bound to *B. brevis* spores correlated with control by the pure gramicidin S. The amount of gramicidin S required in leaves was higher than that in *in vitro* experiments conducted on glass slides due to absorption of the antibiotic on the leaf surface (Edwards and Seddon, 2001). Antibiotic deficient mutants were used by Leifert *et al.* (1995) to demonstrate that antibiosis from both peptide and non-peptide antibiotics was the mode of action for *B. subtilis* CL 27 in control of *B. cinerea*. Cheng *et al.* (2003) used insertional mutagenesis to demonstrate that the biological control activity of *Pseudozyma flocculosa* on *Sphaerotheca fuliginea* was due to the production of an antibiotic octadecyl cellobiose lipid (flocculosin).

Another method of implicating antibiosis as a mechanism of action is microscopic observation of the interaction of the BCA and the target pathogen and comparing these observations with effects of antibiotic compounds in culture filtrates. Based on these methods, *Tilletiopsis palleescens* (Hijwegen, 1989; Klecan *et al.*, 1990; Urquhart *et al.*, 2000) and *Lecanicillium lecanii* (Askary *et al.*, 1998) have been shown to use antibiosis for control of powdery mildew rather than parasitism.

While *Trichoderma* sp. have been shown to produce antibiotic substances and directly parasitise hyphae of fungal pathogens, recent work with mutants has shown that systemic-induced resistance is the primary mode of action (Harman *et al.*, 2004; Woo *et al.*, 2006). Therefore, *Trichoderma* sp. will be discussed under the systemic-induced resistance section and under other modes of action.

Parasitism

Parasitism of powdery mildews by *Ampelomyces quiqualis* is perhaps the classical example of biological control by direct parasitism. Sztejnberg *et al.* (1989) demonstrated parasitism on several powdery mildew genera including: *Oidium, Erysiphe, Sphaerotheca, Podosphaera, Uncinula* and *Leveillula*. This parasite infects the hyphae, conidiophores, conida and cleistothecia and forms pycnidia in or on it's powdery mildew host (Falk *et al.*, 1995). This parasite is now formulated as AQ-10 (Ecogen Inc., Langhorne, PA) and its use has been reviewed recently by Bélanger and Labbé (2002). Other parasites on powdery mildews include *Acremonium alternatum* and *Cladosporium spongiosum* and *C. clodosporioides* that parasitise *Phyllactinia guttata* and *Erysiphe cichoracearum* (Bélanger and Labbé, 2002).

Lysis of pathogen hyphae by hydrolytic enzymes is often characteristic of mycoparasitism. This has been demonstrated for several *Trichoderma* species that control fungal pathogens (Harman, *et al.*, 2004). Other examples include: *Microsphaeropsis* sp. that attack the hyphae and pseudothecia of *Venturia inequalis* such that this pathogen does not overwinter or produces fewer ascospores (Carisse and Rolland, 2004) and *Penicillium purpurgenum* that controls *Monilinia* twig blight on peach (Larena and Melgarejo, 1996).

Sundheim and Tronsmo (1988) reviewed hyperparasites on rusts. They discuss *Eudarluca caricis* (anamorph, *Sphaeropsis filum*) as a hyperparasite on cereal rusts, including: *Puccinia coronata, P. graminis, P. recondita, P. sorghi* and *P. striiformis*. In addition, this hyperparsite has been reported on *Cronartium fusiforme*, *C. strobilinum* and 367 other rust species found in both tropical and temperate climates (Kranz and Brandenburger, 1981). Other rust parasites reviewed by Sundheim and Tronsmo (1988) include: *Verticillium (Lecanicillium) lecani* that parasitises *Uromyces appendiculatus, U. dianthi, P. chrysanthemi* and *Hemiliea vastatrix*; *Tuberculina costaricana* that parasitises *P. penniseti and P. arachidis*; *T. maxima* that parasitises *Cronartium ribicola* and *C. quercuum*; *Scytalidium uredinicola* that parasitises *C. quercuum* f.sp. *fusiforme* and *P. elliotti* var. *elliotti*; *Aphanocladium album* that parasitises *P. graminis* f.sp. *tritici* and several other rust fungi; *Cladosporium hemileiae* that parasitises *H. vastatrix*; *C. gallicola* that parasitises *Endocronartium harknessi*; and *C. uredinicola* that parasitises *P. violae*. More recently, Moricca *et al.* (2001) demonstrated that *C. tenuissimum* provides control of *Cronartium flaccium* and *C. peridermium* by direct parasitism in microscopic studies, although this fungus also produces an antifungal metabolite, cladosporol, that inhibits β-1,3 glucan biosynthesis. In biological control trials with these fungi, control levels varied from insignificant to as much as 80%. In studies with *C. tenuissimum*, control efficiency was dependent on the susceptibility of the host pine, with better control on Austrian, Scots and Larico pine which were less susceptible to *C. flaccidum* than Aleppo, Italian Stone and Maritime pine. Disease reduction ranged from 19-32% on the susceptible varieties and 45-66% on the more resistant varieties. It should be noted that experimental evidence for parasitism that excludes other modes of action are incomplete for most of these rust hyperparasites.

Although the mode of action might more appropriately be described as mycophagy than parasitism, control of grape powdery mildew (*Uncinula nector*) by the tydeid mite, *Orthotydeus lambi* is of considerable interest (English-Loeb *et al.*, 1999). These authors report that the number of mildew colonies per cm^2 and the proportion of leaf infected was reduced by >85% in the presence of the mite relative to non-release controls. In addition, they suggest that *O. lambi* is responsible for preventing major outbreaks of powdery mildew on wild grape (*Vitis riparia*), and that this common and abundant member of leaf fauna on grape and other plants provides significant control of powdery mildew on both wild and cultivated grape. These authors discuss the impact of leaf morphology and note that plants with acaromatia (tufts of hairs or pits located in major leaf vein axes) have higher populations of mycophagous mites. They also note that sulfur and mancozeb fungicides suppress mite populations, and that the fungicides myclobutanil and azoxystrobin had negligible effects on mite populations.

Systemic-Induced Resistance

Effective biological control of plant diseases by BCAs that induce systemic resistance has been described for a wide range of pathosystems. This type of resistance makes an otherwise susceptible plant resistant to a wide array of subsequent pathogen attack (Kloepper *et al.,* 2004; van Loon and Glick, 2004) by activation of host responses that directly attack the pathogen through enzymatic action, production of plant produced antibiotic substances or by lignifying cell walls such that pathogen movement is slowed. In most situations all of these host responses are involved. Implicit to the implication of systemic-induced resistance as a mode of action is that the BCA is applied separately in time and space from the location where pathogen control occurs. Elicitation of systemic resistance can be achieved by several basic types of stimuli: necrotizing pathogens, chemicals, plant growth promoting rhizobacteria, foliar applied bacteria and fungi, oomycete and fungal cell wall fractions and bacterial cell wall fractions (Pieterse *et al.*, 1998; Bargabus *et al.*, 2002; Woo *et al.*, 2006). Several systemic resistance pathways have been described: (i) systemic acquired resistance characterised by salicylic acid signalling and the production of pathogenesis-related (PR) proteins (Delaney, 1997); (ii) systemic acquired resistance characterised by independence from salicylic acid signalling, production of active oxygen species independent of hypersensitive cell death and the production of PR proteins (Bargabus *et al.*, 2003; Bargabus-Larson and Jacobsen., 2006); (iii) induced systemic resistance in which signalling is due to jasmonates and ethylene and is independent of PR proteins (Pieterse *et al.*, 1998). This chapter will use the term systemic-induced resistance to cover all of these phenomena.

Examples of foliar or fruit applied BCAs where systemic-induced resistance has been implicated in disease control include: *Bacillus mycoides* Bac (Bargabus *et al.*, 2002, 2003), *B. mojaviensis* 203-7 (Bargabus *et al.*, 2004), *Bacillus subtilis* Bac B (Collins, *et al.*, 2003a), *Lysobacter enzymogenes* strain C3 (Kilic-Ekici and Yuen, 2003), *Lecanicillium (Verticillium) lecanii* (Benhamou, 2004) and *Trichoderma harzianum* (De Meyer *et al.*, 1998; Woo *et al.*, 2006), *T. virens*, *T. atroviridae,* and *T. asperellum* (Woo *et al.*, 2006). The present author feels that this mode of action is more common than uncommon and that most biocontrol researchers have not appropriately explored this mode of action for BCAs. In addition, it should be noted that there are numerous situations where plant roots colonised by certain rhizobacteria (Kloepper *et al.*, 2004) and *Trichoderma* species (Harman *et al.*, 2004) have induced resistance to bacterial, fungal and viral pathogens and insects expressed on leaves. The effects of induced resistance from either foliar or root colonizing BCAs on resident microbiota has not be studied to date. It would be interesting to determine the effect of induced resistance on the population and distribution of common non-pathogenic phyllosphere residents.

Bacillus mycoides, Bac J, was isolated from the phyllosphere of sugar beet in 1994 and was one of more than 330 isolates evaluated for *in vitro* antibiosis and hydrolytic enzyme attack to *Cercospora beticola,* and *in vivo* control of *Cercospora* leaf spot in glasshouse trials. This bacillus colonises sugar beet leaves as an epiphyte at populations that stabilise at e^{2-3}/cm^2 14 days after application at $e^{4.5}/cm^2$. In field studies it has provided 38-91% control and has been equal statistically to standard fungicide applications in 6 of 10 years of field testing. When used with sugar beet cultivars with moderate levels of resistance, Bac J provided control equal to commercial fungicides, whereas it provided only 41% control afforded by these same fungicides on a susceptible variety (Jacobsen *et al.*, 2004). Control either in the glasshouse or field was unaffected by the addition of the selective nutrient β-glucan presumed to support *Bacillus* species (Jacobsen, unpublished). Control of *Cercospora* leaf spot and induction of resistance does not occur with dead cells or fermentation beer, and systemic resistance is not effectively expressed unless application of the BCA is made 3-6 days prior to inoculation (Bargabus *et al.*, 2002). Induced systemic resistance is expressed for up to 18 days (Jacobsen, unpublished). In addition, disease control in field situations is equal regardless of whether e^4 or e^9cfu/ml spray is used (Jacobsen, unpublished). Induction of resistance as measured by up regulation of the PR protein peroxidases, chitinases and β 1, 3 glucanases, and a biphasic oxidative burst is greater when washed vegetative cells are applied compared to endospore preparations (Bargabus *et al.*, 2002, 2003). Further studies by Baragbus-Larson and Jacobsen (unpublished) using both Bac J and *B. mojaviensis* 203-7 show that the induction of resistance is independent of salicylic acid signalling but is dependent on activation of the NPR-1 gene. This is similar to the induction elicited by acibenzolar-S-methyl (ASM, Actigard, Syngenta, Greensboro, NC) that controls a wide range of diseases (Talley *et al.*, 1999). In addition to control of *Cercospora* leaf spot of sugar beet, Bac J also reduces disease severity of bacterial vascular necrosis of sugar beet caused by *Erwinia carotovora* pv. *betavasculorum* for up to 7 days (Jacobsen, unpublished) and both Bac J and 203-7 control anthracnose of cucumber and cantaloupe caused by *Glomerella cingulata* var *orbicular* and angular leaf spot of cucumber caused by *Pseudomonas syringae* pv. *lachrymans* in the field (Jacobsen, unpublished). An interesting aspect of cucumber anthracnose control by both Bac J and 203-7 is that the latent period is longer, lesion size smaller and the number of viable spores/mm^2 of lesion is reduced by >50% when these BCAs are applied to leaves spatially separating the BCA and the pathogen (Jacobsen, unpublished).

In studies with *B. subtilis* Bac B, biological control of *Cercospora* leaf spot of sugar beet was not as good as that achieved with Bac J, however this organism also has induced systemic resistance as its mode of action and is more amenable to ecological studies due to the fact that single cells can be studied, compared to the problems with chains of cells encountered with *B. mycoides*. In studies with a spontaneous rifampicin resistant mutant, Collins *et al.* (2003a,b), showed that populations declined from e$^{4.5}$/cm^2 to e$^{0.5\text{-}3}$/cm^2 over a 14-day period regardless of whether a selective nutrient β-glucan foodbase was applied at 0, 0.5 or 1.0% of spray formulation. In both field and glasshouse studies, where Bac B populations were modelled on a leaf scale, higher populations of vegetative cells were found after 14 days when the BCA was co-applied with 1% β-glucan than without and that populations were more aggregated without β-glucan. There was no correlation between Bac B population levels or distribution and *Cercospora* leaf spot lesions, thus suggesting that neither antibiosis nor parasitism are likely to be involved in disease control by *B. subtilis*. Collins and Jacobsen (2003a) demonstrated that control in the field was unaffected by application of e$^{6\text{-}9}$ cfu/ml of spray solution. Unlike Bac J, *B. subtilis* Bac B was not more effective when applied as vegetative cells compared to endospore preparations. When applying endospore preparations, application 3-5 days before challenge inoculations was more important than with vegetative cell-based formulations.

Unlike the resistance induced by foliar application of Bac J or 203-7, the resistance induced by *Lysobacter enzymogenes* C3 on tall fescue is localised to the treated leaf with foliar application while systemic resistance occurs with root applications (Kilic-Ekici and Yuen, 2003). Application of C3 live cells resulted in greater inhibition of *Bipolaris sorokiniana* conidial germination and leaf spot development and percent leaf blighted by *Rhizoctonia solani* than dead cells. Another characteristic of induced resistance besides spatial separation is that the BCA must be applied before infection. *L. enzymogenes* C3 provided optimal control when the BCA was applied 1 to 3 days prior to *B. sorokiniana* inoculation.

Biological control fungi in the genus *Trichoderma* are amongst the most studied BCA. Antibiosis, parasitism, competition and systemic-induced resistance have been identified as modes of action in both soil and foliar applications (Harman *et al.,* 2004). Based on use of mutants deficient in antibiotic production or parasitic ability, it appears that systemic-induced resistance is responsible for much of the biological control activity associated with this genus (Harman *et al.,* 2004; Woo *et al.,* 2006). The systemic-induced resistance implicated in *Trichoderma*-plant interactions has recently been reviewed by Woo *et al.* (2006). In this review they document systemic-induced resistance in more than 10 different crop plants including both monocots and dicots. Pathosystems have involved fungal, bacterial and viral pathogens. *Trichoderma* sp. produce three types of systemic-induced resistance elictors. These include; enzymes or peptides such as a 22 kDa xylanase, avirulence proteins with homology to Avr 4 and 9 from *Cladosporium fulvum* and oligosaccharides released by the activity of *Trichoderma* cell wall degrading enzymes on plant or pathogen cell walls. These elictors cause upregulation of several different PR and defense related proteins including chitinases, glucanases, peroxidases and phytoalexins.

Integrated Control Strategies

Because BCAs applied alone are not generally as efficacious or as consistent as fungicides or bactericides, combining them in integrated strategies with other BCAs, systemic resistance inducers, host plant resistance or reduced rates or applications of fungicides offers the opportunity to achieve control that is competitive with the best chemical controls. In theory, integration of several tools brings stability to disease management programmes. Integration of BCAs with other disease management tools may also provide broader crop adaptation In addition, because BCAs typically have different modes of action than fungicides, integrating their use with fungicides that have a high risk for fungicide resistance development may be useful in preventing or delaying resistance to the at-risk fungicide. Use of reduced rated or applications of fungicides is desirable in so far as fungicide residues are reduced and worker exposure is lower. Finally, use of disease prediction systems in conjunction with BCAs should allow them to be applied when conditions are appropriate for the BCA and increase the level of control that likely to be realised by the BCA.

Application of reduced or normal rates of fungicides in combination with BCAs has been shown to give equal or better disease control than either the fungicide or BCA alone, and has also reduced the variability of disease control achieved with BCAs. This latter point is important since a survey of 64 greenhouse experiments showed that *Trichoderma harzianum* T39 controlled *Botrytis cinerea* on tomato and cucumber with efficacy equal to fungicides, but in 20% of these experiments conducted worldwide the BCA was inferior to the fungicide and in 10% of experiments was not different from the untreated control (Guetsky *et al.,* 2001). *Bacillus mycoides* Bac J applied with half the rate of tetraconazole in the first application, followed by three applications of Bac J provided control of *Cercsopora* leaf spot of sugar beet equal to four applications of full rates of tetraconazole and slightly better than four applications of Bac J, (Jacobsen, unpublished). In other work on sugarbeet, Kiewnick *et al.* (2001) reported that the combination of *Bacillus* sp. isolate MSU 127 and a half rate of azoxystrobin had the lowest *Rhizoctonia* crown and root rot index (1.89) and was statistically similar to the full labelled rate of azoxystrobin fungicide applied alone(1.99) while the BCA alone had a root rot index of 2.63 and the half rate of azoxystrobin had a disease index of 2.44. *Rhizoctonia solani* AG 2-2 infects through the crown and azoxystrobin is applied to the crown tissue for control. Qin and Tian (2005) report that application of silicon (Si) with *Cryptococcus laurentii* provided better control of *Penicillium* blue mould and brown rot of cherry than either Si or the BCA alone. These authors attributed the improved control to increased population of the BCA yeast, direct fungitoxicity of Si to the pathogenic fungi and elicitation of biochemical defence responses in the fruit. Korston *et al.* (1997) reported that *Bacillus subtilis* B246 combined with benomyl or copper oxychloride provided better control of avocado black spot than either the BCA or fungicides applied alone. In another example of use of reduced rates of fungicide with BCAs, Buck (2004) reported that *Rhodotorula glutinis* PM4 combined with a 1/10 rate of azoxystrobin or trifloxystrobin provided better control of *Botrytis* grey mould on geranium than either the BCA or full rates of the fungicides used alone. When a vinclozolin resistant isolate of B. cinerea was used, combination of the BCA and either half or full rates of vinclozolin resulted in better control that either the BCA or the fungicide alone. No such interaction was shown for 1/10 rates of copper hydroxide, iprodione, mancozeb or thiophanate methyl. He also showed improved disease control with eight other isolates of *R. glutinis*, two isolates of *R. graminis*, and *R. mucilaginosa* but not *R. minuta*. In other *Botrytis* work, Elad and Stewart (2004) report that *Trichoderma harzianum* T39 mixed with iprodione or copper provided improved control compared to the BCA or the fungicides alone. In an interesting but short paper, Spotts *et al.* (1998) showed that post-harvest application of *Cryptococcus infirmo-miniatus* following preharvest application of iprodione improved control of blue mould and brown rot of sweet cherry compared to the pre-harvest application of the fungicide or the post-harvest application of the yeast alone. In addition they demonstrated that the combination of the fungicide, the BCA and modified atmosphere packaging provided better control than any of these controls used alone or in two-way combinations.

The use of different modes of action in sequential fungicide applications is considered a keystone of resistance management programmes. Since *Bacillus*-based BCAs have modes of action different than synthetic chemical fungicides it is logical that they can be used in fungicide-resistance management programmes. *Bacillus*-based BCAs have modes of action that include antibiosis, parasitism and induced systemic resistance. Larson (2004) studied the prevalence of benomyl, azoxystrobin and tetraconazole sensitivities to *Cercospora beticola* for two years in replicated plots that received no treatments, a rotation of fungicides, *B. mycoides* isolate Bac J, or a programme that integrated Bac J with these fungicides. Data from 2001 and 2002 showed that use of Bac J alone did not reduce the percentage of conidia insensitive to 1 ppm of benomyl, azoxystrobin or tetraconazole. However, the Bac J plus fungicide treatments had a statistically smaller percentage of conidia insensitive to benomyl and tetraconazole than the plots treated with the rotation of fungicides, and was equal to the fungicide rotation plots for azoxystrobin insensitivity. In another set of studies, Buck (2004a) and Buck and Jeffers (2004b) showed that combinations of *Rhodotorula glutinis* PM4 with reduced rates of azoxystrobin, trifloxystrobin and vinclozolin provided improved control and reduced variability of control compared to the fungicide alone against 10 isolates of *B. cinerea* resistant to vinclozolin.

Integration of two or more BCAs could conceptually improve disease control consistency where the BCAs have different modes of action and different ecological requirements (Guetsky *et al.*, 2001). One of the few examples available is the work by Guetsky *et al.* (2001) where they applied the yeast *Pichia guilermondii* and the bacterium *Bacillus mycoides* alone or together for control of *Botrytis cinerea* on strawberry leaves. Control efficacy of the BCAs alone ranged from 38-98%, and the coefficient of variation ranged from 9.7-75% while the BCAs applied in combination had a control efficacy of 80-99.8% and a coefficient of variation of only 0.4-9%. In these trials the authors used temperatures ranging from 10-30^{0}C and relative humidity ranging from 78-100%. The improved control was attributed to the bacterium multiplying faster than the yeast at low temperatures, the production of endospores that provided for survival during stressful conditions, better control of *Botrytis* spore germination by the yeast at temperatures less than 25^{0} C and better control by the bacterium at temperatures greater than 25^{0} C. Thus the combination of organisms allowed better suppression of spore germination over the range of environmental conditions than either BCA alone, and the bacterial BCA compensated for the inability of the yeast BCA to multiply under stressful conditions. In this experiment the modes of action of the two BCAs were not examined. The importance in selecting *Bacillus*-based BCAs adapted to specific environments is shown in the research reported by Leibinger *et al.* (1997) where *Bacillus subtilis* isolates AG704 and HG77 showed good colonisation of apples in the field but poor colonisation and apple fruit rot control in storage compared to two yeast species. These authors also demonstrated that compatibility between BCAs is also important in that the growth of the yeast *Aureobasidium pullans* was reduced by the *Bacillus* BCAs with resultant decrease in efficacy when these BCAs were mixed.

Integration of BCAs with inducers of systemic-induced resistance has not been widely studied, however, Wilson *et al.* (2002) demonstrated reduced incidence and disease severity of bacterial speck of tomato on leaves and incidence on fruit when *P. fluorescens* A506 was combined with acibenzolar-S-methyl (ASM, Actigard, Syngenta, Greensboro, NC), compared to ASM or A506 applied alone. The combination of ASM and A506 also reduced epiphytic populations of *P. syringae* pv. *tomato* more than either ASM or A 506 applied alone. These authors also showed that *P. syringae* Cit 7 combined with ASM provided improved control of bacterial speck when compared to ASM alone or either A 506 or *P syringae* TLP2 combined with ASM at the earliest of two of three sampling dates. However, all treatments were similar at the third sampling date. These authors did not show comparisons with these BCAs applied alone. While antibiosis by A506 cannot be excluded, this set of experiments suggests that combining pre-emptive exclusion and systemic induced resistance provide by ASM may be a promising strategy for management of pathogens that must develop threshold epiphytic populations before pathogenesis begins. In another study, Ji *et al.* (2006) examined the combination of the seed and root applied plant growth promoting rhizobacterium (*P. fluorescens* strain 89B-61) selected for its ability to induce resistance to bacterial speck and *P. syringae* Cit 7 applied to the foliage for control of bacterial speck and spot of tomato. In four field trials, the combination of these BCAs resulted in better control of bacterial speck in one trial and similar control in three trials compared to the BCAs applied alone. In two of three trials, bacterial spot control was increased by the combination of BCAs compared to the BCAs applied alone. These data suggest that the combination of induced resistance and *P. syringae* Cit 7 is not always synergistic but maybe helpful in some situations. It should be noted that the combination of the BCAs was never less efficacious than the BCAs used alone.

Integration of host-plant resistance with BCAs has been demonstrated for *Bacillus mycoides* Bac J (Larson, 2004) for control of Cercospora leaf spot of sugarbeet and for *Pseudozyma flocculosa* control of powdery mildew of cucumber. In a three-year study, Larson showed that *Bacillus mycoides* isolate Bac J provided control of *Cercospora* leaf spot equal to synthetic fungicides as measured by area under the disease progress curve on sugarbeet hybrids that had moderate levels of resistance. Bac J provided only 38-91% of the control afforded by fungicides on more susceptible varieties (Baragbus *et al.,* 2002). The effect of Bac J application was to flatten the disease progress curve such that the economic threshold was exceeded too late to affect yield significantly. Moderately resistant varieties also had flatter disease progress curves such that the economic threshold was exceeded later in crop maturity than with susceptible varieties. This is characteristic of the rate limiting resistance found in these varieties (Rossi *et al.,* 1999). By utilising moderately resistant varieties and Bac J, the level of disease control was equal statistically to four sprays of fungicide on either the susceptible or moderately resistant variety, but yield of extractable sucrose/ha was higher in response to these treatments only on the moderately resistant variety. In their review article, Bélanger and Labbé (2002) report on a trial in the Netherlands where *P. flocculosa* (Sporodex Biological Fungicide, Plant Products Co. Ltd, Dublin, OH) provided season long commercial levels of control of powdery mildew on a semi-tolerant, long English cucumber variety. Integration of host-plant resistance with BCAs may allow BCAs to compete on an even basis with chemical controls in plant disease management programmes.

Finally, the use of disease prediction systems has been shown to improve the performance of BCAs. Johnson *et al.* (2004) showed that BCA treatments could be most effectively timed by using a temperature-based bacterial growth index and disease risk index to allow BCA antagonists (*Pseudomonas fluorescens* A506 and *Pantoea agglomerans* C9-1S populations to grow to effective population sizes that pre-emptively colonise blossoms before the fireblight disease index shifted from low to high. These authors developed a decision matrix based on host bloom stage, 96 h temperature-based bacterial growth index and temperature-based disease risk index that allowed orchard managers to achieve greater consistency in the control achieved by bacterial antagonists. A similar approach was used by Shtienberg and Elad (1997) who used BOTMAN to predict the need for fungicide or *Trichoderma harzianum* T39 application. Fungicides were applied when an outbreak of a *Botrytis* grey mould epidemic on tomato or cucumber was expected. T39 was applied when condition were predicted to be less than favourable for an epidemic but more favourable than when slow or no disease progress was expected during which periods no sprays were applied. In analysis of three years of data encompassing eleven experiments, disease reduction from use of the integrated strategy based on the above parameters was equal to weekly fungicide applications. The mean number of fungicide applications was reduced by the integrated strategy by 6.3 applications and a mean of 5.9 T39 applications were made in the integrated strategy. This is similar to the strategy recommended by Gubler *et al.* (2002) for control of grape powdery mildew where *Bacillus subtilis* QST 713 (Serenade, Agraquest, Davis, CA) and *Ampelomyces quisqualis* are recommended for use when the risk of powdery mildew, disease development is low. Such as strategy limits the use of fungicides and should limit the risk of development of fungicide-resistant strains of powdery mildew.

Conclusions

This review has shown the various modes of action for BCAs applied to the phyllosphere or fruit surface. Several of the BCAs covered have multiple modes of action. Many reported BCAs are not covered in this chapter because a mode of action has not been established. Examples of combining BCAs with other control practices show the potential for increasing BCA control efficacy and consistency, such that the use of BCAs in commercial situations will become more wide spread. Research into how to best integrate BCAs with other disease management tools is needed and this author strongly encourages biocontrol researchers to compare disease control by BCAs not only with chemical standards but to make comparisons that integrate host-plant resistance, cultural controls, other BCAs, reduced use of pesticides, systemic-resistance activators and other BCAs. Research into the effect of BCAs on pest resistance to fungicides and bactericides may suggest a unique role for inclusion of BCAs in many disease control programmes. Successful integration with other BCAs will require research on ecological adaptation and knowledge of the mode of actions of potential BCA partners.

References

Anderson, L.M. and Stockwell, V.O. and Loper, J.E. (2004) An extracellular protease of *Pseudomonas fluorescens* inactivates antibiotics of *Pantoea agglomerans*. *Phytyopathology* 94, 1228-1234.

Andrews, J.H. (1985) Strategies for selecting antagonistic microorganisms from the phylloplane. In: Windels, C.E and Lindow, S.E. (eds) *Biological Control in the Phylloplane.* APS Press, St. Paul, Minnesota, USA. pp. 31-44.

Askary, H., Carriére, Y., Bélanger, R.R. and Brodeur, J. (1998) Pathogenicity of the fungus *Verticillium lecanii* to aphids and powdery mildew. *Biocontrol Science Technology* 8, 23-32.

Bargabus, R.L., Zidack, N.K., Sherwood, J.E. and Jacobsen, B.J. (2002) Characterization of systemic resistance in sugar beet elicited by a non-pathogenic, phyllosphere-colonizing *Bacillus mycoides*, biological control agent. *Physiological and Molecular Plant Pathology* 61, 289-298.

Bargabus, R.L., Zidack, N.K., Sherwood, J.E. and Jacobsen, B.J. (2003) Oxidative burst elicited by *Bacillus mycoides* isolate Bac J, a biological control agent, occurs independently of hypersentive cell death in sugar beet. *Molecular Plant-Microbe Interactions* 16, 1145-1153.

Bargabus, R.L., Zidack, N.K., Sherwood, J.E. and Jacobsen, B.J. (2004) Screening for the identification of potential biological control agents that induce systemic acquired resistance in sugar beet. *Biological Control* 30, 342-350.

Bargabus-Larson, R.L. and Jacobsen, B.J. (2006) Biocontrol elicited systemic resistance in sugarbeet is salicylic acid independent and NPR1 dependent. *Journal of Sugar Beet Research,* in press

Beattie, G.A. (2002) Leaf surface waxes and the process of leaf colonization by microorganisms. In: Lindow, S.E, Hecht-Poinar, E. and Elliott, V.J. (eds) *Phyllosphere Microbiology*. APS Press, St. Paul, Minnesota, USA. pp. 3-26.

Beattie, G.A. and Lindow, S.E. (1995) The secret life of foliar bacterial pathogens on leaves. *Annual Review of Phytopathology* 33, 145-172.

Beattie, G.A. and Lindow, S.E. (1999) Bacterial colonization of leaves: a spectrum of strategies. *Phytopathology* 89, 353-359.

Bélanger, R.R. and Labbé, C. (2002) Control of powdery mildews without chemicals: Prophylactic and biological alternatives for horticultural crops. In: Bélanger, R.R., Bushnell, W.R., Dik, A.J. and Carver, T.L.W. (eds) *The Powdery Mildews: A Comprehensive Treatise.* APS Press, St. Paul, Minnesota, USA. pp. 256-267.

Benhamou, N. (2004) Potential of the mycoparasite, *Verticillium lecanii*, to protect citrus fruit against *Penicillium digitatum*, the causal agent of green mold: a comparison with the effect of chitosan. *Phytopathology* 94, 693-705.

Buck, J.W. (2004a) Combinations of fungicides with phylloplane yeasts for improved control of *Botrytis cinerea* on geranium seedlings. *Phytopathology* 94, 196-202.

Buck, J.W.and Jeffers, S.N. (2004b) Effect of pathogen aggressiveness and vinclozolin on efficacy of *Rhodotorula glutinis* PM4 against *Botrytis cinerea* on geranium leaf discs and seedlings. *Plant Disease* 88, 1262-1268.

Bull, C.T., Stack, J.P. and Smilanick, J.L. (1997) *Pseudomonas* strains ESC-10 and ESC-11 survive in wounds on citrus and control green and blue molds of citrus. *Biological Control* 8, 81-88

Bull, C.T., Wadsworth, M.L., Sorensen, K.N., Takemoto, J.Y., Austin, R.K. and Smilanick, J.L. (1998) Syringomycin E produced by biological control agents controls green mold on lemons. *Biological Control* 12, 89-95.

Carisse, O. and D. Rolland. 2004. Effect of timing of application of the biological control agent *Microsphearopsis ochracea* on the production and ejection pattern of ascospores by *Venturia inequalis. Phytopathology* 94, 1305-1314.

Cheng, Y., McNally, D.J., Labbé, C. Voyer, N., Belzile, F. and Bélanger, R.R. (2003) Insertional mutagenesis of a fungal biocontrol agent led to discovery of a rare cellobiose lipid with antifungal activity. *Applied Environmental Microbiology* 69, 2595-2602.

Collins, D.P. and Jacobsen, B.J. (2003a). Optimizing a *Bacillus subtilis* isolate for control of sugar beet *Cercospora* leaf spot. *Biological Control* 26, 153-161.

Collins, D.P., Jacobsen, B.J. and Maxwell, B. (2003b) Spatial and temporal population dynamics of a phyllosphere colonizing *Bacillus subtilis* biological control agent of sugar beet *Cercospora* leaf spot. *Biological Control* 26, 224-232.

Delaney, T.P. (1997) Genetic dissection of acquired resistance. *Plant Physiology* 113, 5-12.

De Meyer, G., Bigirimana, J., Elad, Y. and Höfte, M. (1998) Induced systemic resistance in *Trichoderma harzianum* T39 biocontrol of *Botrytis cinerea. European Journal of Plant Pathology* 104, 279-286.

Edwards, S.G. and Seddon, B. (2001) Mode of antagonism of *Brevibacillus brevis* against *Botrytis cinerea in vitro. Journal of Applied Microbiology* 91, 652-659.

Elad, Y. and Stewart, A. (2004) Microbial control of *Botrytis* sp. In: Elad, Y., Williamson, B., Tudzynski, P. and Delen, N. (eds) *Botrytis: Biology, Pathology and Control.* Kluwer Academic Publishers, Dordrecht, Netherlands. pp. 223-241.

English-Loeb, G., Norton, A.P., Gadoury, D.M., Seem, R.C. and Wilcox, W.F. (1999) Control of powdery mildew in wild and cultivated grapes by a Tydeid mite. *Biological Control* 14, 97-103.

Falk, S.P., Gadoury, D.M., Cortesi, P., Pearson, R.C. and Seem, R.C. (1995) Parasitism of *Uncinula nector* cleistothecia by the mycoparasite *Ampelomyces quisqualis*. *Phytopathology* 85, 794-800.

Filonow, A.B. (1998) Role of competeition for sugars by yeasts in the biocontrol of gray mold of apple. *Biocontrol Science and Technology* 8, 243-256.

Gubler, W.D., Stapleton, J.J., Leavitt, G.M., Purcell, A.H., Varela, L.G. and Smith, R.J. (2002) UC IPM: UC management guidelines for powdery mildew on grape. http://www.ipm.ucdavis.edu/PMG/r302100311.html

Guetsky, R., Shtienberg, D., Elad, Y. And Dinoor, A. (2001) Combining biological control agents to reduce the variability of biological control. *Phytopathology* 91, 621-627.

Harman, G.E., Howell, C.R., Viterbo, A., Chet, I. and Lorito, M. (2004) *Trichoderma* species-opportunistic, avirulent plant symbionts. *Nature Reviews Microbiology* 2, 43-56.

Hijwegen, T. (1989) Effect of culture filtrate of seventeen fungicolous fungi on sporulation of cucumber powdery mildew. *Netherlands Journal of Plant Pathology* 95, 95-98.

Jacobsen, B.J., Zidack, N.K. and Larson, B.J. (2004) The role of Bacillus-based biological control agents in integrated pest management systems: plant diseases. *Phytopathology* 94, 1272-1275.

Janisiewicz, W.J. and Korston, L. (2002) Biological control of postharvest diseases of fruit. *Annual Review of Phytopathology* 40, 411-441.

Ji, P., Campbell, H.L., Kloepper, J.W., Jones, J.B., Suslow, T.V. and Wilson, M. (2006) Integrated biological control of bacterial speck and spot of tomato under field conditions using foliar biological control agents and plant growth-promoting rhizobacteria. *Biological Control* 36, 358-367.

Johnson, K.B., Stockwell, V.O. and Sawyer, T.L. (2004) Adaption of fire blight forecasting to optimize the use of biological controls. *Plant Disease* 88, 41-48.

Kessel, G.J.T., De Haas, B.H., Lomaers-Van der Plas, C.H., Van den Ende, J.E., Pennock-Vos, M.G., van der Werf, W. and Köhl, J. (2001) Comparative analysis of the role of substrate specificity in biological control of *Botrytis elliptica* in lily and *B. cinerea* in cyclamen with *Ulocladium atrum*. *European Journal of Plant Pathology* 107, 273-284.

Kessel, G.J.T., Köhl, J., Powell, J.A., Rabbinge, R. and van der Werf, W. (2005) Modeling spatial characteristics in the biological control of fungi at leaf scale: competitive colonization by *Botryis cinerea* and the saprophytic antagonist *Ulocladium atrum*. *Phytopathology* 95, 439-448.

Kiewnick, S., Jacobsen, B.J., Braun-Kiewnick, A., Eckhoff, J.L.A. and Bergman, J.W. (2001) Integrated control of *Rhizoctonia* crown and root rot of sugar beet with fungicides and antagonistic bacteria. *Plant Disease* 85, 718-722.

Kiewnick, S. and Jacobsen, B.J. (1998) Control of *Fusarium* dry rot on potato with antagonistic bacteria. *American Journal of Potato Research* 75, 284.

Kilic-Ekici, O. and Yuen, G. (2003) Induced resistance as a mechanism of biological control by *Lysobacter enzymognes* strain C3. *Phytopathology* 93, 1103-1110.

Klecan, A.L. Hippe, S. and Sommerville, S.C. (1990) Reduced growth of *Erysiphe graminis* f.sp. *hordei* induced by *Tilletiopsis pallescens*. *Phytopathology* 80, 325-331.

Kloepper, J.W., Ryu, C. and Zhang, S. (2004) Induced systemic resistance and promotion of plant growth by *Bacillus* sp. *Phytopathology* 94, 1259-1266.

Korsten, L., DeVilliers, E.E., Wehner, F.C. and Kotze, J.M. (1997) Field sprays of *Bacillus subtilis* for control of preharvest fruit diseases of avocado in South Africa. *Plant Disease* 81, 455-459.

Kranz, J. and Brandenburger, W. (1981) An amended host list of the rust parasite *Eudarluca caricis*. *Journal of Plant Disease and Protection*. 88, 682.

Larena, I. and Melgarejo, P. (1996) Biological control of *Monilinia laxa* and *Fusarium oxysporum* f.sp. *lycopersici* by a lytic enzyme-producing *Penicillium purpurogenum*. *Biological Control* 6, 361-367.

Larson, B. (2004) Integrated management of Cercospora leafspot. Masters thesis. Montana State University, Bozeman, Montana, USA.

Leibinger, W., Breuker, B., Hahn, M. and Medgen, K. (1997) Control of postharvest pathogens and colonization of the apple surface by antagonistic microorganisms in the field. *Phytopathology* 87, 1103-1110.

Leifert, C., Li, H., Chidburee, S., Hampton, S., Workman, S., Sigee, D., Epton, H.A.S. and Harbor, A. (1995) Antibiotic production and biological control activity by *Bacillus subtilis* CL27 and *Bacillus pumilis* CL45. *Journal of Applied Bacteriology* 78, 97-108.

Mercier, J. and Lindow, S.E. (2000) Role of leaf surface sugars in colonization of plants by bacterial epiphytes. *Applied Environmental Microbiology* 66, 369-374.

Moricca, S., Ragazzi, A., Mitchelson, K.R. and Assante, G. (2001) Antagonism of the two-needle pine rust fungi *Cronartium flaccidum* and *Peridermium pini* by *Cladosporium tenuissimum in vitro* and *in planta*. *Phytopathology* 91, 457-468.

Paulitz, T.C. and Bélanger, R.R. (2000) Biological control in greenhouse systems. *Annual Review Phytopathology* 39, 103-33.

Pieterse, C.M..J., van Wees, S.C.M., van Pelt, J.A., Knoester, M., Laan, R., Gerrits, H., Weisbeek, P.J. and van Loon, J.C. (1998) A novel signalling pathway controlling induced systemic resistance in Arabidopsis. *Plant Cell* 10, 1571-1580.

Qin, G.Z. and Tian, S.P. (2005) Enhancement of biocontrol activity of *Cryptococcus laurentii* by silicon and the possible mechanisms involved. *Phytopathology* 95, 69-75.

Rossi, V., Giosue, S. and Racca, P. (1999) A model integrating components of rate-reducting resistance to *Cercospora* leaf spot in sugar beet. *Journal of Phytopathology* 147, 339-346.

Scherm, B., Ortu, G., Muzzu, A., Budroni, M., Arras, G. and Migheli, Q. (2003) Biocontrol activity of antagonistic yeasts against *Penicillium expansum* on apple. *Journal of Plant Pathology* 85, 205-213.

Shtienberg, D and Elad, Y. (1997) Incorporation of weather forecasting in integrated, biological-chemical management of *Botrytis cinerea*. *Phytopathology* 87, 332-340.

Spotts, R.A., Cervantes, L.A., Facteau, T.J. and Chand-Goyal, T. (1998) Control of brown rot and blue mold of sweet cherry with pre harvest iprodione, post harvest *Cryptococcus infirmo-miniatus*, and modified atmosphere packaging. *Plant Disease* 82, 1158-1160.

Stockwell, V.O., Johnson, K.B., Sugar, D. and Loper, J.E. (2002) Antibiosis contributes to biological control of fireblight by *Pantoea agglomerans* strain Eh 252 in orchards. *Phytopathology* 92, 1202-1209.

Sundheim, L. and Tronsmo, A. (1988) Hyperparasites in biological control. In: Mekerji, K.G. and Garg, K.L. (eds). *Biological Control of Plant Diseases* Vol. 1. CRC Press, Boca Ratan, FL, USA pp. 53-69.

Sundin, G.W. (2002) Ultraviolet radiation on leaves: Its influence on microbial communities and their adaptation. In: Lindow, S.E, Hecht-Poinar, E.I. and Elliott, V.J. (eds) *Phyllosphere Microbiology*. APS Press, St. Paul, Minnesota, USA. pp. 27-41.

Sztejnberg, A., Galper, S., Mazar, S. and Lisker, N. (1989) *Ampelomyces quisqualis* for biological and integrated control of powdery mildews in Isreal. *Journal of Phytopathology* 124, 285-289.

Tally, A., Oostendorp, M., Lawton, K., Staub, T. and Bassi, B. (1999) Commercial development of elicitors of induced resistance to pathogens. In: Agrawal, A.A., Tuzun, S. and Bent, E. (eds) *Induced Plant Defences Against Pathogens and Herbivores.* APS Press, Saint Paul, Minnesota, USA. pp. 299-318.

Temple, T.N., Stockwell, V.O., Loper, J.E. and Johnson, K.B. (2004) Bioavailabilty of iron to *Pseudomonas fluorescens* A 506 on flowers of pear and apple. *Phytopathology* 94, 1286-1294.

Urquhart, E.J., Li, J. and Punja, Z.K. (2002) Hydrolytic enzymes and antifungal compounds produced by *Tilletiopsis* species, phyllosphere yeasts that are antagonists of powdery mildews. *Canadian Journal of Plant Pathology* 48, 219-229.

van Loon, L.C. and Glick, G.R. (2004) Increased plant fitness by rhizobacteria. In: Sanderman, H. (ed.) *Molecular Exotoxicology of Plants.* Vol. 170. Springer-Verlag, Berlin, Germany. pp. 177-205.

Vanneste, J.L. and Yu, J. (1996) Biological control of fire blight using *Erwinia herbicola* Eh252 and *Pseudomonas fluorescens* A506 separately or in combination. *Acta Horticulturae* 411, 351-353.

Wilson, M. and Lindow, S.E. (1993) Interactions between the biological control agent *Pseudomonas fluorescens* strain A506 and *Erwinia amylovora* in pear blossoms. *Phytopathology* 83,117-123.

Wilson, M., Campbell, H.L., Ji, P., Jones, J.B. and Cuppels, D.A. (2002) Biological control of bacterial speck of tomato under field conditions at several locations in North America. *Phytopathology* 92, 1284-1292.

Wodzinski, R.S., Umholtz, T.E, Rundle, J.R. and Beer, S.V. (1994) Mechanisms of inhibition of *Erwinia amylovora* by *Erw. herbicola in vitro* and *in vivo*. *Journal of Applied Bacteriology* 76, 22-29.

Woo, S.L., Scala, F., Ruocco, M. and Lorito, M. (2006) The molecular biology of the interactions between *Trichoderma,* phytopathogenic fungi and plants. *Phytopathology* 96, 181-185.

Chapter 10

Ecophysiology of Biocontrol Agents for Improved Competence in the Phyllosphere

N. Magan
Applied Mycology Group, Institute of BioScience and Technology, Cranfield University, Silsoe, Bedford, MK45 4DT, UK
n.magan@cranfield.ac.uk

Introduction

The phyllosphere environment is very harsh and subject to significant variations in abiotic factors generally, and these interact with biotic (plant) factors. There have been significant advances in the development of effective biological control agents (BCAs) to control bacterial and fungal diseases (Butt *et al.*, 2001). Post-harvest application of BCAs has been very successful because of the controlled environment which exists during processing, packaging and storage which is conducive to application and establishment of BCAs especially on perishable surfaces. In the phyllosphere, fluctuating abiotic factors, particularly of water availability and temperature, length of dew periods, microclimate and type of canopy, and rainfall events all impact on the use of BCAs. Tolerance to such fluctuations is a prerequisite for the successful development of ecologically competent BCAs for use in the field. Thus a major bottleneck, with regard to successful application of BCAs in the phyllosphere, has been the adequate development of inocula with the required physiological quality characteristics and with the capability for germination and establishment over as wide a relative humidity and temperature range as possible.

With regard to production of BCAs for use generally, economic production of inocula has involved the use of either liquid or solid substrate fermentation systems. While such studies have been numerous, most have concentrated on optimising the quantity of propagules, mycelial fragments or microbial biomass, with practically no consideration given to the quality of inocula. Indeed, few studies have considered the very real practical problem of effective establishment of prospective BCAs in the natural environment, be it for control of pests, diseases or weeds. This aspect needs to be addressed to improve the potential use of BCAs to control diseases in the phyllosphere. Unfortunately, the area of improving ecological fitness of inocula has received very little attention, although some elegant studies of desiccation tolerance on BCAs have been published (Jin *et al.,* 1991; Jackson *et al.*, 1997). Effective environmental stress tolerance of inocula can lead to improved establishment, which could contribute significantly to improving efficacy of BCAs in the phyllosphere environment.

There are thus some key questions which need to be addressed: (a) can the physiology of BCAs be effectively manipulated to accumulate useful endogenous reserves into inocula for improved environmental stress tolerance; (b) would this result in improved germination/growth over a wider relative humidity range; (c) can this lead to improved establishment of BCAs and conserve biocontrol potential in the field; and (d) does ecophysiological manipulation have a role in improving production and quality of BCAs and be useful in the development of ecological competent formulations. This chapter will address these issues and present examples to demonstrate that studies on quality of BCAs may be a valuable approach to improving the formulation and delivery of ecologically competent inocula for use in the phyllosphere to control plant diseases and pests.

Environmental Stress Tolerance - Background

There are a number of fungi which can be classified ecologically based on their water stress tolerance according to Griffin (1981). The xerotolerant and xerophilic fungi are able to tolerate a very wide range of water availability, with *Penicillium*, *Aspergillus* and *Eurotium* species able to grow at between 0.85 (= 85% Equilibrium Relative Humidity, ERH) and 0.70 water activity (a_w, = 70%, ERH) (Magan, 1997). This is significantly wider than most BCA microorganisms used for plant disease and pest control in the phyllosphere, which are able to germinate and grow only very slowly at <0.95-0.93 a_w (= 93-95% ERH). Very few are able to grow at <0.90 a_w (= <90% ERH).

These xerophilic/xerotolerant fungi survive such large fluctuations in external environmental stress due to their ability to synthesise compatible solutes, particularly low molecular weight sugar alcohols such as glycerol and erythritol (polyols) which enable enzyme systems to function effectively (Magan, 1997). For bacteria, this involves the production of glycine betaine, ectoine and proline (Teixido *et al.*, 2005). Recent studies of spoilage fungi such as *Aspergillus flavus* and *Aspergillus ochraceus* have demonstrated that when exposed to osmotic water stress, they accumulate significantly elevated amounts of glycerol and sometimes erythritol in their biomass and also in the conidia when compared to those present under freely available water conditions (Neschi *et al.*, 2004). Fig. 10.1 compares the impact of such treatments on the changes in the relative proportions of sugar alcohols which occur in *A. flavus* (Neschi *et al.*, 2004). Such studies suggest that physiological manipulation of growth conditions can significantly modify the endogenous compounds synthesised and accumulated in the biomass, and also that are channelled into conidia or similar propagules.

There is also some evidence that it may be possible to widen the environmental range for effective growth and biocontrol by producing low a_w mutant strains. Matawele *et al.* (1994) demonstrated that low a_w UV mutant strains of *Metarhizium anisopliae* and *Paecilomyces farinosus* used for pest control germinated over a wider a_w range (minima of 0.957 a_w) and, more importantly, were more virulent at controlling green leafhoppers than the original wild-type strains (germination minima of 0.975 a_w). This represents the type of control needed in the phyllosphere to obtain better disease and pest control. Unfortunately, although this work implied that improved water stress tolerance of the mutants was responsible for the effects observed, no quantification of the endogenous content of the propagules was carried out. Other studies on physiological manipulation of C:N ratios for production of blastospores of entomogenous fungi also showed that increasing the glycogen content resulted in modified and improved virulence against some pest species (Lane *et al.*, 1991). However, glycogen accumulation is more often associated with non-stress conditions and compatible solute accumulation under C and N stress (van Laere, 1989).

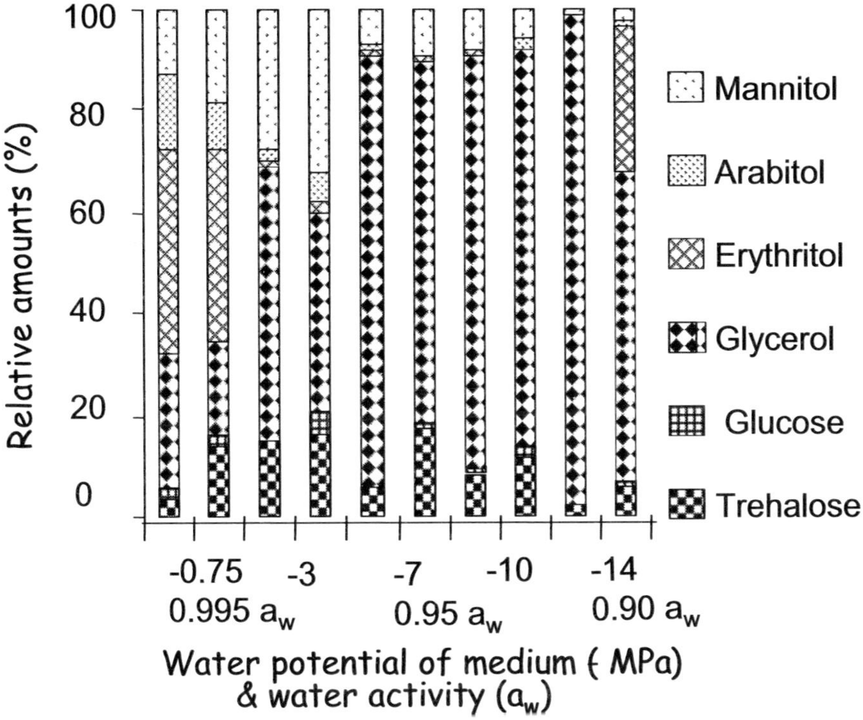

Fig. 10.1. Change in endogenous sugar alcohol content of the xerophilic species *Aspergillus flavus* in relation to water availability (adapted from Neschi *et al.*, 2004).

There have also been few studies that have directly collected spores/biomass of BCAs established on either crop debris or senescing leaves, and where the endogenous compounds present have been analysed. This would indicate the type of mixture of accumulated/synthesised compounds which may contribute to growth under naturally fluctuating environmental conditions. There is evidence that spores of BCAs such as *Ulocladium atrum* can germinate very rapidly during episodes of high humidity, with germ tubes surviving dry periods and being able to continue growing again when humid conditions or dew periods prevail (Köhl and Molhoek, 2001). Similar studies have been carried out with conidia of *Metarhizium anisopliae* sporulating from dead insects. Interestingly, such conidia were also found to be more virulent against pests than those obtained from rich artificial media. Magan (2001) showed that the profiles and quantities of sugars and sugar alcohols found in such conidia were significantly different from those obtained from spores harvested from artificial media. The presence of high amounts of erythritol, mannitol and glucose, with low amounts of trehalose in conidia from killed insects, was very different from that grown on richer artificial media where mannitol, glucose and trehalose were the major reserves. This demonstrated that the endogenous profiles of conidia grown on rich artificial media are significantly different from the profiles present in those obtained directly from insects. This is also indicative that the nutritional status of artificial media commonly used may not be best for the production of such BCAs.

This type of information is very useful. It suggested that ecophysiological systems could be developed to produce BCAs under conditions which may be conducive to the synthesis and accumulation of useful compounds in the inocula/conidia in the laboratory, and then used to screen and quantify environmental stress tolerance and biocontrol capability. This has led to extensive research in the Applied Mycology Group, Cranfield University, in collaboration with a number of other research groups, on a range of fungal biocontrol agents in order to understand the impact that physiological manipulations have on synthesis and accumulation of sugars/sugar alcohols on ecological competence, formulation and biocontrol efficacy. These studies have examined: *Epicoccum nigrum* and *Penicillium frequentans* for control of *Monilinia laxa* on peaches; *Gliocladium roseum* for *B. cinerea* on a variety of crops; *Candida sake* for *Penicillium expansum* on apples; *Ulocladium atrum* for *B. cinerea* on a range of crops; *Pantoea agglomerans* for post-harvest control of citrus and pome fruits; *Pichia anomala* for mould spoilage of moist grain; the use of entomogenous fungi for pest control (Magan, 2001).

Ecophysiological Manipulation of Endogenous Reserves in Inocula

Physiological Water Stress and Temporal Accumulation of Compatible Solutes in BCAs

Initial work to examine the impact of modifications in physiological water stress showed that changes in carbon concentration of media, as well as the addition of solutes, had a significant impact on the behaviour of BCAs (Hallsworth and Magan, 1994a). These studies showed that when culturing entomogenous fungi using trehalose, glycerol, glucose or starch as a major carbon source, both C-concentration and time had a significant effect on accumulation of sugars (glucose, trehalose) and sugar alcohols (polyols: glycerol, erythritol, arabitol and mannitol) in conidia. The time at which optimum quantities of low molecular weight polyols accumulated varied from 7-21 days. Trehalose content increased during the first 5 days and then decreased again. However, these studies were all carried out at one steady state temperature, 25°C. Subsequent studies by Hallsworth and Magan (1996) showed that a_w, temperature, time and pH all affected the synthesis and accumulation of sugars and polyols in conidia of the fungi studied.

Recent studies with bacteria, yeasts and filamentous fungi have all demonstrated that such physiological manipulation does modify endogenous accumulation/synthesis of compatible solutes and sugars. For example, experiments with the BCA *Pantoea agglomerans* CPA-2 showed that modifying the growth medium to 0.98 and 0.97 a_w with the ionic solute NaCl for 24 h at 30°C resulted in a significant increase in glycine betaine and ectoine when compared to unmodified controls. The quantities of these two compatible solutes were increased from 6-7 μmol g^{-1} dry weight to 35-45 μmol in the 0.98 a_w treatment (Teixido *et al.*, 2005). Studies on yeasts such as *Candida sake* and *Pichia anomala* have demonstrated that physiological modifications using ionic solutes, glycerol, sorbitol or proline could have a significant influence on accumulation of trehalose, an important desiccation protectant, and on the ratio of different polyols (Teixido *et al.*, 1998; Ypsilos and Magan, 2004).

A further important consideration is whether these modified endogenous compounds can be conserved effectively in the BCA during harvesting and subsequent processing and formulation prior to application. This aspect of improving the quality of BCAs has received practically no attention. Studies with the cells of *P. anomala* and blastospores of *M. anisopliae* have demonstrated that using isotonic solutions close to the parameters used in the fermentation process can result in a significant improvement in retention of these endogenous compounds. Fig. 10.2 compares the sugar and sugar alcohol accumulation in *P. anomala* cells modified with proline and then harvested with water or isotonic solutions. This clearly shows that there is a significant increase in endogenous amounts of trehalose, arabitol and mannitol when compared with harvesting in either water or isotonic solutions using molasses based media (Mokiou, 2005).

Studies with the filamentous fungi *E. nigrum, U. atrum* and *P. frequentans* have shown that the endogenous contents of polyols can all be significantly increased using this approach with both liquid and solid substrate fermentation systems (Pascual *et al.,* 1999; Frey and Magan, 2000).

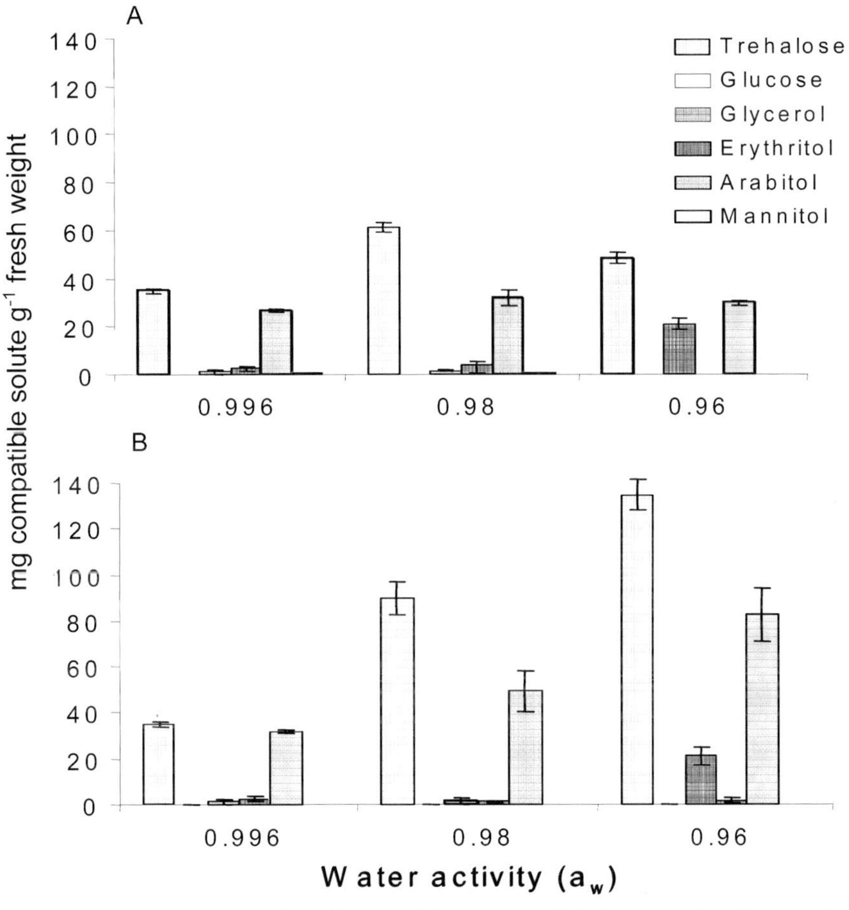

Fig. 10.2. Endogenous concentration of compatible solutes of *Pichia anomala* cells grown in proline modified molasses media after 72 h incubation at 25°C. Cells were washed with water (A) and isotonic solution (B) prior to extraction. Polyol and sugar concentrations are presented with trehalose to the left and mannitol to the right. Bars indicate standard error of the mean (n = 5).

Improvements in Quality of Modified Inocula by Measurements of Viability and Growth

The next key question is whether such modifications can translate into improved germinative capacity over a wider range of environmental stress conditions, and perhaps also more rapid germination than spores from unmodified inocula grown on nutrient-rich unmodified media.

The capacity for withstanding wider water availability ranges was assessed by using weak water-based agar media modified with polyethylene glycol 200/400 and 600 or mixtures thereof or ionic solutes, to avoid using solutes which might be taken up directly by cells or propagules. The inocula were prepared in a diluent of the same a_w as the test conditions to simulate freely available water (0.995 a_w) down to about 0.90a_w (90% ERH), depending on the BCA, to represent conditions that can simulate humidity changes which they might be exposed to during field application. In some cases, richer potato dextrose or yeast/glucose based media were also used.

Studies with *P. agglomerans* showed that the enhanced accumulation of compatible solutes (glycine betaine and ectoine) contributed to a significant improvement in both heat shock tolerance and viability under water stress of 0.96 a_w after 24-36 h. Viability was approximately 1-2 x 10^9 ml^{-1} viable cells for the modified treatments compared to 5 x 10^7 ml^{-1} viable cells in the unmodified controls (Teixido *et al.,* 2005). Additionally, the viability of *P. anomala* cells containing modified endogenous polyols and trehalose was significantly better than unmodified control cells when grown on a molasses-based growth medium, modified with NaCl and proline for 72 h (Fig. 10.3 and Mokiou, 2005). Studies with the cells of the yeast *C. sake* used for control of *Penicillium* rot of apples, modified by culture in weak nutrient yeast broth media modified with either glucose or glycerol to 0.96 a_w also demonstrated that a greater number of modified yeast cells were viable over a range of water availabilities (0.95 to 0.93 a_w) than unmodified yeast cells, which were significantly more sensitive (Teixido *et al*., 1998; Abadias *et al*., 2001).

Studies with entomogenous fungi have also demonstrated that significant improvements in germination could be obtained with conidia with modified endogenous reserves than with unmodified conidia of *Beauveria bassiana*, *M. anisopliae* and *P. farinosus* (Hallsworth and Magan, 1995; Chandler *et al*., 2005). However, by using isotonic solutions to harvest blastospores of *M. anisopliae,* Ypsilos and Magan (2004) were able to show significant improvements in viability, especially at lowered water availabilities. Fig. 10.4 shows the results of such improvement on germination of blastospores harvested in water and with isotonic solutions. In this study, ionic solutes were best, and NaCl and KCl were compared at different concentrations. This showed that under marginal conditions for germination the physiologically modified BCA propagules could survive better than unmodified controls. The interesting finding is that the isotonically harvested, modified spores were more stress tolerant that those harvested in water. This could have significant implications for formulations and final applications of BCAs in the field and give a crucial advantage to facilitating germination and establishment in a competitive environment in the phyllosphere.

In contrast, conidia of dematiaceous BCAs such as *E. nigrum* and *U. atrum*, where significant physiological modifications of endogenous reserves was possible, no improvement in germination was achieved under water stress treatment conditions. This suggested that larger, heavily pigmented spores of fungi originating from the harsher phyllosphere environment may already, to a large extent, have evolved to tolerate environmental stress.

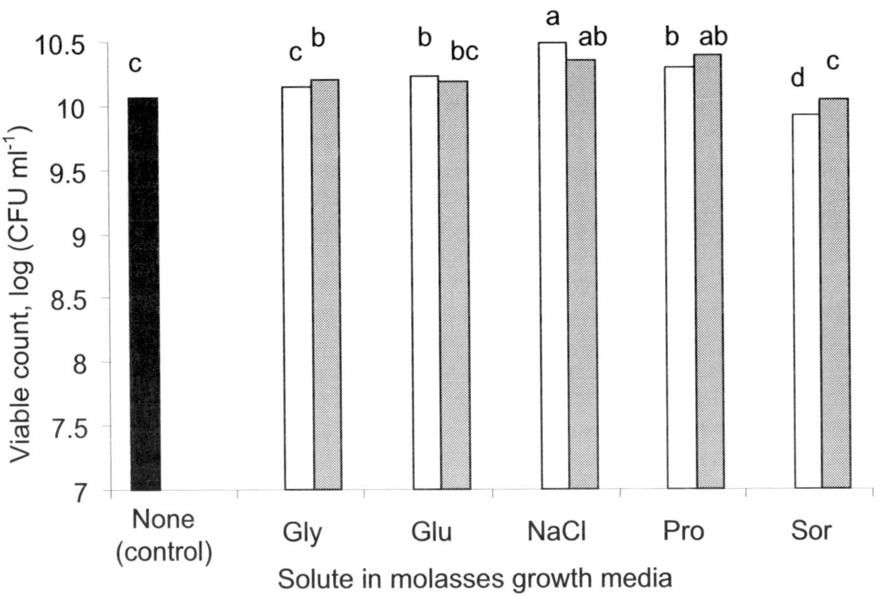

Fig. 10.3. Viability of *Pichia anomala* cells of each (molasses-based media) treatment on water stressed (0.96 aw) nutrient yeast dextrose agar media. Molasses media were unmodified or modified to 0.98 and 0.96 aw with PEG. Cells were incubated at 25°C for 72 h. Different letters indicate statistical differences ($P<0.05$) between means. Key to treatments: control, ■ ; 0.98 aw, □; 0.96 aw, ▒.

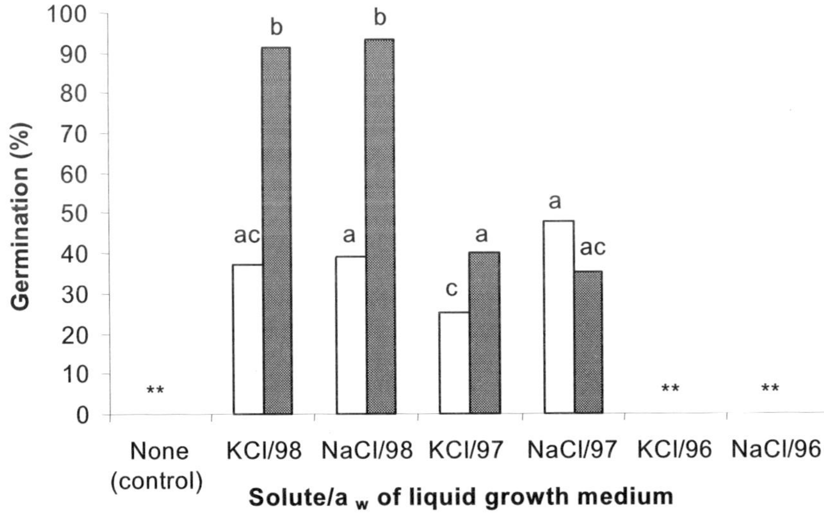

Fig. 10.4. Effect of harvesting with water (□) or isotonic solutions (▒) on germination of blastospores of *Metarhizium anisopliae* on unmodified (control) and water stressed agar media after 84 h at 25°C (from Ypsilos and Magan, 2004). Different letters indicate statistical differences ($P<0.05$) between means. Asterisks indicate no germination.

After germination, the next important phase is germ tube extension and establishment, or appresorium formation for some entomogenous species. Detailed studies were made under the different water stress regimes described previously, which demonstrated that germ tube extension of modified conidia of *E. nigrum* and *U. atrum* could be improved significantly under water stress. Indeed, in the case of *E. nigrum*, in some weak media, this improvement was maintained during *in vitro* mycelial colony development (Pascual, 1998). Table 10.1 shows an example of the improved germ tube extension observed with endogenously characterised conidia of *E. nigrum* obtained from colonies grown on low a_w-stress media.

Trehalose is an important sugar for survival of water stress/desiccation tolerance by replacing water in dehydrated phospholipid membranes. This inhibits transition of the liquid crystalline phase to the gel phase, and in so doing preserves the cell membranes (Crowe *et al.,* 1984). This is critical in desiccation tolerance and maintaining the integrity of cell membranes during wetting and drying cycles. Such tolerance is also an important parameter, especially when BCAs are being formulated as a wettable powder. Studies by Jin *et al.* (1991) with *Trichoderma harzianum* demonstrated that by manipulating the growth conditions with polyethylene glycol, the trehalose content of the conidia could be increased significantly and this enabled the inocula to survive desiccation better than unmodified control conidia. Although the PEG molecular weight and concentrations used unfortunately were not detailed, this study pointed to the critical importance of considering this approach where desiccation tolerance is imperative, especially in the phyllosphere. Trehalose may be more important in some groups of fungi than others. For example, the yeasts *C. sake* and *P. anomala* accumulated and synthesised trehalose rapidly (Teixido *et al.,* 1998; Mokiou, 2005), and this sugar was implicated in improved viability of cells at lowered a_w. Other studies by Jackson and Bothast (1990) and by Jackson and Schisler (1992) also pointed to variation in C:N ratios and limitation as a means of improving desiccation tolerance of mycoherbicides (e.g. *Colletotrichum truncatum*) and entomogenous species (e.g. *Paecilomyces farinosus*). Similar modifications in C:N ratio combined with water stress tolerance with *M. anisoplaea* has also shown similar improvements in ecological competence (Ypsilos and Magan, 2005). It needs to be recognised that the effect of ecophysiological manipulation on endogenous accumulation of reserves varies with fungal species and groups, and that the type of modification must be appropriate for the purpose. For example, if increased glycerol is required, solute stress modifications may be important; whilst C:N limitation and ratios may affect and increase trehalose accumulations within inocula. The use of isotonic solutions to further conserve the endogenous modifications achieved may enable better quality of BCAs to be produced, and which may have more consistent efficacy when applied to control diseases, or indeed pests, in the phyllosphere.

Recent studies by Wilson and Lindow (1994a, b) with bacterial BCAs has also suggested that competitiveness of individual strains could be estimated with the *in vitro* Niche Overlap Index (NOI) derived from the carbon utilisation profiles (NOI is the proportion of the C-compounds utilised by one species that are also utilised by the competing species). They hypothesised that the effectiveness of a strain as a pre-emptive biocontrol agent of epiphytic phytopathogenic bacteria was proportional to the ecological similarity between the BCA and the target pathogen in the phyllosphere of the host plant. They suggested that NOIs >0.9 implied that two competing species could occupy the same niche, while those with <0.9 occupied separate niches. This approach has been used to examine unmodified and modified conidia of *E. nigrum*. Table 10.2 shows the Niche size and NOI for *E. nigrium* against *M. laxa*. At reduced a_w (0.985 a_w treatment), *E. nigrum* was able to utilise a larger percentage of the C-sources used by *M. laxa* than the unmodified conidia (0.996 a_w treatment). NOI values >0.90 are indicative of competitive co-occupation. This suggests that endogenous modifications of the polyols and sugars can also modify competitiveness based on the NOI system previously used so effectively by Wilson and Lindow (1994a) for bacterial BCA selection.

Table 10.1. Mean percentage spore germination of *Epicoccum nigrum* (treatments 0.98, 0.995 a_w) and germ tube extension on water agar at 0.996 a_w and 0.935 a_w at different incubation times. Data are means of three replicate experiments (from Pascual, 1998).

0.996 a_w

	4 h		6 h	
Inoculum	Germination (%)	Germ tube extension (μm)	Germination (%)	Germ tube extension (μm)
E. nigrum 996	87.3	14.4	98.0	32.0
E. nigrum 98	83.0	11.7	97.3	33.7
LSD (P=0.05)	4.0	4.3	4.0	4.3

0.935 a_w

	8 h		12 h		23 h	
Inoculum	Germina -tion (%)	Germ tube extension (μm)	Germina -tion (%)	Germ tube extension (μm)	Germina -tion (%)	Germ tube extension (μm)
E. nigrum 996	51.3	5.4	82.0	9.9	98.7	38.2
E. nigrum 98	32.7	5.5	80.0	12.0	95.3	78.0
LSD (P=0.05)	7.5	1.7	7.5	1.7	7.5	1.7

This approach has been used to understand the interactions between mycotoxigenic *Fusarium* and *Aspergillus* species causing ear rot of maize grain and other mycobiota. Interestingly, it was found that the NOIs varied depending on both water availability and temperature (Marin *et al.,* 1998; Lee and Magan, 2000). However, practically no other studies have examined whether manipulation of endogenous reserves might affect the NOIs of potential BCAs, in relation to environmental factors. This approach may be one method for ensuring that, by simulation of environmental factors relevant to the phyllosphere environment, appropriate BCAs can be chosen for specific targets and for a better chance of success.

Formulations Based on Physiologically Modified BCAs

Predominantly, BCAs are applied as fresh inoculum for control of diseases and pests in terrestrial ecosystems generally. However, for their effective exploitation and commercialisation it is essential that formulations are produced which are as effective as fresh cells. Thus conservation of quality is a critical issue in producing formulations of

BCAs. They also need to not only have a reasonable shelf-life and viability of approximately 6-12 months but also perform at least similarly to fresh inoculum. This area of research has received attention in the last decade but some of the work has neglected the quality of inocula and appropriate formulations which are fit for purpose (see Wraight *et al.*, 2001).

Table 10.2. Niche Overlap Indices (NOI) for *Epicoccum nigrum* paired with *Monilinia laxa* derived from carbon source utilisation patterns at two water activity (a_w) levels at 25°C. Data are means of three experiments (from Pascual, 1998).

		Water activity level				
		0.996			0.985	
Conidial Type	Niche size[a]	NOI *M. laxa*[b]	NOI *E. nigrum*[c]	Niche size	NOI *M. laxa*	NO *E. nigrum*
E. nigrum 996	44	0.85	0.66	34	0.76	0.71
E. nigrum 98	45	0.85	0.64	46	0.90	0.57

[a] Per 96 carbon sources. The niche size for *M. laxa* was 34 at 0.996 a_w and 29 at 0.985 a_w.
[b] NOI $_{M. laxa}$ represents the proportion of the carbon sources on which *M. laxa* grew and were also utilised by *E. nigrum.*
[c] NOI $_{E. nigrum}$ represents the proportion of the carbon compounds on which growth of *E. nigrum* occurred that were also utilised by *M. laxa*.

Studies at Cranfield have concentrated on examining the potential of (a) using wet pastes of modified BCAs in refrigerated storage and (b) the use of fluidised bed drying systems. Initial work with concentrated pastes of *U. atrum* and entomogenous fungi showed some promise as a means of refrigerated storage of endogenously characterised spores/blastospores (Frey, 1998; Ypsilos, 2005). However, viability declined rapidly under both freely available water conditions and under water stress conditions. Thus alternative methods were examined.

Studies were performed with modified BCAs by drying in a fluidised bed at different temperatures (30-80°C) and for different times (10-30 min). This showed that extruded cells or formulations of yeast cells (*P. anomala*, *C. sake*) could survive 50-60°C effectively, while *U. atrum* spores could survive 80°C. Fig. 10.5 shows an example of the data obtained on viability of *P. anomala* cells dried at various temperatures for 10 or 20 min. This shows that viability was significantly decreased only at 70°C (Mokiou, 2005). *U. atrum* could survive 80°C for up to 30 min without loosing significant viability. This may partially be because of the highly pigmented nature of the spores. The question arises as to how effectively these formulations may survive at ambient temperatures. The viability of *U. atrum* spore treatments for 2 (60-80°C) and 6 months (30-50°C) showed that viability of these spores was retained at a level which was as good as fresh spores when compared on water agar at 0.995 a_w and on media with imposed water stress (0.95 a_w). Recent studies have shown that it is possible to combine the use of isotonic solutions for harvesting characterised BCAs with specific additives (e.g. skimmed milk powder; different carbohydrates) and to extrude this prior to fluidised bed drying (Mokiou, 2005). This produces powder formulations which have been shown to survive for up to 12 months with excellent viability of both *P. anomala* and spores of *U. atrum*. After fourteen months refrigerated storage (+4°C) of cottonseed flour based formulations of *P. anomala*, the viability in a pilot scale trial to control spoilage of wheat by *P. roqueforti* was 63% of total cell number, compared to 72% for commercially available dried bakers' yeast (Druvefors *et al.*, 2006). Furthermore, the biocontrol activity of this yeast formulation was also maintained. *P. roqueforti* in the treatment with rehydrated yeast was log 1.9 ± 0.2 mould CFU g^{-1} after 14 days incubation in the presence of air, compared to average values of log 4.5 ± 0.5 mould CFU g^{-1}. Yeast levels reached log 7.9 ± 0.1, which is similar to values obtained using fresh yeast cells.

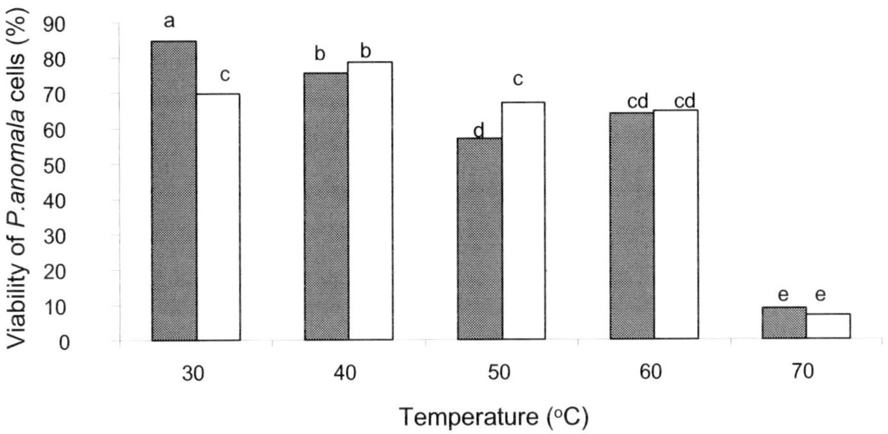

Fig. 10.5. Effect of fluidised bed drying on viability of *Pichia anomala* cells after different time periods at different temperatures (from Mokiou, 2005). Different letters indicate significant (P<0.05) differences between treatments. Key to treatments: 10 min (▩); 20 min (□).

Improvements in Biocontrol in the Field

The critical component of this strategy is whether biocontrol can be either conserved or improved under a range of environmental factors. To this end, a series of studies have tested the low a_w-stress inocula described previously.

Studies with field spraying of different inocula of the yeast *C. sake* on apples in orchards prior to harvest and storage have demonstrated two important things. First, that low-a_w tolerant yeast cells grow better on the apple surface, and by such establishment can give as good if not better post-harvest control of *Penicillium* rot (Fig. 10.6) (Teixido *et al.*, 1998a, b). This work also suggested that the low a_w-tolerant inocula could be applied at a lower concentration to obtain the same efficacy. This could be important for the development of economic production, formulation and application systems for BCAs.

Application of *E. nigrum* inocula to peach twigs in the field has also demonstrated that better control of brown rot could be achieved with endogenously modified spores than with unmodified inocula or the fungicide captan. This BCA was also examined for post-harvest control of *B. cinerea* on cherries under different storage humidity regimes. In this case, good control was achieved at different relative humidities, but there were no statistically significant differences between inocula.

Studies with *U. atrum* for control of *B. cinerea* suggest that manipulation of endogenous nutrient reserves of spores of *U. atrum* can modify the establishment on leaf surfaces, and also affect the levels of biocontrol achieved. Improved suppression of sporulation by *B. cinerea* was achieved by the modified inocula when compared to freshly produced spores.

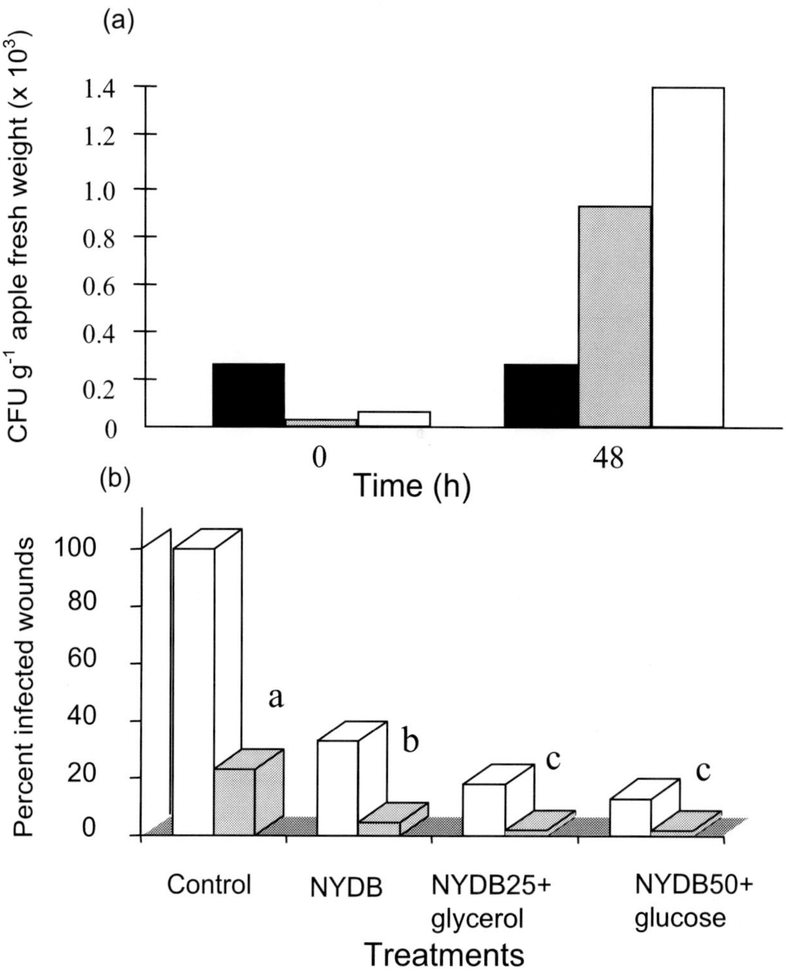

Fig. 10.6. Comparison of the effect of ecologically competent inocula of *Candida sake* CPA1 on (a) establishment of cells on surface of apples pre-harvest and (b) effect on control of *Penicillium* rot in post-harvest storage. In (a) NYDB, nutrient dextrose broth (■); 25% (▩), 50% (□), diluted medium with glucose. In (b) (□), infected wounds and (▩), lesion diameter. Different letters indicate significant differences ($P<0.05$).

Recently formulations of *P. anomala* were applied to moist grain to examine the efficacy for control of a mycotoxigenic spoilage mould (*P. verrucosum*) and a microaerophilic mould (*P. roqueforti*) which can dominate in sealed stored moist wheat grain. Long-term studies showed that the best formulation (based on cottonseed flour) was as effective as fresh cells in controlling total moulds and *P. roqueforti* over a period of 14 months storage. Four different formulations were also examined for controlling *P. verucosum* at different wheat moisture contents, where this mycotoxigenic mould was also used as an inoculant. In these studies, especially at 0.93 a_w in the presence of air some formulations were effective at controlling growth and, more importantly, ochratoxin production (Fig. 10.7; Mokiou and Magan, 2005).

These data have thus demonstrated that the four questions posed can indeed be answered positively, which suggests that potential does exist in this approach. It should be noted, however, that while the examples presented are not numerous, few other groups have examined this approach for improving ecological fitness of BCAs for improved field efficacy.

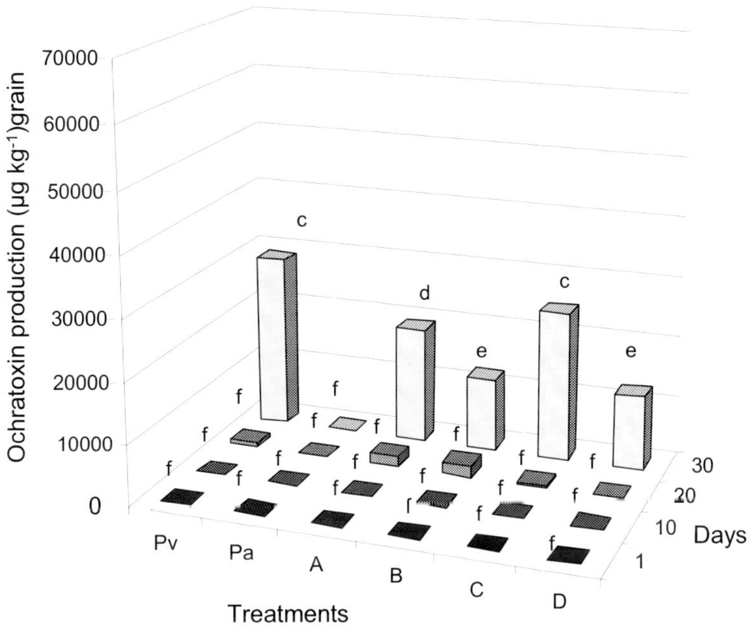

Fig. 10.7. Effect of *Pichia anomala* on ochratoxin A (OTA) contamination of grain at 0.93 a_w co-inoculated (A, B, C, D) with the spoilage mould *Penicillium verrucosum* on wheat grain stored for up to 30 days at 25°C. Pv, *Penicillium verrucosum* alone; Pa, *Pichia anomala* alone; A, unmodified cells; B, proline modified; C, proline + cottonseed flour + skimmed milk in water; D, same as C plus isotonic solutions. Different letters indicate significant differences ($P<0.05$; from Mokiou and Magan, 2005).

Future Prospects for Improving Commercialisation of Ecologically Fit Inocula

Physiological manipulation could have a significant impact on the efficacy and reliability of microbial BCAs. Such BCAs can give better disease and pest control in unfavourable phyllosphere environments, which have significant diurnal fluctuations in humidity. It is important that sufficient attention is paid to the quality of BCAs produced for commercial biopesticides, alongside efforts to maximise yields of inocula in mass production. Further research is required to improve understanding of the ecophysiological behaviour of inocula in the target environment, and to investigate how inoculum quality can be optimised in commercial production systems without adversely affecting yield. In this regard, it may be instructive to draw lessons from work in the plant seed industry designed to improve the uniformity and germination behaviour of seed batches with techniques such as seed priming, and to understand the effects of the environment on seedling emergence using population-based models (Finch-Savage, 2004).

Whilst there is undoubtedly a requirement to improve the quality of inocula used as BCAs, it is important to remember that the active constituents are living organisms. As such it is unrealistic to expect them to perform to the same levels of efficacy as conventional, chemical pesticides. Often the development of BCAs tends to follow the chemical pesticide model, and while this approach has certain benefits, it has some serious negative consequences in that the microbial BCAs are perceived as chemical analogues, with false expectations of chemical-like performance. Thus for enhancing BCA performance in the field an integration of knowledge of technological improvements with microbial ecology is required. The use of isotonic solutions for harvesting BCAs, coupled with improvement of effective formulations, should pay dividends in trying to improve field performance and consistency, especially in the phyllosphere. Formulations must have a shelf-life of 6-12 months and certainly fluidised bed drying offers this option as it is commercially available and is cheaper than freeze drying. A fundamental understanding of the endogenous changes and accumulations of polyols and sugars in inocula of BCAs can also enable information to be obtained on any modifications to internal water and solute potentials. This can be combined with exogenous additions of specific compounds, or mixtures of compounds, to maintain the concentrations in the inocula for conserving viability, and for long-term stability of formulations which have ecological competence. This may be a significant way forward to develop ecologically competent microbial BCAs which can perform effectively in the phyllosphere.

Acknowledgements

I am very grateful to all my past and present research students and collaborators for their commitment and support in this area of research: Dr J.E. Hallsworth, Dr S. Pascual, Dr N. Teixido, Dr M. Abadias, Dr M. Anderson, Dr S. Sancisi-Frey, Dr I. Ypsilos, Dr S. Mokiou, Mrs S. Swanwick, Dr D. Chandler, Dr P. Melgarejo, Prof. I. Vinas and Dr J. Usall.

References

Abadias, M., Teixido, N., Usall, J., Vinas, I. and Magan, N. (2001) Impact of physiological manipulation of endogenous reserves in the biocontrol yeast *Candida sake* on water stress tolerance. *Canadian Journal of Microbiology* 47, 123-129.

Butt, T.M., Jackson, C. and Magan, N. (2001) *Fungi as Biocontrol Agents: Progress, Problems and Potential.* CABI, Wallingford, UK.

Chandler, D., Anderson, M. and Magan, N. (2005) A population based threshold model to demonstrate that physiological manipulation of endogenous reserves can increase the virulence of insect pathogenic fungi. *BCPC Crop Science and Technology* 4B, 309-314.

Crowe, J.H., Crowe, L.M. and Chapman, D. (1984) Preservation of membranes in anhydrobiotic organisms: the role of trehalose. *Science* 223, 701-703.

Frey, S. and Magan, N. (1998) Improving quality and quantity of *Ulocladium atrum* for enhanced biological control of *Botrytis cinerea. Proceedings of Brighton Crop Protection Symposium – Pests and Diseases 4D*, 305-306.

Hallsworth, J.E. and Magan, N. (1994) Effect of carbohydrate type and concentration on polyhydroxy alcohol and trehalose content of conidia of three entomopathogenic fungi. *Microbiology-UK* 140, 2705-2713.

Hallsworth, J.E. and Magan, N. (1995) Manipulation of intracellular glycerol and erythritol enhances germination of conidia at low water availability. *Microbiology-UK* 141, 1109-1115.

Hallsworth, J.E. and Magan, N. (1996) Effect of temperature, pH and culture age on polyol accumulation in conidia of fungal propagules. *Applied and Environmental Microbiology* 62, 2435-2447.

Jackson, M.A. and Bothast, R.J. (1990) Carbon concentration and C-N ration influences submerged culture conidiation by the potential bioherbicide *Colletotrichum truncatum* NRRL 137371. *Applied and Environmental Microbiology* 56, 3435-3438.

Jackson, M.A. and Schisler, D.A. (1992) The composition and attributes of *C. truncatum* spores are altered by the nutritional environment. *Applied and Environmental Microbiology* 58, 2260-2265.

Jackson, M.A., McGuire, M.R., Lacey, L.A. and Wraight, S.P. (1997) Liquid culture production of desiccation tolerant blastospores of the bioinsecticidal fungus *Paecilomyces fumosoroseus*. *Mycological Research* 101, 35-41.

Jin, X., Harman, G.A. and Taylor, A.G. (1991) Conidial biomass and desiccation tolerance of *Trichoderma harzianum* produced at different water potentials. *Biological Control* 1, 237-243.

Köhl, J. and Molhoek, W.M.L. (2001) Effect of water potential on conidial germination and antagonism of *Ulocladium atrum* against *Botrytis cinerea*. *Phytopathology* 91, 485-491.

Lane, B.S., Trinci, A.P.J. and Gillespie, A.T. (1991) Endogenous reserves and survival of blastospores of *Beauveria bassiana* harvested from carbon and nitrogen-limited batch cultures. *Mycological Research* 95, 821-828.

Lee, H.B. and Magan, N. (1999) Environmental factors and nutritional utlization patterns affect niche overlap indices between *Aspergillus ochraceus* and other spoilage fungi. *Letters in Applied Microbiology* 28, 300-304.

Magan, N. (2001) Physiological approaches to improving ecological fitness of biocontrol agents. In: Butt, T., Jackson, C. and Magan, N. (eds), *Fungal Biocontrol Agents: Progress, Problems and Potential*, CABI, Wallingford, UK. pp. 239-252.

Magan, N. (1997) Fungi in extreme environments. In: Wicklow D.T. and Soderstrom, B. (eds) *Mycota* Vol. IV. *Environmental and Microbial Relationships*, Springer Verlag, Berlin, Germany. pp. 99-114.

Marin, S., Sanchis, V., Ramos, A.J., Vinas, I. and Magan, N. (1998) Environmental factors, *in vitro* interspecific interactions, and niche overlap *between Fusarium moniliforme*, *F. proliferatum* and *F. graminearum*, *Aspergillus* and *Penicillium* spp. isolated from maize grain. *Mycological Research* 102, 831-837.

Matawele, P., Trinci, A.P.J. and Gillespie, A.T. (1994) Mutants of entomopathogenic fungi that germinate and grow at reduced water activities and reduced relative humidities are more virulent to *Nephotettix virescens* (green leafhopper) than the parental strain. *Mycological Research* 98, 1329-1333.

Mokiou, S. (2005) Ecophysiological approaches to production and formulation of the biocontrol yeast *Pichia anomala*. PhD Thesis, Institute of BioScience and Technology, Cranfield University, Silsoe, Bedford, UK.

Mokiou, S. and Magan, N. (2005) Control of spoilage and ochratoxin A production in moist grain for animal feed using the biocontrol agent *Pichia anomala*. *BCPC Crop Science and Technology* 2A, 115-120.

Neschi, A., Etcheverry, M. and Magan, N. (2004) Osmotic and matric potential effects on growth and compatible solute accumulation in *Aspergillus* section *Flavi* strains from Argentina. *Journal of Applied Microbiology* 96, 965-972.

Pascual, S. (1998) Studies on production and ecophysiology of fungal antagonists for improved biological control of *Fusarium oxysporum* f.sp. *lycopersici* and *Monilinia laxa*. PhD Thesis, Institute of BioSciences and Technology, Cranfield University, UK.

Pascual, S., Magan, N. and Melgarejo, P. (1998) Improvement of growth and biocontrol ability of *Epicoccum nigrum* under water stress. Proceedings of 7th International Congress of Plant Pathology, Edinburgh. p. 6-52.

Pascual, S., Melgarejo, P. and Magan, N. (2000) Production of the biocontrol agents *Epicoccum nigrum* by solid substrate fermentation: effect of water activity on accumulation of compatible solutes. *Mycopathologia* 146, 83-89.

Teixido, N., Vinas, I., Usall, J. and Magan, N. (1998) Improving ecological fitness and environmental stress tolerance of the biocontrol yeast *Candida sake* by manipulation of intracellular sugar alcohol and sugar content. *Mycological Research* 102, 1409-1417.

Teixidó, N., Cañamás, T., Usall, J., Torres, R., Magan, N. and Viñas, I. (2005) Accumulation of the compatible solutes, glycine-betaine and ectoine, as an osmotic stress adaptation and heat shock cross-protection in the biocontrol agent *Pantoea agglomerans* CPA-2. *Letters in Applied Microbiology* 41, 248-252.

Van Laere, A. (1989) Trehalose, reserve and/or stress metabolite. *FEMS Microbiology Reviews* 63, 201-210.

Wilson, M. and Lindow, S.E. (1994a) Coexistence among epiphytic bacterial populations mediated through nutritional resource partitioning. *Applied and Environmental Microbiology* 60, 4468-4477.

Wilson, M. and Lindow, S.E. (1994b) Ecological similarity and coexistence of epiphytic ice-nucleating (Ice+) *Pseudomonas syringae* strains and a non-ice-nucleating (Ice-) biological control agent. *Applied and Environmental Microbiology* 60, 3128-3137.

Ypsilos, I. (2005). Impact of water relations on the production, ecological fitness and stability of the biocontrol fungus Metarhizium anisopliae. PhD Thesis, Institute of BioScience and Technology, Silsoe, Cranfield University, UK.

Ypsilos, I.K. and Magan, N. (2004) The effect of water activity, modifying solute and culture age on *Metarhizium anisopliae* blastospore production, quality and germinability. *Mycological Research* 108, 1337-1345.

Ypsilos, I.K. and Magan, N. (2005) Characterisation of optimum cultural environmental conditions for the production of high numbers of *Metarhizium anisopliae* blastospores with enhanced ecological fitness. *Biocontrol Science and Technology* 15, 683-699.

Chapter 11

Compost Teas: Alternative Approaches to the Biological Control of Plant Diseases

W.F. Mahaffee[1] and S.J. Scheuerell[2]
[1]United States Department of Agriculture, Agriculture Research Service, Horticulture Crops Research Laboratory, 3420 NW Orchard Ave, Corvallis, Oregon 97330, USA; [2]Evergreen State College, 2700 Evergreen Parkway NW, Olympia, Washington 98505, USA
mahaffee@science.oregonstate.edu

Introduction

Plant disease control in modern agriculture has generally been based on the paradigm of one active ingredient used to target one or multiple pathogens. This approach has generated numerous successes and some failures (e.g. resistance development), while enhancing agricultural productivity. However, there is an increasing concern by the general public over the use of synthetic pesticides and their potential health and environmental impact. This concern has led to very rapid increases in the demand for organic production and foods produced with little or no synthetic pesticides (Dimitri and Greene, 2002).

As a consequence, there have been intense efforts to develop biological control agents for disease control over the past few decades, (Adams, 1990; Johnson and Stockwell, 1998; Sturz *et al.*, 2000; Paulitz and Belanger, 2001; Stewart, 2001; Janisiewicz and Korsten, 2002). In most cases, these research efforts have followed the same paradigm, a single organism is identified and developed for delivery into the agroecosystem. This approach has resulted in several commercial successes (e.g. Galtrol, BlightBan, Kodiak, PlantShield) (Backman *et al.*, 1997; Harman, 2000; Stewart, 2001; McSpadden Gardener, 2005). However, the use of biological control agents has also developed a reputation for inconsistent disease control (Deacon, 1994; Stewart, 2001; Shtienberg and Elad, 2002).

A multitude of reasons have been put forward for this inconsistency, including numerous biotic and abiotic factors that impact the biocontrol agent's ability to survive the application process, attach to, grow, and reproduce on the plant and inhibit pathogenesis. However, they all are related to the inability of a single organism to be all things at all times in an extremely dynamic and harsh environment, and that an organism placed into a competitive environment rarely realises its full niche potential (Deacon, 1994).

Thus, alternative strategies based on the delivery of multiple organisms at the same time have been pursued. Three general types of approaches have been taken: (i) the introduction of uncharacterised microbial communities with no known activity (Litterick *et al.*, 2004); (ii) the mixing of several known biological control agents with different modes of action or that colonise different ecological niches (Raupach and Kloepper, 1998; Guetsky *et al.,* 2002; Jacobsen, this volume); and (iii) the enhancement of indigenous populations existing on or around the plant (Davis *et al.*, 1992; Kokallis-Burelle *et al.*, 1992). Each of these approaches has met with some success, but they still suffer from the inconsistency associated with traditional biological control measures.

This review will focus on the induction of uncharacterised microbial communities through the use of compost teas. Due to the scarcity of peer-reviewed literature on this topic, this review will incorporate non-peer-reviewed literature and popular press accounts to indicate areas that show promise and are in need of scientific investigation.

Compost Teas

Agricultural and horticultural producers, home owners, and landscape managers are looking for viable alternatives for disease control due to the expense, lack of commercial organic formulations, and the inconsistency associated with traditional biological control agents. The introduction of complex microbial communities through approaches such as the application of compost, compost teas, and plant extracts/fermentations is being perceived as providing useful alternatives. The use of compost for the control of diseases has been extensively investigated (reviewed in Hoitink, 2004; Litterick *et al.*, 2004; Noble, 2005). However, the use of compost teas and plant extracts/fermentations have received far less scientific attention (Scheuerell and Mahaffee, 2002; Litterick *et al.*, 2004), despite increased demand for this information. Thus, the research done on this topic has been overshadowed by a proliferation of misinformation arising from the attempt of a few individuals to take advantage of this void.

The use of compost teas for disease control is not new. It has likely been around since the time of the Romans (Koepf, 1992). Despite this apparently long history, recently there has been a rapid increase in popular interest in the use of compost teas for improving plant health and disease control (Scheuerell and Mahaffee, 2002; Litterick *et al.*, 2004). This interest has arisen in two main groups of people, agricultural producers and managers of landscapes at the home or municipal level. The increased interest has been driven by either the lack of certified organic options for disease control, or reduced crop and soil health in conventional fields in spite of increased synthetic inputs. These issues force producers to experiment with any method that shows any sign of promise or that fits into their philosophical framework. For managers of landscapes, including professional landscapers, golf course managers, municipal parks and recreation departments, and home gardeners, interest in the use of compost tea has coincided largely with increased public interest in organically produced food and the desire to live and recreate in non-toxic environments. This shift in public perception has been facilitated further by public exposure to these concepts by garden writers (Robertson, 2000).

From both groups of people, there are an abundant number of anecdotal success stories relating to the utility of compost teas for disease control, but very few scientific investigations. This has led to the development of a rather large compost tea related industry that makes claims of plant disease control at levels that synthetic chemicals have not achieved (Ingham, 2003). Meanwhile, a number of individuals and companies are selling unproven compost tea at farmers' markets, retail outlets, internet sites and through application services.

Terminology

There continues to be a great deal of confusion associated with the terminology used to describe the production of compost teas and related products, resulting in little consistency among the various groups that utilise compost teas. For instance, Scheuerell and Mahaffee (2002) define compost extract as the filtered product of compost mixed with any solvent, but not fermented (i.e. incubated in any manner to allow for microbial growth), generally used for analytical or assay work. Diver (2002), on the other hand, defines compost extract as a compost-water slurry allowed to sit for 7-14 days, which Scheuerell and Mahaffee (2002) term a non-aerated compost tea. There are also a multitude of terms that attempt to divide aerated compost teas into sub-groups (Ingham, 2003) based on individual perceptions of the ability of human pathogens to reproduce during fermentation or their potential for growth under reduced oxygen. In order to avoid confusion, Table 11.1 presents these terms along with the definitions that will be used throughout this chapter and a list of terms considered by the authors to be synonyms. A more through review of these terms and their origins can be found in Scheuerell and Mahaffee (2002).

Table 11.1. Terms and definitions and synonyms used in relation to compost teas and their production.

Term	Definition	Definition
Compost tea	a mixture of water and compost that has been incubated for a period of time to allow for microbial reproduction	watery fermented compost extract, compost extract, organic tea, amended extracts, steepages, slurries
Aerated compost tea	any method in which the compost/water mixture is actively aerated during the fermentation process	put-to-sleep teas
Non-aerated compost tea	methods that do not disturb or only minimally disturb the fermentation after initial mixing	passive brewing or tea, European tea, fermentative compost tea
Compost leachate	water that has been passed through or collected from compost piles	compost runoff
Fermentation	the cultivation of microorganisms	brewing
Nutrient supplements	any compound added to a fermentation to enhance microbial growth	

Methods for Producing Compost Teas

There are almost an infinite number of methods by which compost teas are produced, however, nearly all can be grouped according to whether or not they are actively aerated. All compost teas start by placing compost in contact with a volume of water (1 part compost to 3-50 parts water) in a vessel, and fermenting for a period of time. The compost can be mixed in the water, placed in bags or similar containers and suspended in the water, or have the water passed through it. The amount of compost used is largely dependent on the size of the fermentation vessel. However, based on the work of Yohalem *et al.* (1996b) it appears that a compost sample size of at least 500 g is required to avoid sampling error when making non-aerated compost teas and this minimum is likely to be required for aerated tea production as well.

Non-aerated compost teas, generally of a 1:3-10 v/v ratio of compost to water, are produced using methods that do not attempt to maintain oxygen levels in the water. Typically, these methods allow the vessel to remain minimally disturbed for 1-120 days, typically 7-14 days, at ambient temperature, prior to filtering out the particulate matter and utilizing the tea (Weltzien, 1991). Derivations include periodic stirring of the mixture (e.g. stirring for 5-10 min every 1-5 days (Brinton *et al.*, 1996)) and/or the addition of nutrients (see below). They can be produced very inexpensively anywhere there is access to potable water. However, they can produce offensive odours when simple sugars or similar nutrients are added during fermentation.

Aerated compost teas, generally of a 1:10-50 v/v ratio of compost to water, are produced using a constant mechanical method of aeration to increase the potential of high oxygen levels in the water/compost mixture. A multitude of methods are used to inject air into the mixture, including showering of recirculated water over an open tank or through a porous bag of compost suspended over the tank (Cantisano, 1995; Riggle, 1996; Merril and Mckeon, 2001), injecting air through a hollow propeller shaft (Scheuerell and Mahaffee, 2004; Soilsoup.com), Venturi nozzles (Composttea.com), vortex nozzles (Scheuerell and Mahaffee, 2004), aquarium stones (Ingham, 2000a), fine bubble diffusion mats (Growing Solutions Inc. Compara, Inc.) or perforated tubing (KIS Inc.). Nutrients are generally added at the start of the fermentation. The main advantage afforded by aerated production methods is the reduction of the fermentation to 12-24 h for most situations (Ingham, 2003), although longer fermentations have been used (Cantisano, 1998). There is a growing number of companies offering units designed to produce aerated compost teas (Diver, 2002). Depending on the choice of equipment and nutrients, aerated compost teas can be very expensive to produce, particularly at larger scales with commercially sold brewers.

Control of Foliar and Fruit Diseases

Compost teas have been tested for suppression of a variety of foliar and fruit diseases with mixed results (reviewed in: Scheuerell and Mahaffee, 2002; Litterick *et al.*, 2004). Examples of diseases significantly suppressed *in planta* using compost tea applications include bacterial speck (Zang *et al.*, 1998), bacterial spot (Al-Dahmani *et al.*, 2003), early blight (Tsror, 1999), grey mould (Weltzien, 1989; Ketterer *et al.*, 1992; Elad and Shtienberg, 1994; McQuilken *et al.*, 1994; Scheuerell and Mahaffee, 2006;), leaf blight (Weltzien, 1989), Northern blight (Yohalem *et al.*, 1994), powdery mildew (Weltzien, 1989; Sackenheim, 1993; Elad and Shtienberg, 1994; Scheuerell, 2002;), shoot blight (Yohalem *et al.*, 1994). While many would probably conclude that the disease suppression observed in most of these studies is not sufficient for commercial production, there are numerous non-standard production systems (e.g. organic, biodynamic) for which these results are far better than the alternative of no control options.

Control of Root Diseases

There has been little research conducted on soil-borne disease suppression with compost tea applications for either aerated or non-aerated fermentations. Again, compost teas in soil systems are widely used by tea producers in an attempt to improve "soil health" by enhancing microbial diversity and providing various nutrients (Ingham, 2003; Anon., 2004). Whether the soil microbiota as a whole is altered is highly debatable, but there are very clear indications that the applications of compost tea drenches do have some disease suppressive effects, and therefore, affect at least one member of the soil microbial community. There are only a few studies that have investigated the efficacy of compost tea applications for control of soil-borne diseases.

Tränkner (1992) demonstrated that non-aerated teas significantly reduced *Pythium* damping-off of pea seedlings. They also significantly increased seed germination, root length, and root dry weight when seeds were soaked in the tea for 30 min then re-dried prior to sowing. Ma *et al.* (1999, 2001) demonstrated that *Fusarium* wilt of greenhouse grown pepper (*F. oxysporum* f. sp. *vasinfectum*) and cucumber (*F. oxysporum* f. sp. *cucumerinum*) was controlled significantly by non-aerated compost tea drenches. *In vitro* assays indicated that *Fusarium* microspores and chlamydospores were lysed when exposed to the non-aerated compost tea. In examining the efficacy of aerated and non-aerated compost teas for control of *Pythium* damping-off of cucumber in soil-less media, Scheuerell and Mahaffee (2004) found that both aerated and non-aerated teas significantly reduced disease development, with an aerated compost tea fermented with kelp and humic acid nutrients exhibiting the most consistent disease suppression.

These results indicate that the use of compost teas as soil drenches could be a useful control measure in the management of root diseases in soil-less production systems. However, it is unclear if this will translate to disease control in field soil systems. Since the majority of compost tea use is directed towards to improving "soil health" and control soil-borne diseases in row and turf production systems, this potential deserves far more scientific investigation.

Factors Affecting Disease Control

Aeration

There is debate over whether aeration produces a superior compost tea (Brinton *et al.*, 1996, 2005; Ingham, 1999, 2000b, 2003). Ingham (2000b, 2003) states that non-aeration methods yield a tea that is less effective at disease control. Conversely, the majority of the scientific literature indicates that non-aerated teas effectively control disease (reviewed in Litterick *et al.*, 2004). There have only been a few studies that have directly compared aerated and non-aerated teas made from the same compost source. Cronin *et al.* (1996) found that a non-aerated compost tea inhibited germination of *Venturia inaequalis* conidia while the aerated compost tea did not. However, conidial inhibition was induced after the aerated tea was allowed to incubate for an additional 7 days without aeration. Recently, Scheuerell and Mahaffee compared aerated and non-aerated methods for disease control of grey mould of geraniums (Scheuerell and Mahaffee, 2006), *Pythium* damping-off of cucumber (Scheuerell and Mahaffee, 2004), and powdery mildew, rust and blackspot of rose (Scheuerell, 2002), and found no differences in disease control due to aeration in any of the pathosystems examined. Similarly, Al-Dahmani *et al.* (2003) compared aerated and non-aerated compost tea fermentations to a true anaerobic production method (compost mixed with water and placed in anaerobic jar with O_2 absorbers and CO_2 indicators for 7 days) for control of bacterial spot of tomato and observed no differences in disease control. Thus, it appears that aeration has no effect on disease control.

The popular literature is also replete with assertions that non-aerated methods result in phytotoxicity and are highly conducive for the regrowth of human pathogens. Again, there appears to be no documented evidence to substantiate these claims. In numerous studies (Weltzien and Ketterer, 1986a,b; Weltzien *et al.*, 1987; Weltzien, 1989, 1990, 1991; Al-Dahmani, 2003) there are no reports of phytotoxicity, nor have we observed phytotoxic symptoms when non-aerated compost tea was used as a foliar spray or potting mix drench in over 100 experiments (Scheuerell, 2002; Scheuerell and Mahaffee, 2004, 2006). There are also no biological grounds to substantiate the claim that low oxygen conditions are ideal for the growth of most human pathogens (Ingham, 2003). Actually, the literature indicates the opposite. Facultative anaerobes (e.g. species of *Escherichia*, *Klebsiella*, *Salmonella*,

Shigella), grow more rapidly under aerobic conditions than anaerobic conditions but possess the metabolic capacity to also grow in anaerobic environments (Murray, 1999).

Numerous factors other than aeration (Table 11.2) potentially can influence the efficacy of compost tea applications for disease control. However, few of these factors have been examined systematically for their impact on disease suppression as they relate to both aerated and non-aerated methods for tea production. The factors that appear to impact disease suppression are discussed below.

Table 11.2. Factors affecting disease suppression by compost tea applications

Fermentation Parameters	Application Parameters
Compost source	Adjuvants
- Feed stocks	- Surfactants
- Stability	- Stickers
- Moisture content	- UV stablizers
- Available nutrients	- Other microorganisms
	- Nutrients
Water source	Application Equipment
- Ratio	- Pump type
- Temperature	- Pressure
- Chlorination	- Nozzle type
Nutrient amendments	Concentration
- Type	Rate
- Quantity	Interval or frequency
- Type	
- Quantity	

Adapted from Scheuerell and Mahaffee (2002)

Compost Source

Compost source has, been shown, in most cases to be a significant factor in whether a tea is disease suppressive or not. Investigations of non-aerated production methods indicated that composts containing animal manure where more efficacious than those made with undigested vegetative matter (Weltzien, 1990, 1991; Andrews, 1993), while other research indicated that undigested vegetative matter was equally effective (Elad and Shtienberg, 1994; Scheuerell, 2002; Scheuerell and Mahaffee, 2004, 2006). Scheuerell and Mahaffee (2006) examined aerated and non-aerated compost teas made from 30 different composts for control of grey mould of geranium and found that disease suppression was associated with the particular batch of compost and not necessarily the feed stocks used to create the compost.

Compost Age

Compost age is a component of compost source that has been examined extensively for its impact on disease suppression with non-aerated compost teas. It appears that what constitutes a suitable age is dependent on the feedstocks used to make the compost and storage conditions (Weltzien, 1991; Tränkner, 1992; Andrews, 1993; Brinton *et al*., 1996) and is probably related to the biomass available to support microbial activity (i.e. maturity). When using compost directly as a soil amendment, there is a strong correlation between the level of microbial activity and potential for disease suppression (Chen *et al*., 1988; Inbar *et al.,* 1991; Boehm and Hoitink, 1992; Craft and Nelson, 1996). However, Al-Dahmani *et al,*. (2003) observed no effect of compost age on the disease suppressiveness of manure-based composts when 3, 5, 10, and 16-month-old composts were investigated. In an effort to remove compost age as a factor in disease suppression, Urban and Tränkner (1993) demonstrated that air dried compost was as effective as fresh compost at suppressing *B. cinerea.*

Fermentation Time

For both aerated and non-aerated compost teas, disease suppression tends to increase with fermentation time to a maximum and then declines (Ketterer, 1990; Weltzien, 1990; Ketterer and Schwager,1992; Cantisano, 1998). The time periods involved are likely to be related to the compost source and fermentation methods employed. *Botrytis* suppression on detached grape leaves treated with 1, 3, 7, and 14-day non-aerated compost tea fermentations made from three different composts reached a maximum at 7 days, while grape berry infection was suppressed maximally after 8 days of fermentation (site source). Similarly, Weltzien (1990) demonstrated that late blight (*P. infestans*) lesions on detached tomato leaves were suppressed maximally by 7 and 14-day non-aerated fermentations as compared to 1, 2 and 28-day fermentations. Only anecdotal evidence exists for the disease suppression afforded by aerated compost teas in relation to fermentation time. Ingham (1999, 2000b, 2003) claims that optimum aerated fermentation times coincide with the development of maximum active biomass levels during fermentation, generally achieved after 18-36 h. Conversely, Cantisano (1998) states that while 24 h fermentations are good for fertilisation, 7-14 days are need for maximum disease control. The trend within the user community is to use 18-24 h fermentations (Ingham, 2003).

Fermentation Nutrients

Very few aerated compost teas are produced without the addition of nutrients from sources external to the compost. There are a multitude of recipes that claim to result in growth of various microbial populations or specific bacteria:fungi ratios which, in turn, will supposedly yield specific effects (Ingham and Alms 1999; Ingham, 2003). Several makers of compost tea brew also sell proprietary formulations of various nutrients which tend to contain a sugar source (used to be molasses), soluble kelp, humic acid and other organic materials and minerals.

Numerous nutrients, sucrose (Urban and Tränkner, 1993; Sackenheim *et al.,* 1994; Scheuerell and Mahaffee, 2004;), malt (Ketterer, 1990), yeast extract (Urban and Tränkner, 1993; Scheuerell and Mahaffee, 2004), peptone, starch (Urban and Tränkner, 1993), nutrient broth (Elad and Shtienberg, 1994), humic acid, kelp, molasses, rock dust (Scheuerell and Mahaffee, 2004, 2006), have been examined for their effects on disease suppression when added to compost tea fermentations. The addition of nutrients has been shown to increase (Ketterer, 1990; Urban and Tränkner, 1993; Sackenheim *et al.,* 1994; Scheuerell and Mahaffee, 2004, 2006), decrease (Urban and Tränkner, 1993; Scheuerell and Mahaffee, 2004), or have no effect on disease suppression (Elad and Shtienberg, 1994).

In a systematic evaluation of the effects of nutrient amendments on the control of *Pythium* damping-off of greenhouse grown cucumber seedlings, Scheuerell and Mahaffee (2004) observed over repeated experiments that making compost tea with molasses resulted in erratic disease control. They concluded, based on a series of experiments in which simple sugars were added before and after fermentation, that the loss of disease control was due to residual nutrients in the compost tea. The residual nutrients stimulated *Pythium* propagule germination and/or growth which resulted in loss of disease control. However, compost tea made with kelp, humic acid and rock dust consistently suppressed disease development; further experiments concluded that residual levels of these nutrients did not stimulate disease development (Scheuerell and Mahaffee, 2004). It appears that the interaction between aeration and the addition of fermentation nutrients for optimizing disease suppression has potential, and deserves more investigation.

Dilution of Compost Teas

A major factor contributing to whether or not compost teas can be used economically on a large scale is whether the disease supressiveness is maintained after diluting with water. It would be nearly impossible to produce enough tea to apply, either as foliar or soil applications, to thousands of hectares at the recommended rate of 50L tea/hectare (Ingham, 2003). Again, only a few investigations have been conducted to examine how dilution impacts disease control. In the majority of the published literature, only efficacy against a specific pathogen has been investigated (Elad and Shtienberg, 1994; Yohalem *et al.*, 1994; Scheuerell and Mahaffee, 2004). In a greenhouse study using pepper and tomato foliage, Elad and Shtienberg (1994) showed that the ability to retain disease suppression upon dilution appeared to be dependent on compost source. Scheuerell and Mahaffee (2004) observed that dilution decreased disease suppression of *Pythium* damping-off of cucumber.

Application Frequency

Another factor contributing to whether or not compost teas can be used economically is the application frequency required to yield the desired effect(s). This presents another area for potential research given that the effect of compost tea application frequency has not been evaluated systematically. Significant disease control has only been observed in field studies where compost teas were applied at less than 14-day intervals and 5-10 total applications per year (Samerski, 1989; Ketterer, 1990; Weltzien, 1991; Tsror, 1999). Although application frequency was not a variable examined in these studies, the spray interval was similar to those used for synthetic pesticides and traditional biological control agents.

Adjuvants

The use of various adjuvants to enhance distribution, adherence, and/or stability of synthetic pesticides on plant surfaces is a common practice (Backman, 1978; Steurbant, 1993) and is often the basis for the efficacy of a particular formulation. However, for some reason this same approach has not been pursued in an effort to enhance the efficacy of biological control agents.

There are four general types of adjuvants that have the potential to enhance the disease control afforded by compost tea applications: (i) spreaders, adjuvants that reduce the surface tension of spray droplets, allowing them to spread evenly over plant surfaces as opposed to simply beading up; (ii) stickers, adjuvants that enhance the adherence of spray droplets to plant surfaces; (iii) protectants, adjuvants that protect microbes from desiccation, UV light, and other stresses; (iv) nutrient substrates, adjuvants that can be used as food substrates for beneficial applied microbes. The use of spreaders, stickers and nutrients has been investigated for the enhancement of disease suppression of compost teas (Ketterer and Schwager, 1992; Sackenheim *et al.*, 1994; Yohalem *et al.*, 1996a; Tränkner and Brinton, 1998; Scheuerell, 2002; Scheuerell and Mahaffee, 2006).

In studies on field-grown roses and greenhouse-grown geraniums, Scheuerell and Mahaffee (Scheuerell, 2002; Scheuerell and Mahaffee, 2006) observed a more uniform distribution and adherence of bacteria on leaf surfaces due to the addition of spreader-stickers. Similarly, Sackenheim *et al.* (1994) observed that the addition of methylcellulose, a sticker, to a compost tea fermentation resulted in more recovered organisms per leaf area and significantly increased disease control when compared to the compost tea applied alone. However, Yohalem *et al.* (1996a) did not observe increased disease control of apple scab with the addition of Latron B1956 spreader-sticker or fish oil.

There also exists the potential of increasing disease control through the addition of nutrients immediately prior to application of the compost tea. Tränkner and Brinton (1998) observed the reduction of disease incidence of grape powdery mildew in field plots from 62% to 8% by the addition of 0.5% CASO bouillon and 0.05% rape seed oil to the compost tea just prior to application. Similarly, Ketterer and Schwager (1992) observed enhanced disease control of powdery mildew and late blight on bean and tomato, respectively, with the addition of 0.5% casein. It is clear from this work that the use of spray adjuvants to enhance biological control efficacy warrants further investigation.

Potential for Human Pathogen Regrowth

One of the most controversial issues surrounding the use of compost teas for plant disease management in food crops is the potential of teas to distribute human pathogens from composts made with manure or other materials with the potential to harbour human pathogens (Anon., 2004). The potential for human pathogen regrowth was first raised by Yohlem *et al.* (1994) in discussing the work of Urban and Tränkner (1993), where populations of *Enterobacteriaceae* where indicated to have increased in non-aerated compost teas. Welke (1999) demonstrated that faecal coliforms increased from <3 MPN/g chicken manure compost to 0.8 cfu/ml of non-aerated tea and, after biweekly application to field grown broccoli and leek, to 43 MPN/g plant tissue at harvest. However, they did not observe a similar transfer of *Salmonella* nor did they observe similar results using cattle manure compost. There are claims that human pathogens can not compete in highly aerated compost teas that contain high numbers of beneficial organisms (Ingham, 2003). Ingham (2003) states that facultative anaerobes do not grow well under conditions of high oxygen concentration and nutrient competition, but offers no data to support this position. This statement is also contrary to the scientific understanding of microbial physiology; *E. coli* undergoing aerobic respiration generates 793 kJ from 1 mole of glucose, but produces 122 kJ while undergoing anaerobic respiration (Gottshalk, 1985).

Recently, there have been a few studies that have investigated further this potential for human pathogen regrowth in aerated compost teas (Bess *et al.*, 2002; Brinton, *et al.*, 2004; Duffy *et al.*, 2004). Based on these studies, it can be concluded that pathogen regrowth does not appear to be supported in compost tea fermentations that do not contain added nutrients. Bess *et al.* (2002) utilised incompletely composted manure that had low endemic levels of *E. coli* to examine the effects of various nutrient sources on the regrowth of *E. coli* in aerated compost teas. They observed that the addition of nutrients containing simple sugars (molasses, yeast extract, barley malt) resulted in pathogen regrowth, while more complex nutrient sources (kelp, fish and manure, and fish emulsion) did not result in increased *E. coli* populations. Similarly, Duffy *et al.* (2004) observed a strong positive correlation between the growth of *Salmonella enterica* and that of *E. coli* O157:H7 when flasks containing 20 g compost, 180 ml sterile water, and 0-1% molasses were inoculated with 1cfu/ml bacteria and shaken (100 rpm) at 20°C for 3 days. The results of the study by Brinton *et al.* (2004) appeared to indicate that *E. coli* populations did not increase during aerated compost tea production in the absence of added sugars and that the populations tended to decline the longer the fermentations were conducted. However, the validity of these results is unclear due to the lack of statistical analysis presented.

This issue was recently addressed by the National Organic Standards Board (USDA, 2004). They concluded that potable water should be used, all equipment should be sanitised before use with a certified sanitizing agent, and compliant compost meeting the specifications set out in the National Organic Standards should be used. Using these methods, teas made without added nutrients could be applied without restriction. Compost teas made with nutrients, on the other hand, could be applied without restriction if the production system (i.e. the compost batch, additives and equipment) has been shown twice to produce teas that meet United States Environmental Protection Agency guidelines for recreational water, otherwise their use on non-grain food crops is restricted to the 90/120 day pre-harvest interval rule as dictated for non-composted manure. These standards are likely to eliminate the addition of nutrients to compost teas that will be applied to food crops, since most producers are unlikely to go through this testing process for each batch of compost used for tea production. It also will be likely to result in limited disease suppression from aerated compost tea applications since the most consistent control has occurred with teas made with nutrients (Scheuerell and Mahaffee, 2004, 2006). Given the nature of the studies done thus far, and the apparent importance of nutrient addition to obtain disease suppressive compost teas, considerable research effort is needed to determine when human pathogen regrowth in compost teas can occur. Methods for preventing regrowth are also needed in order to ensure consumer safety and utilisation of compost teas as a potential disease management tool.

Predicting Suppression

Currently, there is a perception perpetuated by some commercial laboratories that various microbial analyses can be used to predict the effectiveness of a compost tea batch for disease control (Ingham, 2000b, 2003). These analyses, which employ various techniques, present quantitative data for components of the microbiota (total bacteria and fungi, active bacteria and fungi, actinomycetes, yeast, nematodes, protozoa and others) of compost teas using either culture media or direct microscopic enumeration of microbes. These data are then related to various thresholds or indices (e.g. 20-150 ug/ml of total bacteria) that supposedly indicate an acceptable range indicative of an effective compost tea. There is, however, no evidence to support these claims. It is questionable as to whether direct counting methods for the enumeration of microbial populations are more representative and predictive than culturing methods.

Scheuerell and Mahaffee (2004) investigated whether there was a relationship between microbial populations in compost teas and disease control using culturing and direct enumeration techniques (per training by Elaine Ingham October 1998, and described in Scheuerell and Mahaffee, 2004). They observed that when all compost tea preparation methods were analysed as a group, there was no clear relationship between disease suppression of *Pythium* damping-off of cucumber and bacterial population of compost tea, regardless of whether microscopic counts or culturing methods were used to access bacterial populations. However, they did observe a positive relationship when teas made with molasses were removed from the analysis. Interestingly, they observed that the culturing methods and direct methods resulted in statistically similar population estimates for aerated compost teas with added nutrients, while statistically different population estimates were obtained from the two methods for unaerated fermentations. This indicated that aerated fermentations selected for culturable bacteria.

In reviewing the literature on the relationship of microbial populations to disease suppression, Scheuerell and Mahaffee (2002) found that disease suppressive compost teas had total bacterial populations ranging from 10^7 to 10^{10} cfu/ml. Similarly, Scheuerell and Mahaffee (2004) observed that suppression of *Pythium* damping-off by a compost tea was related to bacterial populations greater than 10^7 cfu/ml, 10^6 active bacterial cells/ml or 3.16 x 10^7 total bacterial cells/ml.

These data appear to indicate that information on the bacterial populations of a compost tea can be somewhat predictive of its level of disease suppression. However, the practical utility of such a measure is questionable. In all published research, the microbial analyses were conducted within a few hours of sampling and the samples were maintained under similar conditions. In contrast, samples sent to laboratories providing these analyses will incur unknown environmental conditions related to shipping and handling over the 24-48 h period between sample collection and analysis, which will affect the microbial populations present in the sample. For instance, Scheuerell and Mahaffee (unpublished) did not observe active fungal growth in any of the hundreds of samples examined within hours of sampling; however, when samples were stored for 24 h, actively growing fungal hyphae were observed.

Future Direction

We are at the beginning of a paradigm shift where growers, horticulturists, landscape managers and home owners in certain expanding markets are starting to realise that the sustainable management of plant diseases requires more than just the application of synthetic pesticides and fertilisers. They are beginning to realise that the microbiota associated with soil and plant surfaces impact upon disease development and can be manipulated to reduce the use of costly synthetic pesticides, while achieving some level of disease management. However, there is also reluctance to switch to costly biological control agents using a similar "spray and pray" approach. Thus, they are attempting to develop systems for introducing microbes that are cost effective.

Many factors are being considered by these groups in deciding which control measure(s) to utilise. For many, these factors require the use of non-traditional methods for disease management, however, the information available to make well-informed decisions is often inadequate or not available. Due to the lack of information from the scientific community on the topic of compost tea production and utilisation, these groups have sought and utilised information from other sources. This information void has been filled partially by a few individuals who have created and marketed simplistic solutions for plant disease control which are not supported by past and current research, but are appealing to the producer.

There is a pressing need for critical evaluation of various approaches to managing plant diseases in both conventional and organic production. Numerous reports from both practitioners and the scientific community indicate that compost tea applications can suppress diseases in numerous pathosystems. Many consider disease suppression afforded by compost tea applications to be not suitable for commercial production systems. However, there is an ever expanding group of growers working with either limited control options or insufficient return on synthetic inputs, and who consider any level of suppression better than their current situation.

We speculate that achieving consistent disease suppression with compost tea applications will require: modification of compost tea production steps (e.g. nutrient amendments) to ensure the growth of specific groups of microbes; the incorporation of spray adjuvants to enhance attachment and survival of microbes in soil or on plant surfaces, and the development of spray equipment that is more suitable to the delivery of living microorganisms. Ultimately, compost teas represent only one tool in an arsenal of options that can be utilised within an integrated system that incorporates genetic disease resistance, fertility and water management, disease and pest forecasting, and other cultural approaches to enhance plant health.

References

Adams, P.B. (1990) The potential of mycoparasites for biological control of plant diseases. *Annual Review of Phytopathology* 28, 59-72.

Al-Dahmani, J.H., Abbasi, P.A., Miller, S.A. and Hoitink, A.J. (2003) Suppression of bacterial spot of tomato with foliar sprays of compost extracts under greenhouse and field conditions. *Plant Disease* 87, 913-919.

Andrews, J.H. (1993) Compost extracts and the biological control of foliar plant disease. *Pacific Northwest Pollution Prevention Resource Center Grant Report project* # LNC 91-31 Seattle, Washington, USA.

Anonymous (2004) Report of the Compost Tea Task Force, United States Department of Agriculture. http://www.ams.usda.gov/nosb/meetings/CompostTeaTaskForceFinalReport.pdf.

Backman, P.A. (1978) Fungicide formulation: relationship to biological activity. *Annual Review of Phytopathology* 16, 211-237.

Backman, P.A., Wilson, M. and Murphy, J.F. (1997) Bacteria for biological control of plant diseases. In: Rechcigl, N.A. and Rechcigl, J.E. (eds) *Environmentally Safe Approaches to Crop Disease Control.* J.E. Lewis Publishers, New York, USA. pp. 95-109.

Bess, V.H., Manes, R.B.S. and Snodgrass, J.L. (2002) *E. coli* survival in compost tea using different nutrient substrates. In: Michel Jr., F.C., Rynk, R.F. and Hoitink, H.A.J. (eds) *Proceedings, 2002 International Symposium on Composting and Compost Utilization.* JG Press, Emmaus, Pennsylvania, USA. pp. 1150-1157.

Boehm, M.J. and Hoitink, H.A.J. (1992). Sustenance of microbial activity in potting mixes and its impact on severity of *Pythium* root rot of *Poinsettia*. *Phytopathology* 82, 259-264.

Brinton, W.F., Storms, P., Evans, E. and Hill, J. (2004) Compost teas: Microbial hygiene and quality in relation to method of preparation. *Journal of Biodynamics* 2, 36-45.

Brinton, W.F., Trankner, A., and Droffner, M. (1996) Investigations into liquid compost extracts. *Biocycle* 37, 68-70.

Cantisano, A. (1995) The biological control of plant pathogens in raspberries using beneficial microorganisms, compost teas and nutritional management. *Organic Farming Research Foundation - Project Report 94-01.* Santa Cruz, Califonia, USA.

Cantisano, A. (1998) *Compost teas.* Organic Agricultural Advisors letter, Colfax, California, USA.

Chen, W.D., Hoitink, H.A.J., Schmitthenner, A. F. and Tuovinen, O.H.1988. The role of microbial activity in suppression of damping-off caused by *Pythium ultimum. Phytopathology* 78, 314-322.

Craft, C.M. and Nelson, E.B. (1996) Microbial properties of compoststhat suppress damping-off and root rot of creeping bentgrass caused by *Pythium graminicola. Applied Environmental Microbiology* 62, 1550-1557.

Cronin M.J, Yohalem D S, Harris R.F. and Andrews J H. (1996) Putative mechanism and dynamics of inhibition of the apple scab pathogen *Venturia inaequalis* by compost extracts. *Soil Biology and Biochemistry* 28, 1241-1249.

Davis, R.F., Backman, P.A., Rodriguez-Kabana, R. and Kokalis-Burelle, N. (1992) Biological control of apple fruit diseases by *Chaetomium globosum* formulations containing cellulose. *Biological Control* 2, 118-123.

Deacon, J.W. (1994) Rhizosphere constraints affecting biocontrol organisms applied to seed. In: BCPC Monograph No. 57*: Seed Treatment: Progress and Prospect.* British Crop Protection Council, Thornton Heath, UK. pp. 315-326.

Dimitri, C. and Greene, C. (2002) Recent Growth Patterns in the US Organic Foods Market, Economic Research Service – United States Department of Agriculture. *Agriculture Information Bulletin No. AIB777.*

Diver, S. (2002) Notes on Compost Teas: A 2002 Supplement to the ATTRA Publication Compost Teas for Plant Disease Control. Approriate Technology Transfer for Rural Areas Publication, Fayetteville, Arkansas, USA.

Duffy, B., Sarreal, C. Subbarao, R. and Stanker, L. (2004) Effect of molasses on regrowth of *E. coli* O157:H7 and *Salmonella* in compost teas. *Compost Science and Utilization* 12, 93-96.

Elad, Y. and Shtienberg, D. (1994) Effect of compost water extracts on gray mould *(Botrytis cinerea*). *Crop Protection* 13, 109-114.

Gottschalk, (1985) *Bacterial Metabolism.* Springer-Verlag, NY, USA.

Guetsky, R., Shtienberg, D., Elad, Y., Fischer, E. and Dinoor, A. (2002) Improving biological control by combining biocontrol agents each with several mechanisms of disease suppression. *Phytopathology* 92, 976-985.

Harman G.E. (2000) Myths and dogmas of biocontrol: changes in perceptions derived from research on *Trichoderma harzianum* T-22. *Plant Disease* 84, 377-393.

Hoitink, H.A.J. (2004) Production and utilization guidelines for disease suppressive composts. *Acta Horticulturae* 635, 87-92.

Inbar, Y., Boehm, M.J. and Hoitink, H.A.J. (1991) Hydrolysis of fluorescein diacetate in sphagnum peat container media for predicting suppressiveness to damping-off caused by *Pythium ultimum. Soil Biology and.Biochemistry* 23, 479-483.

Ingham, E.R. (1999) Making a high quality compost tea, Part II. *BioCycle* 40, 94.

Ingham, E.R. (2000a) Brewing compost tea. *Kitchen Gardener* 29, 16-19.

Ingham, E.R. (2000b) *The Compost Tea Brewing Manual.* Unisun Communications, Corvallis, Oregon, USA.

Ingham, E.R. (2003) *The Compost Tea Brewing Manual.* Soil Food Wed, Inc, Corvallis, Oregon, USA.

Janisiewicz, W.J. and Korsten, L. (2002) Biological control of postharvest diseases of fruits. *Annual Review of Phytopathology* 40, 411-441

Johnson, K.B. and Stockwell, V.O. (1998) Management of fire blight: a case study in microbial ecology. *Annual Review of Phytopathology* 36, 227-248.

Ketterer, N. (1990) Untersuchungen zur Wirkung von Kompost-Extraken auf den Blattbefall ser Kartoffel und Tomate durch *Phytophthora infestans* sowie auf den Befall der Weinrebe durch *Plasmopara viticola, Psuedopziza tracheiphila* und *Uncinula necator.* Dissertation, University of Bonn, Germany.

Ketterer N., Fisher B. and Weltzien H.C. (1992) Biological control of *Botrytis cinerea* on grapevine by compost extracts and their microorganisms in pure culture. In: Verhoeff, K., Malathrakis, N. E and Williamson, B. (eds), *Recent Advances in* Botrytis *Research. Proceedings of the 10th International Botrytis Symposium.* Heraklion, Crete, Greece, 5-10 April 1992. Pudoc Scientific Publishers, Wageningen, The Netherlands. pp. 179-186.

Ketterer, N. and Schwager, L. (1992) Effect of compost extracts on the disease incidence and the phyllosphere flora of bush bean and tomato leaves. *Mededelingen van de Faculteit-Landbouwwetenschappen, Universiteit Gent* 57, 411-421.

Koepf, H.H. (1992) *Biodynamic Farming: Principles and Practice.* Anthro Press, NY, USA.

Kokalis-Burelle, N., Backman, P.A., Rodriguez-Kabana, R. and Ploper, L.D. (1992) Potential for biological contol of early leafspot of peanut using *Bacillus cereus* and chitin as foliar amendments. *Biological Control* 2, 321-328.

Litterick, A.M., Harrier, L., Wallace, P., Watson, C.A. and Wood, M. (2004) The role of uncomposted materials, composts, manures, and compost extracts in reducing pest and disease incidence and severity in sustainable temperate agricultural and horticultural crop production - a review. *Critical Reviews in Plant Sciences* 23, 453-479.

Ma, L.P., Gao F. and Qiao X.W. (1999) Efficacy of compost extracts to cucumber wilt (*Fusarium oxysporum* f. sp. *cucumerinum*) and its mechanisms. *Acta Phytopathologica Sinica* 29, 270-274.

Ma, L.P., Qiao ,X.W., Gao, F. and Hao, B.Q. (2001) Control of sweet pepper fusarium wilt with compost extracts and its mechanism. *Chinese Journal of Applied and Environmental Biology* 7, 84-87.

McQuilken, M.P., Whipps, J.M. and Lynch, J.M. (1994) Effects of water extracts of composted manure-straw mixture on the plant pathogen *Botrytis cinerea. World Journal of Microbiology and Biotechnology* 10, 20-26.

McSpadden Gardener, B. (2005) Commercial biocontrol products available for use against plant pathogens. [Accessed 2 March, 2006] Available from http:// http://www.oardc.ohio-state.edu/apsbcc/productlist.htm

Merrill, R. and McKeon, J. (2001) Apparatus design and experimental protocol for organic compost teas. *Organic Farming Research Foundation Information Bulletin* 9, 9-15.

Murray, P.R. (1999) *Manual of Clinical Microbiology*. American Society of Microbiolgoy Press, Washington, D.C, USA.

Noble, R. (2005) Suppression of soil-borne plant diseases with composts: a review. *Biocontrol Science and Technology* 15, 3-20.

Paulitz, T.C. and Belanger, R.R. (2001) Biological control in greenhouse systems. *Annual Review of Phytopathology* 39, 103-133.

Raupach, G.S. and Kloepper, J.W. (1998) Mixtures of plant growth-promoting rhizobacteria enhance biological control of multiple cucumber pathogens. *Phytopathology* 88, 1158-1164.

Riggle, D. (1996) Compost teas in agriculture. *Biocycle* 37, 65-67.

Robertson, S. (2000) *Summer is Feeding Time for your Plants*. Democrat-Herald Newspaper, Albany, OR, USA.

Sackenheim, R. (1993) Untersuchungen ber Wirkungen von wassrigen, mikrobiologisch aktiven Extrakten aus kompostierten Substraten auf den Befall der Weinrebe (*Vitis vinifera*) mit *Plasmopara viticola, Uncinula necator, Botrytis cinerea* und *Pseudopezizia tracheiphila*. PhD Dissertation. University of Bonn, Germany.

Sackenheim, R., Weltzien, H.C. and Kast, W.K. (1994) Effects of microflora composition in the phyllosphere on biological regulation of grapevine fungal diseases. *Vitis* 33, 235-240.

Samerski, C. (1989) Untersuchungen zu Wirkung und Wirkungsmechanismen von Kompostextrakten im Pathosystem Gurke (*Cucumis sativus* L.) Echter Gurkenmehltau (*Sphaerotheca fuliginea*) (Schlechtendal ex. Fr.) Pollaci. PhD Dissertation, University of Bonn, Germany.

Scheuerell, S.J. (2002) Compost teas and compost amended container media. PhD Dissertation, Oregon State University, Corvallis, Oregon, USA.

Scheuerell, S.J. and Mahaffee, W.F. (2006) Variability associated with suppression of gray mold (*Botrytis cinerea*) on geranium (*Pelargonium* x *hortorum*) by non-aerated and aerated compost teas. *Plant Disease* 90, 1201-1208.

Scheuerell, S.J. Sullivan, D. and Mahaffee, W.F. (2005) Suppression of seedling damping-off caused by *Pythium ultimum, Pythium irregulare*, and *Rhizoctonia solani* in compost amended container media. *Phytopathology* 95, 306-315.

Scheuerell, S.J. and Mahaffee, W.F. (2004) Compost tea as a container media drench for suppressing seedling damping-off caused by *Pythium ultimum*. *Phytopathology* 94, 1156-1163.

Scheuerell, S.J. and Mahaffee, W.M.(2002) Compost tea: principles and prospects for plant disease control. *Compost Science* 10, 313-338.

Shtienberg, D. and Elad, Y. (2002) Is it possible to cope with variability of biocontrol? *Bulletin International Organization for Biological and Integrated Control of Noxious Animals and Plants*, West Palearctic Regional Section 25, 1-4.

Steurbaut, W. (1993) Adjuvants for use with foliar fungicides. *Pesticide Science* 38, 85-91.

Stewart, A. (2001) Commercial biocontrol reality or fantasy? *Australasian Plant Pathology* 30, 127-131.

Sturz, A.V., Christie, B.R., and Nowak, J. (2000) Bacterial endophytes: potential role in developing sustainable systems of crop production. *Critical Reviews in Plant Science* 19, 1-30.

Tränkner, A. (1992) Use of agricultural and municipal organic wastes to develop suppressiveness to plant pathogens. In: Tjamos, E.S., Papavizas G.C. and Cook R.J (eds) *Biological Control of Plant Diseases:Progress and Challenges for the Future*. Plenum Press, New York, USA. pp. 35-42.

Tränkner, A. and Brinton, W.B. (1998) Compost practices for control of grape powdery mildew (*Uncinula necator*). Woods End Research Laboratory, PO Box 297, Mt Vernon, Maine, USA.

Tsror, L. (1999) Biological control of early blight in tomatoes. *Acta Horticulturae* 487, 271-273.

Urban, J. and Trankner, A. (1993) Control of gray mold (*Botrytis cinerea*) with fermented compost/water extracts. *Bulletin International Organization for Biological and Integrated Control of Noxious Animals and Plants*, West Palearctic Regional Section 16, 8-11.

Welke, S. (1999) Report: Effectiveness of Compost Extracts as Disease Suppressants in Fresh Market crops in BC. Organic Farming Research Foundation, Santa Cruz, California, USA.

Weltzien, H.C. (1989) Some effects of composted organic materials on plant health. *Agriculture, Ecosystems and Environment* 27, 439-446.

Weltzien, H.C. (1990) The use of composted materials for leaf disease suppression in field crops. *Monograph British Crop Protection Council* 45, 15-120.

Weltzien, H.C. (1991) Biocontrol of foliar fungal disease with compost extracts. In: J.H. Andrews and S.S. Hirano (eds), *Microbial Ecology of Leaves*. Springer-Verlag, New York, USA. pp. 430-450.

Weltzien, H.C. and Ketterer, N. (1986a) Control of downy mildew, *Plasmopara viticola* (de Bary) Berlese et de Toni, on grapevine leaves through water extracts from composted organic wastes. *Phytopathologische Zeitschrift-Journal of Phytopathology* 116, 186-188.

Weltzien, H.C. and Ketterer, N. (1986b) Control of *Phytophthora infestans* on tomato leaves and potato tubers through water extracts of composted organic wastes. *Phytopathology* 76, 1104.

Weltzien, H.C., Ketterer, N., Samerski, C., Budde, K. and Medhin, G. (1987) Untersuchungen zur Wirkung von Kompostextrakten auf die Pflanzengesundheit. *Nachrichtenblatt des Deutschen Pflanzenschutzdienstes* (Braunschweig) 39, 25-28.

Yohalem, D.S., Harris, R.F. and Andrews, J.H. (1994) Aqueous extracts of spent mushroom substrate for foliar disease control. *Compost Science and Utilization* 2, 67-74.

Yohalem, D.S., Nordheim, E.V. and Andrews, J.H. (1996a) The effect of water extracts of spent mushroom compost on apple scab in the field. *Phytopathology* 86, 914-922.

Yohalem, D.S., Voland, R., Nordheim, E.V., Harris, R.F. and Andrews, J.H. (1996b) Sample size requirements to evaluate spore germination inhibition by compost extracts. *Soil Biology and Biochemistry* 28, 519-525.

Zhang, W.D., Han, Y., Dick, W.A. Davis, K.R. and Hoitink, H.A.J. (1998) Compost and compost water extract-induced systemic acquired resistance in cucumber and *Arabidopsis*. *Phytopathology* 88, 450-455.

Chapter 12

Molecular Interactions at the Leaf Surface: *Xanthomonas* and its Host

M. Wilson[1], W. Moss[2], Y. Zhang[3] and J. Jones[3]
[1]Biology Department, The Colorado College, Colorado Springs, CO 80903, USA; [2]Department of Entomology and Plant Pathology, Auburn University, Auburn, AL 36830, USA; [3]Department of Plant Pathology, University of Florida, Gainesville, FL 32611, USA
mwilson@coloradocollege.edu

Introduction

Bacterial pathogens in the genus *Xanthomonas* cause diseases in a wide range of hosts. This chapter will focus on the pathogen *Xanthomonas campestris* pv. *vesicatoria* (also known as *Xanthomonas euvesicatoria* (Jones *et al.*, 2004)) which is one of the causal agents of the disease known as bacterial spot of tomato (*Lycopersicon esculentum*). While much of the study of this pathosystem has focused on the incompatible reaction where a pathogen interacts with a resistant host, this chapter will discuss the compatible reaction where a pathogen interacts with a susceptible host. Specifically, the chapter will address the molecular interactions between pathogen and host occurring in the first 24 h after immigration of cells onto the leaf, or in the experimental situation, following spray-inoculation onto the leaf. This is the period of early colonisation, sometimes referred to as the "epiphytic" phase, prior to entry into the leaf mesophyll via the stomates or hydathodes of an intact tomato leaf. This early interaction at the leaf surface has been long neglected, but raises the important and topical issues of pathogen recognition of a host plant; host recognition of a potential pathogen; elicitors and pathogen-associated molecular patterns (PAMPs); and the early, non-specific defence responses, recently likened to "innate immunity" observed in mammalian systems. It is hoped that this chapter will highlight the importance of understanding this understudied stage of the infection process.

The Pathogen

The *Hrp* cluster and the Type III Secretion System (TTSS)

The genome of the pathogen *X. campestris* pv. *vesicatoria* strain 75-3 carries a 23-kb "*hrp* cluster" consisting of 6 operons with 22 genes: 11 hypersensitive response conserved (*hrc*) genes; 7 hypersensitive response and pathogenicity (*hrp*) genes; and 4 *hrp*-associated (*hpa*) genes (Bonas *et al.*, 2000; Buttner and Bonas, 2002a). In addition, two important regulatory genes, *hrpG* (Wengelnik *et al.*, 1996; Wengelnik *et al.*, 1999; Noel *et al.*, 2001) and *hrpXv* (Wengelnik and Bonas, 1996), occur outside the *hrp* cluster. The *hrc* gene products are involved in the structure and function of the type III secretion system (TTSS) (Buttner and Bonas, 2002b). The function of this TTSS also involves the protein HrpF as part of the translocon (Huguet and Bonas, 1997; Rossier *et al.*, 2000; Buttner *et al.*, 2002); and HrpE, recently shown to be the main protein of the *X. campestris* pv. *vesicatoria* Hrp-pilus (Weber *et al.* 2005). The role of this TTSS is believed to be delivery of effector proteins (e.g. Avr, Hpa, and Xop) from the cytoplasm of the bacterial cell into the cytoplasm of the mesophyll cells of the leaf (Lahaye and Bonas, 2001; Buttner and Bonas, 2002; Alfano and Collmer, 2004). It is now becoming apparent that at least a subset of these effectors are necessary for pathogenicity, may alter gene expression in the host, and may be involved in suppressing host basal defence responses (Lahaye and Bonas, 2001; Buttner and Bonas, 2002; Marois *et al.*, 2002; Alfano and Collmer, 2004; Espinosa and Alfano, 2004).

Regulation of the Pathogenicity Locus and Induction of *hrp* Genes

Expression of the pathogenicity locus is controlled by two regulatory genes *hrpG* (Wengelnik *et al.*, 1996, 1999; Noel *et al.*, 2001) and *hrpXv* (Wengelnik and Bonas, 1996). HrpG is proposed to be a response regulator in a putative two-component system that detects environmental signals typical of the plant apoplast (Wengelnik *et al.*, 1996). As yet, however, the cognate sensor has not been identified, so the mechanism by which the pathogen senses the host environment remains speculative. In the current model (Wengelnik *et al.*, 1996; Buttner and Bonas, 2002a), environmental signals are detected by a membrane bound sensor which leads to the phosphorylation and activation of HrpG. HrpG then activates transcription of *hrpXv* and *hrcC* (previously known as *hrpA*) (Wengelnik *et al.*, 1996; Buttner and Bonas, 2002a). HrpXv (Wengelnik and Bonas, 1996), which acts like an *E. coli* AraC-type transcriptional activator, activates transcription of the operons *hrpB* through *hrpF* (Wengelnik *et al.*, 1996; Buttner and Bonas, 2002a), and also other genes in the *hrpG* regulon (Noel *et al.*, 2001).

Relatively little has been published regarding the nature of the environmental signals involved in induction of the *hrp* gene cluster in *X. campestris* pv. *vesicatoria* or the timing of induction during colonisation and infection. Using *hrp*::Tn*3*-*gus* fusions, Schulte and Bonas (1992a) showed that the *hrp* gene cluster was plant-inducible, but these experiments employed syringe inoculation or vacuum infiltration into the leaf tissue of pepper (*Capsicum annuum*), rather than a 'natural' spray-inoculation onto the leaf surface. Schulte and Bonas (1992b) also showed that the pathogenicity locus was induced *in vitro* in XVM1 medium by sucrose and sulfur-containing amino acids, while Wengelnik *et al.* (1996) developed the defined *hrp*-inducing medium XVM2 which is free of plant-derived molecules, both of which were based on the minimal medium MM1. It was assumed that the composition of the medium mimicked that of the plant apoplast, since that is where *hrp*-induction had been observed (Schulte and Bonas, 1992a) and where the TTSS was presumed to function.

The recent availability of the green fluorescent protein (GFP) and other fluorescent proteins combined with confocal laser scanning microscopy (CLSM) has revolutionised the study of *in vivo* pathogenesis in phytobacteriology. Aldon *et al.* (2000) were among the first to publish a study in which GFP was employed as a reporter of *hrp* gene expression *in vivo*, in this case with the pathogen *Ralstonia solanacearum*. In this pathosystem, the *hrpB-gfp* reporter was specifically induced on physical contact of pathogen cells with the plant cell wall (Aldon *et al.*, 2000). This was followed by a study by Boureau *et al.* (2002) of the foliar pathogen *Pseudomonas syringae* pv. *tomato* DC3000. In their experiments, fluorescence from the *hrpA-gfp* reporter was observed as soon as 3 h after infiltration into the leaf interior (Boureau *et al.*, 2002). Commendably, these authors also spray-inoculated the pathogen onto the tomato leaf surface in an effort to simulate the natural infection process. In this scenario, however, it is notable that no fluorescence was observed until 72 h after inoculation and this was only in pathogen cells that had invaded the interior of the leaf (Boureau *et al.*, 2002). In other words, in *P. syringae* pv. *tomato* no *hrp* gene induction was observed on the tomato leaf surface.

In our study, *hrp-gfp* reporters were used in *X. campestris* pv. *vesicatoria* to try to ascertain the earliest time and the location at which *hrp* gene induction could be observed *in planta* following a spray-inoculation of the pathogen onto the leaf surface (Zhang *et al.*, 1999, 2006; Bussey *et al.*, 2001; Wilson *et al.*, 2002). Initially a constitutive P_{trp}-*gfp* reporter construct was used to determine the distribution of cells following inoculation and the location of multiplication on the leaf during colonisation. Following this, tomato leaves were mist-inoculated with strains carrying one of two reporter constructs, *hrpG-gfp* or *hrpF-gfp*, and examined by CLSM at 24 h intervals (Zhang *et al.*, 1999, 2006; Bussey *et al.*, 2001; Wilson *et al.*, 2002). Recall that *hrpG* and *hrpX* are regulatory genes and that *hrpF* is a structural gene. As observed with the *hrpA-gfp* reporter in *P. syringae* pv. *tomato* (Boureau *et al.*, 2002), high-level expression of the *hrpG-gfp* and *hrpF-gfp* reporters occurred in cells that had entered stomates and colonised the interior of the leaf between 48 h and 72 h after inoculation (Zhang *et al.*, 1999, 2006; Bussey *et al.*, 2001; Wilson *et al.*, 2002). What was unexpected, was that low-level *hrpG-gfp* expression could be observed in cells on the leaf surface as early as 24 h after mist-inoculation, at a time at which there was no evidence of colonisation of the interior of the leaf (Zhang *et al.*, 1999, 2006; Bussey *et al.*, 2001; Wilson *et al.*, 2002). This contrasts with the observations for *P. syringae* (Boureau *et al.*, 2002), however, it should be noted that the low-level expression is less obvious and could have been missed in the *P. syringae* work.

Recently, the *hrpG*-regulon has been shown to include genes throughout the genome of *X. campestris* pv. *vesicatoria* (Noel *et al.*, 2001). The observation of *hrpG-gfp* expression on the leaf surface, only 24 h after inoculation, albeit at a low level, strongly suggests that some genes of the *hrpG*-regulon are expressed on the leaf surface prior to invasion of the leaf interior. We will speculate on the possible reasons for this expression of the *hrpG*-regulon after a discussion of the host response to colonisation by the pathogen.

The Host

Studies of the effects of pathogens on patterns of host gene expression have concentrated primarily on the incompatible interaction between a pathogen and a resistant host; specifically, on the signal transduction pathways following the Avr-R interaction and the molecular nature of the hypersensitive response (HR). Many others have examined the phenomenon of systemic acquired resistance (SAR), where a pathogen inoculated onto a lower leaf induces a systemic defence response that protects distal leaves against subsequent challenge inoculation. Far fewer studies, however, have examined the host response to inoculation with a virulent, compatible pathogen (e.g. Santaella *et al.*, 2004) or the host response to non-pathogenic *hrp* mutants of a compatible or even incompatible pathogen (Jakobek and Lindgren, 1993; Newman *et al.*, 1994; Brown *et al.*, 1995; Faize *et al.*, 1999; Newman *et al.*, 2001; Venisse *et al.*, 2002).

The Basal Defence Response

One of the first studies of this nature was published by Jakobek and Lindgren (1993). In that study, a *hrp* mutant of *Pseudomonas syringae* pv. *tabaci* infiltrated into the leaves of the non-host common bean (*Phaseolus vulgaris*) resulted in accumulation of transcripts of chalcone synthase (CHS), chalcone isomerase, chitinase (CHT) and phenylalanine ammonia lyase (PAL). While bean is a non-host for *P. syringae* pv. *tabaci*, their observation that "there is a general, nonspecific mechanism for the induction of defence transcripts and phytoalexins by pathogenic bacteria that is distinct from the more specific mechanism associated with the induction of the hypersensitive response" seems to have been borne out by subsequent observations. Similarly, a *hrp* mutant of the pathogen *X. campestris* pv. *campestris* induced β-1,3-glucanase (BGL) transcript accumulation in cabbage (*Brassica campestris*) (Newman *et al.*, 1994); *hrpA*, *B*, *C*, *D*, *E*, and *F* mutants of *X. campestris* pv. *vesicatoria* induced papilla deposition in pepper (*C. annuum*) mesophyll cells (Brown *et al.*, 1995); a *hrp* mutant of *X. campestris* pv. *vesicatoria* induced accumulation of phenolic conjugates in pepper (Newman *et al.*, 2001); in apple (*Malus* sp.) *hrp* mutants of *Erwinia amylovora* induced enzymatic activity of PAL and guaiacol peroxidase in one study (Faize *et al.*, 1999) and increased abundance of transcripts of CHS, dihydroflavonol reductase and flavonol synthase in another study (Venisse *et al.*, 2002). All of these studies suggest strongly that the host plant recognises a potential pathogen, even one that is non-pathogenic by virtue of a mutation in the *hrp* genes, and mounts a defence response. In the case of *hrp* mutants this probably contributes to the limitation on internal colonisation of the leaf that is usually observed (Bonas *et al.*, 1991), but in the case of pathogenic compatible bacteria, it now appears that this defence response is suppressed by secretion of effectors into the plant cells (Lahaye and Bonas, 2001; Marois *et al.*, 2002; Buttner and Bonas, 2002; Alfano and Collmer, 2004; Espinosa and Alfano, 2004). Recently, this non-specific defence response, independent as it is of an R-avr interaction, has been referred to as the "basal defence" (Keshavarzi *et al.*, 2004) or "innate immunity" (Cohn *et al.*, 2001).

Our own interest in the host response to pathogen colonisation arose from experiments using *hrp* mutants as potential biological control agents of bacterial spot of tomato. Having obtained only limited control of the disease using saprophytic bacteria applied to the foliage in advance of challenge with the wild-type pathogen (Byrne *et al.*, 2005), near-isogenic non-pathogenic *hrp* mutants of *X. campestris* pv. *vesicatoria* strain 75-3 (courtesy of Ulla Bonas) were employed in an attempt to employ pre-emptive competitive exclusion, previously shown to be effective against bacterial speck, caused by *Pseudomonas syringae* pv. *tomato* (Wilson *et al.*, 2002). Surprisingly, the *hrp* mutants exhibited differential activity, with increasing protective effect observed in the order *hrpE*, *hrpF*, *hrpX*, and *hrpG* (Moss *et al.*, 1997, 1998 a and b, 1999; Moss, 2000). Indeed, inoculation of a tomato plant with the *hrpG* mutant 24-48 h prior to challenge with the wild-type pathogen provided a greater protective effect than observed with any of the saprophytic bacteria or other *hrp* mutants (Moss *et al.*, 1997, 1998 a,b, 1999; Moss, 2000). Further, there was no evidence that this protective effect was due to pre-emptive competitive exclusion, as had been demonstrated to be the case with Ice$^-$ and Ice$^+$ *P. syringae* (Wilson *et al.*, 1994) or non-pathogenic saprotrophs with *P. syringae* pv. *tomato* (Wilson *et al.*, 2002). Induction of a host defence response by the *hrp* mutants seemed the most likely explanation.

As one might expect given the absence of an HR, there was no evidence of a systemic protective effect or SAR; the defence response was limited to the leaf or leaflet inoculated with the *hrp* mutants. Experiments with transgenic tomato plants altered in signal transduction pathways based upon salicylic acid (SA) (Brading *et al.*, 2000), jasmonic acid (JA) (Lightner *et al.*, 1993), or ethylene (ETH) (Lanahan *et al.*, 1994) demonstrated that sensing of and/or response to the *hrpG* mutant required the SA, JA, and ETH-mediated signal transduction pathways, while sensing of and/or response to the *hrpX* and *hrpE* mutants appeared not to require any of the pathways or possibly only the SA-mediated pathway (Moss *et al.*, 1999; Moss, 2000). Northern analyses of pathogenesis-related (PR) protein and PAL gene transcript abundance at 24 h after inoculation of tomato plants with the *hrp* mutants, wild-type pathogen, or deionised water, clearly showed differential responses to the *hrpG* and *hrpE* mutants (Wilson *et al.*, 2000 a,b; 2005). Semi-quantitative RT-PCR analyses of PR protein, PAL, CHT, BGL, and lipoxygenase gene expression similarly showed differential responses to the *hrpG* and *hrpE* mutants and further, suggested that there may be a "burst" of gene expression as early as 5-8 h after inoculation (Wilson *et al.*, 2005). These data suggest that the *X. campestris* pv. *vesicatoria* 75-3 *hrp* mutants induced a non-specific defence response in tomato and that the response was both qualitatively and quantitatively different in response to the *hrpG* and *hrpE* mutants. The magnitude of the defence response or level of induction of 'innate immunity' by the *hrpG* regulatory mutant (as measured by the protective effect against subsequent challenge with the wild-type pathogen) was significantly greater than the magnitude of the response or level of induction by the *hrpE* mutant. It should be noted that we are not suggesting that a "burst" of a product such as BGL or CHT, which likely have no antibacterial activity, is the cause of this protection, but they are indicative that the plant is responding, and is responding differentially. Hence, we have a scenario in which the tomato leaf is responding to the presence of a *potential* pathogen on the surface, possibly as early as 5-8 h after inoculation, but certainly within 24 h, with the production of defensive products able to prevent or minimise invasion and/or infection. How the plant senses cells of the pathogen on the leaf and distinguishes those cells from cells of saprotrophs is another question.

Elicitors and PAMPs

It has long been known that molecules derived from bacterial pathogens, such as lipopolysaccharide (LPS) or extracellular polysaccharide (EPS), can act as general elicitors of defence responses. Newman *et al.* (1995) demonstrated that LPS from *X. campestris* pv. *campestris* induced defence-related gene expression in cabbage (*B. campestris*) and subsequent studies by this and other groups have consistently shown that LPS from *X. campestris* not only induces defence gene expression but also modifies the response of the plant to subsequent inoculation with pathogen cells (Romeiro and Kimura, 1997; Newman *et al.*, 2000; Meyer *et al.*, 2001; Newman *et al.*, 2002; Erbs and Newman, 2003; Braun *et al.*, 2005). EPS from *X. campestris* has also been shown to induce resistance mechanisms and provide protection against subsequent pathogen challenge (Romeiro and Kimura, 1997; Guzzo *et al.*, 1993). In addition to LPS and EPS, plants can also respond to flagellin (Boller, 1995; Gomez-Gomez and Boller, 2002; Zipfel *et al.*, 2004). The suite of general elicitors produced during growth of the pathogen have come to be known as pathogen-associated molecular patterns (PAMPs) (Parker, 2003; Zipfel *et al.*, 2004). It is likely that plants have evolved to specifically recognise the EPS, LPS, flagellin, and secreted products of pathogenic bacteria, such that these PAMPs elicit a defence response, while EPS, LPS, flagellin, and secreted products from saprophytic bacteria either elicit no response or elicit a lesser response.

In preliminary experiments addressing the PAMP hypothesis, we observed that the relative magnitudes of the defence response of tomato plants to either cells or *cell-free supernatant* from culture broth of the *hrpG* and *hrpE* mutants were similar (Wilson *et al.*, 2005). In other words, the magnitude of the defence response (as measured by protection during subsequent challenge with the wild-type pathogen) mounted by tomato was similar to both cells of the *X. campestris* pv. *vesicatoria* 75-3 *hrpG* mutant and to cell-free culture supernatant from that mutant, and this was significantly larger than the magnitude of the defence response to either cells or cell-free supernatant of the *hrpE* mutant. Undoubtedly the cell-free culture supernatant in this preliminary study contained waste products in addition to LPS, EPS and sheared-flagella, however, the data suggest that the differential response of tomato to the *hrpG* and *hrpE* mutants described above is due to different PAMPs (i.e. the quality and/or quantity of the LPS, EPS, or flagellin, or some other as yet unidentified compound was different for the *hrpG* and *hrpE* mutant). Here we return to the pathogen and a possible reason for the activation of the *hrpG*-regulon in the early stages of colonisation while the cells are still on the leaf surface.

Speculation

We have data so far that support the assertions that the *hrpG*-regulon is induced within 24 h of arrival on the leaf surface; that the host plant detects and responds to the presence of a potential pathogen on the leaf surface within 24 h, and possibly as early as 5-8 h; that the plant responds differentially to the *hrpG* regulatory mutant and the *hrpE* structural mutant; and that the differential response probably results from different PAMPs in the *hrpE* and *hrpG* mutants. To add to this, we have observed that the *hrpG* mutant synthesises considerably more EPS on LB agar than does the *hrpE* mutant and we presume that the difference in quantity and possibly quality of the EPS is one component contributing to different PAMPs for these two mutants. We hypothesise that when the wild-type pathogen initially lands on a susceptible host plant the environmental cues lead to activation of the *hrpG*-regulon and that this triggers a change in the phenotype or PAMP of the pathogen to a "pathogenic" form which is less likely to induce the basal defence responses of the host. In other words, the pathogen undergoes a "phenotype-conversion" (a phenomenon previously described for *R. solanacearum* (Brumbley *et al.*, 1993)) or "PAMP-conversion" from a "saprophytic" form, evolved for epiphytic colonisation of non-host plants, to a "pathogenic" form, evolved for minimisation of elicitor-induced basal defence responses. In the case of the *hrp* mutants, we hypothesise that the *hrpG* mutant, which lacks the functional HrpG response regulator for the putative two-component system, is unable to activate the *hrpG*-regulon; unable to undergo PAMP-conversion from "saprophytic" form to "pathogenic" form; unable to avoid eliciting the full basal defence response; and, in the absence of a functional TTSS, cannot suppress the defence responses. The *hrpE* mutant, on the other hand, retains an intact environmental sensing system and is able to activate the *hrpG*-regulon; able to undergo PAMP-conversion; and able to avoid eliciting the full basal defence response. However, since the *hrpE* mutant cannot synthesise the HrpE protein of the Hrp pilus, the TTSS is non-functional and it cannot suppress even the minimal defence response observed.

Conclusions

The few hours immediately following the arrival of a cell of a phytopathogenic bacterium on the leaf surface of a potential host deserve much greater attention and numerous questions remain. Does the pathogen distinguish between a host and non-host while on the surface and if so, how? Does the plant distinguish between the elicitor profile (PAMP) of a pathogen and the elicitor profile of a saprotroph while the organism is still on the surface and if so, how? Are pathogens, in fact, able to undergo a phenotype- or PAMP-conversion on the leaf surface? Is this conversion triggered only on a host and not on a non-host? Is the Hrp pilus elaborated by cells on the leaf surface and if so, what function does it serve? The answers to these questions will likely be different among the foliar pathogens, which seem to follow quite different colonisation and invasion strategies. So, for example, the observation that the *hrp*-regulon in *P. syringae* is not induced on the leaf surface (Boureau *et al.*, 2002) is contradictory to the observation that the *hrpG*-regulon in *X. campestris* pv. *vesicatoria* is active on the leaf surface. Even among pathovars of *P. syringae* there appears to be a range of colonisation and invasion strategies (Wilson *et al.*, 1999). Therefore, we urge our colleagues to devise experiments to try to answer these questions in as wide a range of phytopathogenic genera, species and pathovars as possible.

References

Aldon, D., Brito, B., Boucher, C. and Genin, S. (2000) A bacterial sensor of plant cell contact controls the transcriptional induction of *Ralstonia solanacearum* pathogenicity genes. *The EMBO Journal* 19, 2304-2314.

Alfano, J.R. and Collmer, A. (2004) Type III secretion system effector proteins: double agents in bacterial disease and plant defence. *Annual Review of Phytopathology* 42, 385-414.

Boller, T. (1995) Chemoperception of microbial signals in plant cells. *Annual Review of Plant Molecular Biology* 46, 189-214.

Bonas, U., Schulte, R., Fenselau, S., Minsavage, G.V., Staskawicz, B.J. and Stall, R.E. (1991) Isolation of a gene cluster from *Xanthomonas campestris* pv. *vesicatoria* that determines pathogenicity and the hypersensitive response on pepper and tomato. *Molecular Plant-Microbe Interactions* 4, 81-88.

Bonas, U., Van den Ackerveken, G., Buttner, D., Hahn, K., Marois, E., Nennstiel, D., Noel, L., Rossier, O. and Szurek, B. (2000) How the bacterial plant pathogen *Xanthomonas campestris* pv. *vesicatoria* conquers the host. *Molecular Plant Pathology* 1, 73-76.

Boureau, T., Routtu, J., Roine, E., Taira, S. and Romantschuk, M. (2002) Localization of *hrpA*-induced *Pseudomonas syringae* pv. *tomato* leaves. *Molecular Plant Pathology* 3, 451-460.

Brading, P.A., Hammond-Kossack, K.E., Parr, A. and Jones, J.D.G. (2000) Salicylic acid is not required for *Cf-2*-and *Cf-9*-dependent resistance of tomato to *Cladosporium fulvum*. *The Plant Journal* 23, 305-318.

Braun, S.G., Meyer, A., Holst, O., Puhler, A. and Niehaus, K. (2005) Characterization of the *Xanthomonas campestris* pv. *campestris* lipopolysaccharide substructures essential for elicitation of an oxidative burst in tobacco cells. *Molecular Plant-Microbe Interactions* 18, 674-681.

Brown, I., Mansfield, J. and Bonas, U. (1995) *hrp* genes in *Xanthomonas campestris* pv. *vesicatoria* determine ability to suppress papilla deposition in pepper mesophyll cells. *Molecular Plant-Microbe Interactions* 8, 825-836.

Brumbley, S.M., Carney, B.F. and Denny, T.P. (1993) Phenotype conversion in *Pseudomonas solanacearum* due to spontaneous inactivation of PhcA, a putative LysR transcriptional regulator. *Journal of Bacteriology* 175, 5477-5487.

Bussey, M., Callaway, E., Zhang, Y., Jones, J.B. and Wilson, M. (2001) Host-pathogen interactions between *Xanthomonas campestris* pv. *vesicatoria* and *tomato. 10th International Congress on Molecular Plant-Microbe Interactions*, Madison, Wisconsin, USA.

Buttner, D. and Bonas, U. (2002a) Getting across - bacterial type III effector proteins on their way to the plant cell. *The EMBO Journal* 21, 5313-5322.

Buttner, D. and Bonas, U. (2002b) Port of entry – the type III secretion translocon. *Trends in Microbiology* 10, 186-192.

Buttner, D., Nennstiel, D., Klusener, B. and Bonas, U. (2002) Functional analysis of HrpF, a putative type III translocon protein from *Xanthomonas campestris* pv. *vesicatoria. Journal of Bacteriology* 184, 2389-2398.

Byrne, J.M., Campbell, H.L., Dianese, A.C., Ji, P., Cuppels, D.A., Louws, F.J., Miller, S.A., Jones, J.B. and Wilson, M. (2005) Biological control of bacterial spot of tomato under field conditions at multiple locations in North America. *Biological Control* 32, 408-418.

Cohn, J., Sessa, G. and Martin, G.B. (2001) Innate immunity in plants. *Current Opinion in Immunology* 13, 55-62.

Erbs, G. and Newman, M.-A. (2003) The role of lipopolysaccharides in induction of plant defence responses. *Molecular Plant Pathology* 4, 421-425.

Espinosa, A. and Alfano, J.R. (2004) Disabling surveillance: bacterial type III secretion system effectors that suppress innate immunity. *Cellular Microbiology* 6, 1027-1040.

Faize, M., Brisset, M.-N., Paulin, J.-P. and Tharaud, M. (1999) Secretion and regulation Hrp mutants of *Erwinia amylovora* trigger different responses in apple. *FEMS Microbiology Letters* 171, 173-178.

Gomez-Gomez, L. and Boller, T. (2002) Flagellin perception: a paradigm for innate immunity. *Trends in Plant Science* 7, 251-256.

Guzzo, S.D., Bach, E.E., Martins, E.M.F. and Moraes, W.B.C. (1993) Crude exopolysaccharide (EPS) from *Xanthomonas campestris* pv. *manihotis*, *Xanthomonas campestris* pv. *campestris* and commercial xanthan gum as inducers of protection in coffee plants against *Hemileia vastatrix. Journal of Phytopathology* 139, 119-128.

Huguet, E. and Bonas, U. (1997) *hrpF* of *Xanthomonas campestris* pv. *vesicatoria* encodes and 87-kDa protein with homology to NolX of *Rhizobium fredii. Molecular Plant-Microbe Interactions* 10, 488-498.

Jakobek, J.L. and Lindgren, P.B. (1993) Generalized induction of defence responses in bean is not correlated with the induction of the hypersensitive reaction. *The Plant Cell* 5, 49-56.

Jones, J.B., Lacy, G.H., BOuzar, H., Stall, R.E. and Schaad, N.W. 2004. Reclassification of the xanthomonads associated with bacterial spot disease of tomato and pepper. *Systematic and Applied Microbiology* 27, 755-762.

Keshavarzi, M., Soylu, S., Brown, I., Bonas, U., Nicole, M., Rossiter, J. and Mansfield, J. (2005) Basal defences induced in pepper by lipopolysaccharides are suppressed by *Xanthomonas campestris* pv. *vesicatoria. Molecular Plant-Microbe Interactions* 17, 805-815.

Lahaye, T. and Bonas, U. (2001) Molecular secrets of bacterial type III effector proteins. *Trends in Plant Science* 6, 479-485.

Lanahan, M.B., Yen, H.-C., Giovannoni, J.J. and Klee, H.J. (1994) The Never Ripe mutation blocks ethylene perception in tomato. *Plant Cell* 6, 521-530.

Lightner, J., Pearce, G., Ryan, C.A. and Browse, J. (1993) Isolation of signaling mutants of tomato (*Lycopsersicon esculentum*). *Molecular and General Genetics* 241, 595-601,

Marois, E., Van den Ackerveken, G. and Bonas, U. (2002) The *Xanthomonas* type III effector protein AvrBs3 modulates plant gene expression and induces cell hypertrophy in the susceptible host. *Molecular Plant-Microbe Interactions* 15, 637-646.

Meyer, A., Puhler, A. and Niehaus, K. (2001) The lipopolysaccharides of the phytopathogen *Xanthomonas campestris* pv. *campestris* induce an oxidative burst reaction in cell cultures of *Nicotiana tabacum. Planta* 213, 214-222.

Moss, W.P. (2000) Interactions of *Xanthomonas campestris* pv. *vesicatoria hrp* mutants with the pathogenic parent and the host plant leading to biological control of bacterial spot disease of tomato. PhD.Thesis, Auburn University, Auburn, Alabama, USA.

Moss, W., Byrne, J.M. and Wilson, M. (1997) Interactions between *Xanthomonas axonopodis* pv. *vesicatoria hrp* mutants and the pathogenic parent. *Phytopathology* 87, S68.

Moss, W.P., Jones, J.B. and Wilson, M. (1999) Differential induction or suppression of host defence responses by *Xanthomonas campestris* pv. *vesicatoria*. *9th International Congress on Molecular Plant-Microbe Interactions*, Amsterdam, NL.

Moss, W.P., Bonas, U., Jones, J.B. and Wilson, M. (1998) Interactions of *Xanthomonas axonopodis* pv. *vesicatoria* 75-3*hrp* mutants, the pathogenic parent, and the host plant. Abstract 1.7.2. *ICPP 98, 7th International Congress of Plant Pathology*, Edinburgh, UK.

Moss, W.P., Bonas, U., Jones, J.B. and Wilson, M. (1998) Interactions of *Xanthomonas axonopodis* pv. *vesicatoria* 75-3*hrp* mutants, the pathogenic parent, and the host plant. Abstract N29. American Society for Microbiology, 98th General Meeting, Atlanta, Georgia, USA.

Newman, M.-A., Conrads-Strauch, J., Scofield, G., Daniels, M.J. and Dow, J.M. (1994) Defence-related gene induction in *Brassica campestris* in response to defined mutants of *Xanthomonas campestris* with altered pathogenicity. *Molecular Plant-Microbe Interactions* 7, 553-563.

Newman, M.-A., Daniels, M.J. and Dow, J.M. (1995) Lipopolysaccharide from *Xanthomonas campestris* induces defence-related gene expression in *Brassica campestris*. *Molecular Plant-Microbe Interactions* 8, 778-780.

Newman, M.-A., von Roepenack, E., Daniels, M. and Dow, M. (2000) Lipopolysaccharides and plant responses to phytopathogenic bacteria. *Molecular Plant Pathology* 1, 25-31.

Newman, M.-A., von Roepenack-Lahaye, E., Parr, A., Daniels, M.J. and Dow, J.M. (2001) Induction of hydroxycinnamoyl-tyramine conjugates in pepper by *Xanthomonas campestris*, a plant defence response activated by *hrp* gene-dependent and *hrp* gene-independent mechanisms. *Molecular Plant-Microbe Interactions* 14, 785-792.

Newman, M.-A., von Roepenack-Lahaye, E., Parr, A., Daniels, M.J. and Dow, J.M. (2002) Prior exposure to lipopolysaccharides potentiates expression of plant defences in response to bacteria. *The Plant Journal* 29, 487-495.

Noel, L., Thieme, F., Nennstiel, D. and Bonas, U. (2001) cDNA-AFLP analysis unravels a genome-wide *hrpG* regulon in the plant pathogen *Xanthomonas campestris* pv. *vesicatoria*. *Molecular Microbiology* 41, 1271-1281.

Parker, J.E. (2003) Plant recognition of molecular patterns. *Trends in Plant Science* 8, 245-247.

Romeiro, R.S. and Kimura, O. (1997) Induced resistance in pepper leaves infiltrated with purified bacterial elicitors from *Xanthomonas campestris* pv. *vesicatoria*. *Journal of Phytopathology* 145, 495-498.

Rossier, O., Van de Ackerveken, G. and Bonas, U. (2000) HrpB2 and HrpF from *Xanthomonas* are type III-secreted proteins and essential for pathogenicity and recognition by the host plant. *Molecular Microbiology* 38, 828-838.

Santaella, M., Suarez, E., Lopez, C., Gonzalez, C., Mosquera, G., Restrepo, S., Tohme, J., Badillo, A. and Verdier, V. (2004) Identification of genes in cassava that are differentially expressed during infection with *Xanthomonas axonopodis* pv. *manihotis*. *Molecular Plant Pathology* 5, 549-558.

Schulte, R. and Bonas, U. (1992a) Expression of *Xanthomonas campestris* pv. *vesicatoria hrp* gene cluster, which determines pathogenicity and hypersensitivity on pepper and tomato, is plant inducible. *Journal of Bacteriology* 174, 815-823.

Schulte, R. and Bonas, U. (1992b) A *Xanthomonas* pathogenicity locus is induced by sucrose and sulfur-containing amino acids. *Plant Cell* 4, 79-86.

Venisse, J.-S., Malnoy, M., Faize, M., Paulin, J.-P. and Brisset, M.-N. (2002) Modulation of defence responses of *Malus* spp. during compatible interactions with *Erwinia amylovora*. *Molecular Plant-Microbe Interactions* 15, 1204-1212.

Weber, E., Ojanen-Reuhs, T., Huguet, E., Hause, G., Romantschuk, M., Korhonen, T.K., Bonas, U. and Koebnik, R. (2005) The type III-dependent *hrp* pilus is required for productive interaction of *Xanthomonas campestris* pv. *vesicatoria* with pepper host plants. *Journal of Bacteriology* 187, 2458-2468.

Wengelnik, K. and Bonas, U. (1996) HrpXv, an AraC-type regulator, activates expression of five of the six loci in the *hrp* cluster of *Xanthomonas campestris* pv. *vesicatoria*. *Journal of Bacteriology* 178, 3462-3469.

Wengelnik, K., Marie, C., Russelm M. and Bonas, U. (1996) Expression and localization of HrpA1, a protein of *Xanthomonas campestris* pv. *vesicatoria* essential for pathogenicity and induction of the hypersensitive reaction. *Journal of Bacteriology* 178, 1061-1069.

Wengelnik, K., Rossier, O. and Bonas, U. (1999) Mutations in the regulatory gene *hrpG* of *Xanthomonas campestris* pv. *vesicatoria* result in constitutive expression of all *hrp* genes. *Journal of Bacteriology* 181, 6828-6831.

Wengelnik, K., Van den Ackerveken, G. and Bonas, U. (1996) HrpG, a key *hrp* regulatory protein of *Xanthomonas* pv. *vesicatoria* is homologous to two-component response regulators. *Molecular Plant-Microbe Interactions* 9, 704-712.

Wilson, M., Bussey, M., Garton, R., Moss, W.P. and Jones, J.B. (2000) Response of tomato (cv. Agriset) to inoculation with wild-type or *hrp* mutants of *Xanthomonas campestris* pv. *vesicatoria* 75-3. *Phyllosphere 2000, 7th International Symposium on the Microbiology of Aerial Plant Surfaces.* University of California, Berkeley, USA.

Wilson, M., Bussey, M., Garton, R., Martin, R., Moss, W., Smith, L. and Staley, P. (2005) Mechanism of biological control of tomato by *Xanthomonas campestris* pv. *vesicatoria hrp* mutants. *12th International Congress on Molecular Plant-Microbe Interactions*, Cancun, Mexico.

Wilson, M., Callaway, E.M., Zhang, Y. and Jones, J.B. (2002) Induction of *hrp* genes of *Xanthomonas campestris* pv. *vesicatoria* on the tomato leaf surface. Abstract B-258. *American Society for Microbiology, 102nd General Meeting*, Salt Lake City, Utah, USA.

Wilson, M., Campbell, H.L., Ji, P., Jones, J.B. and Cuppels, D.A. (2002) Biological control of bacterial speck of tomato under field conditions at several locations in North America. *Phytopathology* 92, 1284-1292.

Wilson, M., Hirano, S.S. and Lindow, S.E. (1999) Location and survival of leaf-associated bacteria in relation to pathogenicity and potential for growth within the leaf. *Applied and Environmental Microbiology* 65, 1435-1443.

Wilson, M. and Lindow, S.E. (1994) Ecological similarity and coexistence of epiphytic ice-nucleating (Ice^+) *Pseudomonas syringae* and a non-ice-nucleating (Ice-) biological control agent. *Applied and Environmental Microbiology* 60, 3128-3137.

Wilson, M., Moss, W.P., Garton, R., Zhang, Y. and Jones, J.B. (2000) Induction of a local defence response in tomato by a *hrpG* mutant of *Xanthomonas campestris* pv. *vesicatoria*. *10th International Conference on Plant Pathogenic Bacteria*, Prince Edward Island, Canada.

Zhang, Y., Callaway, E., Jones, J.B. and Wilson, M. (2006) Unpublished data.

Zhang, Y., Jones, J.B. and Wilson, M. (1999) Use of green fluorescent protein to monitor gene expression and colonization by *Xanthomonas campestris* pv. *vesicatoria*. *Phytopathology* 89, S89

Zipfel, C., Robatzek, S., Navarro, L., Oakeley, E.J., Jones, J.D.G., Felix, G. and Boller, T. (2004) Bacterial disease resistance in *Arabidopsis* through flagellin perception. *Nature* 428, 764-767.

Chapter 13

Erwiniae: Genomics and the Secret Life of a Plant Pathogen

I.K. Toth[1], L. Moleleki[1], L. Pritchard[1], H. Liu[1], S. Humphris[1], L. Hyman[1], G.W. Axelsen[2], M.B. Brurberg[3], M. Ravensdale[1], E. Gilroy[1] and P.R.J. Birch[1]

[1]Scottish Crop Research Institute, Invergowrie, Dundee, DD2 5DA, Scotland; [2]Department of Chemistry, Biotechnology and Food Science, Norwegian University of Life Sciences, Chr. M. Falsensvei 1, PO Box 5003, 1432 As, Norway; [3]Department of Biotechnological Sciences, Agricultural University of Norway, Norway

ian.toth@scri.ac.uk

Introduction

Erwinia carotovora subsp. *atroseptica* (*Eca*), also referred to as *Pectobacterium atrosepticum*, is a bacterial pathogen of potato, causing blackleg of potato plants in the field (Fig. 13.1) and soft rot of tubers in storage (Perombelon, 2002; Gardan *et al.*, 2003). It is of special research interest for two main reasons; first because it causes major crop losses in temperate regions, and secondly because it is a plant pathogenic member of the same family (*Enterobacteriaceae*) as important human pathogens such as *E. coli*, *Salmonella* spp., *Yersinia* spp. and *Shigella* spp. *Eca* is particularly well known for its production of large quantities and several types of plant cell wall degrading enzymes (PCWDEs) during the latter stages of disease development (Toth *et al.*, 2003). As a consequence it has been described as a necrotrophic pathogen, causing death of host tissue and colonisation of dead substrate. However, relatively little is known about the mechanisms used by *Eca* to initiate infection, how this leads to the regulation and expression of virulence factors or how and where *Eca* exists when not causing disease.

In an attempt to find out more about these processes, the fully annotated genome sequence of *Eca* was obtained through collaboration between the Scottish Crop Research Institute (SCRI) and the Sanger Institute, Cambridge, UK. A detailed investigation was then undertaken using computational analyses to compare the *Eca* genome with other bacterial genomes, to identify interesting and novel genes involved in disease development and in the pathogen's existence away from the diseased plant (Bell *et al.*, 2004). The type III secretion system (T3SS) is important in the initiation of disease caused by a number of pathogens of both plants and animals (He *et al.*, 2004) and was thus an important area of focus in this work. In addition, known regulatory processes involved in infection were also investigated in an attempt to identify novel virulence-associated genes within these key regulons.

Fig. 13.1. Potato plants showing disease symptoms following infection by *Erwinia carotovora* subsp. *atroseptica* (left) next to a healthy plant (right).

Experimental Approach

Genome Comparisons

Annotation of the *Eca* genome sequence placed coding sequences (CDSs) into a number of functional groups, including regulators, pathogenicity determinants, phage/IS elements and osmotic adaptation etc. *Eca* CDSs were also compared at the protein level (using FASTA analysis) with those of other fully sequenced bacterial genomes to identify reciprocal best hits (RBH), indicating potential orthologous sequences. The RBH results, comparing *Eca* against all other sequenced genomes, were displayed using the GenomeDiagram software package developed at SCRI (http://bioinf.scri.sari.ac.uk/lp/programs.html#genomediagram; Fig. 13.2) (Pritchard *et al.,* 2005). RBHs were also differentiated on the basis of their presence in enteric and/or non-enteric organisms (Bell *et al.,* 2004).

Mutation Grid Screening

Following the above analyses, a number of CDSs were selected for further study based both on their putative role in disease and life in the environment. To study these genes more closely, a pooled transposon mutation grid for the large-scale production of mutants was developed. This grid consisted of around 10,000 Tn*5* mutants (approx. 2-fold genome coverage) in 384-well microtitre plates. The mutants were pooled separately by plate, sector, quadrant, row and column and DNA extracted such that only a handful of PCR reactions were required to identify the presence and position of mutants for isolation and further study (Holeva *et al.,* 2004) (Fig. 13.3). This method has allowed isolation of over 40 mutants of interest from the *Eca* genome, and is expected to prove similarly useful when applied to other genetically amenable bacteria (Holeva *et al.*, 2004). Three mutants isolated from the grid are described in detail below and include a *hrpL* mutant (a gene encoding a regulatory protein involved in the coordinate induction of genes that make up the type III secretion system [T3SS] and its secreted/translocated proteins); a putative T3SS-translocated protein (designated *geneX*) mutant; and an *expI* mutant involved in quorum sensing regulation. The location of the insertions in the mutants was confirmed by sequencing, and all mutant strains were transduced into a wild type *Eca* background prior to pathogenicity assays (Toth *et al.*, 1997). Where mutants were found to be reduced in virulence, the disrupted gene was complemented by inserting a PCR product containing the wild-type gene into the plasmid pGEM-T Easy (Promega) and transforming this plasmid into the appropriate mutant by electroporation. Self-ligated plasmids were transformed separately as controls.

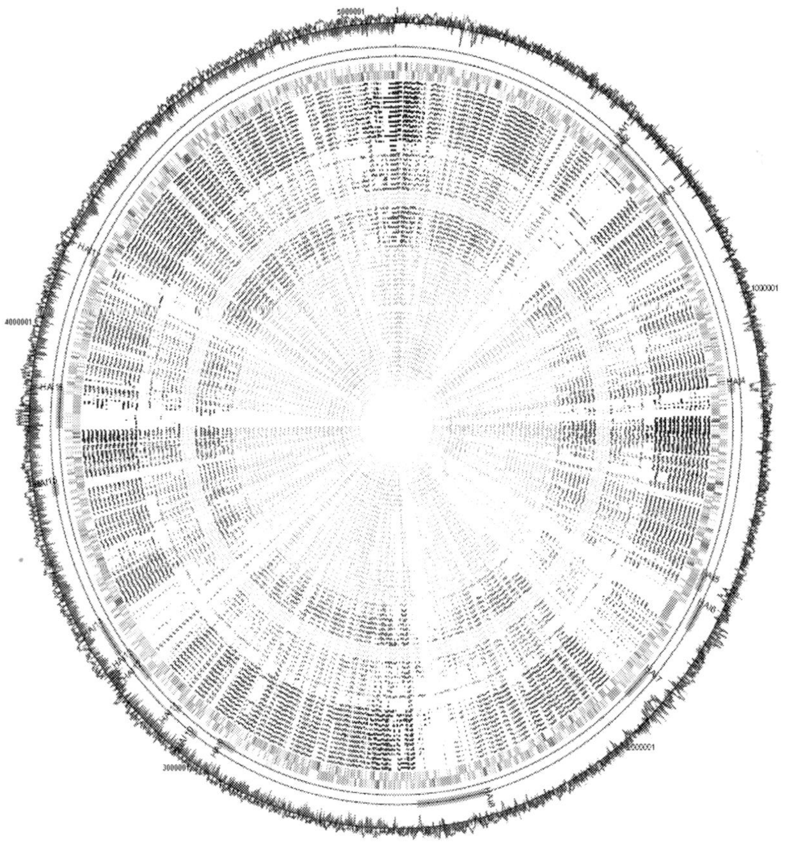

Fig. 13.2. Image generated by GenomeDiagram showing comparison of the *Erwinia carotovora* subsp. *atroseptica* genome sequence with other bacterial genome sequences found by reciprocal FASTA of coding sequences. Inner tracks show locations of reciprocal best hits, while outer tracks show plots of percentage G+C and G+C skew. Two further tracks indicate possible positions of horizontally acquired islands.

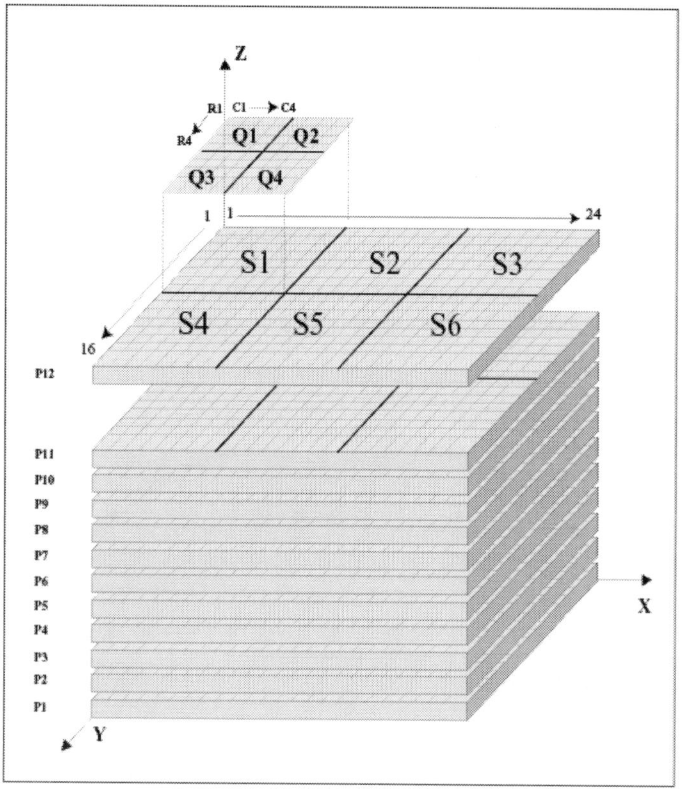

Fig. 13.3. Representation of a pooled mutation grid with 12 x 384-well microtitre plates (P). Each plate is divided into sectors (S), each sector into quadrants (Q), and each quadrant into four rows (R) and four columns (C).

Pathogenicity Assays

Micropropagated potato plants (cultivar Estima; Gentech, Dundee, Scotland) were grown in a glasshouse at 22°C. Six-week-old plants were transferred to a controlled environment room at 20°C, 16 h day/light and high humidity for pathogenicity assays. Overnight cultures were inoculated into potato stems (to give a final inoculum in the plant of 10^2 colony-forming units - CFU) under the second fully expanded leaf and the wound sealed with Parafilm. Twenty replicates were inoculated per strain and disease lesions measured (mm) over a period of 12 days post-inoculation (dpi). Cells (10^4 CFU final concentration) were stab inoculated into potato tubers (cultivar Maris Piper), the wound sealed and rot weight measured (g) after 96 h post-inoculation (hpi). The statistical significance of the results was analysed using a T-test or ANOVA (Genstat v6.1.0.200) (Holeva *et al.*, 2004).

Expression of TF1 Transcription Factor

Both wild-type and mutant strains (10^2 CFU) were vacuum infiltrated into detached leaves before incubating at 18°C and 70% humidity in the light. Samples were obtained at 0.5, 2, 4, 7 and 10 hpi, RNA extracted using a Qiagen RNA plant extraction kit and cDNA synthesised using superscript™ II RT (Invitrogen). Primers were designed to amplify *Solanum tuberosum* (*St*)TF1, together with *St*Tubulin as a constitutive endogenous control in a real-time PCR reaction mix containing SYBR green (Qiagen). Relative expression levels of TF1 were calculated according to an ABI PRISM 7700 Sequence Detection System (Applied Biosystems).

Microarray Analysis

The wild-type strain (*Eca* SCRI1043) and two mutant strains (with disrupted *hrpL* and *expI* genes) were used in this study. Previously, Tn*5* mutants of *Eca* carrying mutations in the regulatory gene *hrpL* and quorum sensing regulon gene *expI* were isolated, and it was shown that they were both significantly reduced in pathogenicity. To determine differences in transcription across the whole genomes of the wild-type and mutant strains, total RNA was prepared from bacterial strains inoculated into potato tubers. Each strain was inoculated into three separate tubers and RNA was extracted from each. For cDNA synthesis, total RNA was mixed with ten 11mer primers as described by Dellagi *et al.* (2000). WT cDNA was labelled fluorescently with Cy3 and mutant cDNA with Cy5 from each pair of RNA, and hybridised to an *Eca*-specific microarray slide, representing over 98% of *Eca* CDSs, in accordance with the manufacturer's protocol (Agilent Technologies). Tuber inoculation, cDNA synthesis and hybridisations were repeated to produce a total of three biological replicates.

Results and Discussion

Reciprocal best hit analyses were used to compare approximately 250 published bacterial genome sequences, including plant pathogens and other plant-associated bacteria, and the information displayed in GenomeDiagram (Fig. 13.2). These analyses revealed that approximately 66% of *Eca* CDSs were shared with at least one other enterobacterium. The majority of these CDSs are likely to represent genes shared through common ancestry with this group of organisms. A number of horizontally-acquired islands (HAIs), in some cases thought to have been acquired from organisms outside of the enterobacteria, were also revealed by GenomeDiagram. These islands contain a higher proportion of pathogenicity or putative pathogenicity genes than the enterobacterial-common backbone sequences, and many of these genes show a high degree of similarity to other plant pathogens (Fig. 13.4), e.g. genes similar to the phytotoxins syringomycin and coronafacic acid from *Pseudomonas syringae*. The coronatine-related toxin now has been shown to have a role in pathogenicity (Bell *et al.*, 2004).

The T3SS, and proteins thought to be secreted through this system, such as the effector protein DspE, were also shown to have a role in pathogenicity (Holeva *et al.*, 2004). A mutant with a disruption in a gene (designated *geneX*) encoding a putative Type III secreted protein was found to be reduced in pathogenicity when compared to the wild-type strain, thus indicating that the product of this gene is required for virulence in *Eca*. To further assess the role of *geneX* in pathogenicity, a plasmid carrying *geneX* was used to complement the mutant. In *Eca*, as *geneX* is part of a two-gene operon together with its chaperone, it was necessary to include this chaperone on the complementing plasmid to restore virulence to wild-type levels. It is likely therefore that the *geneX* chaperone plays an important role possibly in the stabilisation of *geneX* during pathogenesis.

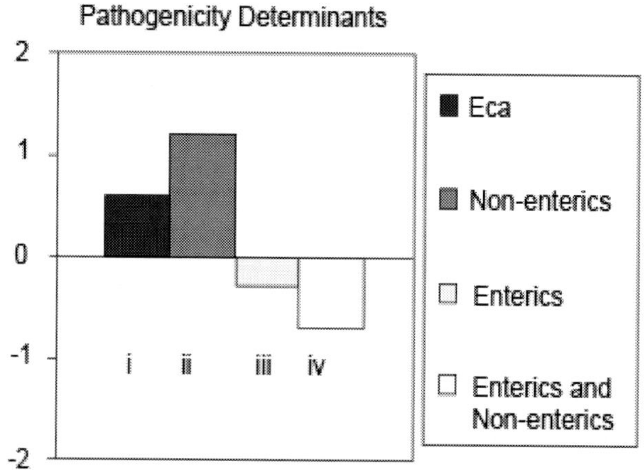

Fig. 13.4. Log_2 ratio of observed to expected coding sequences associated with pathogenicity in each of four bacterial groups; coding sequences in *Erwinia carotovora* subsp. *atroseptica* shared with: (i) no other species; (ii) non-enteric only; (iii) enteric only; (iv) enteric and non-enteric species.

To study the effect of *geneX* product on the plant response during infection, expression of a transcription factor, designated TF1, was assessed in potato leaves using quantitative RT-PCR. TF1 was selected as it has been shown previously to be up-regulated in response to *Eca* (Dellagi *et al.,* 2000). Using detached potato leaf assays (cultivar Desiree) vacuum infiltrated with *Eca* wild-type or the *geneX* mutant, low levels of TF1 were observed initially (0.5 hpi) in response to the wild-type strain (Fig. 13.5). One possible explanation for this may be that during early infection the pathogen suppresses plant defences to allow time for bacterial multiplication. However, at 10 hpi when the infection process is largely completed, TF1 transcription increased dramatically (over 200 fold) as its suppression may no longer be of benefit to the pathogen. In the *geneX* mutant, the transcription factor was up-regulated highly (175 fold) during the early stages of the infection process (0.5 hpi) (Fig. 13.5). This lends support to the hypothesis above and suggests that this protein may be involved in suppression of the TF1 transcription factor and thus in overcoming plant resistance. If this is the case, these results are the first clear indication that *Eca* may be interacting with living potato cells as part of the infection process in a biotrophic interaction. Thus *Eca* appears to behave as a hemibiotroph, requiring living cells as part of its life cycle during the early infection process, much like *Pseudomonas syringae*, but acting as a necrotroph (with the production of PCWDEs) towards the end of infection (Toth and Birch, 2005).

Microarrays with probes designed to over 98% of CDSs within the *Eca* SCRI1043 genome were used to analyse the expression of genes in the *hrpL* mutant versus the wild type strain following infection of potato tubers. The data revealed that the genes in *Eca* controlled by *hrpL* (T3SS and its secreted proteins) were switched on during the early stages of the infection process (after only 30 minutes), supporting the role of the T3SS and its translocated proteins in the early biotrophic stages of infection. A bioinformatics approach was used to examine both the regulatory region of these potential HrpL-regulated genes to identify potential hrp-boxes (HrpL specific regulatory regions), and their N-terminal sequences to find evidence of a type III secretion signal (Petnicki-Ocwieja *et al.,* 2002). This approach led to the identification of 27 new potential effector proteins in *Eca* that are being characterised further.

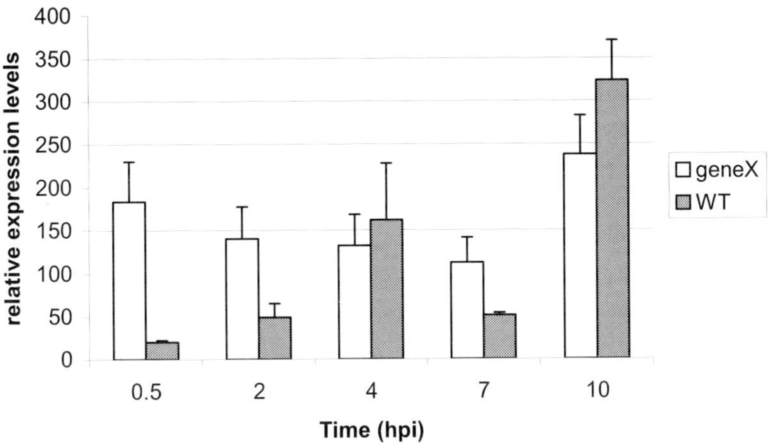

Fig. 13.5. Relative expression levels of the TF1 transcription factor in detached potato leaves infiltrated with wild-type *Erwinia carotovora* subsp. *atroseptica* strain (SCRI1043) or *geneX* mutant at sampling times up to 10 h post inoculation.

Many bacteria use a process called quorum-sensing (QS), which involves a hormone called acyl-homoserine lactone (AHL) that switches on genes involved in disease and other processes (McGowan *et al.*, 2005). In *Eca*, at high bacterial populations towards the end of infection, the AHLs are synthesised via the *expI* gene and are responsible for regulating the production of secreted PCWDEs. In studies comparing the *Eca* wild-type and *expI* mutant strains following potato tuber inoculation, the microarray data revealed that PCWDEs, other genes known to be involved in the latter stages of infection and novel genes potentially involved at this latter stage, were no longer switched on in the *expI* mutant (Fig. 13.6). These genes are now being investigated further to identify their roles in both the QS process and disease development.

Based on the above analyses, microarrays have enabled the expression of the majority of *Eca* genes to be examined in different regulatory backgrounds. New putative virulence factors that may be involved in suppressing plant defences through regulation by the HrpL protein, and in attacking the plant at the latter stages of infection through QS regulation, have been identified. These microarrays have demonstrated that HrpL-regulated genes are expressed at an early stage of infection, before the onset of quorum-sensing. The quorum-sensing regulators then activate genes after 12 hpi, leading to the high expression of PCWDEs and other disease-inducing proteins.

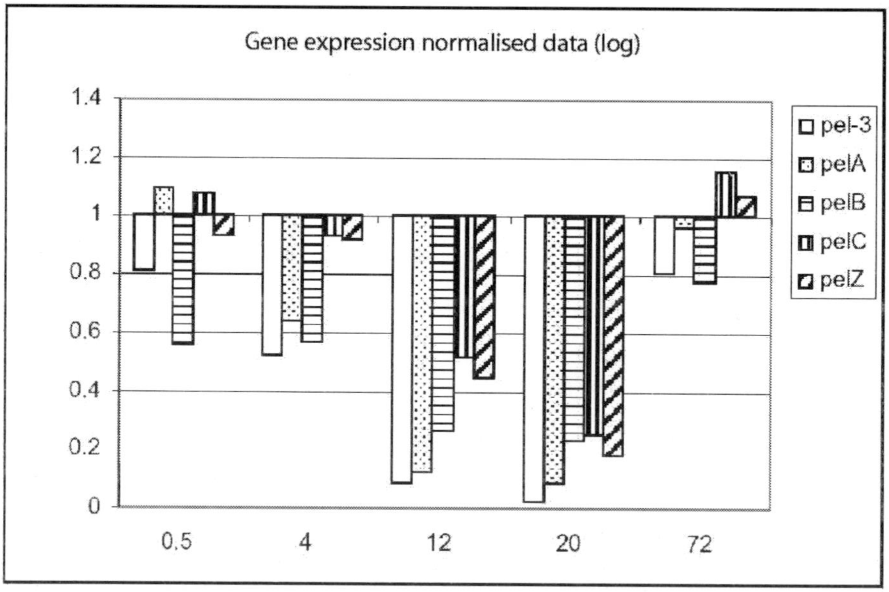

Fig. 13.6. Five genes encoding plant cell wall degrading enzymes are no longer induced 12-20 h post inoculation in the *expI* mutant strain of *Erwinia carotovora* subsp. *atroseptica*.

Little is known about *Eca* in the environment but it is thought that it survives for a short period in soil and ground water following potato infection (Pérombelon 2002). However, the genome sequence, particularly through the HAIs, has revealed genes that suggest a more persistant existence on or around the roots of plants. For example, the HAIs contain CDSs similar to those of other plant-associated bacteria but thought not to be directly involved in pathogenicity. These include genes involved in nitrogen fixation, opine catabolism, antibiotic production, root adhesion and heavy metal resistance, together with a greater number of methyl-accepting chemotaxis proteins and ABC transport uptake systems than any other sequenced enterobacterial genome.

Eca possesses genes for converting atmospheric nitrogen into ammonium, nitrites and nitrates, which may help to provide the necessary nitrogen compounds for successful plant growth. This suggests a very different life style to disease on potato, and offers clues to the role of *Eca* on the roots of other plant hosts. The possibility of root association with non-potato hosts is supported by the discovery of genes within the *Eca* genome involved in root attachment. Although mutants in these genes were unaffected in their ability to bind to potato roots, leaves or tubers, they were reduced in their ability to bind to the roots of alternative crops and/or weeds, suggesting that *Eca* may interact with such plants in the environment. These findings may have important implications for disease management, e.g. by reducing the contamination of seed potato stocks through appropriate crop rotation and weed management strategies.

Conclusion

Genome-scale approaches have provided an important way forward in understanding of the biological processes involved in disease initiation, plant resistance and the ecology of *Eca*.

However, there are still many issues to be resolved. Future work will seek to identify novel Type III-secreted effector proteins, to better understand regulatory processes involved during the *Eca*-plant interaction, to more fully understand the nature of plant defences against *Eca* infection, and to determine whether plants other than potato might act as reservoirs for contamination and subsequent infection. Where possible, these approaches will be used to develop control strategies for this economically-important plant pathogen.

References

Bell, K.S., Sebaihia, M., Pritchard, L., Holden, M., Hyman, L.J., Holeva, M. C., Thomson, N.R., Bentley, S.D., Churcher, L.J.C., Mungall, K. *et al.* (2004) Genome sequence of the enterobacterial phytopathogen *Erwinia carotovora* subsp. *atroseptica* and characterisation of novel virulence factors. *Proceedings of the National Academy of Sciences, USA* 101, 11105-11110.

Dellagi, A., Helibronn, J., Avrova, A. O., Montesano, M., Palva, E.T., Stewart, H.E., Toth, I.K., Cooke, D.E., Lyon, G. D. and Birch, P.R.J. (2000) A potato gene encoding a WRKY-like transcription factor is induced in interactions with *Erwinia carotovora* subsp. *atroseptica* and *Phytophthora infestans* is coregulated with class I endochitinase expression. *Molecular Plant-Microbe Interactions* 13, 1092-1101.

Gardan, L., Gouy, C., Christen, R. and Samson, R. (2003) Elevation of three subspecies of *Pectobacterium carotovorum* to species level: *Pectobacterium atrosepticum* sp. nov., *Pectobacterium betavasculorum* sp. nov. and *Pectobacterium wasabiae* sp. nov. *International Journal of Systematic and Evolutionary Microbiology* 53, 381-391.

He, S-Y., Nomura, K. and Whittam, T.S. (2004) Type III protein secretion mechanisms in mammalian and plant pathogens. *Biochemica et Biophysica Acta* 1694, 181-206.

Holeva, M.C., Bell, K.S., Hyman, L.J., Avrova, A.O., Whisson, S.C., Birch, P.R.J. and Toth, I.K (2004) Use of a pooled transposon mutation grid to demonstrate roles in disease development for *Erwinia carotovora* subsp. *atroseptica* putative Type III secreted effector (DspE/A) and helper (HrpN) proteins. *Molecular Plant-Microbe Interactions* 17, 943-950.

McGowan, S.J., Barnard, A.M.L., Bosgelmez, G., Sebaihia, M., Simpson, N.J.L., Thomson, N.R., Todd, D.E., Welch, M., Whitehead, N.A. and Salmond, G.P.C. (2005) Carbapenem antibiotic biosynthesis in *Erwinia carotovora* is regulated by physiological and genetic factors modulating the quorum sensing-dependent control pathway. *Molecular Microbiology* 55, 526-545.

Pérombelon, M.C.M. (2002) Potato diseases caused by soft rot *Erwinias*: an overview of pathogenesis. *Plant Pathology* 51, 1-12.

Petnicki-Ocwieja T., Schneider D.J., Tam V. C., Chancey S.T., Shan L., Jamir Y., Schechter L.M., Janes M.D., Buell C.R., Tang X.Y., Collmer A. and Alfano J.R. (2002) Genome-wide identification of proteins secreted by the Hrp type III protein secretion system of *Pseudomonas syringae* pv. *tomato* DC3000. *Proceedings of the National Academy of Sciences, USA* 99, 7652-7657.

Pritchard, L., White, J.A., Birch, P.R. and Toth, I.K. (2005) GenomeDiagram: a Python package for the visualisation of large-scale genomic data. *Bioinformatics* Epub Dec 23.

Toth I.K., Bell, K.S., Holeva, M.C. and Birch, P.R.J. (2003) Soft rot *Erwiniae*: from genes to genomes. *Molecular Plant Pathology* 4, 17-30.

Toth, I.K. and Birch, P.R J. (2005) Rotting softly and stealthily. *Current Opinion in Plant Biology* 8, 424-429.

Toth, I.K., Mulholland, V., Cooper, V., Bentley, S., Shih, Y-L., Perombelon, M.C.M. and Salmond, G.P.C. (1997) Generalised transduction in the blackleg pathogen *Erwinia carotovora* subsp. *atroseptica* by bacteriophage ϕM1. *Microbiology* 143, 2433-2438.

Chapter 14

Host-Pathogen Interactions of Relevance to the Phyllosphere

G.W. Sundin
Department of Plant Pathology, Centre for Microbial Ecology, Michigan State University, 103 Center for Integrated Plant Systems, East Lansing, MI 48824-1311, USA
sundin@msu.edu

Introduction

The phyllosphere houses a diversity of microbes, some resident, some transient. A small proportion of phyllosphere residents are capable of invading leaves and initiating disease. Over the last 10 years, tremendous advances have been witnessed in the understanding of host-pathogen interactions, particularly of bacterial pathogens such as *Pseudomonas syringae*. However, knowledge of the relevance of the phyllosphere milieu to host-pathogen interactions is still quite limited. Since the last International Symposium on the Microbiology of Aerial Plant Surfaces held in 2000, genomic analyses of bacterial plant pathogens, including three pathovars of *P. syringae*, have provided a framework from which research to elucidate mechanisms of epiphytic fitness and the inter-relationships between epiphytic fitness, population size, and virulence, can begin. In this chapter, I will review our knowledge of the effect of host genotype on the phyllosphere population size of bacterial pathogens and outline the environmental, organismal and host-mediated cues that *P. syringae* will sense and respond to in the phyllosphere. This chapter expands on one from the meeting in 2000 in which Hirano and Upper (2000) provide excellent background on requirements for pathogenicity in *P. syringae* and the effects of some of these genes on field fitness. In addition, my hope is that during the next few years researchers will address broad questions in phyllosphere microbiology as part of their host-pathogen interactions research portfolio because there are an abundance of questions. Is the situation of a microbial propagule in the phyllosphere merely a prelude to the intimate host-pathogen interplay that occurs within the interior of leaves? Or is the phyllosphere itself a location of importance to early events in pathogenesis and pathogen recognition? Do pathogens behave differently and is there an underlying differential genetic response when colonising the phyllosphere of susceptible vs. resistant hosts or host vs. nonhost plants? How does possession of a functional type III secretion system influence bacterial population size in the phyllosphere?

Factors that Affect Bacterial Colonisation of the Phyllosphere

The leaf surface is a living entity and is part of a plant host that typically possesses a battery of constitutive and inducible traits aimed at limiting the ability of pathogens to cause disease. Phyllosphere geography, particularly the location of structures such as trichomes and veins, clearly influences the localised growth of pathogens. In addition, the nonrandom availability of nutrients influences bacterial growth, and probably enables the growth of immigrating cells into aggregates (Monier and Lindow, 2003b, 2004). While much is known about the behaviour of *P. syringae* and some other epiphytes on leaf surfaces, the effect of host genotype on this behaviour is not well known.

Three broad groups of ecological factors (environmental, organismal and plant host) affect the epiphytic population size of bacterial pathogens. Not coincidentally, these three factor groups also represent the three points of the plant disease triangle model which states that the occurrence of disease is predicated upon the coincident occurrence of a virulent pathogen, susceptible plant host and disease-conducive environment (Agrios, 1997).

The environment imposes stresses upon microbes in most terrestrial habitats; in the phyllosphere, these stresses can be compounded by rapid fluctuations. The types and effects of environmental stress in the phyllosphere have been reviewed (Beattie and Lindow, 1995) and desiccation and exposure to ultraviolet radiation may be the two most important. Strategies of tolerance or avoidance of stress are two possible fitness strategies of foliar pathogens and colonists in response to environmental stress (Beattie and Lindow, 1995). The strategy of avoidance, through endophytic colonisation or through colonisation of external "protected sites" is important in this regard because of the presumed requirement of pathogenicity for widespread endophytic colonisation.

The role of the organism is clearly important in successful epiphytic colonisation: this becomes more significant from a disease standpoint because of the known association of epiphytic population size and disease incidence in several *P. syringae* pathosystems (reviewed in Hirano and Upper, 2000). The sequence of events resulting in successful epiphytic colonisation by *P. syringae* has been carefully and extensively examined by Lindow's group. Immigration to a leaf results in the placement of a *P. syringae* cell in a stressful environment; cells initially respond by reducing cell size, which is a known physiological response to stress (Monier and Lindow, 2003a). Growth is dependent upon the availability of water and sugars, can be associated with certain favourable physical locations on leaves such as broken trichomes, and results in cellular aggregation. Cells in aggregates are increased in survival compared to solitary cells and eventually comprise the majority of a population on an individual leaf (Monier and Lindow 2003b, 2004). Such aggregates may form biofilms, which is a microbial growth habit of significant environmental importance, and the importance of genes regulated by population density-dependent regulatory systems is beginning to become known in *P. syringae* (Quinones *et al.*, 2005). The organismal component of phyllosphere colonisation also includes that of competitors and other organisms located in the phyllosphere geographical context. I will not discuss this directly in this chapter, but researchers must always be cognizant of the potential for other coexisting organisms to influence pathogen dynamics in phyllosphere systems.

Finally, the host plant plays an important role in modulating the population size of bacterial pathogens. Susceptible genotypes are compatible with pathogen growth and are predicted to favour internal colonisation and pathogen growth, increased egress back to the surface, increased dissemination, as well as the possibility of disease incidence. Resistant genotypes are potentially incompatible with internal pathogen growth; although pathogens will colonise the phyllosphere of resistant hosts, their populations are reduced in many cases, presumably due to reduced internal growth and subsequent egress. In addition, plants have evolved a number of pathogen surveillance mechanisms and can recognise pathogen-associated molecular patterns (PAMPs) which can limit the growth of pathogens on non-host plants.

Differential Colonisation of Resistant and Susceptible Host Plants by Bacterial Pathogens

Field observations have shown an effect of host genotype on the epiphytic populations of certain bacterial pathogens. There are two broad observations of significance in this regard. The first is that *P. syringae* pathogens are better epiphytic colonists of host plants than non-host plants (for example, see Ercolani and Crosse, 1966). The second observation is that epiphytic populations of bacterial pathogens tend to be larger on susceptible cultivars than resistant cultivars of host plants. For example, Crosse (1963) in a study of epiphytic populations of the bacterial canker of cherry pathogen *P. syringae* pv. *morsprunorum*, found that the sensitive cherry cultivar 'Napoleon' consistently supported larger epiphytic populations than the resistant cultivar 'Roundel'. Likewise, the snap bean cultivar 'Eagle', which is susceptible to the bacterial brown spot pathogen *P. syringae* pv. *syringae* (Pss), supported epiphytic Pss populations that were up to 1,000-fold larger than those on the resistant cultivar WBR 133 (Daub and Hagedorn, 1981). With *P. syringae*, these population trends are not universal, for example, equal populations of Pss were recovered from susceptible and resistant wheat cultivars (Fryda and Otta, 1978). Host effects on epiphytic population size have also been observed with *Xanthomonas campestris* pv. *vesicatoria* on pepper, as 100 to 1,000-fold larger epiphytic populations are recovered from the susceptible cultivar 'Jupiter' compared to the resistant cultivar 'Boynton Bell' (Pernezny and Collins, 1997).

In phytopathogenic bacteria commonly found as epiphytes on leaves, pathogenicity is dependent upon the presence of a functional type III secretion system. Type III secretion is a mechanism whereby bacterial pathogens establish an intimate association with cells of their respective host and inject effector proteins directly into these host cells (for a general review, see Lahaye and Bonas, 2001). The function of individual effector proteins may be varied, but their delivery to host cells ultimately results in the manifestation of disease symptoms. The influence of type III secretion on epiphytic population size of *P. syringae* has been examined by the group of Hirano and Upper. Two field studies using defined mutants have shown that a functioning type III secretion system is necessary for *P. syringae* to attain maximal epiphytic population size (Hirano *et al.*, 1997; 1999). In another recent study that also incorporated field experiments, it was demonstrated that the lack of a critical effector protein (AvrBs2) resulted in reduced epiphytic populations of *X. axonopodis* pv. *vesicatoria* on pepper (Wichmann and Bergelson, 2004).

Bacterial-induced modifications of the leaf surface resulting in increased host cell leakage of water and nutrients represent a potential mechanism whereby these cells can enhance their colonisation of the phyllosphere environment. For example, bacterial production of indole-3-acetic acid (IAA) may contribute to nutrient leakage (Brandl and Lindow 1998; Manulis *et al.*, 1998). This type of modification or induction of host cell leakage is akin to that accomplished by pathogens colonizing leaves internally in the apoplast. In these cases of course, pathogenesis mediated by type III secretion leads to cell collapse that macroscopically is termed watersoaking. The only difference between external and internal habitat modification is that external modification activities, such as IAA-induced host cell leakage, are performed by both pathogens and non-pathogens.

The ability of pathogen populations to grow internally within susceptible host plants is thought to contribute to larger epiphytic populations due in part to increased egress onto the leaf surface (Beattie and Lindow, 1995). Growth within internal leaf locations would provide protection for bacterial pathogens from the rapidly fluctuating environmental stress conditions present in the phyllosphere (reviewed in Lindow and Brandl, 2003). The ability to access internal leaf sites appears to be dependent upon pathogenicity (Wilson *et al.*, 1999).

Identification of Traits Important to Epiphytic Fitness

Although the epiphytic population size of *P. syringae* type III secretion mutants is significantly reduced in the field, mutations in other loci can also reduce epiphytic population size. For example, traits that facilitate survival in response to environmental stress such as motility, tolerance of ultraviolet radiation, or EPS production are important to epiphytic population size (Haefele and Lindow, 1987; Sundin and Murillo, 1999; Yu *et al.*, 1999).

The Lindow laboratory has been involved for several years in attempts to identify genes of importance to epiphytic fitness in *P. syringae* (Lindow *et al.*, 1993; Marco *et al.*, 2003, Marco *et al.*, 2005). A novel genetic strategy termed habitat-inducible rescue of survival (HIRS) was utilised by Marco *et al.* (2005) and, coupled with the recent completion of the genome sequence of Pss B728a (Feil *et al.*, 2005), resulted in the identification of a number of genes that were induced in epiphytically growing cells. The HIRS study resulted in an estimate that approximately 2% of the Pss B728a genome was induced during epiphytic growth (Marco *et al.*, 2005). An examination of a compilation of traits known to influence epiphytic fitness or to be induced during epiphytic growth reveals a variety of phenotypes (Table 14.1). Of significance, only the syrE syringomycin synthetase and a gene encoding an ApbE-family protein were also identified in an *in vivo* expression technology examination of genes induced in *P. syringae* pv. *tomato* DC3000 during plant infection (Table 14.1; Boch *et al.*, 2002).

The contrasting genes identified in the HIRS study and the IVET study of Pst DC3000 are interesting. The HIRS study examined Pss genes induced on the leaf surface following a shift from moist permissive growth conditions to dry conditions usually only exploited by pathogenic Pss strains (Marco *et al.*, 2005). The Pst DC3000 study identifed genes induced following bacterial infiltration of *Arabidopsis* leaves (Boch *et al.*, 2002). The distinct lists of identified genes could lead one to speculate that the epiphytic lifestyle and the pathogenic lifestyle are completely different. However, the requirement of a functional type III secretion system for the maintenance of population size in Pss B728a partially refutes that speculation because epiphytic populations were sampled in this study (Hirano *et al.*, 1999). Marco *et al.* (2005) suggested that there was a continuum between epiphytic growth and pathogenesis; similarly, Beattie and Lindow (1999) suggested that external and internal leaf-associated populations probably form a continuum due to the processes of ingress and egress. It would be interesting to track the expression of genes identified in both the HIRS and IVET studies both throughout an infection event and in a type III secretion mutant for the same time frame following leaf surface inoculation. Additional studies could be done tracking the expression of those same genes following inoculation to a resistant host or to a non-host. These types of studies would help to delineate the role of pathogenicity phenotype in phyllosphere fitness and the over-riding effect of host genotype.

Table 14.1. Listing of genes and phenotypes of known importance for epiphytic fitness, and selected genes (7 of 59) identified in an experiment using the novel method habitat-inducible rescue of survival (HIRS).

Gene(s) or trait	Function	Reference
Alginate production	biofilm formation	Yu *et al.*, 1999
gacS	global regulator of virulence factors	Hirano *et al.*, 1997
hrpW	type III secretion	Hirano *et al.*, 1999
metXW	methionine biosynthesis	Andersen *et al.*, 1998
Motility	cell movement to protected sites	Haefele and Lindow, 1987
rulAB	UVR survival	Sundin and Murillo, 1999
HIRS study		Marco *et al.*, 2005
syrE	syringomycin synthetase	
CEL orf6	putative virulence gene	
ApbE-family protein	putative virulence gene	
ssuE	member of sulfate starvation regulon	
betC	sulfatase family enzyme	
corA	putative magnesium transporter	
sylE	putative syringolin A exporter	

Activities of Pathogens in the Phyllosphere and in the Field

If the phyllosphere was a nutrient-rich, stress-free environment, the diversity of plant pathogens currently known may not have evolved. However, since the phyllosphere is relatively nutrient poor, water depleted and influenced by environmental stress, bacterial pathogens have evolved a plethora of functions to survive and overcome the stresses of this habitat. The ability of growing cells of *P. syringae* to aggregate appears to contribute significantly to survival and ultimately to pathogenesis. Aggregation likely affects the cell density-dependent expression of genes regulated by quorum-sensing (Quinones *et al.*, 2005). Quorum-sensing is an elegant regulatory system in bacteria that controls the expression of genes that are apparently more important to a group or population of cells than to solitary cells. Since aggregated Pss cells survive better than solitary cells in the phyllosphere (Monier and Lindow 2003b), it seems likely that quorum sensing influences traits that directly contribute to survival or indirectly through facilitating internal invasion of leaves.

Our understanding of *P. syringae* biology, epiphytic colonisation and pathogenesis has increased tremendously over the past 10 years. Field analyses of pathogen activities have proceeded more slowly. Genomic studies will be quite helpful in assessing the impact of environmental stress or the impact of host genotype on gene expression in pathogens. Initial genomic studies of stress responses are revealing interesting effects on global gene expression. For example, a recent study of the effects of solar radiation exposure on global gene expression in the enterobacterium *Shewanella oneidensis* MR-1 revealed that 29% of the genome was differentially regulated with an exceptionally large number of genes repressed by this stress (Qiu *et al.*, 2005). The biology of the phyllosphere system encompasses the natural variability of the environment. We will continue to require field understanding and validation of new concepts of importance to phyllosphere microbiology and I hope that more of these new concepts will originate from field studies.

I will briefly summarise what I think we know and don't know about host-pathogen interactions of relevance to the phyllosphere:

1 Following immigration to a leaf, *Pss* cells in favourable locations will grow and form aggregates (Monier and Lindow 2003b).

1a This work was done with a virulent *Pss* strain on a compatible host. What result would be seen if these same experiments were done using a resistant host or a *Pss* type III secretion mutant?

2 Mutants with defects in genes that are thought to affect surface population size (such as UVR-sensitive mutants) can also exhibit reduced epiphytic population size.

2a Is the population size reduced because more cells are initially killed on leaf surfaces, thus leaving fewer cells available to grow internally within leaves?

3 Pathogenic *Pss* strains access protected sites on and within leaves and grow internally to a greater extent than nonpathogenic *Pss* strains (Wilson *et al.,* 1999).

3a Would the same results be observed for a nonpathogenic type III secretion mutant?

4 The carriage of a functional type III secretion system is known to influence *Pss* epiphytic population size on apparently healthy leaves (Hirano *et al.,* 1999).

4a What is the contribution of this pathogenicity determinant to growth in the phyllosphere?

Pathogenicity appears to promote longer-term survival and growth of *P. syringae* strains in the favourable environment of a susceptible host plant. On a resistant host, the plant turns the tables on the pathogen through recognition and a cascade of events resulting in a hypersensitive response that ultimately reduces pathogen population size. The lower pathogen populations observed on resistant hosts in previous field studies may be similar to those seen for type III secretion mutants, i.e. the surviving epiphytic population on resistant hosts consists of those cells that have not established an intimate interaction with the host through type III secretion. Studies are needed to further define the role of pathogenicity and virulence determinants in promoting epiphytic fitness, however, it will be difficult if not impossible to obtain accurate data if studies are done in growth chambers. The reason for this is that a greater proportion of the population of pathogenicity-deficient mutants is expected to be on the surface, exposed to environmental stress and nutrient depletion, etc.

and the rapid fluctuations of these stresses, conditions that would be difficult to simulate in chambers.

Acknowledgements

Work in my laboratory is supported by the USDA National Research Initiative, USDA-CSREES, and the Michigan Agricultural Experiment Station.

References

Andersen, G.L., Beattie, G.A. and Lindow, S.E. (1998) Molecular characterisation and sequence of a methionine biosynthetic locus from *Pseudomonas syringae. Journal of Bacteriology* 180, 4497-4507.

Agrios, G.N. (1997) *Plant Pathology*. Academic Press, San Diego, California, USA.

Beattie, G.A. and Lindow, S.E. (1995) The secret life of foliar bacterial pathogens on leaves. *Annual Review of Phytopathology* 33, 145-172.

Beattie, G.A. and Lindow, S.E. (1999) Bacterial colonization of leaves: a spectrum of strategies. *Phytopathology* 89, 353-359.

Boch, J., Joardar, V., Gao, L., Robertson, T.L., Lim, M. and Kunkel, B.N. (2002) Identification of *Pseudomonas syringae* pv. *tomato* genes induced during infection of *Arabidopsis thaliana. Molecular Microbiology* 44, 73-88.

Boureau, T., Routtu, J., Roine, E., Taira, S. and Romantschuk, M. (2002) Localization of hrpA-induced *Pseudomonas syringae* pv. *tomato* DC3000 in infected tomato leaves. *Molecular Plant Pathology* 3, 451-460.

Brandl, M.T. and Lindow, S.E. (1998) Contribution of indole-3-acetic acid production to the epiphytic fitness of *Erwinia herbicola. Applied and Environmental Microbiology* 64, 3256-3263.

Crosse, J.E. (1963) Bacterial canker of stone fruits. V. A comparison of leaf-surface populations of *Pseudomonas morsprunorum* in autumn on two cherry varieties. *Annals of Applied Biology* 52, 97-104.

Daub, M.E. and Hagedorn, D.J. (1981) Epiphytic populations of *Pseudomonas syringae* on susceptible and resistant bean lines. *Phytopathology* 71, 547-550.

Ercolani, G.L. and Crosse, J.E. (1966) The growth of *Pseudomonas phaseolicola* and related plant pathogens *in vivo. Journal of General Microbiology* 45, 429-439.

Feil, H., Feil, W.S., Chain, P., *et al.*, (2005) Comparison of the complete genome sequences of *Pseudomonas syringae* pv. *syringae* B728a and pv. *tomato* DC3000. *Proceedings of the National Academy of Sciences, USA* 102, 11064-11069.

Fryda, S.J. and Otta, J.D. (1978) Epiphytic movement and survival of *Pseudomonas syringae* on spring wheat. *Phytopathology* 68, 1064-1067.

Haefele, D.M. and Lindow, S.E. (1987) Flagellar motility confers epiphytic fitness advantages upon *Pseudomonas syringae. Applied and Environmental Microbiology* 53, 2528-2533.

Hirano, S.S., Charkowski, A.O., Collmer, A., Willis, D.K. and Upper, C.D. (1999) Role of the Hrp type III protein secretion system in growth of *Pseudomonas syringae* pv. *syringae* B728a on host plants in the field. *Proceedings of the National Academy of Sciences, USA* 96, 9851-9856.

Hirano, S.S., Ostertag, E.M., Savage, S.A., Baker, L.S., Willis, D.K. and Upper, C.D. (1997) Contribution of the regulatory gene lemA to field fitness of *Pseudomonas syringae* pv. *syringae. Applied and Environmental Microbiology* 63, 4304-4312.

Hirano, S.S. and Upper, C.D. (2000) Bacteria in the leaf ecosystem with emphasis on *Pseudomonas syringae* - a pathogen, ice nucleus, and epiphyte. *Microbiology and Molecular Biology Reviews* 64, 624-653.

Hirano, S.S. and Upper, C.D. (2002) Effect of pathogenicity-associated genes on phyllosphere fitness of *Pseudomonas syringae* pv. *syringae.* In: S.E. Lindow, E.I. Hecht-Poinar and V.J. Elliott (eds), *Phyllosphere Microbiology*, American Phytopathological Society, St. Paul, Minnesota, USA. pp. 81-99.

Lahaye, T. and Bonas, U. (2001) Molecular secrets of bacterial type III effector proteins. *Trends in Plant Science* 6, 479-485.

Lindow, S.E., Andersen, G. and Beattie, G.A. (1993) Characteristics of insertional mutants of *Pseudomonas syringae* with reduced epiphytic fitness. *Applied and Environmental Microbiology* 59, 1593-1601.

Lindow, S.E. and Brandl, M.T. (2003) Microbiology of the phyllosphere. *Applied and Environmental Microbiology* 69, 1875-1883.

Manulis, S., Haviv-Chesner, A., Brandl, M.T., Lindow, S.E. and Barash, I. (1998) Differential involvement of indole-3-acetic acid biosynthetic pathways in pathogenicity and epiphytic fitness of *Erwinia herbicola* pv. *gypsophilae*. *Molecular Plant-Microbe Interactions* 11, 634-642.

Marco, M.L., Legac, J. and Lindow, S.E. (2003) Conditional survival as a selection strategy to identify plant-inducible genes of *Pseudomonas syringae*. *Applied and Environmental Microbiology* 69, 5793-5801.

Marco, M.L., Legac, J. and Lindow, S.E. (2005) *Pseudomonas syringae* genes induced during colonization of leaf surfaces. *Environmental Microbiology* 7, 1379-1391.

Monier, J.M. and Lindow, S.E. (2003a) *Pseudomonas syringae* responds to the environment on leaves by cell size reduction. *Phytopathology* 93, 1209-1216.

Monier, J.M. and Lindow, S.E. (2003b) Differential survival of solitary and aggregated bacterial cells promotes aggregate formation on leaf surfaces. *Proceedings of the National Academy of Sciences, USA* 100, 15977-15982.

Monier, J.M. and Lindow, S.E. (2004) Frequency, size, and localization of bacterial aggregates on bean leaf surfaces. *Applied and Environmental Microbiology* 70, 346-355.

Morris, C.E. and Monier, J.M. (2003) The ecological significance of biofilm formation by plant-associated bacteria. *Annual Review of Phytopathology* 41, 429-453.

Pernezny, K. and Collins, J. (1997) Epiphytic populations of *Xanthomonas campestris* pv. *vesicatoria* on pepper: relationships to host-plant resistance and exposure to copper sprays. *Phytopathology* 81, 791-794.

Qiu, X.Y., Tiedje, J.M. and Sundin, G.W. (2005) Genome-wide examination of the natural solar radiation response in *Shewanella oneidensis* MR-1. *Photochemistry and Photobiology* 81, 1559-1568.

Quinones, B., Dulla, G. and Lindow, S.E. (2005) Quorum sensing regulates exopolysaccharide production, motility, and virulence in *Pseudomonas syringae*. *Molecular Plant-Microbe Interactions* 18, 682-693.

Sundin, G.W. and Murillo, J. (1999) Functional analysis of the *Pseudomonas syringae* rulAB determinant in tolerance to ultraviolet B (290 to 320 nm) radiation and distribution of rulAB among *P. syringae* pathovars. *Environmental Microbiology* 1, 75-87.

Wichmann, G. and Bergelson, J. (2004) Effector genes of *Xanthomonas axonopodis* pv. *vesicatoria* promote transmission and enhance other fitness traits in the field. *Genetics* 166, 693-706.

Wilson, M., Hirano, S.S. and Lindow, S.E. (1999) Location and survival of leaf-associated bacteria in relation to pathogenicity and potential for growth within the host. *Applied and Environmental Microbiology* 65, 1435-1443.

Yu, J., Penaloza-Vasquez, A., Chakrabarty, A.M. and Bender, C.L. (1999) Involvement of the exopolysaccharide alginate in the virulence and epiphytic fitness of *Pseudomonas syringae* pv. *syringae*. *Molecular Microbiology* 33, 712-720.

Chapter 15

Effects of Endophytes on Colonisation by Leaf Surface Microbiota

R.W.S. Weber[1] and H. Anke[2]
[1] Department of Biotechnology, University of Kaiserslautern, Erwin-Schroedinger-Str. 56, 67663 Kaiserslautern, Germany; [2] Institute of Biotechnology and Drug Research (IBWF), Erwin-Schroedinger-Str. 56, 67663 Kaiserslautern, Germany
rwsweber@rhrk.uni-kl.de

Introduction

Endophytes

The original definition of the term 'endophyte', like so many important terms in plant pathology, was created by de Bary (1866) who distinguished phytopathogenic fungi colonising the interior of plant organs (endophytes) from epiphytes growing mainly on the leaf surface, such as powdery mildews (Erysiphales). The current view of endophytes is restricted to that of an organism spending a sizeable part of its life cycle inside a healthy looking (asymptomatic) host plant from which it can be isolated after rigorous surface sterilisation. Although different kinds of microbe may be involved in endophytic associations, most of the literature on this topic deals with fungi, and the term 'endophyte' is therefore generally taken to imply fungi unless otherwise stated. It is not possible unequivocally to distinguish endophytes from epiphytes because even obligate endophytes belonging to the genus *Neotyphodium* are capable of growing from the interior onto the surface of the living host, forming epiphytic mycelium and conidia (Moy *et al.*, 2000). Conversely, predominantly epiphytic fungi such as *Alternaria* or *Cladosporium* are capable of endophytic colonisation of living plants, especially under adverse conditions.

Endophytes are ubiquitous, and probably all vascular and non-vascular terrestrial plants harbour them. They have also been isolated from freshwater and marine plants, and from lichens. Endophytes may be distributed in a hyperdiverse pattern especially in tropical trees, of which single leaves may be colonised by numerous discrete individual mycelia, some occurring commonly, others being rare. Hyde *et al.* (1997) have shown that there are considerably more species of endophytes than pathogens or saprotrophs associated with tropical palms, and this may also hold for other plants. The conclusion by Dreyfuss and Chapela (1994) that there may be 1.3 million species of endophytes worldwide would render this group the largest contributing factor to the total worldwide fungal biodiversity, which has been conservatively estimated at 1.5 million species (Hawksworth, 2001). In general, most plant species are thought to harbour a number of specific endophytes, with infrequent isolates probably representing chance infections by species specialised on other hosts. It is therefore prudent for ecophysiological studies to focus on fungi isolated as a dominant component of the endophyte population of their host.

The Origin of Endophytism

Two broad ecological groups of endophytes may be distinguished. Many endophytes of grasses (Poaceae) grow systemically throughout above-ground host organs and spread mainly by clonal means, surviving in perennial culms or as seed-borne infections. They are noted as producers of biologically active secondary metabolites, especially alkaloids, in the living host. Their toxicity at physiological concentrations to grazing insects and mammals has led to the view of the endophyte-grass association as one of mutualistic symbiosis. Typical endophytes of Poaceae showing the above characteristics belong to the Clavicipitaceae (Pyrenomycetes, Ascomycota), and the assignment of both hosts and endophytes to narrow taxonomic groupings underlines the specialised nature of their association.

Endophytes belonging to the second category infect a much broader range of hosts, including trees, shrubs and herbaceous plants. Some of these resemble the clavicipitalean endophytes of grasses in showing active systemic intercellular growth in the living plant, whereas many others are confined to small discrete patches of hyphae between or within one or a few individual plant cells. However, they may display rapid colonisation following death of the host plant or host organ. Non-clavicipitalean endophytes are recruited from a wide taxonomic background, with a range of ascomycete orders in the Pyrenomycetes, Loculoascomycetes and inoperculate discomycetes predominating. Their production of metabolites inside the living host is far less well documented, and suggestions of mutualism have been made in only a few cases. Benefits to the host plant are mainly indirect and, where characterised, chiefly concern the production of substances inhibitory to herbivorous insects (Miller *et al.*, 2002).

The carefully balanced nature of the grass/endophyte symbiosis suggests a long evolutionary history shared between host and fungus, and it is possible, using extant members of Clavicipitales as examples, to reconstruct a gradual transition from pathogenicity as found for example in *Epichloë* spp. to the seemingly obligate endophytism of *Neotyphodium* (White and Morgan-Jones, 1996). Such evolutionary trends may be even more straightforward in the case of endophytes of woody or herbaceous plants, with a single mutation event shown to be sufficient to convert the anthracnose pathogen *Colletotrichum magna* into a symptomless systemic endophyte (Redman *et al.*, 1999). Further, *Colletotrichum* strains fully pathogenic on a given host plant may survive in other plant species for prolonged periods as symptomless endophytes (Freeman *et al.*, 2001). Either or both of the above phenomena may be responsible for the fact that numerous fungi con-specific or con-generic with plant pathogens can be isolated as endophytes from healthy-looking host tissues; examples include *Fusarium* spp., *Phomopsis* spp. and *Colletotrichum* spp. Many of these phytopathogens show a hemibiotrophic mode of infection in which penetration of the host cuticle leads to a brief (24-48 h) biotrophic phase of limited intracellular growth, which then gives rise to the necrotrophic phase of aggressive tissue colonisation and destruction. Simplistically, we may view endophytes of this kind as would-be pathogens stalled at the biotrophic stage.

Epiphyte-Endophyte Interactions Involving the Living Host

Because of their economic importance in affecting grazing livestock, endophytes of poaceous grasses are the best examined in terms of secondary metabolite production *in planta*. Various alkaloids have been detected in living host plants (see Lane *et al.*, 2000). Insecticidal toxins include lolines at 0.1-10 mg/g dry weight (Bush *et al.*, 1993) and peramine at up to 40 µg/g (Keogh *et al.*, 1996; Spiering *et al.*, 2002, 2005), whereas compounds primarily toxic to grazing mammals are ergovaline at up to 4 µg/g (Spiering *et al.*, 2002, 2005) and lolitrem B at or below 10 µg/g (Keogh *et al.*, 1996; Spiering *et al.*, 2005). These alkaloids have only low antimicrobial activities. Indeed, even sclerotia of the ergot fungus *Claviceps purpurea*, containing alkaloids at concentrations of 0.2-4.0 mg/g dry weight (Pažoutová *et al.*, 2000), are commonly overgrown by moulds when incubated in a moist chamber.

Whilst endophytes of Poaceae may produce other compounds with activity against fungi colonising from air-borne inoculum, these have not generally been identified or quantified *in planta*. However, Wicklow *et al.* (2005) showed that the antimicrobial metabolites pyrrocidines A and B are produced by the maize endophyte *Acremonium zeae* in pure culture as well as in mature kernels at harvest, and although no quantitative data were provided, the authors suggested that pyrrocidines are responsible for inhibiting the colonisation of developing kernels by *Fusarium verticillioides* (= *F. moniliforme*) and *Aspergillus flavus*. Both of these are producers of potent mycotoxins (fumonisins and aflatoxins, respectively) which are of relevance to human and animal health as carcinogens. Whereas *A. flavus* is an epiphytic species, *F. verticillioides* is itself a systemic endophyte of maize spreading vertically by infected seeds, and horizontally by conidial inoculum produced on dead plant tissue (Oren *et al.*, 2003). Like *A. zeae*, *F. verticillioides* can inhibit colonisation by *A. flavus* (Wicklow *et al.*, 1988). Under certain circumstances, *F. verticillioides* may become pathogenic, rotting roots, stalks and ears. Fumonisins are accumulated mainly during saprotrophic or pathogenic growth, reaching levels of 100 µg/g or more (Marasas, 2001). Bacon *et al.* (2001) have provided evidence that endophytic hyphae may also produce fumonisins, although the concentrations measured (up to 10 µg/g dry weight) were much lower than those required to inhibit other fungi (around 0.2-3.6 mg/ml). Fumonisins are therefore more likely to play a role in killing host tissue and/or in colonisation during the following saprotrophic growth phase.

An interesting case is that of taxol, a microtubule-stabilizing cytotoxic compound of medicinal value as an anti-cancer drug. Taxol has antifungal activities, and its concentrations in bark and needles of yew trees (*Taxus* spp.) at 10-500 µg/g would be sufficient to inhibit fungi attempting to colonise these host tissues (Wagner and Flores, 1994). This is true especially for oomycetes which are particularly sensitive to taxol (Strobel, 2003). Intriguingly, taxol and related metabolites (taxanes) are produced both by yew trees and by a wide range of endophytic fungi and even actinomycetes isolated from them (Tan and Zou, 2001). In pure suspension culture, taxol production by *Taxus* cells can be stimulated by adding fungal cell wall material as an elicitor (Wang *et al.*, 2001), supporting the notion that taxol may be a defence-related compound. It is conceivable that the endophyte population in living yew trees contributes to their antifungal defence. Unfortunately, neither the extent of endophyte colonisation within *Taxus* trees nor the relative biosynthetic contributions by tree host and endophytes to the overall taxol pool in natural trees have been quantified as yet.

It is well known that the presence of non-pathogenic endophytic fungi can protect their hosts against colonisation by taxonomically related (and sometimes also unrelated) plant pathogens. In the case of *Colletotrichum*, this phenomenon seems to be due to a stimulating effect of the endophyte on the host's immune system. Redman *et al.* (1999) have explained their findings by the assumption that the endophyte produces high constitutive levels of antigenic metabolites, so that only low additional amounts need to be released by an incoming pathogen to trigger the immune response of the host plant. This would explain the observation that endophyte-containing hosts are able to launch a more rapid defence response than plants free from endophytes (Redman *et al.*, 1999). Similar phenomena seem to account for the protection of tomato roots by apathogenic *Fusarium* strains against *Fusarium* wilt pathogens (Larkin and Fravel, 1999). In contrast, suppression of *Fusarium verticillioides* by endophytic bacteria (*Bacillus subtilis*) in maize plants has been explained by the concept of competitive exclusion in which *B. subtilis* occupies the same intercellular spaces colonised by *F. verticillioides* (Bacon *et al.*, 2001).

To summarise, whilst endophytes may well act directly or indirectly against penetrating plant pathogens, to the best of our knowledge there is no conclusive evidence of their direct effect on the population of epiphytic microbes on healthy plant surfaces. Such an effect may not even be necessary if we assume that most epiphytic filamentous fungi are present as spores (Andrews and Harris, 2000), whereas epiphytic yeasts may show active vegetative growth but are unable to penetrate the epidermis because they lack the specialised hyphal structures required for this task.

Epiphyte-Endophyte Interactions in Senescent Host Tissue

Endophyte populations in living plant shoots can be quite stable, and in plants other than grasses the density of discrete endophyte colonies may increase gradually with repeated successful infection events from air-borne inoculum. This situation may last throughout the growing season or even for several years, for example in the case of conifer needles or tree branches. Petrini (1991) has considered such populations to be in an equilibrium which is upset by the onset of senescence. This concerns especially the balance between endophytes and epiphytes; unless an endophyte possesses a high competitive potential, it will be replaced by fungi colonizing from the outside. Not surprisingly, many endophytes react strongly to the onset of senescence, as the following examples demonstrate.

Phoma medicaginis

Medicago spp. are of significance in agriculture as a fodder crop, and in research as a genetically tractable leguminous plant. Although *Medicago* spp. may be attacked by a range of hemibiotrophic pathogens in agricultural situations, wild populations of *M. lupulina* and *M. sativa* observed for several growing seasons on the University of Kaiserslautern Campus were entirely free from visible disease symptoms or showed only limited lesions in the shape of confined melanised necrotic spots (see Fig. 15.2A) which did not affect overall plant performance. Nonetheless, a characterisation of the endophyte population within these plants revealed several dominant species, all of which were closely related to hemibiotrophic plant pathogens. There was a sharp demarcation between the roots colonised predominantly by *Fusarium* spp. and the stems and leaves in which *Phoma medicaginis*, *Alternaria* sp. and *Colletotrichum trifolii* predominated (Fig. 15.1).

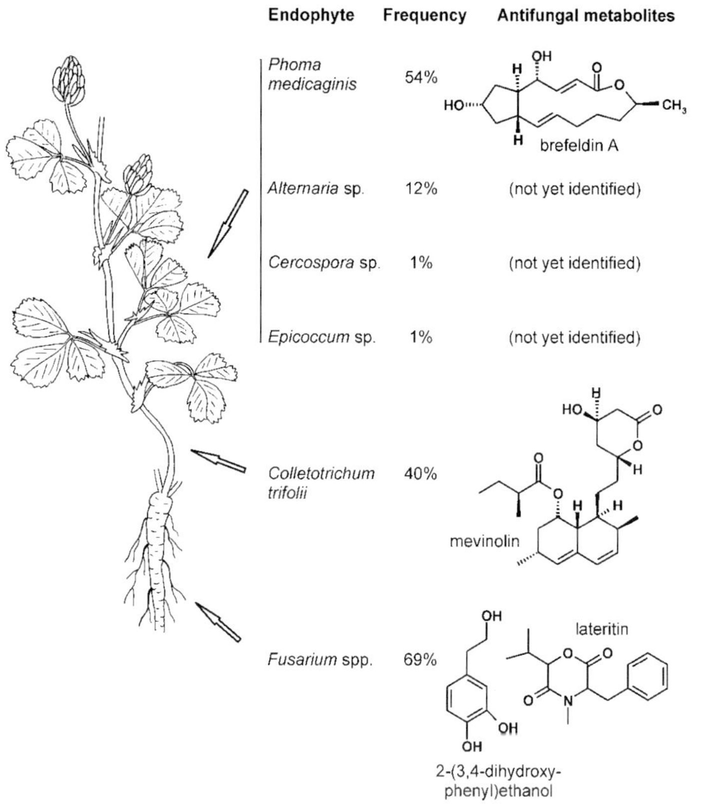

Fig. 15.1. The most abundant endophytes isolated from *Medicago lupulina* and *M. sativa*, and the percentage of surface sterilised root, shoot base or shoot and leaf segments from which they were isolated. Identified metabolites with antifungal activity are also indicated (data from Weber *et al.*, 2004; R.W.S. Weber and E. Stenger, unpublished).

Since *P. medicaginis* was the dominant aerial endophyte, isolated from over 50% of surface-sterilised leaf or stem segments (1-2 mm length), this fungus was examined in detail (Weber *et al.*, 2004). Spray-inoculation experiments with *P. medicaginis* conidia on *M. sativa* and *M. truncatula* gave rise to limited infection counts similar to those observed in the field. No further development during this endophytic phase was observed, and leaves unfolding subsequent to the conidial spray remained free from endophyte infections, indicating that no systemic growth had taken place. However, when leaves collected in the field (Fig. 15.2A) or from artificially inoculated plants were dried and then incubated in a moist chamber, a saprotrophic phase of rapid tissue colonisation was observed, originating from the limited endophytic infection sites. Within 5-10 days the leaf surface became littered with sporulating pycnidia of *P. medicaginis* (Fig. 15.2B,C). Spores of common phylloplane fungi such as *Cladosporium*, *Alternaria* or *Penicillium* sprayed onto such leaves formed mycelium only well outside the areas colonised by *P. medicaginis* (Fig. 15. 2C,D), suggesting antibiosis by *P. medicaginis* against its competitors during the saprotrophic phase.

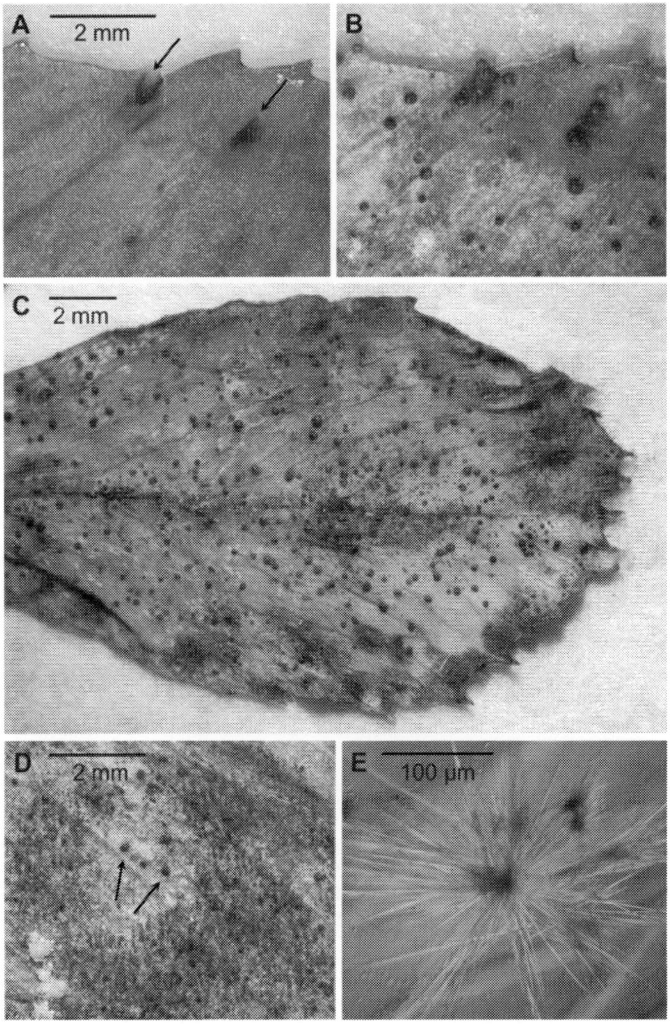

Fig. 15.2. Colonisation of *Medicago sativa* by endophytic *Phoma medicaginis.* (A) Two clearly visible infection foci (arrows) after rehydration of a dried leaf. (B) The same leaf after 5 days; numerous pycnidia indicate that *P. medicaginis* has colonised the leaf tissue. (C) Competition on a *Medicago* leaf. Conidia of *Cladosporium cladosporioides* were sprayed onto a *P. medicaginis*-infected leaf 5 days after rehydration. The photograph, taken a further 3 days later, shows that *C. cladosporioides* has grown only at the extreme margins of the leaf (dark areas). (D) As for C, except that here *C. cladosporioides* was sprayed onto the leaf at the time of rehydration. A few pycnidia of *P. medicaginis* (arrows) are in the centre of a halo of plant tissue not colonised by *C. cladosporioides*. (E) Putative brefeldin A crystals in a conidial suspension of *P. medicaginis* harvested from a malt-agar plate.

The only antifungal metabolite produced in pure culture by our isolates of *P. medicaginis* was brefeldin A (Fig. 15.1; Weber *et al.*, 2004), previously shown by Noordeloos *et al.* (1993) to be deposited in a crystalline form in agar cultures of *P.*

medicaginis. We also observed needle-shaped crystals in malt-extract agar (Fig. 15.2E), although at 20 µg ml^{-1} the production of brefeldin A in liquid culture would have been too low to inhibit phylloplane fungi. However, when *P. medicaginis* was allowed to colonise autoclaved *Medicago* shoots, brefeldin A was produced at up to 3 mg/g dry weight (or 920 µg/ml, assuming a density of 1 g/ml for fully hydrated plant tissue). This is well beyond the toxicity threshold of 50-250 µg/ml against a range of competing phylloplane fungi tested (Weber *et al.*, 2004). No trace of brefeldin A was detected in living leaf tissue colonised endophytically by *P. medicaginis*, suggesting that this metabolite is produced only during or after the switch from the latent endophytic stage to the saprotrophic phase of tissue colonisation, and that it may be a means by which *P. medicaginis* maintains its advantage of early arrival on the substrate against incoming air-borne spores of epiphytic fungi. The putative role of brefeldin A as a habitat defence compound is supported by its detection in numerous strains of *P. medicaginis* (Noordeloos *et al.*, 1993), its secretion into the extracellular medium, and its antibiotic activity against a broad range of fungi as well as other organisms (Betina, 1992; Weber *et al.*, 2004). The excessive concentration of brefeldin A in colonised *Medicago* shoots ties in with a widespread phenomenon that fungi produce higher amounts of secondary metabolites in solid-substrate fermentation than in submerged culture.

The wood-inhabiting ascomycete *Hypoxylon fragiforme* is present in a latent (endophytic) state in healthy sapwood especially of beech, and Chapela and Boddy (1988) have shown that the switch to the actively colonizing saprotrophic stage leading to fructification is triggered by drying, such as is experienced when a living branch colonised by *H. fragiforme* breaks off the main tree. The situation may be more complex in *P. medicaginis*, with physiological changes, for example the host death-associated lifting of growth restrictions imposed by phytoalexins, also likely to play a part (see Fig. 15.4). We have found that brefeldin A biosynthesis can be stimulated by adding spores of competing phylloplane fungi to liquid cultures (Fig. 15.3). These results indicate that *P. medicaginis* may be able to react to the presence of phylloplane fungi by enhanced brefeldin A synthesis, although it is unclear to which extent such a regulation operates in nature, given that the astonishing brefeldin A concentration measured in sterile colonised tissue (920 µg/ml) would be sufficiently toxic against most phylloplane fungi, including the producer itself.

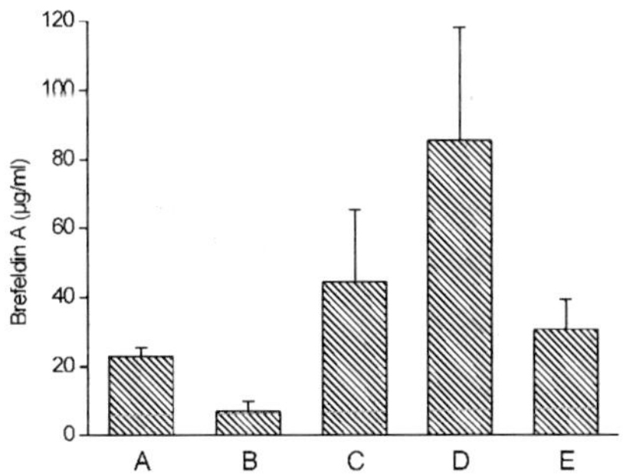

Fig. 15.3. Brefeldin A production by *Phoma medicaginis* in liquid culture. (A) Control culture (4% malt extract, shaken flask). (B) Dried *Medicago sativa* shoots added at 10 g/l prior to inoculation. (C) Yeast cells of *Sporobolomyces roseus* added at 5 × 10^5/ml after 5 days. (D) As for C, but *Mucor mucedo* sporangiospores added at 10^5 spores/ml. (E) As for D, but conidia of *Penicillium citrinum* added. Data are presented as av. ± S.D. after 14 days' cultivation (R.W.S. Weber, unpublished).

Bioactive Metabolites in Other Latent Pathogens

In addition to *P. medicaginis*, several further endophytes which we isolated from *Medicago* spp. have been found to produce metabolites detectable by their antifungal activity in pure culture, including lateritin produced by *Fusarium* spp. and mevinolin produced by *Colletotrichum trifolii* (Fig. 15.1). Like brefeldin A, mevinolin possesses activity against a wide range of organisms, including plants and fungi. The healthy appearance of *Medicago* infected by a battery of these endophytes would argue against the production of such toxic substances during the endophytic phase. It is therefore probable that they have a similar role to brefeldin A for the producing fungi, i.e. that they are synthesised only during saprotrophic growth.

Phomopsis leptostromiformis (teleomorph *Diaporthe toxica*) uncannily resembles *Phoma medicaginis* in many ways, although there is no close phylogenetic relationship between them. It is associated with legumes (lupins), and the endophytic phase in living plants is characterised by limited growth to form coralloid hyphae between the cuticle and the epidermis (Williamson *et al.*, 1991). This latent phase may last for several months as long as the plants remain alive, and animals grazing on infected living plants seem to suffer no ill effects. Senescence of tissue brought about by climate, removal of shoots or herbicide treatment triggers an extensive colonisation by *P. leptostromiformis* (Cowling *et al.*, 1984). Saprotrophic growth is accompanied by the production of phomopsin A which remains in colonised plant material for several weeks in the field (Allen *et al.*, 1980). Phomopsin A is a hexapeptide highly toxic to animals grazing on lupin stubble, causing a disease known as lupinosis. Unfortunately, no information seems to be available on antimicrobial activities of phomopsin A. However, given that it interferes with microtubule assembly/disassembly, we assume that it is fungitoxic, like other mycotoxins such as taxol (see above) which also acts against microtubules, or cytochalasins which inhibit actin polymerisation. In fact, cytochalasin derivatives (chaetoglobosins) are also produced by *P. leptostromiformis* growing saprotrophically on dead plant material (Burlot *et al.*, 2003). Lupin seeds may become colonised by *P. leptostromiformis* either asymptomatically or showing discolourations and cracks or crinkles. Discoloured seeds may contain phomopsin concentrations of up to 3.6 μg/g (Wood and Petterson, 1986). There would appear to be no obvious advantage to the host plant in harbouring an endophyte which produces toxic substances only upon host death. Seeds are unlikely to benefit from *P. leptostromiformis* infections because infected symptomatic seeds are non-viable (Wood and Sivasithamparam, 1989). Further, although data for lupins are not available, seeds of other legumes have been shown to survive passage through the ruminant gut (Gardener *et al.*, 1993), and in any case the amount of phomopsin A present in seeds is not generally sufficient to cause lupinosis in grazing animals (Wood and Petterson, 1986; Wood and Sivasithamparam, 1989). As in the case of *Phoma medicaginis*, territorial behaviour of *P. leptostromiformis* against competing epiphytic fungi during the saprotrophic phase is perhaps the simplest explanation for the observed pattern of secondary metabolite production. That this strategy may be of general relevance is shown by the study of Faifer and Bertoni (1988) in which a range of endophytic fungi from *Eucalyptus* leaves were shown to be more potent producers of antifungal metabolites than epiphytic fungi from the same leaves.

Antifungal Metabolites in Pure Culture

In pure culture, endophytes produce a myriad of metabolites with antifungal activities (reviewed by Tan and Zou, 2001; Strobel, 2003). Some examples are listed in Table 15.1.

With the notable exception of taxol (see above), few reports exist in which their presence in the living plant at biologically active concentrations has been demonstrated. Apart from the technical challenges involved in quantifying such compounds in complex plant extracts, this may simply reflect the confinement of their biosynthesis to the saprotrophic phase.

Table 15.1. Some examples of antifungal metabolites produced by endophytic fungi under saprotrophic growth conditions (for references, see Tan and Zou, 2001; Strobel, 2003; and the current text).

Compound	Producers	Cellular target
Cytochalasins	numerous species	actin filaments (especially in filamentous fungi)
Taxol	numerous species	microtubules (especially in Oomycota)
Fumonisin B1	*Fusarium verticillioides*	several, including sphingolipid biosynthesis (generally toxic)
Khafrefungin	unidentified (tropical)	sphingolipid biosynthesis (especially in yeasts)
Aureobasidins	*Aureobasidium*	sphingolipid biosynthesis (especially in yeasts)
Cryptocandins, echinocandins	*Cryptosporiopsis*	regulation of β-(1,3)-glucan synthase (generally antifungal)
Mevinolin	*Fusarium* endophytes; various saprotrophic fungi	HMG-CoA reductase, *i.e.* terpenoid biosynthesis (generally toxic)
Mellein	Xylariaceae, *Phomopsis*	? (generally toxic against fungi)
Chokols	*Epichloë typhina*	? (especially filamentous fungi)

Endophytes in Dead Host Tissue

The succession on fallen leaves and wood

Endophytes as well as epiphytes will fruit rapidly on dead host tissue, and both groups of fungi have been shown to be replaced after leaf fall by saprotrophic species specialised in exploiting pre-colonised substrata, including micro-fungi and basidiomycetes. However, there is one taxonomic group of fungi capable of prolonging its colonisation of endophytically infected fallen leaves and wood, the Xylariaceae (Pyrenomycetes, Ascomycota). There are several features contributing to this phenomenon. One is the ability of *Xylaria* or *Hypoxylon* spp. to produce cellulolytic and ligninolytic enzymes, the latter being rare among Ascomycota. A second is the production of numerous antimicrobial compounds, many of which are considered to be characteristic of the group. Typical antifungal metabolites of Xylariaceae include cytochalasins and mellein (Whalley and

Edwards, 1995), sordarin derivatives such as xylarin (Schneider *et al.*, 1995), and xylaramide (Schneider *et al.*, 1996), although some of these are also produced by other endophytes, e.g. *Phomopsis* (Tan and Zou, 2001). Although critical studies concerning the production of xylariaceous metabolites in the environment have not yet been performed, we would not be surprised if some of them were synthesised there at biologically active concentrations, like they are by specialised saprotrophic fungi.

Saprotrophic Fungi and their Metabolites

If we view endophytes as latent saprotrophs employing antibiosis to defend their substrate when this becomes available upon host senescence, it may be useful to consider briefly the role of secondary metabolites in purely saprotrophic fungi. Of the numerous species with antibiotic activity in pure culture, only a few have been examined in their natural substrates. The selectively antifungal strobilurins have been shown to be present at relevant concentrations in wood colonised by various members of the Basidiomycota (Engler *et al.*, 1998). Further, strobilurin biosynthesis may be stimulated by potential competitors, such as *Penicillium* spp. (Kettering *et al.*, 2004). There is therefore good evidence that the strobilurins fulfil an important ecophysiological role for their producers (Anke, 1995). The coprophilous fungus *Stilbella erythrocephala* (Pyrenomycetes, Ascomycota) produces antiamoebins, a family of biocidal peptaibols, in its natural substrate at concentrations sufficient to inhibit other coprophilous fungi, and antibiosis can explain why this species excludes all competitors from the dung pellet it has colonised in natural situations (Lehr *et al.*, 2006). Production of antiamoebins in the natural substrate was found to be enhanced considerably in comparison with liquid culture. This would appear similar to the situation in *P. medicaginis*, although the physiological basis is likely to be different, with dung being rich in nitrogen which is required for the biosynthesis of linear oligopeptides such as peptaibols. Together with secreted cell wall-degrading enzymes, peptaibols have also been shown to be responsible for the antifungal activity of *Trichoderma harzianum*, a soil fungus used extensively in biological control (Lorito *et al.*, 1996).

However, the situation is not always as clear-cut. For example, the coprophilous fungus *Podospora pleiospora* was shown to produce the antifungal compound sordarin and two of its derivatives in pure culture and in dung, but the concentration was far too low to account for the exclusion of competing fungi from dung colonised by *P. pleiospora* (Weber *et al.*, 2005).

Conclusions

A Hypha in the Soup?

In his compelling review of potential ecophysiological roles of alkaloids in the grass/endophyte symbiome, Richardson (2000) has called for a biological rather than ecological consideration of clavicipitalean endophytes. It is well documented that mammalian and insect herbivores show a strong avoidance of endophyte-infected material and favour endophyte-free food, but even if this is due to the high alkaloid content it is an open question whether these herbivores place a sufficient strain on grasses to enforce a selection pressure towards the endophyte/grass symbiome (Lane *et al.*, 2000). Perhaps we should look more closely at any direct beneficial effect which the production of secondary metabolites might provide for their producers, rather than at indirect effects on the host plant. There are few examples among non-clavicipitaceous endophytes in which high concentrations of secondary metabolites have been detected in the living plant. Further, when saprotrophic cultures of endophytic *Fusarium*, *Geniculosporium* or *Coniothyrium* spp. were confronted with callus cultures of their host plants, the latter were killed by phytotoxic metabolites (Peters *et al.*, 1998). Such observations indicate fundamental differences between the subdued secondary metabolism of the endophytic phase and the more active saprotrophic physiology. As the examples given in the current article have indicated, in most endophytic associations (excluding the ones between grasses and clavicipitaceous fungi) the over-riding interest must be for the fungus to complete sporulation within the short time-frame opened up by host senescence and closed by the arrival of more competitive epiphytes and saprotrophs. Any significant extension of that window of time by antibiosis would be of direct selective advantage to the producing organism, as summarised in Fig. 15.4.

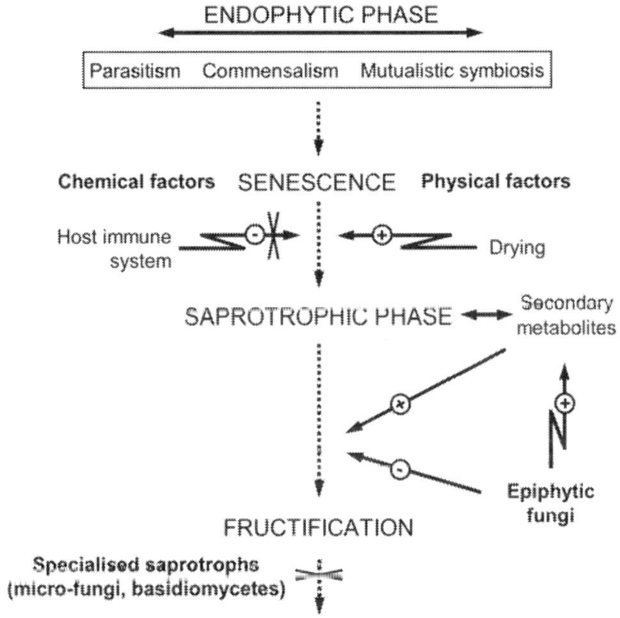

Fig. 15.4. Summary of constraints faced by endophytes relying on horizontal dispersal by spore inoculum. Signals terminating the balanced endophytic phase upon senescence may be chemical factors (e.g. the failing immune system of the host) and physical parameters such as alterations in the water status of the host tissue. Stimulating and inhibitory signals are labelled '+' and '-', respectively. Rapid saprotrophic growth is viewed as a race towards fructification and is accompanied by the production of antimicrobial antibiotics delaying colonisation by competing epiphytes which, in turn, may stimulate antibiotics production. Ultimately, however, both endophytes and epiphytes are displaced by specialised saprotrophs such as soil or humus micro-fungi and basidiomycetes.

Acknowledgements

We are grateful to Anja Meffert for her help with the quantification of fungal secondary metabolites, and to Eva Stenger for performing bioactivity tests.

References

Allen, J.G., Wallace, S.R. and Wood, P.McR. (1980) A study of the period for which lupin stubbles infected with *Phomopsis leptostromiformis* remain toxic. *Australian Journal of Experimental Agriculture and Animal Husbandry* 20, 166-169.

Andrews, J.H. and Harris, R.F. (2000) The ecology and biogeography of microorganisms on plant surfaces. *Annual Review of Phytopathology* 38, 145-180.

Anke, T. (1995) The antifungal strobilurins and their possible ecological role. *Canadian Journal of Botany* 73, S940-S945.

Bacon, C.W., Yates, I.E., Hinton, D.M. and Meredith, F. (2001) Biological control of *Fusarium moniliforme* in maize. *Environmental Health Perspectives* 109 (Supplement 2), 325-332.

Betina, V. (1992) Biological effects of the antibiotic brefeldin A (decumbin, cyanein, ascotoxin, synergisidin): a retrospective. *Folia Microbiologica* 37, 3-11.

Burlot, L., Cherton, J.-C., Convert, O., Correia, I. and Dennetiere, B. (2003) New chaetoglobosins from maize infested by *Phomopsis leptostromiformis* fungi. Production, identification, and semi-synthesis. *Spectroscopy* 17, 725-734.

Bush, L.P., Fannin, F.F., Siegel, M.R., Dahlman, D.L. and Burton, H.R. (1993) Chemistry, occurrence and biological effects of saturated pyrrolizidine alkaloids associated with endophyte-grass interactions. *Agriculture, Ecosystems and Environment* 44, 81-102.

Chapela, I.H. and Boddy, L. (1988) Fungal colonization of attached beech branches. II. Spatial and temporal organization of communities arising from latent invaders in bark and functional sapwood under different moisture regimes. *New Phytologist* 110, 47-57.

Cowling, W.A., Wood, P. McR. and Brown, A.G.P. (1984) Use of a paraquat-diquat herbicide for the detection of *Phomopsis leptostromiformis* infection in lupins. *Australasian Plant Pathology* 13, 45-46.

de Bary, A. (1866) *Morphologie und Physiologie der Pilze, Flechten und Myxomyceten.* Wilhelm Engelmann. Leipzig, Germany.

Dreyfuss, M.M. and Chapela, I.H. (1994) Potential of fungi in the discovery of novel, low-molecular weight pharmaceuticals. In: Gullo, V.P. (ed.) *The Discovery of Natural Products with Therapeutic Potential.* Butterworth-Heinemann, London, UK. pp. 49-80.

Engler, M., Anke, T. and Sterner, O. (1998) Production of antibiotics by *Collybia nivalis*, *Omphalotus olearius*, a *Favolaschia* and a *Pterula* species on natural substrates. *Zeitschrift für Naturforschung* 53c, 318-324.

Faifer, G.C. and Bertoni, M.D. (1988) Interactions between epiphytes and endophytes from the phyllosphere of *Eucalyptus viminalis* III. *Nova Hedwigia* 47, 219-229.

Freeman, S., Horowitz, S. and Sharon, A. (2001) Pathogenic and non-pathogenic lifestyles in *Colletotrichum acutatum* from strawberry and other plants. *Phytopathology* 91, 986-992.

Gardener, C.J., McIvor, J.G. and Jansen, A. (1993) Passage of legume and grass seeds through the digestive tract of cattle and their survival in faeces. *Journal of Applied Ecology* 30, 63-74.

Hawksworth, D.L. (2001) The magnitude of fungal diversity: the 1.5 million species estimate revisited. *Mycological Research* 105, 1422-1432.

Hyde, K.D., Fröhlich, J. and Taylor, J. (1997) Diversity of ascomycetes on palms in the tropics. In: Hyde, K.D. (ed.) *Biodiversity of Tropical Microfungi.* Hong Kong University Press, Hong Kong. pp. 141-156.

Keogh, R.G., Tapper, B.A. and Fletcher, R.H. (1996) Distributions of the fungal endophyte *Acremonium lolii*, and of the alkaloids lolitrem B and peramine, within perennial ryegrass. *New Zealand Journal of Agricultural Research* 39, 121-127.

Kettering, M., Sterner, O. and Anke, T. (2004) Antibiotics in the chemical communication of fungi. *Zeitschrift für Naturforschung* 59c, 816-823.

Lane, G.A., Christensen, M.J. and Miles, C.O. (2000). Coevolution of fungal endophytes with grasses: the significance of secondary metabolites. In: Bacon, C.W. and White, J.F. (eds) *Microbial Endophytes*. Marcel Dekker, New York, USA. pp. 341-388.

Larkin, R.P. and Fravel, D.R. (1999) Mechanisms of action and dose-response relationships governing biological control of Fusarium wilt of tomato by nonpathogenic *Fusarium* spp. *Phytopathology* 89, 1152-1161.

Lehr, N.-A., Meffert, A., Antelo, L., Sterner, O., Anke, H. and Weber, R.W.S. (2006) Antiamoebins, myrocin B and the basis of antifungal antibiosis in the coprophilous fungus *Stilbella erythrocephala* (syn. *S. fimetaria*). *FEMS Microbiology Ecology* 55, 105-112.

Lorito, M., Farkas, V., Rebuffat, S., Bodo, B. and Kubicek, C.P. (1996) Cell wall synthesis is a major target of mycoparasitic antagonism by *Trichoderma harzianum*. *Journal of Bacteriology* 178, 6382-6385.

Marasas, W.F.O. (2001) Discovery and occurrence of the fumonisins: a historical perspective. *Environmental Health Perspectives* 109 (Supplement 2), 239-243.

Miller, J.D., MacKenzie, S., Foto, M., Adams, G.W. and Findlay, J.A. (2002) Needles of white spruce inoculated with rugulosin-producing endophytes contain rugulosin reducing spruce budworm growth rate. *Mycological Research* 106, 471-479.

Moy, M., Belanger, F., Duncan, R., Freehoff, A., Leary, C., Meyer, W., Sullivan, R. and White, J.F. (2000) Identification of epiphyllous mycelial nets on leaves of grasses infected by clavicipitaceous endophytes. *Symbiosis* 28, 291-302.

Noordeloos, M.E., de Gruyter, J., van Eijk, G.W. and Roeijmans, H.J. (1993) Production of dendritic crystals in pure cultures of *Phoma* and *Ascochyta* and its value as a taxonomic character relative to morphology, pathology and cultural characteristics. *Mycological Research* 97, 1343-1350.

Oren, L., Ezrati, S., Cohen, D. and Sharon, A. (2003) Early events in the *Fusarium verticillioides*-maize interaction characterized by using a green fluorescent protein-expressing transgenic isolate. *Applied and Environmental Microbiology* 69, 1695-1701.

Pažoutová, S., Olšovská, J., Linka, M., Kolínská, R. and Flieger, M. (2000) Chemoraces and habitat specialization of *Claviceps purpurea* populations. *Applied and Environmental Microbiology* 66, 5419-5425.

Peters, S., Draeger, S., Aust, H.-J. and Schulz, B. (1998) Interactions in dual cultures of endophytic fungi with host and non-host plant calli. *Mycologia* 90, 360-367.

Petrini, O. (1991) Fungal endophytes of tree leaves. In: Andrews, J.A. and Hirano, S.S. (eds) *Microbial Ecology of Leaves*. Springer Verlag, New York, USA. pp. 179-197.

Redman, R.S., Freeman, S., Clifton, D.R., Morrel, J., Brown, G. and Rodriguez, R.J. (1999) Biochemical analysis of plant protection afforded by a non-pathogenic endophytic mutant of *Colletotrichum magna*. *Plant Physiology* 119, 795-804

Richardson, M.D. (2000) Alkaloids of endophyte-infected grasses: defence chemicals or biological anomalies? In: Bacon, C.W. and White, J.F. (eds) *Microbial Endophytes*. Marcel Dekker, New York, USA. pp. 323-340.

Schneider, G., Anke, H. and Sterner, O. (1995) Xylarin, an antifungal *Xylaria* metabolite with an unusual tricyclic uronic acid moiety. *Natural Product Letters* 7, 309-316.

Schneider, G., Anke, H. and Sterner, O. (1996) Xylaramide, a new antifungal compound, and other secondary metabolites from *Xylaria longipes*. *Zeitschrift für Naturforschung* 51c, 802-806.

Spiering, M.J., Davies, E., Tapper, B.A., Schmid, J. and Lane, G.A. (2002) Simplified extraction of ergovaline and peramine for analysis of tissue distribution in endophyte-infected grass tillers. *Journal of Agricultural and Food Chemistry* 50, 5856-5862.

Spiering, M.J., Lane, G.A., Christensen, M.J. and Schmid, J. (2005) Distribution of the fungal endophyte *Neotyphodium lolii* is not a major determinant of the distribution of fungal alkaloids in *Lolium perenne* plants. *Phytochemistry* 66, 195-202.

Strobel, G.A. (2003) Endophytes as sources of bioactive products. *Microbes and Infection* 5, 535-544.

Tan, R.X. and Zou, W.X. (2001) Endophytes: a rich source of functional metabolites. *Natural Product Reports* 18, 448-459.

Wagner, L.J. and Flores, H.E. (1994) Effect of taxol and related compounds on growth of plant pathogenic fungi. *Phytopathology* 84, 1173-1178.

Wang, C., Wu, J. and Mei, X. (2001) Enhancement of taxol production and excretion in *Taxus chinensis* cell culture by fungal elicitation and medium renewal. *Applied Microbiology and Biotechnology* 55, 404-410.

Weber, R.W.S., Stenger, E., Meffert, A. and Hahn, M. (2004) Brefeldin A production by *Phoma medicaginis* in dead pre-colonized plant tissue: a strategy for habitat conquest? *Mycological Research* 108, 662-671.

Weber, R.W.S., Meffert, A., Anke, H. and Sterner, O. (2005) Production of sordarin and related metabolites by the coprophilous fungus *Podospora pleiospora* in submerged culture and in its natural substrate. *Mycological Research* 109, 619-626.

Whalley, A.J.S. and Edwards, R.L. (1995) Secondary metabolites and systematic arrangement within the Xylariaceae. *Canadian Jornal of Botany* 73, S802-S810.

White, J.F. and Morgan-Jones, G. (1996) Morphological and physiological adaptations of Balansieae and trends in the evolution of grass endophytes. In: Redlin, S.C. and Carris, L.M. (eds) *Endophytic Fungi in Grasses and Woody Plants*. APS Press, St. Paul, Minnesota, USA. pp. 133-154.

Wicklow, D.T., Horn, B.W., Shotwell, O.L., Hesseltine, C.W. and Caldwell, R.W. (1988). Fungal interference with *Aspergillus flavus* infection and aflatoxin contamination of maize grown in a controlled environment. *Phytopathology* 78, 68-74.

Wicklow, D.T., Roth, S., Deyrup, S.T. and Gloer, J.B. (2005) A protective endophyte of maize: *Acremonium zeae* antibiotics inhibitory to *Aspergillus flavus* and *Fusarium verticillioides. Mycological Research* 109, 610-618.

Williamson, P.M., Sivasithamparam, K. and Cowling, W.A. (1991) Formation of sub-cuticular coralloid hyphae by *Phomopsis leptostromiformis* upon latent infection of narrow-leafed lupins. *Plant Disease* 75, 1023-1026.

Wood, P.McR. and Petterson, D.S. (1986) *Phomopsis leptostromiformis* infection and phomopsin A content of lupin seed in Western Australia. *Australian Journal of Experimental Agriculture* 26, 583-586.

Wood, P.McR. and Sivasithamparam, K. (1989) *Diaporthe woodii* (anamorph *Phomopsis leptostromiformis*) – a toxigenic fungus infecting cultivated lupins. *Mycopathologia* 105, 79-86

Chapter 16

Plant Control of Phyllosphere Diversity: Genotype Interactions with Ultraviolet-B Radiation

A.E. Stapleton[1] and S.J. Simmons[2]
[1]Department of Biology and Marine Biology, University of North Carolina at Wilmington, Wilmington, NC 28403, USA; [2]Department of Mathematics and Applied Statistics, University of North Carolina at Wilmington, Wilmington, NC 28403, USA
stapletona@uncw.edu

Introduction

Leaf surfaces are a key terrestrial habitat for epiphytic microbes (Lindow and Brandl, 2003). Microbial species found on leaf surfaces include members of all the major bacterial and fungal groups (Andrews and Harris, 2000), with sphingomonads, pseudomonads and other Proteobacteria frequently detected, along with large numbers of species identified only from sequences (Kim *et al.*, 1998; de Jager *et al.*, 2001; Yang *et al.*, 2001; Idris *et al.*, 2004). These phyllosphere populations compete for limited resources and are exposed to high levels of ultraviolet radiation and high levels of visible radiation. The habitat can fluctuate to extremes of temperature and UV exposure on very short time scales (Lindow and Brandl, 2003). Ultraviolet radiation is known to alter phyllosphere bacterial communities in both groundnut and maize (Jacobs and Sundin, 2001; Kadivar and Stapleton, 2003).

The microstructure of leaves can affect the type and pattern of colonisation by bacteria (Beattie and Lindow 1999; Kinkel 2000), so there are expected to be different communities and possibly different community environmental responses in different plant species. Both correlation studies and quantitative trait locus (QTL) mapping experiments have shown that bacterial populations can be controlled by plant genotypes (Smith *et al.*, 1999; Gao *et al.*, 2003). Cultured rhizobacteria differ in maize parent and hybrids, with the hybrid having a more diverse collection of pseudomonads (Picard and Bosco, 2005). The details of plant-microbe communication are understood in individual species, such as *Agrobacterium* and nitrogen-fixing symbiotic bacteria (Brencic and Winans, 2005). It is clear that there is cross-talk between pathogens and plants (Ryu *et al.*, 2003; Mayak *et al.*, 2004), and in a few cases there are known biotic effects on one or both partners that could affect environmental responses. One recent example of cross-talk is tomato plant induction of a fungal UV defence enzyme (Alejandre-Duran *et al.*, 2003).

Microbial effects on plants have been documented both for individual microbial species and for mixed microbial populations. Plant-associated bacteria are known to produce the plant growth regulator auxin (Fett *et al.,* 1987; Glickmann *et al.,* 1988; Lindow *et al.,* 1998), affect water and salt stress resistance (Mayak *et al.,* 2004), alter plant cuticular layers and water permeability (Krimm *et al.,* 2005; Schreiber *et al.,* 2005), and promote plant growth via bacterial volatiles (Ryu *et al.,* 2003).

Typical ecological sampling includes recording of the number of each species in the sampled area. For microbes, gene sequences are now commonly used as the proxy for species; species are defined as individual gene sequences, or operational taxonomic units (OTUs). Libraries of gene sequences can be made from each sample and used to fit diversity curves, however, these are unwieldy for quantitative analysis and are impractical for large numbers of different samples (Bohannan and Hughes, 2003). Denaturing gradient gel electrophoresis (DGGE) and terminal fragment length polymorphism (T-RFLP) are widely used for broader surveys (Muyzer, 1999; Kitts, 2001; Blackwood *et al.,* 2003). DGGE, unlike T-RFLP, allows further analysis by hybridisation or sequencing of unknown OTUs. DGGE, however, does have limited resolution, which reduces the ability to detect differences in complex communities. DGGE is also relatively difficult to standardise; it is easiest to make comparisons between the lanes within a single gel. Analysis of band fingerprint patterns requires a multivariate approach. Rees *et al.* (2004) discuss using a non-metric multidimensional scaling (nMDS) and analysis of similarity (ANOSIM) for analysing T-RFLP data. However, ANOSIM is limited by the requirement that the variation between each group be similar (Rees *et al.,* 2004). A robust test for multivariate dispersion is required to determine if samples are different due to dispersion or location (Anderson, 2006).

We used DGGE, sequencing of selected bacterial species, and non-parametric multivariate ecological data analysis methods to examine host plant control of phyllosphere microbial community diversity, and to explore how that role interacts with direct and indirect effects of abiotic ultraviolet radiation stress. In field experiments, we have found that genotype is more important than UV-B in explaining patterns of bacterial rDNA microbial diversity on *Zea mays* (maize) leaves. The patterns of host control were examined in archaeal populations in addition to bacteria; archaeal populations show a different pattern of effects than the bacterial community. For these analyses, we chose plant genotypes that were genetically informative—the two parents of a commonly used mapping population and the F_1 hybrid. This allows us to determine the inheritance of host control of diversity, as well as the ultraviolet radiation effects.

Materials and Methods

Plant Material

Maize seeds of the B73 (abbreviated as B) and Mo17 (abbreviated as M) genotype were generously provided by M. Lee, ISU, USA. Seed stocks were increased in the field nursery, Clayton, NC, USA; pollen from the B73 line was crossed onto Mo17 ears to create the F, which is abbreviated as X.

Growth Conditions

Experiments were conducted essentially as previously described (Kadivar and Stapleton, 2003). Seeds of each line were planted into tilled soil at the University of North Carolina

Wilmington field site in 2004 and grown for 20 days, to the five-leaf stage. Water was provided as needed by drip hose. Plastic shelters of a standard composition (Caldwell, 1971; Kadivar and Stapleton, 2003) were placed over each set of the three lines. One shelter (abbreviated UV) was constructed of cellulose diacetate (United Plastics, Lima, OH USA), which transmits UV-B and visible radiation, while the control shelter was constructed of MylarD (United Plastics, Lima, OH USA), which does not transmit UV-B radiation.

Microbial Community Sampling

Pilot experiments to test sampling and DNA extraction methods were conducted in 2002 and 2003. For the full experiment, three plants of each genotype were harvested for each DNA sample, on 9 July 2004 under clear skies. Plants were cut, submerged in buffer (0.2 mM Tris pH 7.5, 0.02 mM EDTA, 0.00012X? Triton X100), agitated vigorously and removed, as described previously (Kadivar and Stapleton, 2003). Mock samples (with no plants but otherwise following the harvest protocol) were collected at random points during harvesting. The microbial fraction was then collected onto 0.02 micron sterile CA filters (VWR Scientific, Atlanta, GA, USA) and eluted by reverse filtration of 0.2 ml buffer through the filter. The filtrate was frozen at -70 °C until DNA extraction. Four samples of each line-treatment combination were collected. This level of replication (n = 4), was chosen to avoid having ‘gel’ as a factor in the data analysis, as one DGGE gel could accommodate all 24 samples.

DNA Preparation, PCR and DGGE Conditions

Frozen samples were thawed on ice and centrifuged at 10,000 g for 5 min. The pellet (containing cells not lysed by the freeze-thaw) was collected, re-suspended in microbead (supplier) solution, and DNA extracted using an UltraClean Microbial kit according to the manufacturer’s instructions (MoBio, Carlsbad, CA, USA). The supernatant (containing lysed microbial cells) was precipitated with ethanol to recover DNA. The DNA from the pellet and supernatant were then combined. Blank samples (with extraction reagents but no leaf wash) were collected throughout the harvest and DNA preparation process. Ribosomal DNA segments were amplified by PCR using primers and reaction conditions listed in Table 16.1.

Bacterial amplifications were performed with AccutaqLA (Sigma, St. Louis, USA) and archaeal amplifications with KlentaqDV (Sigma, St. Louis, USA). Denaturing gradient gel electrophoresis (DGGE) was carried out using a DCode system according to the manufacturer’s instructions, except that a 32-well comb and 10X loading dye were used and the gel was stained with SYBRGold (Molecular Probes, Eugene, OR, USA). PCR amplifications and DGGE separations were repeated with similar results.

Gel analysis

Gels were imaged on a BioRad GelDoc2000 and bands identified and quantified using QuantityOne software, according to the manufacturer’s instructions (BioRad, Hercules, CA, USA). Each separable band was considered to be a separate operational taxonomic unit (OTU). The number of separable bands is limited by the amount of DNA in each band and by the resolution of the band quantitation software. In our system each DGGE lane could be divided into no more than 72 bands by the software at the resolution that allowed viewing of the entire gel. Sixty-three different size bands were identified in at least one sample.

Table. 16.1 PCR primer sequences and PCR and DGGE conditions.

	Primers Name, sequence, citation	PCR conditions	DGGE
bacterial	B518R 5'- ATTACCGCGGCTGCTG-3' B357Fclamp 5'- CGCCCGCCGCGCGCGGCGGGC GGGGCGGGGGCCCTACGGGAG GCAGCAG-3' B357F 5'- CCTACGGGAGGCAGCAG-3' (Muyzer *et al.*, 1993, Yu and Morrison, 2004)	Cycle 1, 1X 94°C 2 min Cycle 2, 20X Step 1, 94°C 1 min Step 2, 65°C 1 min with 0.5°C decrease each cycle Step 3, 72°C 1 min Cycle 3, 25X Step 1, 94°C 1 min Step 2, 55°C 1 min Step 3, 72°C 1 min Cycle 4, 1X 72°C 7 min	8% bis-acrylamide 30% to 55% denaturant Stacking gel (0% denaturant) 14 h, 70 V, 60°C
archaeal	UA1204R 5'- TTMGGGGCATRCNKACCT-'3 A571Fclamp 5'- CGCCCGCCGCGCGCGGCGGGC GGGGCGGGGGCGCYTAAAGSR NCCGTAGC-3' A571F 5'- GCYTAAAGSRNCCGTAGC-'3 (Baker *et al.*, 2003; Baker and Cowan, 2004)	Cycle 1, 1X 94°C 2 min Cycle 2, 20X Step 1, 94°C 1 min Step 2, 65°C 1 min with 0.5°C decrease each cycle Step 3 72°C 1 min Cycle 3, 40X Step 1, 94°C 1 min Step 2, 55°C 1 min Step 3, 72°C 1 min Cycle 4 1X 72°C 7 min	8% bis-acrylamide 35% to 85% denaturant Stacking gel (0% denaturant) 14 h 70 V 60°C

Individual samples had 7 to 23 bands present. For each sample bands were tabulated into three variables: the number of bands per lane, number of novel bands and number of bands not shared.

Sequence Analysis

Bands present in some treatment/genotype combinations but not others were excised from the bacterial DGGE gel, reamplified using the forward primer without the clamp, cloned into a plasmid vector using a TOPO 2.1 kit (Invitrogen, Carlsbad, CA, USA) and sequenced using an ABI BigDye kit on an ABI3100 capillary sequencer (ABI, Foster City, CA, USA). Sequences were identified by submission to BLAST (Altschul *et al.*, 1990) and RDP (Cole *et al.*, 2005). Genbank accession numbers for the bacterial sequences are DQ131797 and DQ131798.

Statistical analysis

The experimental design includes two factors, the two environments with and without UV-B and the three plant genotypes, inbred lines B and M plus the F_1 cross (X). Three correlated responses were analysed: the number of OTUs shared, the number of OTUs total, and the number of novel OTUs. P values less than 0.05 are considered significant while P values less than 0.10 were deemed to be suggestive.

The incidence matrix (as 1 or 0 for band present versus absent) for each OTU was used to calculate the Shannon diversity index with the program EstimateS (Colwell, 2005). The incidence matrix (0's and 1's) was tabulated by identifying whether a species was present or absent. Using this information, the abundance of species (counts) can be determined and used to calculate the Shannon index.

Canonical variates are useful in reducing the dimensionality of a multivariate response set and can be plotted to visualise the effect of a factor. We used Genstat (VSN International Ltd, Herts, UK) to calculate and plot canonical variates and 95% confidence limits.

The multivariate responses were reduced to one dimension by non-metric nMDS, as previously described (Powell *et al.,* 2003). The nMDS reduced the dimensionality of the multivariate data while preserving the distances between the data values. We reduced the dimensionality of the responses to one dimension by using the MDS function in R (http://www.r-project.org/) (Minchin 1987). The one dimensional scaling variable produced by the nMDS procedure was used to visualise the data and calculate an estimate of heritability (see below).

In order to maintain all the information contained within the correlated responses, the data were analysed for significant differences by using a multivariate approach. The analysis examined significant differences due to genotype, treatment, and the genotype *x* treatment interaction in a multivariate analysis of variance (MANOVA). The small sample size made use of the traditional MANOVA questionable and the non-parametric MANOVA by Anderson (2001) was considered to be more appropriate. The non-parametric MANOVA avoided the assumption of normality, and used permutations to obtain p-values for the analysis. Anderson's FORTRAN code was used to calculate p-values in the non-parametric MANOVA and posterior multiple comparisons for significant differences in genotype, treatment and the genotype *x* treatment interaction.

For each analysis, 5000 permutations were run with the Bray-Curtis measure of dissimilarity. Since the scales of the responses were different, a square root transformation was used to create more equivalent scales. Bray-Curtis distances and square root transformations are commonly used in ecology and have been suggested for use with molecular fingerprint data (Grant and Ogilvie, 2003).

Similar to the case of the univariate ANOVA, the non-parametric MANOVA assumed the variances within each group were the same. Violation of this assumption confounds results produced by the non-parametric MANOVA. In other words, a significant p-value in the analysis indicates a significant difference in location (mean), dispersion, or both. Therefore, we tested for significant differences in dispersion by using the FORTRAN program of PERMDISP developed by Anderson (2004, 2006). The program performed a non-parametric multivariate version of Levene's test of dispersion for significant differences in dispersion due to genotype, treatment, and the genotype x treatment interaction. The Bray-Curtis measure of dissimilarity was used on the square root transformation of the data with 5000 permutations.

An estimate of heritability of microbial community structure, which indicates the extent of plant genetic control of microbial assemblages, can be obtained by first performing an ordination of the variables. We used the non-metric multidimensional scaling (nMDS) values. The heritability was then estimated as the ratio of variation in genotype to the total variation (Lynch and Walsh, 1998). The procedure PROC VARCOMP in SAS was used to estimate the components of variation involved in estimating the heritability, with the following model:

$$y = \mu + G + E + GxE + \varepsilon$$

where μ is the overall mean, G is the genotype and considered to be a random effect, E denotes the fixed effect of UV-B treatment, $(G \times E)$ is the interaction between genotype and UV-B treatment, and ε is the residual error.

Results

Bacterial Community Diversity is Structured by Plant Genotype

A plot of the Shannon index (a commonly used measure of diversity) of each sample suggests that there is a difference in diversity in the three genotypes; the B parental inbred microbial community has lower diversity than the M inbred or the F_1 cross (Fig. 16.1) using this measure. Indices of diversity are not designed for factorial comparisons. We used a multivariate method, non-parametric MANOVA (Anderson, 2001), to examine diversity of the samples and effect of the genotype and treatment factors on diversity.

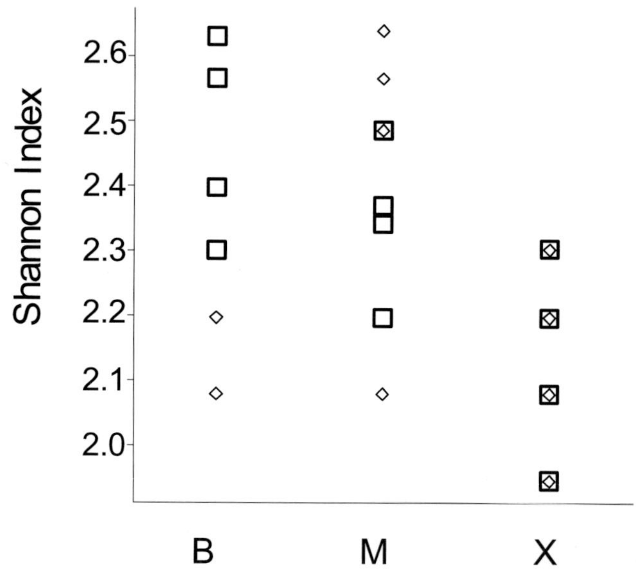

Fig. 16.1. Bacterial community Shannon diversity indices for parent line B73 (B), Mo17 (M) and the F_1 cross (X). The Shannon diversity index was calculated using the EstimateS program for each sample and plotted. The UV-B treatment samples are indicated by squares (□) and the control samples are indicated by diamonds (◊).

Bacterial community multivariate diversity on leaves was significantly affected by plant genotype (Table 16.2, $P = 0.0066$ using permutational MANOVA tests), with the F_1 cross significantly different from the parent lines (Table 16.2). This difference could be due to difference in dispersion (the distance of each observation from the mean) or location (i.e.

the mean of the sample). We used a robust test for dispersion that is suitable for multivariate data (Anderson, 2006). We compared the multivariate dispersion of the bacterial samples (Table 16.2); there was no significant difference in dispersion for the genotypes. Thus, there are significant differences in the amount of overall diversity in the three genotypes. Tests among genotypes showed the F_1 to be significantly different from the parental inbreds (Table 16.2). Examination of the raw data (not shown) suggests that the F_1 cross microbial community generally has fewer OTUs than the parental inbreds, and that the F_1 OTUs tend to be novel.

Table 16.2. Bacterial community diversity significance tests P values from non-parametric MANOVA and from permutational tests for dispersion on bacterial operational taxonomic unit square-root transformed data, with analysis based on Bray-Curtis dissimilarities. Comparisons significant at threshold of P<0.05 are shown in bold and suggestive comparisons at P<0.10 are shown in italics.

Permutational multivariate analysis of variance	
Factor	P value (permutational)
Treatment (UV or control)	0.165
Genotype (B73, Mo17, F1)	**0.006**
Treatment x genotype	0.165
Tests among genotypes	
B73, F1	**0.045**
B73, Mo17	0.229
F1, Mo17	**0.003**
Permutational test of multivariate dispersion	
Factor	P value (permutational)
Treatment (UV or control)	*0.085*
Genotype (B73, Mo17, F1)	0.964
Treatment x genotype	0.617

Canonical correspondence analysis is widely used to compare species lists (ter Braak, 1986). We calculated the canonical variate values and 95% confidence limits to compare the bacterial community diversity in the UV treatment and control treatment (Fig. 16.2). The F_1 genotype groups separately from the parental inbreds in UV (Fig. 16.2a) but not in the control (Fig. 16.2b). In the control sample, the M parental inbred was in a separate group (Fig. 16.2b). Again, the bacterial community diversity was affected by the genotype of the host plant.

The difference in treatment effects can be seen in another ordination comparison, the non-metric multidimensional scaling variates that are plotted in Fig. 16.3. The B parental inbred has clearly separated points for UV and control samples, whereas the M inbred and the F_1 cross show no clear separation by UV-B treatment (Fig. 16.3). The difference in UV and control for the B line was also visible in the Shannon index plot (Fig. 16.1). Thus there was weak evidence for a difference in diversity in UV versus control in the B genotype. The difference in the treatments in the nMDS suggests that the B parental inbred may influence its microbial community differently in UV-B than control, whereas this differential response is not present in the M inbred parent or the F_1 cross.

The dispersion of bacterial community measurements (multivariate distance from centroid) has a trend towards difference in treatment (P = 0.0856), with no difference in dispersion in the three genotypes (Table 16.2). The rank order of the three genotypes' dissimilarity in the UV treatment is different from the rank order in control conditions, with the M inbred average dispersion higher than the other genotypes in UV and lower in the control, with the F1 having the lowest average dispersion in UV and the highest in the

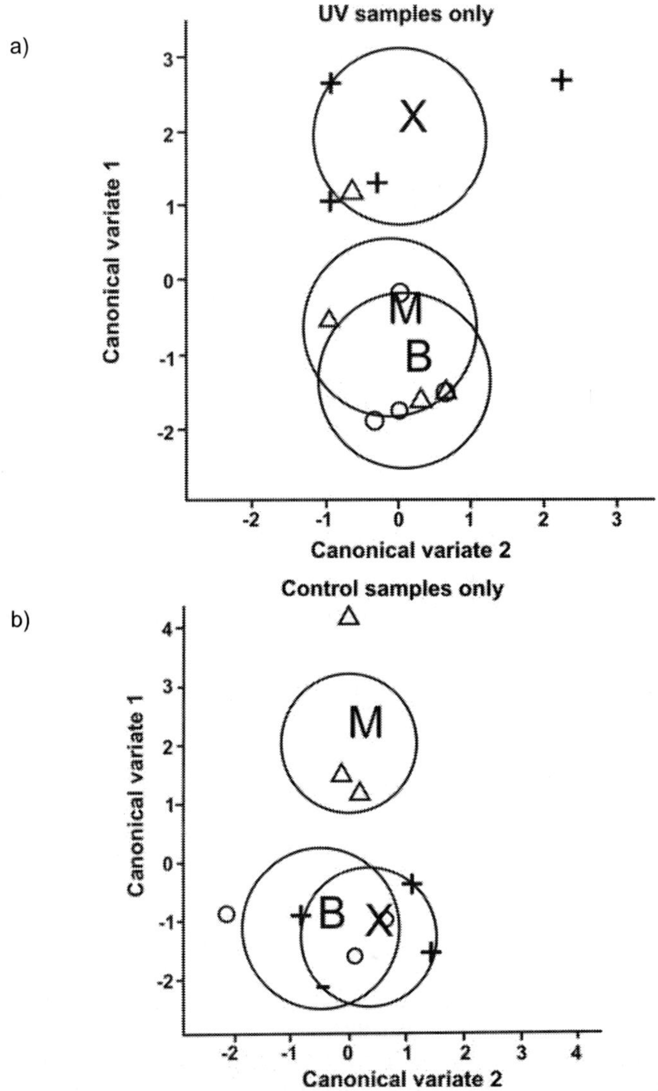

Fig. 16.2. Plot of canonical variates from bacterial fingerprint DGGE band counts (operational taxonomic units). Circles indicate 95% confidence limits. Parent inbred lines are labelled with the letters B and M, and the F1 cross is indicated with the letter X. Symbols indicate the data points for the replicates, with line B as ο, line M as Δ and the cross X as +. a) Canonical variate plot for UV samples; these plants and the associated microbial leaf community were grown under cellulose acetate film, which transmits UV-B. b) Canonical variate plot for control samples; these plants and the associated microbial community were grown under MylarD film, which blocks UV-B radiation transmission.

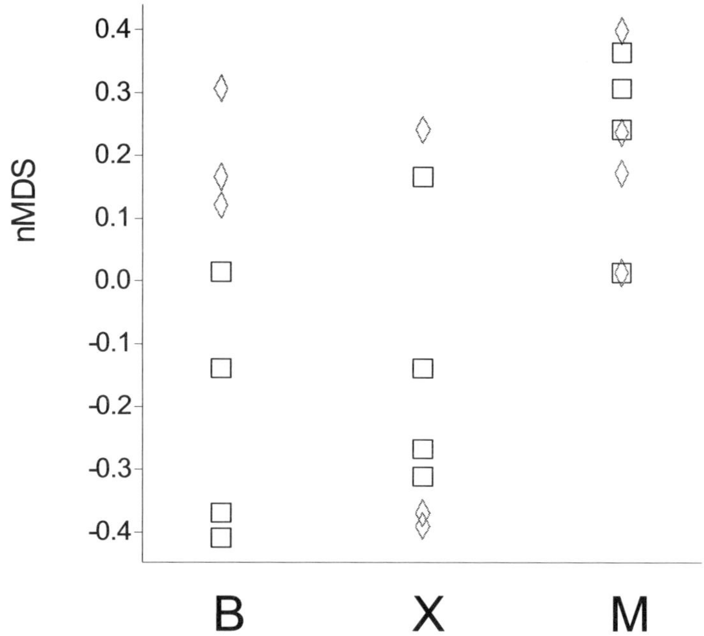

Fig. 16.3. Comparison of bacterial genotypes (plant inbred lines and the F_1 cross) and treatments using the non-metric multidimensional scaling response variable. Parent inbred lines are labelled with the letters B and M, and the F_1 cross is indicated with the letter X. Square symbols (□) indicate values from UV-B exposed samples (grown under cellulose diacetate) and diamonds (◊) indicate values from control samples (grown under MylarD and thus not exposed to UV-B).

control. The confidence limits in the correspondence analysis were larger in control samples than UV for the B parental line (compare the size of the circles in Fig. 16.2a with Fig. 16.2b), which also indicates that dispersion is affected by treatment and interacts with the genotype. This suggests that UV-B is a stress for this microbial community.

We sequenced two DGGE OTUs that were present in the parental inbreds and not in the F_1. The first sequence had most similarity to *Sphingomonas* spp., with the best RDP matches (0.914) to *S. parapaucimobilis* and *S. yabuuchiae* and unclassified isolate AKYG937. The second sequence had most similarity to *Pseudomonas* spp., with the best RDP matches (0.915) to *P. oleovorans, P.* sp. M13, *P. psychrotolerans* and uncultured isolate KUMy1.

Pilot bacterial primer DGGE experiments from the 2002 field samples also showed clear differences between lines and small differences between treatments (data not shown).

Archaeal Phyllosphere Communities are Highly Diverse

No significant effect of the different genotypes or treatments on archaeal diversity was seen with permutational MANOVA tests (Table 16.3), although there is a suggestive difference in the comparison of genotypes (P = 0.069). The canonical variates' confidence intervals overlap for both UV (Fig. 16.4a) and control (Fig. 16.4b) and nMDS has no separation (Fig. 16.5). Thus, genotype and treatment have no striking effect on archaeal diversity. This could be due to differences in the amount of diversity, or to differences in the dispersion.

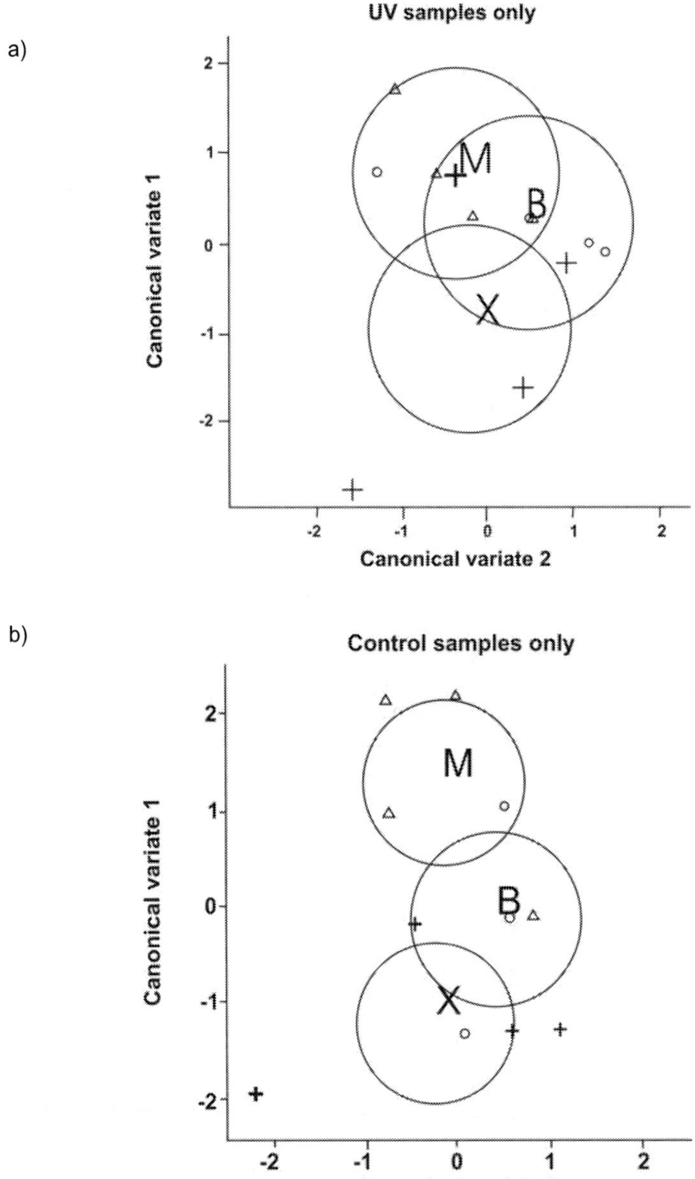

Fig. 16.4. Plot of canonical variates from archaeal fingerprint DGGE band counts (OTUs). Circles indicate 95% confidence limits. Parent inbred lines are labelled with the letters B and M, and the F_1 cross is indicated with the letter X. Symbols indicate the data points for the replicates, with line B as ○, line M as Δ and the cross X as +. a) Canonical variate plot for UV samples; these plants and the associated microbial leaf community were grown under cellulose acetate film, which transmits UV-B. b) Canonical variate plot for control samples; these plants and the associated microbial community were grown under MylarD film, which blocks UV-B radiation transmission.

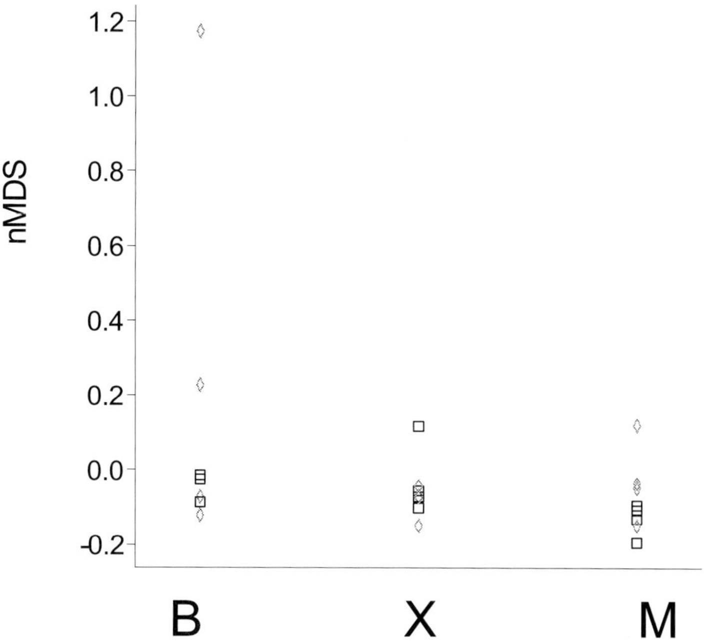

Fig. 16.5. Comparison of bacterial genotypes (plant inbred lines and the F_1 cross) and treatments using the non-metric multidimensional scaling response variable. Parent inbred lines are labelled with the letters B and M, and the F_1 cross is indicated with the letter X. Square symbols (□) indicate values from UV-B exposed samples (grown under cellulose diacetate) and diamonds (◊) indicate samples from control samples (grown under MylarD and not thus exposed to UV-B).

When we examined the dispersion using a permutation test, a significant effect of treatment (P = 0.003) in the overall model (Table 16. 3) was seen. Average dispersion increases in the UV-B exposed archaeal community as compared to the control community. There was a trend toward higher dispersion in the UV-B treatment in the F_1 cross, but not in the parental lines (Table 16.3). The differences in dispersion in the treatment and control can also be seen in the size of confidence limit circles, which are larger in UV than control (compare Fig. 16.4a to Fig. 16.4b).

Examination of the raw data (not shown) supports the statistical analysis and suggests that the archaeal samples are very diverse; the set of OTUs in each sample has little in common with other samples. Each fingerprint has novel bands not found in any other fingerprint.

Table 16.3. Archaeal community diversity significance tests P values from non-parametric MANOVA and from permutational tests for dispersion on archaeal OTU square root-transformed data, with analysis based on Bray-Curtis dissimilarities. Significant differences at a threshold of P<0.05 are shown in bold; suggestive comparisons at P<0.10 are shown in italic.

Permutational multivariate analysis of variance	
Factor	P value (permutational)
Treatment (UV or control)	0.172
Genotype (B, M, X)	*0.069*
Treatment x genotype	0.311
Tests among genotypes	
B, X	0.296
B, M	0.509
X, M	0.188
Permutational test of multivariate dispersion	
Factor	P value (permutational)
Treatment (UV or control)	**0.004**
Genotype (B, M, X)	0.168
Treatment x genotype	0.247
Tests within genotypes, B	
UV, control	0.229
Tests within genotypes, X	
UV, control	*0.085*
Tests within genotypes, M	
UV, control	0.173

Heritability of Host Control Differs

We calculated the heritability (proportion of variation explained by genotype) for bacterial and archaeal diversity using the nMDS scaled numbers (Table 16.4). Our estimate is conservative and number of genotypes is small, so we may be underestimating the extent of genetic host plant control. Host plant genotype has a small effect on bacterial diversity (2%). There was no genotype effect in the archaeal samples, and thus we have no evidence from this measure that the host plant genetically controls the diversity of the archaeal community.

Table 16.4. Heritability proportion of variation explained by genotype estimates from variance components.

Bacterial	
Variance component	Estimate
Variance due to genotype	0.0219
Variance due to interaction	0.0266
Variance due to error	0.0418
Archaeal	
Variance component	Estimate
Variance due to genotype	0
Variance due to interaction	0.0052
Variance due to error	0.0658

Discussion

We used the visualisation and statistical analysis methods that are widely used in terrestrial and marine ecological studies to examine microbial diversity in the phyllosphere. These methods are especially useful for larger designed experiments, such as our factorial design, or for experiments over long temporal scales. The number of samples makes sequence-library analysis impractical for larger experiments.

Maize phyllosphere bacterial diversity, as measured by molecular fingerprinting, is significantly different in the different genotypes used in our experiments, with suggestive differences in the effect of UV-B radiation on individual genotypes. Thus, the bacterial community structure is responsive to genetic differences in the leaf chemistry or structure of the maize host plants. Host control of colonisation has been studied intensively for individual bacteria such as the Rhizobiaceae and *Agrobacterium* spp. (Brencic and Winans, 2005), and host-controlled differences in pseudomonad diversity from culture collections have been described (Picard and Bosco, 2005); however host control of changes in diversity of non-cultured organisms on the leaf surface has not been reported in earlier work.

We previously found increased diversity in UV-B-exposed phyllosphere microbial samples on the B inbred parent (Kadivar and Stapleton, 2003), and we see similar results in comparison of the Shannon index in this experiment (Fig. 16.1), despite the differences in field location (Tennessee vs. North Carolina) and differences in processing of the DNA samples.

We examined some key bands from our DGGE gels by sequencing. We were especially interested in OTUs that were present in the parent inbred but not in the F_1 cross, as if these OTU are pathogens they could contribute to hybrid vigour in the F_1. The bands we sequenced are variants of common phyllosphere species (Pseudomonads and Sphingomonads), although they are not a perfect sequence match to any known species. Sphingomonads are numerically dominant on plant surfaces (Kim *et al.*, 1998; Idris *et al.*, 2004) and have been reported on many plant species, including many Gramineae (Yun *et al.*, 2000; Adhikari *et al.*, 2001; Rivas *et al.*, 2004). Pseudomonads on plant surfaces have been studied extensively and are widespread and abundant (Lindow and Brandl, 2003). There is no indication from sequence comparisons and comparisons with the literature that the two OTUs present only in the inbred parents in our study are pathogens.

We were able to detect archaeal OTU signals by increasing the cycle number of the touchdown PCR. The archaeal community on the leaf surface is very diverse, with no significant effect of plant genotype or UV-B exposure on community diversity, but with a significant effect of the abiotic stress, UV-B exposure, on multivariate dispersion (Table 16.3). Large-scale glacial-succession surveys of soil crenarchaea found no effect of plant species on archaeal communities (Nicol *et al.*, 2005), so it may not be surprising that we found no effect of genotype within a species. A survey of archaeal diversity across a desiccation gradient did show an effect on archaeal diversity (Rothrock and Garcia-Pichel, 2005), possibly suggesting archaeal communities are more strongly affected by abiotic factors than biotic. We were unable to detect fungal PCR signals in our samples; this may be due to our relatively mild leaf wash protocol. Fungal hyphae often extend into the interior of the stomata of leaves, and may not be removed by the buffer wash we used to sample the microbial community. More abrasive methods could be used for future samples to access this portion of the microbial community in addition to the bacterial and archaeal portions.

Bacterial and archaeal phyllosphere communities differ in our experiments, both in their alterations (or lack of change) in response to genotype and UV, as discussed above, and in multivariate dissimilarity. Dissimilarity is affected by treatment for archaeal community (Table 16.3), and there is a suggestive difference in the bacterial dispersion (Table 16.2). Differences in dispersion can indicate environmental stress (Warwick and Clarke, 1993). Average dispersion is higher in UV for both bacterial and archaeal samples, with a suggestive P value for bacterial dispersion and a highly significant P value for archaea. In our study, the effect of UV on dissimilarity comparisons among genotypes is opposite for bacterial and archaeal species, with lower F_1 dispersion in UV in the bacterial community versus higher F_1 dispersion in UV in the archaeal community (archaeal average dissimilarity for F_1 UV = 17.980 and for control = 5.296). Overall, UV radiation shifts the community structure in opposite directions for bacteria versus archaea in the F_1-cross host.

We used non-metric scaled variables to determine how much of the diversity difference can be explained by the plant genotype (Lynch and Walsh, 1998). The heritability is low (less than 10%), so future experiments directed at mapping the plant controlling loci will need large samples sizes and powerful mapping methods. Heritability of archaeal diversity is very low, which makes mapping experiments impractical.

In conclusion, we have identified important host genetic control of the diversity of maize phyllosphere bacterial communities, along with the lack of host control of archaeal community structure. Mapping of maize genes that control microbial community structure is in progress.

Acknowledgements

The experimental design and molecular analysis was conceived and completed by A. Stapleton, with data analysis by S. Simmons and A. Stapleton. Thanks are due to Robert L. Bryden for excellent technical assistance with plant growth and J. Holland for assistance with seed increase. We appreciate funding provided to this project by UNCW International Programs (Faculty Travel Grant) and by the NSF International Directorate. This material is based on work supported by the National Science Foundation under Grant No. OISE-0436491.

References

Adhikari, T., Joseph, B.C.M., Yang, G.,. Phillips, D.A. and Nelson, L.M. (2001) Evaluation of bacteria isolated from rice for plant growth promotion and biological control of seedling disease of rice. *Canadian Journal of Microbiology* 47, 916-924.

Alejandre-Duran, E., Roldan-Arjona, T., Ariza, R.R. and Ruiz-Rubio, M. (2003) The photolyase gene from the plant pathogen *Fusarium oxysporum* f. sp. *lycopersici* is induced by visible light and alpha-tomatine from tomato plant. *Fungal Genetics and Biology* 40, 159-165.

Altschul, S.F., Gish, W., Miller, W., Myers, E.W. and Lipman, D.J. (1990) Basic local alignment search tool. *Journal of Molecular Biology* 215, 403-10.

Anderson, M.J. (2001) A new method for non-parametric multivariate analysis of variance. *Australian Ecology* 26, 32-46.

Anderson, M.J. (2004) PERMDISP: a FORTRAN computer program for permutational analysis of mutivariate dispersions (for any two-factor ANOVA design) using permutation tests. Department of Statistics, University of Auckland, Auckland, New Zealand.

Anderson, M.J. (2006) Distance-based tests for homogeneity of multivariate dispersions. *Biometrics* 62, 245-253.

Andrews, J.H. and Harris, R.F. (2000) The ecology and biogeography of microorganisms on plant surfaces. *Annual Review of Phytopathology* 38, 145-180.

Baker, G.C. and Cowan, D.A. (2004) 16S rDNA primers and the unbiased assessment of thermophile diversity. *Biochemical Society Transactions* 32, 218-221.

Baker, G.C., Smith, J.J. and Cowan, D.A. (2003) Review and re-analysis of domain-specific 16S primers. *Journal of Microbiological Methods* 55, 541-555.

Beattie, G.S. and Lindow, S.E. (1999) Bacterial colonization of leaves: a spectrum of strategies. *Phytopathology* 89, 353-359.

Blackwood, C.B., Marsh, T., Kim, S.H. and Paul, E.A. (2003) Terminal restriction fragment length polymorphism data analysis for quantitative comparison of microbial communities. *Applied Environmental Microbiology* 69, 926-32.

Bohannan, B.J.M. and Hughes, J. (2003) New approaches to analyzing microbial biodiversity data. *Current Opinion in Microbiology* 6, 282-287.

Brencic, A. and Winans, S.C. (2005) Detection of and response to signals involved in host-microbe interactions by plant-associated bacteria. *Microbiology and Molecular Biology Reviews* 69, 155-194.

Caldwell, M.M. (1971) Solar UV irradiation and the growth and development of higher plants. In: A.C. Giese (ed.) *Photophysiology*. Academic Press, New York, USA. pp. 131-177.

Cole, J.R., Chai, B., Farris, R.J., Wang, Q., Kulam, S.A., McGarrell, D.M., Garrity, G.M. and Tiedje, J.M. (2005) The Ribosomal Database Project (RDP-II): sequences and tools for high-throughput rRNA analysis. *Nucleic Acids Research* 33, D294-296.

Colwell, R.K. (2005) EstimateS: Statistical estimation of species richness and shared species from samples. User's Guide and application published at http://viceroy.eeb.uconn.edu/estima

de Jager, E.S., Wehner, F.C. and Korsten, L. (2001) Microbial ecology of the mango phylloplane. *Microbial Ecology* 42, 201-207.

Fett, W.F., Osman, S.F. and Dunn, M.F. (1987) Auxin production by plant-pathogenic pseudomonads and xanthomonads. *Applied and Environmental Microbiology* 53, 1839-1845.

Glickmann, E., Gardan, L., Jacquet, S., Hussain, S., Elasri, M., Petit, A. and Dessaux, Y. (1988) Auxin production is a common feature of most pathovars of *Pseudomonas syringae*. *Molecular Plant-Microbe Interactions* 11, 156-162.

Grant, A. and Ogilvie, L.A. (2003) Terminal restriction fragment length polymorphism data analysis. *Applied and Environmental Microbiology* 69, 6342-6343.

Idris, R., Trifonova, R., Puschenreiter, M., Wenzel, W.W. and Sessitsch, A. (2004) Bacterial communities associated with flowering plants of the Ni hyperaccumulator *Thlaspi goesingense*. *Applied and Environmental Microbiology* 70, 2667-2677.

Jacobs J.L. and Sundin, G.W. (2001) Effect of solar UV-B radiation on a phyllosphere bacterial community. *Applied and Environmental Microbiology* 67, 5488-5496.

Kadivar, H. and Stapleton, A.E. (2003) Ultraviolet radiation alters maize phyllosphere bacterial diversity. *Microbial Ecology* 45, 353-361.

Kim, H., Nishiyama, M., Kunito, T., Senoo, K., Kawahara, K., Murakami, K. and Oyaizu, H. (1998) High population of *Sphingomonas* species on plant surface. *Journal of Applied Microbiology* 85, 731-736.

Kinkel, L., Wilson, L.M. and Lindow, S.E. (2000) Plant species and plant incubation conditions influence variability in epiphytic bacterial population size. *Microbial Ecology* 39, 1-11.

Kitts, C.L. (2001) Terminal restriction fragment patterns: a tool for comparing microbial communities and assessing community dynamics. *Current Issues International Microbiology* 2, 17-25.

Krimm, U., Abanda-Nkpwatt, D., Schwab, W. and Schreiber, L. (2005) Epiphytic microorganisms on strawberry plants (*Fragaria ananassa* cv. Elsant): identification of bacterial isolates and analysis of their interaction with leaf surfaces. *FEMS Microbiology Ecology* 53, 483-492.

Lindow, S.E. and Brandl, M.T. (2003) Microbiology of the phyllosphere. *Applied and Environmental Microbiology* 69, 1875-1883.

Lindow, S.E., Desurmont, C., Elkins, R., McGourty, G., Clark, E. and Brandl, M.T. (1998) Occurence of indol-3-acetic acid-producing bacteria on pear trees and their association with fruit russet. *Phytopathology* 88, 1149-1157.

Lynch, M. and Walsh, B. (1998) *Genetics and Analysis of Quantitative Traits*. Sinauer Associates, Inc., Sunderland, MA, USA.

Mayak, S., Tirosh, T. and Glick, B.R. (2004) Plant growth-promoting bacteria that confer resistance to water stress in tomatoes and peppers. *Plant Science* 166, 525-530.

Minchin, P.R. (1987) An evaluation of relative robustness of techniques for ecological ordinations. *Vegetatio* 71, 145-156.

Muyzer, G. (1999) DGGE/TGGE a method for identifying genes from natural ecosystems. *Current Opinion in Microbiology* 2, 317-322.

Muyzer, G., De Waal, E.C. and Uitterlinden, A.G. (1993) Profiling of complex microbial populations by denaturing gradient gel electrophoresis analysis of polymerase chain reaction-amplified genes coding for 16S rRNA. *Applied and Environmental Microbiology* 59, 695-700.

Nicol, G.W., Tscherko, D., Embley, T.M. and Prosser, J.I. (2005) Primary succession of soil Crenarchaeota across a receding glacier foreland. *Environmental Microbiology* 7, 337-347.

Picard, C. and Bosco, M. (2005) Maize heterosis affects the structure and dynamics of indigenous rhizospheric auxins-producing *Pseudomonas* populations. *FEMS Microbiology Ecology* 53, 349-357.

Powell, S.M., Bowman, J.P., Snape, I. and Stark, J.S. (2003) Microbial community variation in pristine and polluted nearshore Antarctic sediments. *FEMS Microbiology Ecology* 45, 135-145.

Rees, G.N., Baldwin D.S., Watson G.O., Perryman S. and Nielsen D.L. (2004) Ordination and significance testing of microbial community composition derived from terminal restriction fragment length polymorphisms: application of multivariate statistics. *Antonie van Leeuwenhoek* 86, 339-347.

Rivas, R., Abril, A., Trujillo, M.E. and Velásquez, E. (2004) *Sphingomonas phyllosphaerae* sp. no., from the phyllosphere of *Acacia caven* in Argentina. *International Journal of Systematic and Evolutionary Microbiology* 54, 2147-2150.

Rothrock, M.J. and Garcia-Pichel, F. (2005) Microbial diversity of benthic mats along a tidal desiccation gradient. *Environmental Microbiology* 7, 593-601.

Ryu, C.-M., Farag, M., Hu, C.-H., Reddy, M.S., Wei, H.-X., Pare, P.W. and Kloepper, J.W. (2003) Bacterial volatiles promote growth in *Arabidopsis*. *Proceedings of the National Academy of Sciences, USA* 100, 4927-4932.

Schreiber, L., Krimm, U., Knoll, D., Sayed, M., Auling, G. and Kroppenstedt, R.M. (2005) Plant-microbe interactions: identification of epiphytic bacteria and their ability to alter leaf surface permeability. *New Phytologist* 166, 589-594.

Smith, K.P., Handelsman, J. and Goodman, R.M. (1999) Genetic basis in plants for interactions with disease-suppressive bacteria. *Proceedings of the National Academy of Sciences, USA* 96, 4786-4790.

ter Braak, C.J.F. (1986) Canonical correspondence analysis: A new eigenfactor technique for multivariate direct gradient analysis. *Ecology* 67, 1167-1179.

Warwick, R.M. and Clarke, K.R. (1993) Increased variability as a symptom of stress in marine communities. *Journal of Experimental Marine Biology and Ecology* 172, 215-226.

Yang, C.-H., Crowley, D.E., Borneman, J. and Keen, N.T. (2001) Microbial phyllosphere populations are more complex than previously realized. *Proceedings of the National Academy of Sciences, USA* 98, 3889-3894.

Yu, Z. and Morrison, M. (2004) Comparisons of different hypervariable regions of *rrs* genes for use in fingerprinting of microbial communities by PCR-denaturing gradient gel electrophoresis. *Applied and Environmental Microbiology* 70, 4800-4806.

Yun, N.R., Shin, Y.K., Hwang, S.Y., Kuraishi, H. and Sugiyama, J. (2000) Chemotaxonomic and phylogenetic analyses of *Sphingomonas* strains isolated from ears of plants in the family Gramineae and a proposal of *Sphingomonas roseaoflava* sp. nov. *Journal of General and Applied Microbiology* 46, 9-18.

1

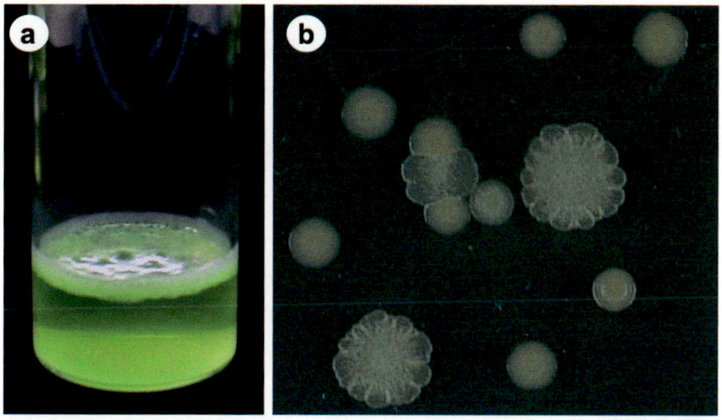

2

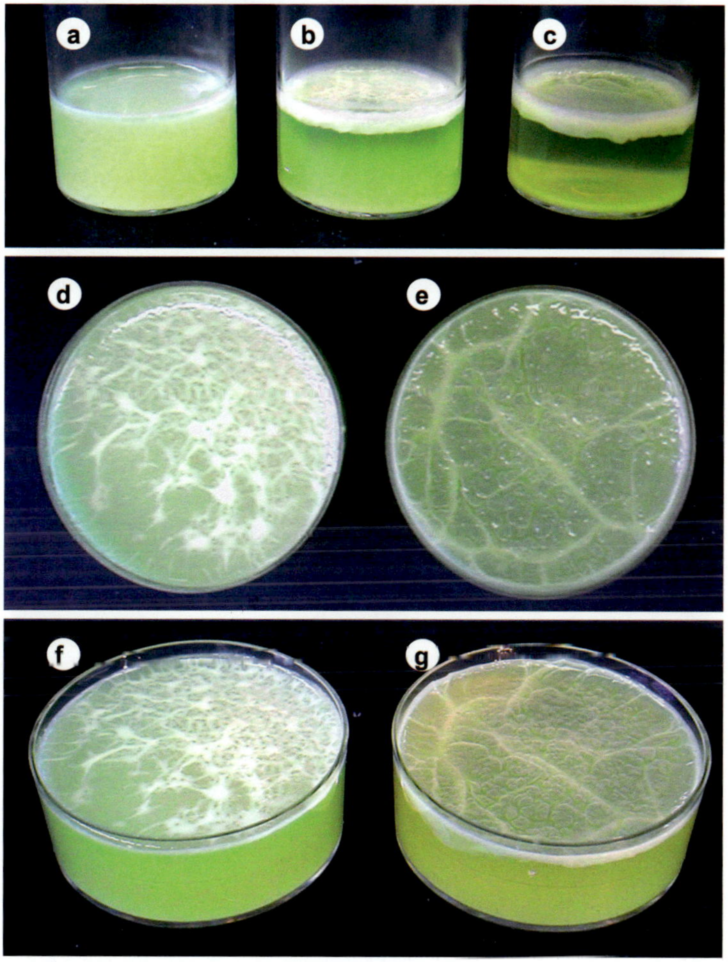

Plate 1. The Wrinkly Spreader (WS) mutant of *P. fluorescens* SBW25 was selected in static microcosms where it forms a biofilm of substantial size and robustness at the air-liquid (A-L) interface. Shown are **(a)** the WS biofilm *in situ* (the biofilm covers 3.8 cm^2 with a depth of 1–1.5 mm); and **(b)** smooth and circular wild-type colonies, and flat and wrinkled WS colonies.

Plate 2. A-L biofilms produced by *Pseudomonas* spp. selected in static microcosms show considerable phenotypic variation which falls into four broad types: waxy aggregation (WA), viscous mass (VM), floccular material (FM) and physically cohesive (PC)-type biofilms. Shown are **(a)** *P. putida* PH6 (WA-type) biofilm with little evident surface topography (even by eye); **(b, e and g)** *P. marginalis* CR30 (VM-type) biofilm incorporating lumps of viscous material at the surface; (**c**) *P. fluorescens* 54/96 (FM-type) biofilm with obvious surface topography but little connectivity; and **(d and f)** *P. putida* A1 (PC-type) biofilm with obvious surface topography as well as physical strength.

3

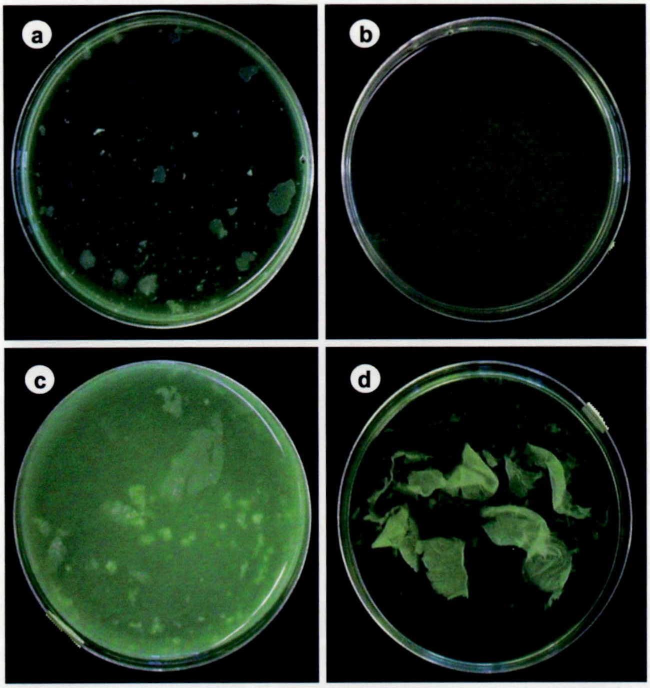

4

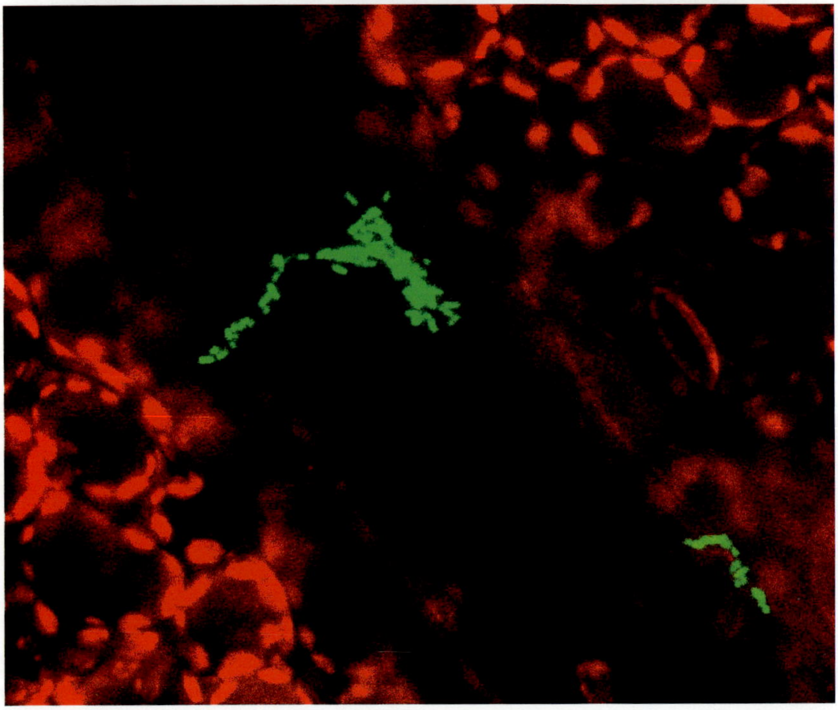

Plate 3. The physical integrity of A-L biofilms varies considerably and is reflected by the fragmentation pattern produced when transferred from microcosms to Petri dishes. Shown are **(a)** *P. putida* PH6 (WA-type biofilm); **(b)** *P. marginalis* CR30 (VM-type biofilm); **(c)** *P. fluorescens* 54/96 (FM-type); and **(d)** *P. putida* A1 (PC-type).

Plate 4. Confocal micrograph of *S. enterica* cells labelled with the green fluorescent protein, 9 days after inoculation on to the leaves of cilantro plants incubated at 28°C under high humidity conditions.

5

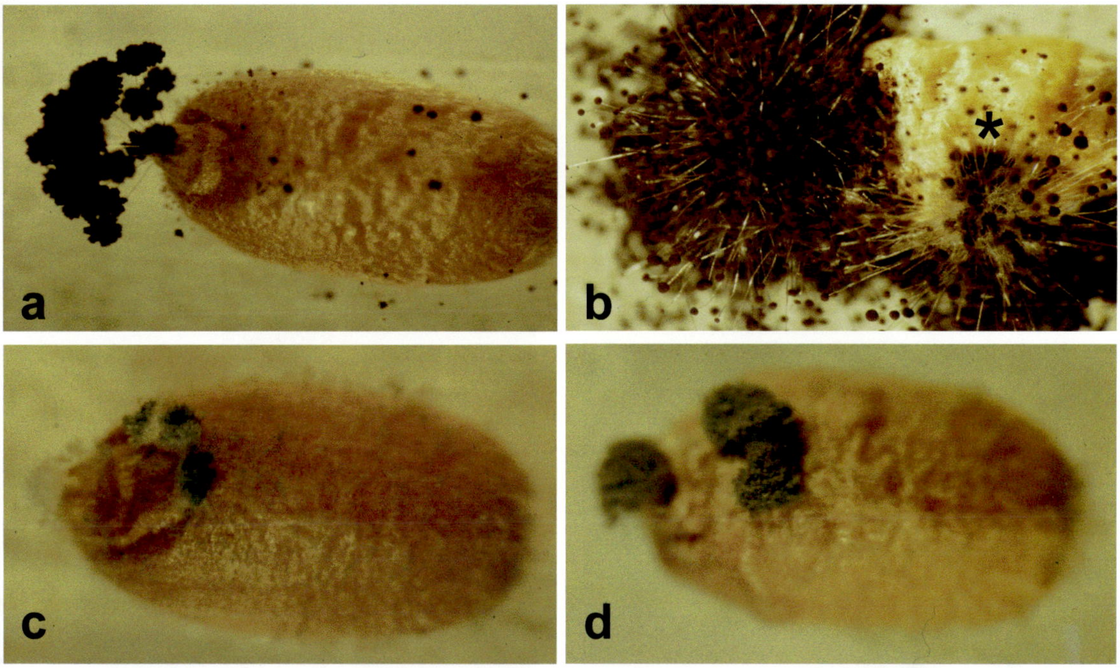

6

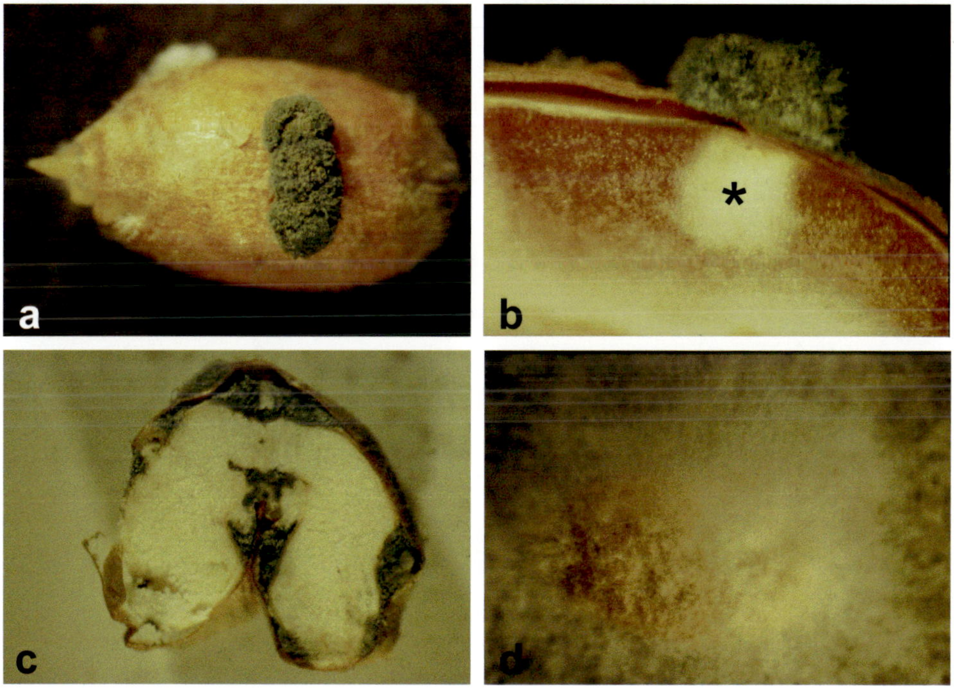

Plate 5. a and b. Wheat grain showing development of sporulation at 13 **(a)** and 30 **(b)** days post-inoculation (dpi) with spores of *Aspergillus niger*. Most profuse sporulation occurred from the embryo end of the grain, with small sporophores also visible as dark specks on the grain surface after 13 days. A second area of profuse sporulation developed after 30 days, with sporophores erupting through the seed coat (*). **c and d.** Wheat grain showing sporulation at 13 **(c)** and 30 **(d)** dpi with spores of *Penicillium aurantiogriseum*. Characteristic "blue eye" symptoms developed as sporophores emerged most profusely from the embryo end of the grain.

Plate 6. a and b. *P. aurantiogriseum* conidiophores at 13 dpi erupting through the seed coat **(a)** are seen in transverse section **(b)** to be associated with a discrete area of mycelium (*) in the endosperm. **c.** Transversely sectioned grain with *P. aurantiogriseum* conidiophores formed under the seed coat, causing bulges at the grain surface. **d.** Wheat grain colonised by *F. culmorum* at 21 dpi barely visible beneath dense hyphal growth.

7

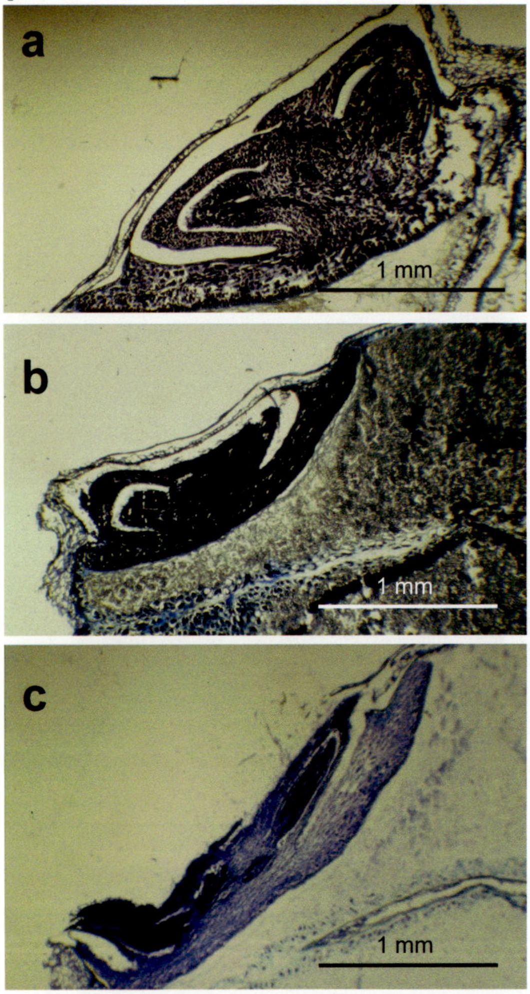

8

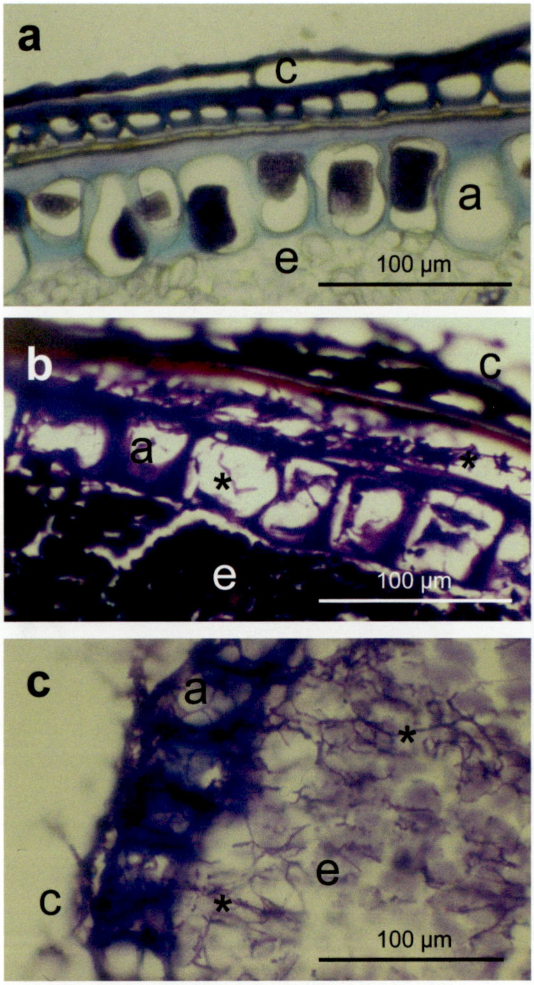

Plate 7. Transverse sections of wheat grains stained with Toluidine Blue. **a.** Control grain showing intact embryo structure (14 dpi with sterile distilled water). **b and c.** Embryos of grains infected by *P. aurantiogriseum* **(b)** and *A. niger* **(c)** have lost cellular structure and collapsed (14 dpi).

Plate 8. Transverse sections of wheat grains stained with Toluidine Blue. **a.** Uninfected grain showing structure of seed coat (c), aleurone layer (a) and endosperm (e). **b and c.** *P. aurantiogriseum* infected grains (29 dpi). Hyphae (*) have penetrated through the seed coat (c) and aleurone layer (a) to colonise the endosperm (e).

9

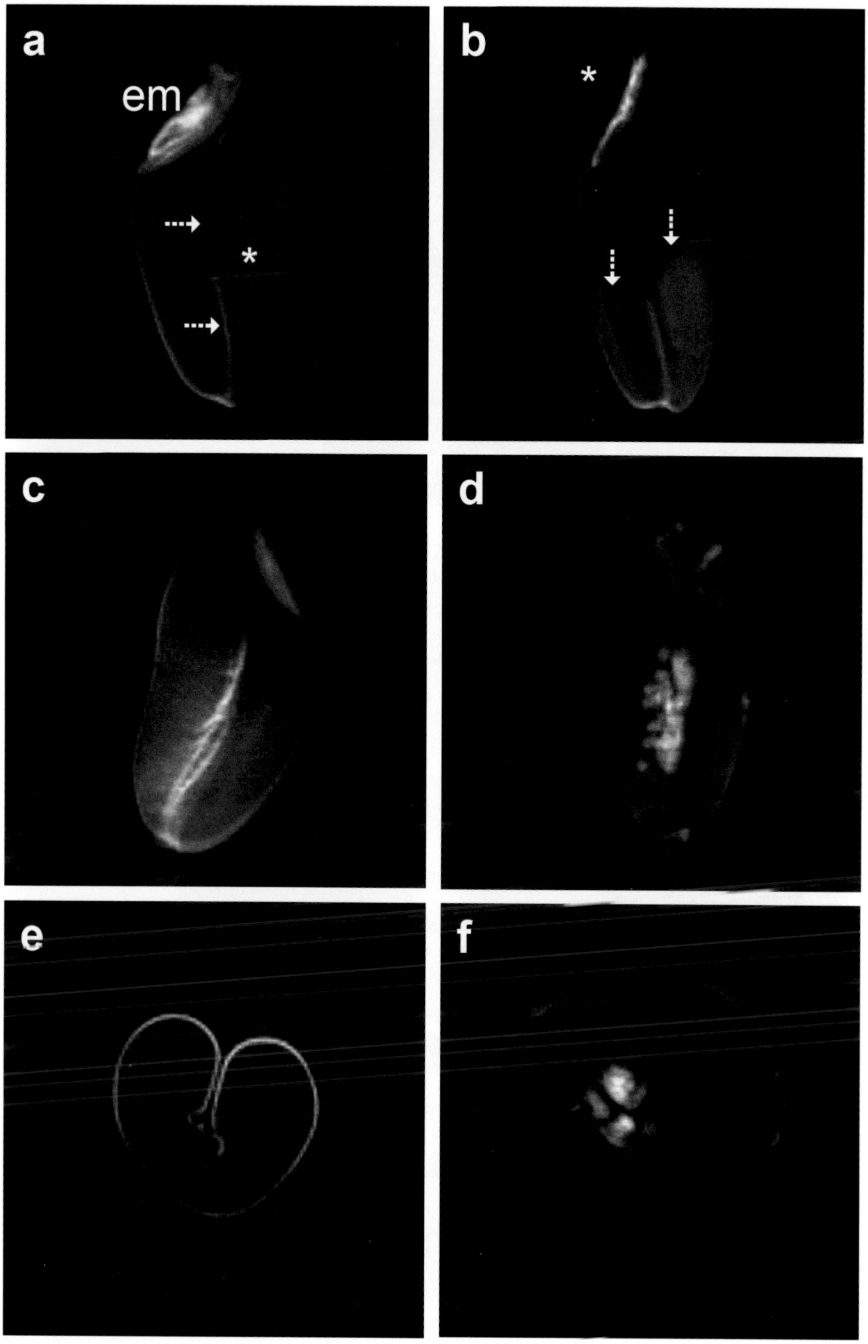

Plate 9. Nuclear magnetic resonance micro-imaging of wheat grains. **a and e.** The embryo (em) of uninfected grains (longitudinal scan in **a**) gave a high intensity signal, whilst the endosperm (transverse scan in **e**) gave a relatively low intensity, uniform signal. Higher signals were associated with the nucellar projection (arrows) and the crease side of the grain (*) due to the double layer of seed surface imaged. **b**. A. *niger* infection has caused collapse of the embryo and reduced signal in the adjacent endosperm, with a clear demarcation zone (arrows). Tips of conidiophores arising from the infected embryo are visible as bright flecks (*). **c, d and f.** *P. aurantiogriseum* infected grains showing reduced signal adjacent to the embryo but enhanced signal at the crease in **c** at 14 dpi. Sporulation from the embryo **(d)** and within the crease (longitudinal scan in **d**, transverse scan in **f**) at 30 dpi are visible as areas of relatively high signal intensity. Areas of reduced intensity are apparent immediately adjacent to these regions.

10

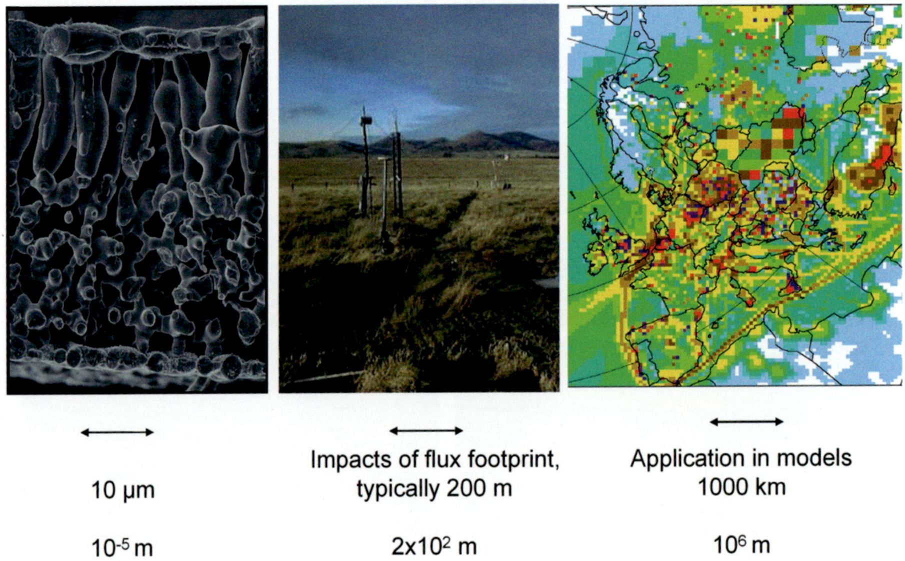

11

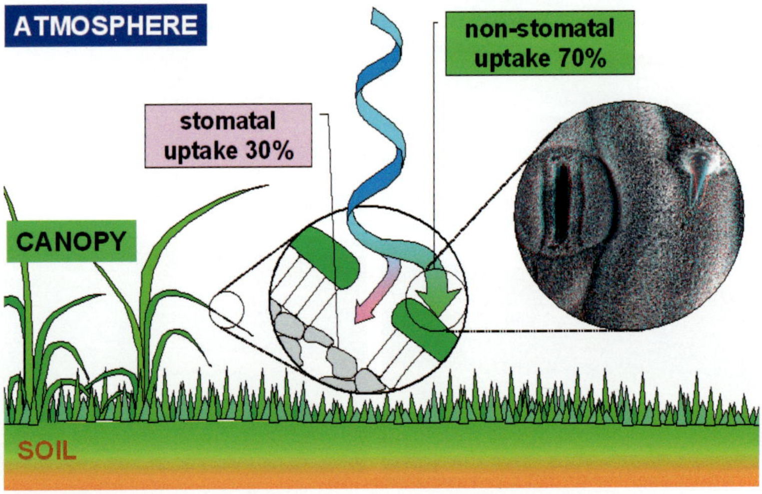

Plate 10. Three spatial scales for studies of land–atmosphere exchange of reactive trace gases. From left to right, the microscopic scale, with a cross-section of a bean leaf showing pallisade and mesophyll cells bounded by epidermal cells and the airspace for internal exchange within the leaf. The field scale at which most measurements are made by micro-meteorological methods, and the continental scale at which the land–atmosphere knowledge is applied using long-range transport models.

Plate 11. Schematic illustration of the turbulent transfer of ozone towards leaf surfaces and the molecular diffusion at the leaf surface and within sub-stomatal cavities.

12

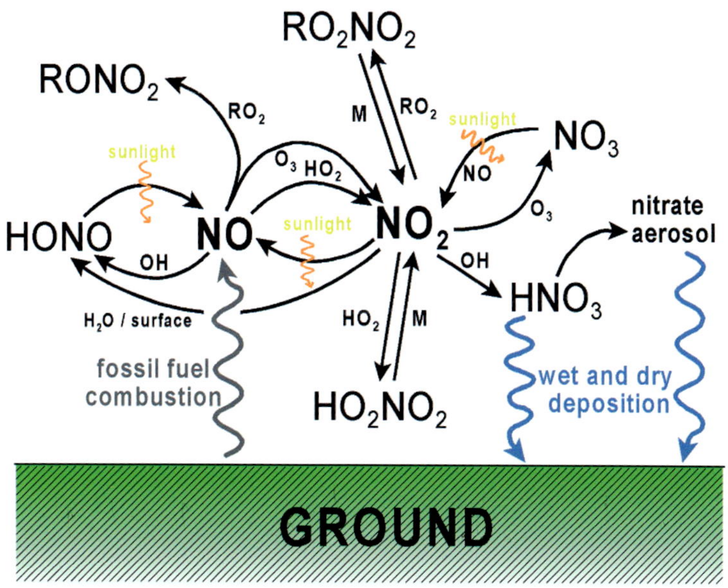

13

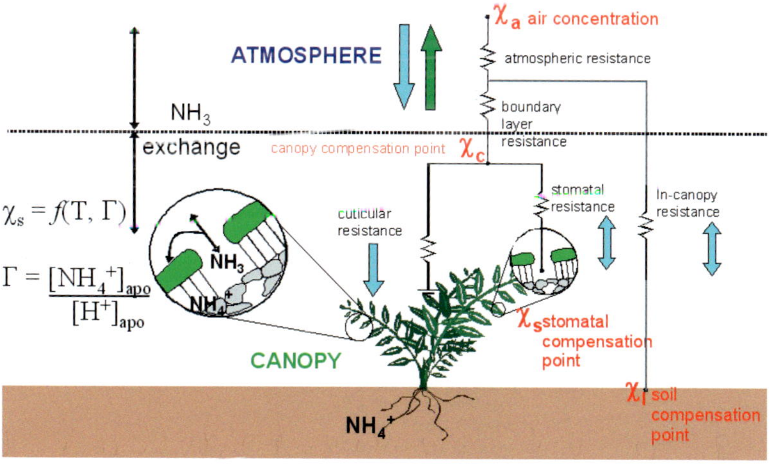

Plate 12. Air chemistry of nitrogen oxide (from PORG, 1998).

Plate 13. A schematic diagram of the compensation point for NH_3 exchange between plant canopies and the atmosphere (after Sutton *et al.*, 1995 and Nemitz *et al.*, 2001).

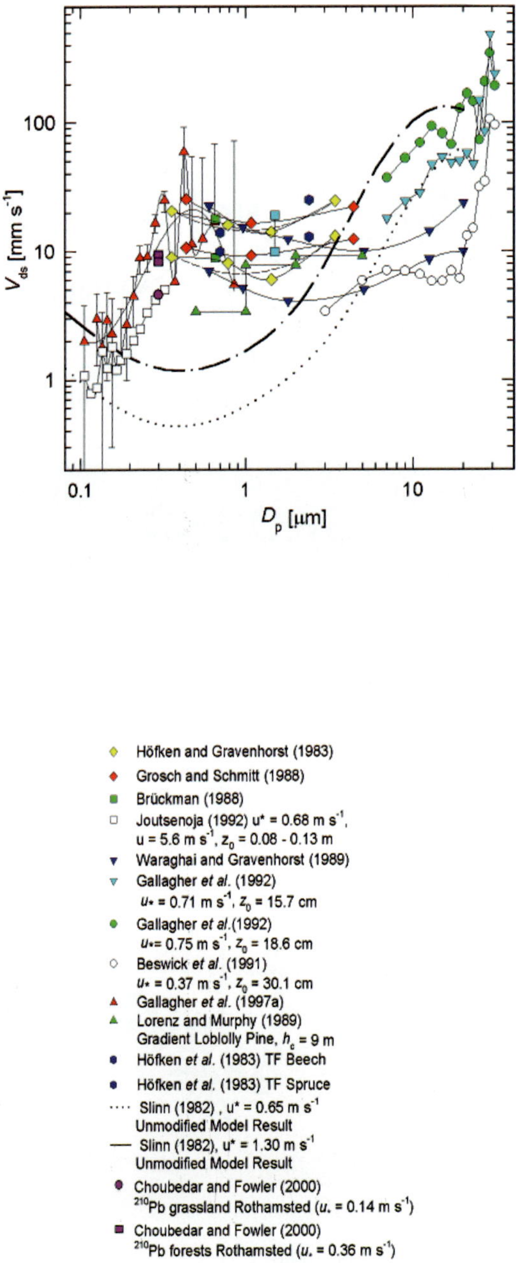

Plate 14. The size dependence of particle deposition on to woodland canopies from a compilation of different experimental studies (from Gallagher *et al.*, 2002).

Chapter 17

Population Growth and the Landscape Ecology of Microbes on Leaf Surfaces

J.H. Andrews
Plant Pathology Department, 1630 Linden Dr., University of Wisconsin, Madison, WI 53706, USA
jha@plantpath.wisc.edu

Introduction

The leaf surfaces of terrestrial plants, with a total surface area estimated at 4-6 x 10^8 km^2 (Morris and Kinkel, 2002) represent one of the largest and most significant microbial habitats. Epicuticular colonists, including yeasts, filamentous fungi and protists, live on nutrients that are either deposited as fallout from the atmosphere or exuded from within the leaf. The epiphytic community is remarkably variable in several ways and at many levels of scale. This reflects the unknown but presumably relatively large influence of underlying habitat heterogeneity. Virtually from the emergence of phyllosphere microbiology as a discipline more than 50 years ago, research has focused on the well known but unexplained phenomenon of extreme leaf-to-leaf variability in microbe population densities (Woody *et al.,* 2003). Here we consider how this arises and how it is shaped by microscale processes. There are clear analogies and implications to other microbial systems and to macro-ecology.

Phyllosphere Dynamics and Landscape Ecology

Microbial ecologists, as well as plant and animal ecologists, have tended until fairly recently to assume spatial uniformity below an arbitrarily selected level of scale. The chosen level varies, but in microbial ecology it has frequently been taken to be a discrete habitat patch such as a branch or rock surface. In macro-ecology, the unit is often represented as an early successional or old-growth field, forest patch, or tree. Assumptions of homogeneity facilitate modelling and comparisons across sites in assessments of population dynamics. These premises obscure local effects, however, such as patchy colonisation due to dispersal (recruitment) limitation, variation in terrain, resources and interactions among neighbours (Tilman and Kareiva, 1997). Neighbourhood effects are particularly important for sessile organisms such as plants, benthic marine invertebrates, and microorganisms (Tilman and Kareiva, 1997; Andrews, 1998). For example, spatial patterns of vegetation correspond broadly to temperature and moisture distribution, and more locally to gradients in soils and exposure, and to local topographic variation in temperature and moisture. Analogously, microbial abundance patterns, such as those on a leaf are likely to correlate with spatially determined factors such as nutrient pools (Mercier and Lindow, 2000) or physical features (Beattie and Lindow, 1995; Lindow and Brandl, 2003). Among these are topographic landmarks such as trichomes, stomata, veins and depressions above the anticlinal walls of epidermal cells (Andrews and Harris, 2000; Lindow and Brandl, 2003).

Landscape ecology concerns spatial heterogeneity and its implications to biotic and abiotic processes across broad scales (Tilman and Kareiva, 1997; Turner *et al.*, 2001). There are at least three processes governing the distribution and abundance of microbes on leaf surfaces that involve the tenets of landscape ecology. First, leaves are colonised by propagules (spores, vegetative cells, etc.) from sources mainly off the leaf. Variation in leaf topography and cuticle characteristics is likely to create a heterogeneous rather than a random distribution of initial colonists. If this distribution of colonists is not matched to the distribution of resources (nutrients, water), the growth of colonies will be less than that which would occur if there were a perfect correlation between resources and colonists (Ives and May, 1985). In other words, the initial distribution of colonists is likely to be patchy and this pattern probably intensifies through time due to habitat irregularities and local disturbances.

Second, colonies arising from the immigrants will grow under favourable conditions as allowed by the availability of nutrients. This growth depends not only on the local concentration of nutrients, but also on the (micro-) spatial extent of the nutrients. Colonies might be expected to grow in spatial extent until the local concentration of nutrients drops below some threshold. However, the ability of a colony to spread depends also on how the nutrients are distributed. This phenomenon is consistent with percolation theory (Stauffer and Aharony, 1992; see section on Prospective analyses and population models, below).

Third, due to the landscape patterns of resource distribution, the total population growth on a leaf is the average taken over a wide variation in growth rates of many individual colonies at distinct sites. A result from theoretical ecology is that the average growth rate of many populations (colonies) depends not only on the mean resource abundance, but also on the variation in resource abundance (Caswell, 1989; Yoshimura and Jansen, 1996). Rich and complex population dynamical patterns emerge when there is both spatial and temporal variation in resources, and when there is movement of individuals (Hanski *et al.,* 1995; Hastings, 1991). Later, we present an example showing how variation in resources across the surface of a leaf may affect the aggregate population growth rate of an epiphytic microbial colony. This characteristic illustrates one of the basic themes of landscape ecology: processes investigated at a fine spatial scale (in our case, the growth rate of individual colonies) can be used to understand emergent patterns at a coarser spatial scale (the population dynamics of *A. pullulans* on a leaf or the metapopulation among leaves). It also has practical (e.g. biocontrol) implications and is of great theoretical interest for estimating overall population growth in any heterogeneous habitat.

Indeed, there is now evidence from an assay based on repeated samples of the same leaf over time in the field that *A. pullulans* populations are consistently higher on some leaves of a plant than on others (Woody, *et al.,* 2003) (Fig. 17.1). This is probably due to persistent differences in leaf habitat quality. Furthermore, changes in density of *A. pullulans* tend to be synchronous among leaves, such that rank order of leaves is largely maintained through time (Woody, *et al.*, 2003). Previously such information could not be obtained directly in phyllosphere microbiology because of the destructive nature of conventional leaf assays. Though population densities clearly varied dramatically among leaves sampled over time, because of inherent experimental design limitations it could never be ascertained whether this was due to innate and consistent leaf differences or, alternatively, because populations were simply gyrating over time on any given leaf. So the question now becomes why leaves are different and to what extent micro-scale processes are responsible for the differences.

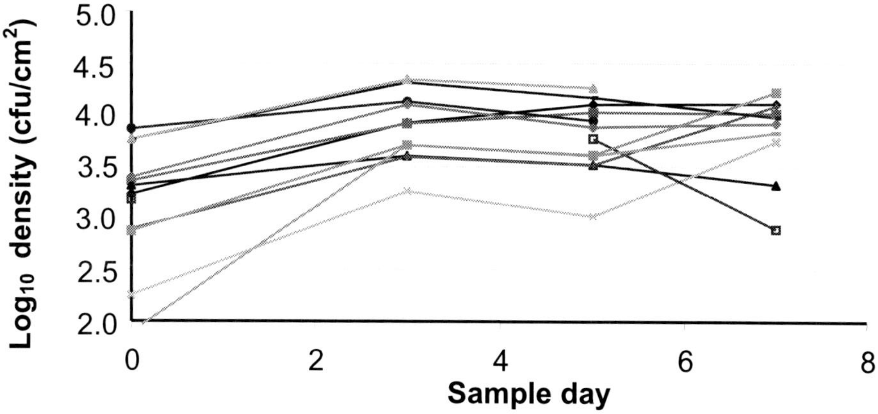

Fig. 17.1. The leaf resampling assay (Woody *et al.,* 2003) shows that *Aureobasidium pullulans* densities on apple leaves in the field are different among leaves harvested over time but populations on the individual leaves tend to track synchronously (10 leaves).

Leaves and *Aureobasidium pullulans* as a Model System

Aureobasidium pullulans (de Bary) Arnaud is a polymorphic deuteromycete. In many respects it behaves as a yeast and is informally classified as a member of the so-called "black yeasts" (De Hoog, 1999). *A. pullulans* is ubiquitous in nature (Andrews *et al.*, 1994, 2004; De Hoog, 1999) and is one of the major colonists of the phyllosphere (Andrews *et al.*, 2002; Wiens, 1989), reaching densities that are comparable to common epiphytic bacterial species. It occurs on leaves primarily in the unicellular form either as individual or budding yeast cells (blastospores; ca. 3-5 μm), or as swollen cells/chlamydospores. At least among the wild-type strains and habitats that we study, hyphae and pseudohyphae are rare in nature. Unlike individual bacterial cells, yeasts can be seen easily and enumerated microscopically, particularly on an uneven landscape such as a leaf. *A. pullulans* on leaves would be described best as a commensal, though it may operate mutualistically with the plant as it has demonstrable biocontrol efficacy against pathogens (Andrews *et al.*, 1994; De Hoog, 1999). Additionally, the fungus has several interesting properties, including secondary metabolic compounds, among them antibiotics and extracellular polysaccharides (Pouliot *et al.*, 2005). For all of these reasons, we have focused on it as a model organism to study the population biology (birth, death, immigration and emigration) of a resident phyllosphere colonist.

Spatial Approaches and Results

Over approximately the past decade, we have invested much time and resources to develop the quantitative tools to study the population biology of *A. pullulans*. These include primarily the following components: a fluorescent molecular probe (18S rRNA) (Li *et al.*, 1997; Spear *et al.*, 1999a,b); a GFP-marked strain (Vanden Wymelenberg *et al.*, 1997) (and, more recently a line transformed to express a red fluorescent protein [DsRed] from the soft coral *Discosoma* sp.); an *in situ* live/dead cell staining procedure based on the combined stains CellTracker Blue™ (chloromethyl hydroxycoumarin) and DEAD Red™ (ethidium homodimer 1; both reagents from Molecular Probes Inc., Eugene, OR, USA) (Nelson *et al.*, 2000); a semi-destructive apple leaf assay system (Woody *et al.,* 2003); and quantitative/modelling approaches (Li *et al.*, 1997; Andrews *et al.*, 2002; Woody *et al.*, 2003; Ives *et al.,* 2004; Woody *et al.*, submitted). By applying these methods in multi-year field studies we are learning some intriguing things about natural (unmanipulated) populations of the fungus in nature.

Where, specifically, on the phylloplane is *A. pullulans* found? To focus on spatial (specifically veinal) relationships between the fungal cells and the plant surface, we use site-specific mapping of the leaf landscape facilitated by Geographic Information Systems (GIS) software. In brief, the procedure is as follows: (i) Field leaves are divided transversely into typically four, ~1 cm wide segments and processed through fixatives and various reagents for fluorescence *in situ* hybridisation (FISH) (Li *et al.*, 1997; Andrews *et al.*, 2002). (ii) Adjacent, non-overlapping transects 500 μm wide as determined by the field diameter of a 40 x long-working distance objective lens (total magnification 400 x; area of field, 0.196 mm^2) are run along the adaxial surface of each apple leaf segment. As four such segments are normally taken at approximately equidistant intervals per leaf, 20 transects for about 15-30% of the adaxial surface area are routinely examined per leaf. Transect data include spatial information as to whether *A. pullulans* cells are located over the midvein, the smaller veins, or the interveinal region. We concentrate on veins because they are important landmarks for this fungus and seemingly so for microbes in general (Andrews *et al.,* 2002). An operational advantage of using the adaxial rather than the abaxial surface, at least for apples, is that there are relatively few stomata and trichomes; the latter complicate efficient quantification. (iii) The observer then records the presence and number of *A. pullulans* cells and leaf landscape features of interest (veins versuss no veins; wounds) along transects using a tape recorder. These data are transcribed into spreadsheets when convenient and summarised. We are able to count up to about 50 separated cells per field accurately; when this number is exceeded or cells are tightly aggregated, estimates are made based on the area covered and thickness of the cluster as determined by optical sectioning (deconvolution or confocal laser scanning microscopy). In the unusual situation where hyphae or pseudohyphae are present, the cell number recorded reflects the number of cells in the filament in the microscope field. The data are expressed spatially by use of ArcView™ GIS software (ESRI, Redlands, California, USA). We have prepared macro programs for Microsoft Excel™ that stack the transects for GIS presentation on a "prime meridian" based on the midvein as the vertical reference point. The macros plot cell number and leaf features for each microscope field outward from this longitudinal divide onto an artificial grid of latitude and longitude compatible with the format of ArcView. Additional macros sort and arrange the data for analysis with SAS software.

We assay colonisation both qualitatively (as occupancy, i.e. either a site has one or more *A. pullulans* cells present or it does not) and quantitatively (as cell density at sites that are occupied). Overall, when expressed as mean occupancy per date of sample collection, a range of about 20-40% of the microscope fields with at least one cell is observed.

Occupancy on a microscope field basis is always greater over the midvein (typically about 45-90%) or smaller veins (~50-65%) than over interveinal regions (~10-20%). An example of the data in GIS fomat is shown in Fig. 17.2). Intensity of colonisation, whether expressed as a percentage of total *A. pullulans* cells associated with a particular leaf feature, or as cell density per unit area, is also significantly greater over the veinal than the non-veinal areas. Locally high concentrations of cells are seen associated with veins (up to about 250 cells per field) or in association with micro-wounds in interveinal regions (up to about 1,500 cells per field). The morphotypes usually seen are blastospores and swollen cells or chlamydospores; hyphae and pseudohyphae, alluded to above, rarely occur.

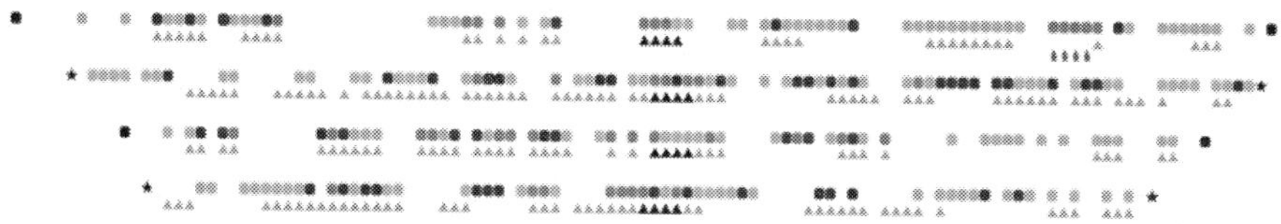

Fig. 17.2. *A. pullulans* densities are highly variable by microscopic position on adaxial leaf surfaces. Each circle of different intensity represents a microscopic field of different density (the darker the shade, the greater the cell density); gaps between circles represent zero counts. The main vein and smaller veins are indicated by the triangles (dark and light, respectively); arrowheads are wound microsites. The star symbols represent the ends of each of three transects. From Andrews *et al.* (2002).

We conclude from these assessments that *A. pullulans* colonises the phylloplane (at least the phylloplane of apple in cool temperate climactic regions) predominantly as single cells and small groups thereof in highly heterogeneous fashion. Leaf sites evidently exist that are relatively conducive for epiphytic fungal growth (veins; micro-wounds) or non-conducive (unwounded interveinal areas). Much space is unoccupied by *A. pullulans* and perhaps also by other microbes. On a microscope field basis (0.196 mm^2), this is up to about 70% of the fields overall and up to 85% in the interveinal regions. Where colonisation does occur, it can be locally dense.

This colonisation pattern prompts many biological questions, perhaps the most obvious are why is not more of the seemingly colonisable space occupied and why is the veinal system so intensely colonised. One hypothesis is that this is attributable to higher carbohydrate concentrations on the apple phylloplane above the veins than above the interveinal areas. While it is intuitive that levels of photoassimilates would be higher on the epidermis in the proximity of underlying veins, due to leakage, there is no direct physiological evidence for this in the literature to our knowledge. Nevertheless, phloem tissues in veins conduct concentrated sugars (e.g. sucrose at ~0.5 M; Salisbury and Ross, 1992). There are parenchymal vein extension cells connecting the vasculature to the epidermis (Salisbury and Ross, 1992), and assimilates may move along a symplastic or apoplastic route from vein to epidermis (dyes move quickly by this route; Salisbury and Ross, 1992). Indeed, we have some preliminary evidence from microassays on field leaves to support this hypothesis (McGrath and Andrews, unpublished). What is perhaps even more interesting is that also in preliminary controlled laboratory experiments where glucose is added incrementally at physiological levels to interveinal areas, *A. pullulans* populations do not reach the densities characteristic of veins. This intriguing result suggests that other factors, such as possibly different nutrients or chemicals, or the microtopography of the leaf landscape, are involved. For instance, the anticlinal walls of the veinal cells, and hence the grooves above those walls, tend to run in parallel. A consequence of this may be that yeast cells can bud unrestrictedly in linear fashion along these canal-like channels. In contrast, epidermal cells in the interveinal area join frequently and irregularly; possibly the arrangement of the grooves or cell topography impedes budding.

Prospective Analyses and Population Models

Here we discuss, first, how it would be possible to use data such as the foregoing to characterise microbe population growth on leaves in spatially explicit terms and, second, how to model overall population growth on the leaf landscape. We are exploring these models and approaches in our ongoing research and present them as being potentially useful to phyllosphere microbiologists. The overall goal of the models is to simultaneously explore the spatial patterns of *A. pullulans* dynamics (or any other microbe of interest) on leaves and to integrate these fine-scale patterns to understand the observed population dynamics at the scale of whole leaves. The bulk of microbial research has addressed coarse-scale spatial patterns, such as differences among leaves in microbial abundance, whereas here the focus is on explaining these coarse-scale patterns by understanding the finer-scale processes that underlie them. For example, it is instructive to know how the occurrence of "hot spots" and "cold spots" for localised growth influence estimation of overall population growth in a heterogeneous system.

Statistical Model for *A. pullulans* Spatial Distribution on a Leaf

Resource selection functions (RSF) and similar logistic, regression-type models are frequently employed by macro-ecologists to explain habitat use patterns by vertebrates (Johnson *et al.*, 2002) or plants (Tilman and Kareiva, 1997; Turner *et al.*, 2001), but the approach can be applied at finer spatial scales. RSFs are any functions that are proportional to the probability of use for a resource unit. They can be utilised in our context to investigate the role of leaf features such as veinal structure on colony growth of *A. pullulans*.

A common approach for estimating RSFs is to compare specific environmental co-variates at a particular location to those potentially available, for example, a set of random landscape locations. Co-variates for a leaf could include physical leaf features, for example veinal or interveinal region; distance to the nearest leaf vein or other landmark such as a glandular trichome; and other properties such as concentration of particular nutrients. The values of the co-variates at sites occupied by colonies can be contrasted with an equal number of randomly chosen sites on the same leaf; this process is repeated for each of the leaves designated for spatial analysis. The model can then be used to predict colonisation patterns on new leaves. In such analyses, typically a two-way contingency matrix is constructed, contrasting observed values of *A. pullulans* occurrence with predicted occurrence. This table can be evaluated by using the Tau_p statistic, which quantifies the improvement in a model's predictive power over a random assignment of values to grid cells (Ma and Redmond, 1995). The RSF approach could be expanded by comparing the distribution of *A. pullulans* colonies, say, on leaves from different plant species varying in their geometry. A RSF could be estimated for each plant species, and cross-validated to quantify the model's predictive power. Differences in the RSF models fitted to different plant species would reveal potential differences in processes dictating the distribution of colonies. Conversely, if the same RSF model fits all plant species tested, then the same processes are likely to be operating despite the very different host architecture and vein geometries of these plants.

Simulation Model of Colony Growth and Spread

Simulations derived from neutral landscape models (NLMs) (Gardner *et al.*, 1987) can be used to predict the growth patterns of colonies upon leaf surfaces. These spatially explicit models were originally developed from percolation theory (Stauffer and Aharony, 1992) and are useful for predicting how processes such as animal movement and seed dispersal respond to spatial heterogeneity (Gardner *et al.*, 1987; Gardner and Gustafson, 1996). The essence of percolation theory (Stauffer and Aharony, 1992; Andren, 1994) can be understood by imagining a sheet of paper ruled into an infinite number of squares, an array known to physicists as a square lattice. Some of the squares contain sufficient nutrients to support cell growth; others do not. The ability of a colony to grow into a region depends on the relative abundances of nutrient-filled or nutrient-empty squares. A well-known result from percolation theory is that if squares are randomly assigned to contain nutrients with probability p, then colony growth can only spread through space if $p > 0.41$ (assuming an "8-neighbour" rule). For values of p less than 0.41, there is a good chance that the growing colony will encounter a barrier of squares all of which contain no nutrients, thereby limiting further spatial spread of the colony. If, however, squares are not randomly assigned to having nutrients or not, but instead the squares with nutrients are clustered, then colony spread could occur with fewer squares containing nutrients. This result from percolation theory has been used to explain the spread of such diverse phenomena as forest fires and oil inside porous rock in oil reservoirs (Stauffer and Aharony, 1992).

The implications of different spatial arrangements of resources; in particular those owing to venation patterns, for colony growth and expansion can be explored by use of NLMs. The spatial distribution of habitat is depicted in a 2-dimensional array in which positions, or grid sites, are considered as suitable or unsuitable for the species of interest Both the amount of suitable habitat (e.g. for *A. pullulans*, the proportion of a leaf occupied by veins) and its spatial arrangement (e.g. venation pattern) can be varied separately. Individual leaves can be represented as a 2-dimensional array with a typical objective lens field as site size (0.196 mm^2; Andrews *et al.*, 2002) in which the amount and arrangement of veins are varied. Initial colonisation by *A. pullulans* would be taken to occur at random (at one or multiple locations on the leaf), and population spread can be simulated as the probability of colony growth into nearby sites. Fecundity and mortality can be represented by using probabilities derived from colony growth dynamics under controlled conditions. For instance, we might hypothesise that the growth and expansion of *A. pullulans* colonies would be most rapid and population sizes the greatest when leaf vein sites are nutrient-rich, abundant, and well connected. Furthermore, if nutrients are evenly distributed, a single colonisation event would be adequate to initiate extensive spatial spread, i.e. the system would percolate (Gardner *et al.,* 1989, Stauffer and Aharony, 1992). In contrast, if nutrient-rich sites were not well connected spatially (i.e. in the hypothetical absence of veins), then colony expansion would be slower, and multiple colonisation events will be required for *A. pullulans* populations to occupy a substantial proportion of the nutrient-rich sites on the leaf surface.

The advantage of the simulation approach is that the full range of parameter space can be explored and the conditions under which spread does or does not occur can be identified. The effect of disturbance events, such as rainfall, on microbe colonisation can also be simulated. The hypothesised effect of disturbance could be simulated by pulsing the leaf with a greater number of suitable sites, and then imposing intervening periods in which only a few of the highest quality sites allow growth. These pulsed periods in which suitable sites are more common may greatly speed the expansion of colonies if this allows them to jump a barrier of sites which under normal conditions would be unsuitable. By varying the magnitude and rate of pulsed disturbances, one can ask how disturbances may affect the expansion of colonies.

Probabilistic Model of Colony Growth and Spread

A probabilistic model could be used to explain the growth of many colonies located in different regions of the leaf surface with different nutrient concentrations. The model would complement the simulation model (above) of colony growth and expansion. While the simulation provides a detailed, mechanistic description of the colony growth process, the probabilistic approach offers a more qualitative description that makes it easier to extract general results pertinent to a range of biological systems beyond the organism of interest, represented here by *A. pullulans*. Bringing different modelling approaches to bear on the same problem is a powerful technique because it capitalises on the strengths of different approaches and potentially stimulates new ideas.

The growth of colonies can be modelled as a stochastic branching process in which cell division and death are described in terms of probabilities (Feller, 1968). Specifically, let X_t denote the random variable giving the number of cells in a colony at time t, with $P_k(t)$ giving the probability that $X_t = k$. Assume that cell division and death can be summarised by the probabilities $P_k(0) = p_k$, such that p_k gives the probability that a given cell is replaced by k cells in the following generation. For example, if the cell dies before reproducing with probability 0.2, then $p_0 = 0.2$. Assuming that birth and death are independent of the size of the colony and time, then it is possible by standard probability theory to calculate the distribution $p_k(t)$ for X_t. This makes it possible to calculate the growth of the expected colony size and the chance that a colony will go extinct over some time period. Note that, rather than using a branching process formulation of colony growth, it is also possible to model colony growth as a stochastic birth-death process (Nisbet and Gurney, 1982).

The most biologically interesting questions arise when the standard assumptions of branching processes are violated, in particular, the assumptions that cell division and death are independent of colony size and time. To illustrate the consequences of colony size-dependence, consider the process in which cells never die ($p_o = 0$), and in each time period the probability they divide is $p_2 = r(1 - X_t/K)$, leaving $p_1 = 1 - p_2$. These assumptions give a stochastic branching process comparable to logistic growth, in which K gives the colony carrying capacity as the maximum colony size that can be achieved. A particularly pertinent question that can be asked with this model is how spatial variation in nutrient concentration might affect the growth of the *A. pullulans* population over an entire leaf. To model spatial variation in nutrients, suppose that the values of K's differ for the collection of colonies on a leaf. Fig. 17.3A illustrates the stochastic growth of ten colonies when values of K range over two orders of magnitude. This is compared to the growth of ten colonies which have the same value of K, set as equal to the average K in Fig. 17.3B. Although the number of cells on the leaf eventually reach the value in both cases A and B, they do so more slowly in case A (Fig. 17.3C), indicating that variation in nutrient concentration slows the overall population growth of *A. pullulans* on leaves. If a branching process model is fitted directly to data collected on colony growth, the population growth trajectory for an entire leaf can be predicted. These predictions can be tested by comparing the estimated population growth trajectories to the observed trajectories obtained from field-collected leaves evaluated by direct microscopy (Andrews *et al.*, 2002), an *in situ* live/dead cell assay (Nelson *et al.*, 2000), or other suitable assay.

The branching-process models can further be used to explore the consequences of

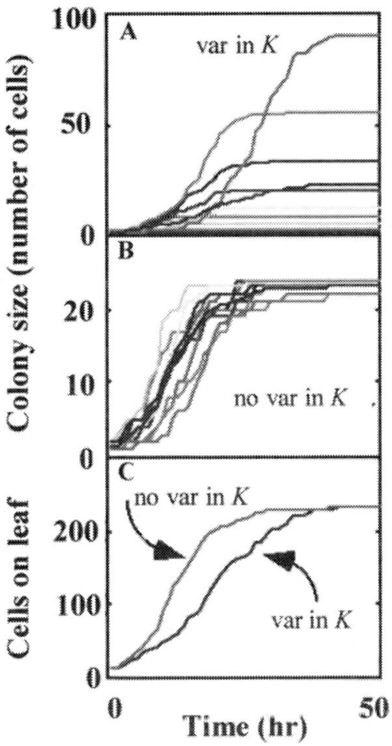

Fig. 17.3. Growth of ten hypothetical microbial colonies by a stochastic branching process (A) with and (B) without variation in location-specific carrying capacities, *K*. (C) gives the total number of cells on the leaf for the cases in (A) and (B). From A.R. Ives, M.G. Turner, E.V. Nordheim and J.H. Andrews (unpublished).

environmental variation, parallelling similar investigations with the simulation model (above). An interesting aspect of *A. pullulans* and yeast biology in general is that when phylloplane conditions are favourable, cells can divide fairly rapidly (blastospore phase), but when conditions are poor, cells can survive for extended periods (chlamydospore phase). This leads to an asymmetry in which population increase is rapid, yet population decline is slow. This asymmetry can lead to a strong buffering effect against environmental variation, allowing colonies to rapidly capitalise on good conditions and ride out bad conditions. In the branching process model, this behaviour can be incorporated by allowing values of p_k to depend on time, with the asymmetry in increase versus decline incorporated as high probabilities of division that depend strongly on the environment and low probabilities of death that depend weakly on the environment.

The probabilistic model can also be used to explore the consequences of cell movement on leaf surfaces. If rainfall events move cells across leaf surfaces, then new micro-colonies can be initiated; this is one hypothesis for the surge of population growth on leaves following rainfall events (McGrath and Andrews, unpublished). Thus, the overall population growth on a leaf depends both on the growth of individual colonies and the rate at which new colonies are founded by cell movement. The combined effects of these processes can be modelled by allowing a possible transition in the branching process from 0 cells to 1.

Conclusion

This chapter has considered how a dominant member of the epiphytic microbial community grows and colonises leaves. The situation can be generalised to elucidate the spatially explicit role nutrients and other factors play in microbe population dynamics and density. Intensity of occupation (cell packing) has numerous consequences, including the ability of a benign microbe to exclude a pathogenic microbe, cell-to-cell signalling, density-dependent regulation (quorum sensing), genetic exchange and communication with the plant (Beattie and Lindow, 1995; Andrews and Harris, 2000; Lindow and Brandl, 2003). Conversely, the apparent absence of colonisation over vast reaches of the phylloplane is also likely to have consequences. One of these may be vulnerability to infection or the epiphytic establishment of plant pathogens.

The spatial configuration of individuals is an important consideration in understanding their ecology. It has been suggested that landscape ecology represents the relationship between spatial heterogeneity and ecological processes that can be translated across a wide range of scales and applied to any population or process if it is scaled appropriately (Turner *et al.*, 1989, 2001; Turner and Gardner, 1991). However, the consistency of relationships observed between habitat heterogeneity and population growth has been tested over relatively few taxa (Turner *et al.*, 1989, 2001; Turner and Gardner, 1991; Fischer *et al.*, 2004) and apparently no studies have tested such relationships by using microbial communities. The influence of dendritic habitat structure, which is similar to leaf venation, on populations has been considered (Fagan, 2002). Application of the principles and approaches of landscape ecology to microbial ecology contributes useful new insights. Conversely, a microbe system offers a means to explain the processes underlying a particular landscape pattern, such as the dendritic network of veins. Mechanistic explanations and the ability to do multiple replications have often eluded landscape ecologists (Turner and Gardner, 1991; Turner *et al.*, 2001). Thus, there is considerable potential for complementation across levels of scale, with landscape ecology providing the theoretical principles and phyllosphere microbiology providing the robust experimental units.

Acknowledgements

I thank Russell N. Spear for assistance in many ways and M.J. McGrath, A.R. Ives, M.G. Turner and E.V. Nordheim for collaboration. Fig. 17.3 and the associated text were prepared by A.R. Ives. Research reported here was supported by USDA-Hatch grants 142P446 and 142P4873 to J.H. Andrews, and NSF grant DEB-0075358 to J.H. Andrews, A.R. Ives and E.V. Nordheim.

References

Andren, H. (1994) Effects of habitat fragmentation on birds and mammals in landscapes with different proportions of suitable habitat: a review. *Oikos* 71, 355-366.

Andrews, J.H. (1998) Bacteria as modular organisms. *Annual Review of Microbiology* 52, 105-126.

Andrews, J.H. and Harris, R.F. (2000) The ecology and biogeography of microorganisms on plant surfaces. *Annual Review of Phytopathology* 38, 145-180.

Andrews, J.H. Spear, R.N. and Nordheim, E.V. (2002) Population biology of *Aureobasidium pullulans* on apple leaf surfaces. *Canadian Journal of Microbiology* 48, 500-513.

Andrews, J.H., Harris, R.F., Spear, R.N. and Lau, G.W. (1994) Morphogenesis and adhesion of *Aureobasidium pullulans*. *Canadian Journal of Microbiology* 40, 6-17.

Beattie, G.A. and Lindow, S.E. (1995) The secret life of foliar bacterial pathogens on leaves. *Annual Review of Phytopathology* 33, 145-172.

Caswell, H. (1989) *Matrix Population Models*. Sinauer Associates Inc., Sunderland, MA, USA.

De Hoog, G.S. (ed.) (1999) Ecology and evolution of the black yeasts and their relatives. Centraalbureau voor Schimmelcultures, *Studies in Mycology* 43, 1-208.

Fagan, W.F. (2002) Connectivity, fragmentation, and extinction risk in dendritic metapopulations. *Ecology* 83, 3243-3249.

Feller, W. (1968) *An Introduction to Probability Theory and its Applications*. Vol. 1. John Wiley & Sons, NY, USA.

Fischer, J., Lindenmayer, D.B. and Fazey, I. (2004) Appreciating ecological complexity: habitat contours as a conceptual landscape model. *Conservation Biology* 18, 3243-3249.

Gardner, R.H. and Gustafson, E.J. (1996) The effect of landscape heterogeneity on the probability of patch colonisation. *Ecology* 77, 94-107.

Gardner, R.H., Milne, B.T., Turner, M.G. and O'Neill, R.V. (1987) Neutral models for the analysis of broad-scale landscape patterns. *Landscape Ecology* 1, 19-28.

Gardner, R.H., O'Neill, R.V., Turner, M.G. and Dale, V. H. (1989) Quantifying scale-dependent effects with simple percolation models. *Landscape Ecology* 3, 217-227.

Hanski, I., Poyry, J., Pakkala, T. and Kuussaari, M. (1995) Multiple equilibria in metapopulation dynamics. *Nature* 337, 618-621.

Hastings, A. (1991) Structured models of metapopulation dynamics. *Biological Journal of the Linnean Society* 42, 57-71.

Ives, A.R. and May, R.M. (1985) Competition within and between species in a patchy environment: relations between microscopic and macroscopic models. *Journal of Theoretical Biology* 115, 65-92.

Ives, A.R., Woody, S.T., Nordheim, E.V., Nelson, C. and Andrews, J.H. (2004) The synergistic effects of stochasticity and dispersal on population densities. *American Naturalist* 163, 375-387.

Johnson, C.J., Parker, K.L., Heard, D.C. and Gillingham, M.P. (2002) A multiscale behavioral approach to understanding the movements of woodland caribou. *Ecological Applications* 12, 1840-1860.

Li, S., Spear, R.N. and Andrews, J.H. (1997) Quantitative fluorescence *in situ* hybridization of *Aureobasidium pullulans* on microscope slides and leaf surfaces. *Applied Environmental Microbiology* 63, 3261-3267.

Lindow, S.E. and Brandl, M.T. (2003) Microbiology of the phyllosphere. *Applied and Environmental Microbiology* 69, 1875-1883.

Ma, Z. and Redmond, R.L. (1995) Tau coefficients for accuracy assessment of classification of remote sensing data. Photogram. *Engineering and Remote Sensing* 61, 435-439.

Mercier, J. and Lindow, S.E. (2000) Role of leaf surface sugars in colonization of plants by bacterial epiphytes. *Applied and Environmental Microbiology* 66, 369-374.

Morris, C.E. and Kinkel, L.L. (2002) Fifty years of phyllosphere microbiology: significant contributions to research in related fields. In: S.E. Lindow, E.I. Hecht-Poinar and V.J. Elliott (eds) *Phyllosphere Microbiology*. APS Press, St. Paul, Minnesota, USA. pp. 365-375.

Nelson, C.D., Spear, R.N. and Andrews, J.H. (2000) Automated image analysis of live/dead staining of the fungus *Aureobasidium pullulans* on microscope slides and leaf surfaces. *BioTechniques* 29, 874-882.

Nisbet, R.M. and Gurney, W.S.C. (1982) *Modeling Fluctuating Populations*. John Wiley & Sons, Chichester, UK.

Pouliot, J.M, Walton, I., Parkhouse, M.N., Abu-Lail, L.I. and Camesano, T.A. (2005) Adhesion of *Aureobasidium pullulans* is controlled by uronic acid based polymers and pullulan. *Biomacromolecules* 6, 1122-1131.

Salisbury, F.B. and Ross, C.W. (1992) *Plant Physiology*, 14th ed. Wadsworth, Belmont, CA, USA.

Spear, R.N., Cullen, D. and Andrews, J.H. (1999a) Fluorescent labels, confocal microscopy, and quantitative image analysis in the study of fungal biology. *Methods in Enzymology* 307, 607-623.

Spear, R.N., Li, S., Nordheim, E.V. and Andrews, J.H. (1999b) Quantitative imaging and statistical analysis of fluorescence *in situ* hybridization (FISH) of *Aureobasidium pullulans. Journal of Microbiological Methods* 35, 101-110.

Stauffer, D. and Aharony, A. (1992) *Introduction to Percolation Theory*, 2nd edition. Taylor & Francis, London, UK.

Tilman, D. and Kareiva, P. (eds) (1997) *Spatial Ecology*. Princeton University Press, Princeton, NJ, USA.

Turner, M.G. and Gardner, R.H. (eds) (1991) *Quantitative Methods in Landscape Ecology*. Springer Verlag, New York, USA.

Turner, M.G., Gardner, R.H., Dale, V.H. and O'Neill, R.V. (1989) Predicting the spread of disturbance across heterogeneous landscapes. *Oikos* 55, 121-129.

Turner, M.G., Gardner, R.H. and O'Neill, R.V. (2001) *Landscape Ecology in Theory and Practice*. Springer-Verlag, New York, USA.

Vanden Wymelenberg, A.J., Cullen, D., Spear, R.N., Schoenike, B. and Andrews, J.H. (1997) Expression of green fluorescent protein in *Aureobasidiuim pullulans* and quantification of the fungus on leaf surfaces. *BioTechniques* 28, 686-690.

Wiens, J. A. (1989) Spatial scaling in ecology. *Functional Ecology* 3, 385-397.

Woody, S.T., Ives, A.R., Nordheim, E.V. and Andrews, J.H. (2006) Dispersal, density-dependence, and dynamics of the microbe *Aureobasidium pullulans* on apple leaves. *Ecology* (submitted).

Woody, S.T., Spear, R.N., Nordheim, E.V., Ives, A.R. and Andrews, J.H. (2003) Single-leaf resolution of the temporal population dynamics of *Aureobasidium pullulans* on apple leaves. *Applied and Environmental Microbiology* 69, 4892-4900.

Yoshimura, J. and Jansen, V.A.A. (1996) Evolution and population dynamics in stochastic environments. *Research in Population Ecology* 38, 165-182.

Chapter 18

What DNA Microarrays Can Tell Us About Bacterial Diversity: A New Light on an Old Question

G.L. Andersen, Y.M. Piceno, T.Z. DeSantis and E.L. Brodie
Center for Environmental Biotechnology, Lawrence Berkeley National Laboratory, 1 Cyclotron Road, Berkeley, CA 94720, USA
glandersen@lbl.gov

Introduction

A limited number of studies on the types of microorganisms that are routinely present in air have been completed, which is surprising considering their importance for a host of biological processes. At any given location, bacterial diversity and concentration is believed to be dependent on both local and long-range sources, as well as current meteorological conditions. Bacteria from many different environments, such as sewage treatment plants, air-conditioning ducts, vegetation and soil, have the potential to move into the atmosphere for periods of minutes to days, and potentially even longer. Once the bacteria have been aerosolised, meteorological conditions are the dominant influence determining the dispersal and survival of organisms from a particular source (Fig. 18.1) (Mohr, 2002). Thus, the impact of humans on the microbial composition of the atmosphere can either be direct, by influencing the sources of bacteria in the air, or indirect, by affecting global climate patterns.

Prior to the advent of molecular techniques for bacterial identification, sampling methods typically consisted of culturing bacteria from concentrated air samples or from direct deposition onto media plates. To facilitate classification, visualisation of agar plates containing nutrients to preferentially grow certain organisms, sometimes under different temperature regimes, has been used. Among the traits that have been observed are macro- and microscopic morphology, colour and physiological growth requirements. There have been limits to this process, particularly for organisms unable to grow under the conditions offered (Buttner *et al.*, 2002). Often less than 1% of the bacteria present in a sample are cultured readily (Staley and Konopka, 1985). Although limited to specific classes of organisms, the traditional culture studies provided valuable information about bacterial transport through the air, particularly as it pertained to plant and animal disease.

As early as 1837 it was shown that microorganisms were in the air and soon after that diseases could spread from person to person through the air (Lighthart and Shaffer, 1995; Stites, 2002). Culture-based studies have been used to demonstrate that plant epiphytes and animal disease agents can be transported through the air (Lindemann *et al.*, 1982; Lindemann and Upper, 1985; Weir-Brush *et al.*, 2004), some impacting ecosystems thousands of miles away from the source systems (Griffin *et al.*, 2001). The effect of long-range wind dispersal of microorganisms into established communities is still relatively unknown, but immigration can be an important source of epiphytic bacteria, as various studies have shown. Deposition or short-range immigration of bacteria onto leaves has been studied for important plant pathogens (e.g. *Pseudomonas syringae*: Hirano and Upper, 2000; Upper and Hirano, 2002) and for ice nucleation-active bacteria (Lindow and Andersen, 1996).

Fig. 18.1. The microbial composition in the atmosphere is highly dynamic. Organisms are released into the air from both local and long-range sources. Meteorological and electrostatic forces influence microbial dispersal and eventual deposition. Vegetation and other localised reservoirs may influence concentration, types and distribution of organisms in outdoor environments.

The type of leaf surface also has a large impact on the amount of bacteria present from immigration and whether bacteria grow on the leaf surface. Lindow and Andersen (1996) found immigrating bacteria to be the main source for the epiphytic bacterial communities on navel orange leaves, whereas *in situ* bacterial growth contributed more epiphytic bacteria to bean leaves than immigration (Upper and Hirano, 2002 and references therein). In part, physical and nutritional differences may determine if growth or immigration dominate phyllosphere bacterial populations. Thick, waxy cuticles may inhibit the diffusion of nutrients from leaf surfaces, thereby limiting bacterial growth (Lindow and Andersen, 1996). Also, the waxy cuticle may reduce the number of crevices available for bacterial colonisation and greatly reduce the amount of available water, again limiting bacterial growth. Regardless of the causes, there are lower population numbers on waxy leaves than on non-waxy (more wettable) leaves (Lindow and Andersen, 1996 and references therein).

Plants are not only influenced by bacteria deposited on their surfaces, they are also contributors to aerosolised bacterial populations. Lindemann *et al.* (1982) demonstrated that in addition to bacterial fluxes into the atmosphere (upward flux) being greater over vegetated areas than over bare soils, the type of plant present also affects the total bacterial load in the air. This was confirmed by Lindow and Andersen (1996) while examining bacterial immigration onto citrus tree leaves. Trees growing at the edge of the grove near grasses and broadleaf plants had significantly greater population sizes on their leaves than did leaves of trees closer to the middle of the orchards. Furthermore, Shaffer and Lighthart (1997) demonstrated that the stage of crop development and processing affected the total number of culturable bacteria over the same type of plant. They found the greatest number of bacteria present (in upward flux) after rye grass had its seed harvested and was cut short and left on the ground to dry. Interestingly, even the electrostatic charge of plant leaves may affect bacterial release from the plant canopy (Lindemann and Upper, 1985) and deposition of spores onto surfaces (Levetin, 2002).

Diurnal Cycles and Transport Mechanisms of Bioaerosol Populations

Larger-scale environmental forces also have notable influence on bioaerosols. Several investigations have shown daily fluxes in the concentration of airborne bacteria (Hirano and Upper, 1989; Lighthart and Shaffer, 1995; Shaffer and Lighthart, 1997). At a high desert location, Lighthart and Shaffer (1994) found a mid-day low corresponding to a solar radiation peak, a rise in temperature, and decreased relative humidity compared to early morning and evening samples. The opposite pattern was demonstrated over a bean plant crop, but it was restricted to days when wind speeds were greater than 1 m/s (Lindemann and Upper, 1985) (i.e. a mid-day peak when wind speeds were high enough to contribute to an upward flux of bacteria from plant leaves). In a river valley affected by strong, mid-afternoon, onshore winds, the viable airborne bacterial load had three peaks: sunrise, early afternoon and late afternoon (Lighthart and Shaffer, 1995). The first increase was accompanied by increased winds and solar heating. Solar heating destabilised the atmosphere near the earth's surface and dried the soil. Drier surfaces release more bacteria than wet surfaces, and the wind likely increased aerosolisation also. The second peak occurred from noon to 3.00 pm (roughly), and ended abruptly with the onset of onshore winds. The third peak had a sharp rise corresponding to slacking onshore winds and tapered off as evening set in. The viable counts for that study had been collected over a grass seed field.

On a much larger scale, African dust storms are known to carry dust and associated microorganisms to the Americas, the Caribbean and the Sargasso Sea (Griffin *et al.*, 2001; McCarthy, 2001; Weir-Brush *et al.*, 2004). From February to April, the northeastern Amazon Basin is affected by the trans-oceanic dust, while Central and North America (mostly Florida) is affected from June to October (Griffin *et al.*, 2001). Weir-Brush and colleagues showed an important relationship between fungi cultivated during African dust events in the Caribbean and fungal disease agents of corals there. Dust storms within Africa (Mali) are also associated with meningococcal meningitis, the outbreaks of which correspond directly with the Harmattan winds, blowing from February to May (Sultan *et al.*, 2005). Kellogg *et al.* (2004) have characterised the culturable airborne bacteria and fungi in Mali, West Africa, which is an important first step in understanding the potential for plant and human pathogen transport from one continent to the next.

Impact and Importance of Molecular Methods for the Microbial Characterisation of Bioaerosols

As important as culture-based studies have been in elucidating bioaerosol transport, the small percentage of cultivated bacteria compared to total bacteria in the environment has resulted in a large gap in the knowledge of what organisms are predominant in the air. A much-needed revolution in microbial detection and identification took place with the recognition of the phylogenetic importance of house-keeping genes and the advent of the polymerase chain reaction (PCR). Zuckerkandl and Pauling (1965) formally introduced the idea of using biopolymers for developing systematics that were universally applicable. Carl Woese and colleagues were instrumental in constructing the framework for phylogenetic studies of bacteria, fungi and other organisms (Pace, 1997). This represented a particularly important step given the inherent difficulty of detecting and identifying microorganisms with their often limited morphology, reluctance to grow on standard laboratory media, and abundant genetic diversity. Ribosomal RNA became the molecule of choice for constructing a universal phylogeny and now provides a large platform from which to ask questions about phylogeny and evolution (Liu and Stahl, 2002). Working with molecular structures allows us to learn about bacteria and fungi present in soils, lakes, rivers, forest litter, plant tissues and many other places without having to culture them first. This vastly improves our ability to learn about ecosystems and microbial niches.

The most common molecular method for detecting and identifying bacteria from environmental samples is PCR followed by a technique that separates the PCR products for identification. The most critical step in PCR, then, is to identify primers (short pieces of DNA) corresponding to gene regions shared by many or all microbes of interest. DNA sequencing of cloned, 16S ribosomal fragments is the current gold standard for classifying bacteria into recognised groups. A pool of 16S rDNA PCR products is generated from the various bacteria present in a sample and then separated to determine the individual bacterial identities. Cloning, gel-based separations (e.g. DGGE, TGGE), and chromatography-based separation (e.g. HPLC) strategies have all been used. Yang and colleagues (2001) found greater bacterial diversity present on the phyllosphere of citrus leaves using PCR and denaturing gradient gel electrophoresis (DGGE) analysis than had been noted previously by culturing. Soils, which are assumed to provide a major component of microbes in the air, have been found to possess very high levels of diversity when assessed using molecular methods (Torsvik *et al.*, 1990; Dunbar *et al.*, 1999). Air samples, too, are suspected to be more diverse than previously demonstrated by culturing. Radosevich and colleagues (2002) constructed a clone library from outdoor air showing most of the 319 clones analysed to have distinct sequences. Not only were the more common Gram-positive bacteria detected (e.g. *Bacillus*, *Streptococcus*) as would be expected in culture-based studies, but bacterial groups less likely to be cultured from air samples also were detected (e.g. *Alphaproteobacteria, Cytophaga, Nitrospira, Clostridium*).

Molecular methods improve our ability to study bioaerosols for another important reason: sampling obstacles caused by the low relative concentrations of bacteria in the air. A variety of samplers have been designed in efforts to trap as many kinds of organisms (bacteria, fungi, viruses) as possible and, more importantly, to keep them viable long enough for them to grow and be counted. A major difficulty for culture-based studies is drying and other stresses caused by sampling itself, especially if long collection times are required (Piceno and Davis, 2002). Using molecular methods to analyse bioaerosol samples does not require that the microorganisms remain viable, since the analysis is performed on cellular components. This allows the use of harsher, yet potentially more inclusive, sampling strategies to increase the chance of detecting less abundant or less hardy bacteria.

The disadvantage is not knowing if the cells were viable even before collection onto the filter. For example, filtering air onto any kind of membrane will reduce the viability of vegetative cells due to the stress of desiccation, and yet, high flow filtration provides an opportunity to sample much larger volumes of air or to sample for much longer times than is possible even with liquid impinger samplers. Radosevich and colleagues (2002) used a high-volume air sampler that filtered 1000 liters of air per minute and collected air for 24 hours. By comparison, typical agar impaction or liquid impingement samplers operate at 12-15 L/min for 1-2 minutes (impaction) or up to 90 minutes (impingement).

An Introduction to DNA Microarrays

Although many molecular methods show promise in the identification of microorganisms that are present in complex environments, such as the earth's atmosphere, there remain roadblocks to the adoption of these techniques for routine study. Notably, methods such as DNA sequencing of cloned library fragments are expensive and time consuming, so that only a limited number of samples may be analysed at the depth of coverage required for diversity assessments. Conversely, methods for the electrophoretic separation of amplified 16S rDNA fragments (such as T-RFLP or DGGE) are useful for determining changes in bacterial diversity over multiple samples but do not have the resolution to identify individual components. We have explored the use of phylogenetic 16S rDNA microarrays for the categorisation of bacterial composition in complex environmental samples, such as bioaerosols, because of their ability to both identify individual components and assess multiple samples simultaneously. A DNA microarray often consists of large numbers of individual short DNA segments called probes bound to a two-dimensional surface (e.g. glass) with specific sequences located at defined x,y coordinates. Nucleic acids from environmental samples are labelled as targets so that hybridisation of specific sequences will be visualised by the number of labelled molecules attached to the probes. 16S rDNA microarrays have been used successfully to differentiate bacteria in specific groups (Lehner *et al.*, 2005 (*Enterococcus*); Castiglioni *et al.*, 2004 (Cyanobacteria); Kelly *et al.*, 2005 (nitrifying bacteria); Warsen *et al.*, 2004 (fish pathogens)). In order to extend the technique's capability towards detecting all known bacteria, a greater number of total probes are employed and several probes are designed for each taxon to increase confidence via redundancy. This limits the appearance of false positives for it is less likely that all the probes directed at a single taxon will be bound by non-target organisms. As microarray manufacturing processes improve, the resolution of organism identification will be likely to increase because of the higher density of probes placed on each array.

The First Generation DNA Microarray

We constructed our first generation high-density small subunit (SSU) rDNA microarray (G1) containing 62,358 16S rDNA probes for parallel, multi-species identification from complex samples. At the time the probes were chosen, the public databases contained complete and aligned SSU sequences from about 3000 organisms. A highly variable, 100 base pair region near the 3' gene terminus bounded on both ends by universally conserved segments was targeted for differentiation of prokaryotes and eukaryotes. Multiple probes were unique for each subdivision and were synthesised directly onto the microarray's glass surface. A control probe (or mismatch probe, MM) was paired with each complimentary probe. The MM probe was complimentary to the target 20-mer except at the central position so that it would have a lower hybridisation intensity score than the perfectly matched probe. The MM probes helped to ensure that the probe-target interaction was sequence dependent.

The initial environmental test samples came from air samples filtered onto polyester filters (Wilson *et al.*, 2002). The filters were washed to recover the bacteria. After extracting the DNA from the filter washes, PCR primers corresponding to conserved segments were used to generate a pool of partial length SSU rDNA amplicons. Microarray hybridised target DNA was end-labelled with biotin, stained with a streptavidin phycoerythrin-R conjugate, and scanned with a laser to measure fluorescence intensity. The fluorescence from the control probes and the target probes was compared, and probe pairs in which the mismatch probe fluorescence was higher than the target probes were removed from further analysis. Having multiple probes for each phylogenetic group allowed confident (reproducible) presence/absence scoring for each order. Fig. 18.2 shows the results for hybridisation scores (fluorescence intensities) for bacterial and fungal orders detected in the air samples. The results depicted are replicates of one sample; they demonstrate overall array-to-array reproducibility, even across different manufacturing lots of microarrays.

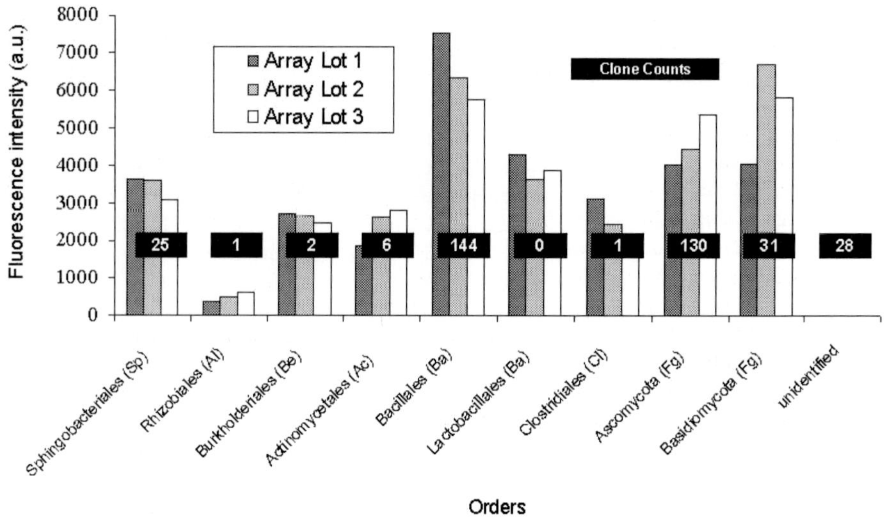

Fig. 18.2. Fluorescence intensities (in arbitrary units) for three replicate samples analysed by DNA microarrays (different manufacturing lots) using DNA amplified from outdoor air samples. The number of clones found in each category is shown in the boxes above the replicate bars of each category.

The G1 microarray provided a low resolution (Order level) depiction of the microorganisms present in aerosols. The bacterial orders *Vibrionales*, *Flavobacteriales*, *Clostridiales*, *Rhizobiales*, *Lactobacillales*, *Bacillales*, *Acholeplasmatales*, *Deferribacterales*, *Mycoplasmatales*, *Burkholderiales*, *Sphingomonadales* and *Sphingobacteriales* were commonly observed in air samples, as well as the fungal phyla *Ascomycota* and *Basidiomycota*, and the brown algae *Phaeophyceae* (DeSantis *et al.*, 2003, 2005). Importantly, the array provided, for the first time, a method to categorise a blended pool of DNA from diverse organisms in as little as 9 hours without the need for labourious isolation of each sequence type. Just as importantly, replicates could be obtained for each location, sampling method, or season, thus allowing greater statistical accuracy when determining quantitative trends. At this phylogenetic resolution, the G1 microarray and the clone library methods produced similar results. The cloning library yielded some DNA sequences not detected on the G1 microarray because they were found to be novel (not represented in the RDP database at that time) and therefore could not have been represented on the arrays. However, the microarray detected sequences not recovered from the sequencing efforts and also indicated a discrepancy in the prevalence of certain sequences. That is to say, the number of clones in a bacterial order did not always correlate with the fluorescence intensities of the probes in the same order on the microarray (see Fig. 18.2). This hinted at possible cloning biases, but more work was needed to determine where the bias lay.

Progress and Promise of DNA Microarrays

Microarray manufacturing technology has borrowed from computer chip technology to allow increasingly higher densities of probes to be arrayed onto one chip. This has allowed an increase in taxonomic resolution by increasing the number of unique DNA segments that can be assayed. The general microarray design approach is described in detail in DeSantis *et al.* (2003) as it was applied to one bacterial family. Briefly, *Escherichia coli* 16S rRNA sequences (positions 47 to 1473) were extracted from approximately 30,000 gene sequences at least 600 nucleotides in length from the 16S rDNA database Greengenes (http://greengenes.lbl.gov). This region was selected because it is bounded on both ends by universally conserved segments that can be used as PCR priming sites to amplify bacterial or archaeal (Dojka *et al.*, 1998) genomic material using only 2 to 4 primers. The sequences were filtered using the software package Bellerophon (Huber *et al.*, 2004) to remove putative chimeric sequences from the data set. This was necessary because chimeras in the public databases are aberrations that do not reflect DNA sequences found in nature. They have been misconstrued as novel organisms (Hugenholtz and Huber, 2003) and can complicate phylogenetics. The filtered sequences are considered to be the set of putative 16S rDNA amplicons. Sequences were clustered to enable each sequence of a cluster to be complementary to a set of perfectly matching (PM) probes. Putative amplicons were placed in the same cluster as a result of common 17-mers found in the sequence. The resulting 8,988 clusters, each containing less than 5% sequence diversity, were considered operational taxonomic units (OTUs) representing all 91 known prokaryotic orders.

The objective of the probe selection strategy was to obtain an effective set of probes capable of correctly categorising mixed amplicons into their proper OTU. For each OTU, a set of 11 or more specific 25-mers (probes) were sought that were prevalent in members of a given OTU but were dissimilar from sequences outside the given OTU. In the first step of probe selection for a particular OTU, each of the sequences in the OTU was separated into overlapping 25-mers, the potential targets. Then each potential target was matched to as many sequences of the OTU as possible. Since each sequence in an OTU could range from 600 to 1,500 nucleotides, it was not sufficient to use a simple text search. Therefore, the sequences had to be aligned. Sequence alignment provided discrete measurement of group size at each potential probe site. For example, if an OTU containing seven sequences possessed a probe site where one member was missing data, then the site-specific OTU size was only six. In ranking the possible targets, those having data for all members of that OTU (very prevalent) were preferred over those found only in a fraction of the OTU members. In the second step, a subset of the prevalent targets was selected and complemented for probe construction, avoiding those capable of cross-hybridising to an unintended amplicon.

Probes presumed to have the ability to cross-hybridise were those 25-mers that contained a central 17-mer matching sequences in more than one OTU (Urakawa *et al.*, 2002). Thus, probes that were unique to an OTU solely due to a distinctive base in one of the outer four bases were avoided. As each PM probe was chosen, it was paired with a control 25-mer (mismatching probe, MM), identical in all positions except the thirteenth base. The MM probe did not contain a central 17-mer complimentary to sequences in any OTU. Similar to the G1 microarray, the PM and MM probes constitute a probe pair and are analysed together.

Probe Set Scoring

To estimate the likelihood that an organism from an OTU is present in a sample and to quantify the collective hybridisation intensity from the probes assigned to that OTU, each probe pair (PM and MM) within a probe set is scored as positive or not positive. Pairs scored positive are those that produce significantly more fluorescence from the PM probe compared to the MM probe. The positive fraction is calculated for each OTU as the number of positive probe pairs divided by the total number of probe pairs in a probe set. An OTU (typically consisting of organisms with 95-99% rRNA gene sequence identity) is considered present in the sample when over 92% of its assigned probe pairs are positive (PosFrac >0.92). Therefore, having one or a few unresponsive probes will not preclude detection of a taxon.

The hybridisation score is used to quantify the collective intensity from the whole set of probe pairs for each OTU. Hybridisation scores are expressed in arbitrary units (a.u.) for each OTU as the trimmed average of the PM minus MM intensity differences across the probe pairs for a given OTU (the trimmed average disregards the maximum and minimum difference). To allow meaningful chip-to-chip comparisons given that there is variation in target labelling efficiency, stain brightness, and scanner intensity, the hybridisation scores can be normalised so that the average of the scores corresponding to the spiked-in DNA was a constant 2500 a.u.

We speculated there would be an advantage in analysing both PM and MM probes, especially when detecting taxa in low abundance within a complex sample. A low intensity value from a PM probe itself can indicate either a low abundance of the perfectly matching target or a high abundance of a non-perfectly matching target that has cross-hybridisation potential. A comparison to the MM control probe intensity could reveal that the PM and MM have similar intensities, denoting non-specific hybridisation to both probe types, or that a significant difference exists, denoting the PM target is more likely in the sample compared to the MM target. We regularly require a positive fraction of >92% for detection of a taxon, which was determined empirically to provide stringency while maintaining sensitivity to the amplicons known to be present from cloning experiments (unpublished data). The advantage of having multiple probes instead of one for identification is that it allows detection of environmental amplicon sequences that do not perfectly match a database sequence. A high cutoff value of >92% increases the confidence of a correct call, but also increases the possibility that some organisms may be overlooked. A lower cut-off value increases the number of organisms that can be identified, but will also increase the number of false positives that result from cross-hybridisation. The array scoring procedure permits flexibility in stringency, enabling the researcher to decide how much information to gather. In any case, having multiple probes for each OTU results in increased potential for accurate identification and is necessary for sampling organisms in uncharacterised environments.

To validate results showing positive probe sets, PCR primers can be designed to target a bacterial family or genus of interest. This strategy has been used in conjunction with clone library analysis for three different matrices (air, soil, and water) and is proving the strategy described above to be reliable and comprehensive (based on currently used universal bacterial PCR primers) (unpublished data).

Applying Microarray Technology for Bioaerosol Detection

Outdoor air samples from cities across the USA are routinely collected and analysed for bioterrorism surveillance measures. A US Department of Homeland Security programme, BioWatch, has provided the main focus of air samples analysed in our lab using the latest version of our high-density microarrays to categorise the individual bacterial components. These samples are providing a baseline for the types of bacterial organisms travelling through the air and how they may impact pathogen detection devices. The data may also provide valuable information about the bacteria present that are potential animal or plant pathogens, addressing bioterrorism concerns (Sheeran, 2002), as well as basic scientific questions.

Air samples collected onto porous filters were initially processed by a bead mill homogenisation technique for DNA extraction, followed by PCR amplification, and DNA microarray hybridisation and analysis. Genomic DNA extraction techniques have been optimised for air filters and air filter washes and can be viewed online (http://greengenes.lbl.gov/Download/Protocols/).

Filtered air samples from multiple sites from throughout each city were pooled for each sampling day to provide a more uniform picture of aerosol microbial composition. After DNA was extracted from the pooled filters, 16S rRNA gene PCR products were generated using the universal bacterial primers 27F and 1492R. PCR product pools were generated by combining the PCR products from four days selected for each week; this sample constituted a 'city-week' sample that was used as the basic measure for changes in microbial composition over time and location. Nucleic acid controls for sample processing, hybridisation, and x,y probe position determination (scanning) were added to each sample prior to placement on the microarray. After overnight hybridisation, the microarrays were then washed, stained and scanned, similar to the G1 arrays, to collect information about the bacteria present.

A Specific Example

Two cities, Detroit and Philadelphia, sampled through the BioWatch national biosurveillance programme were chosen for paired comparisons of microarray results from air filters processed as described above. Our goal was to determine the variability of microbial composition over time and location, as well as to identify the common bacterial groups that were present. Specifically, two seasons (spring and summer) from 2004 were analysed, where each season comprised 3 consecutive weeks. A background sample was established by using a blank DNA extraction performed along with the DNA extractions from one of the cities. There was no PCR product visible by gel electrophoresis for the blank sample extracted, but any OTU called present in the blank microarray sample was removed from further analysis. We used positive fraction values equal to and above 0.9 to indicate presence and anything below that to indicate inconclusive or negative results (absence).

For a greater understanding of how bacterial diversity might differ by city or season, the samples were compared using pair-wise T-tests of the fluorescence intensities of the microarray probes. Results are shown in Table 18.1. The greatest differences were between cities, but this appeared to be driven by the Philadelphia spring sample values. When comparing the composition of OTUs considered present between spring and summer in each city, samples from Philadelphia were found to have the greatest differences, while Detroit samples had fewer differences between seasons. In general, there were fewer OTUs called present in the Philadelphia spring samples than in any of the other samples, which contributed to these large differences. Work is ongoing to determine why the Philadelphia spring samples had a lower level of diversity than the other samples even though the overall quantity of DNA extracted, amplified, and hybridised to the microarray was similar to the other samples.

Table 18.1. Pair-wise comparison of the number of OTUs significantly different between cities and seasons based on fluorescence intensity

	Detroit Spring	Detroit Summer	Philadelphia Spring
Detroit Spring	0	-	-
Detroit Summer	42	0	-
Philadelphia Spring	1431	1244	0
Philadelphia Summer	64	163	612

Very little is known about how the microbial composition of aerosols varies over location or season. As an initial attempt to understand if urban areas share similar types of organisms in the air or, conversely, if airborne bacteria are distinctive for each city, we determined which OTUs were present in Detroit and Philadelphia samples in both spring and summer seasons using our phylogenetic microarrays. To be considered present in both cities, a bacterial OTU had to have an average positive fraction value greater than or equal to 0.92 (the average pf was calculated from the three individual city-week samples processed for each city from each season). A list of all OTUs detected within the taxonomic classification of 'Order' is presented in Table 18.2. A more detailed description of the taxonomic classification used in this study is available at http://greengenes.lbl.gov/Download/Taxonomic_Outlines/ in the file called G2_chip_SeqDescByOTU_tax_outline.txt.

There were 34 named phyla and 109 named families present in both cities in both seasons. The total number of families detected in both cities and both seasons, including the named and currently unclassified families, was 144. Some of the unclassified families belonged to phyla such as TM7, which are not among the readily culturable bacteria. Orders containing anaerobes (e.g. *Clostridiales*) were detected, supporting the use of molecular methods to detect bacteria not readily cultured under standard aerosol sampling and culture methods. Culture-based studies often have far fewer families reported, most likely due to the inability to grow members of more families, the problem of over- or under-sampling (leading to few colonies or too many colonies to identify), and/or the enormous amount of time and effort required to characterise more colonies (Shaffer and Lighthart, 1997; Griffin *et al.*, 2001; Kellogg *et al.*, 2004).

Table 18.2. The number of families and OTUs detected in Detroit and Philadelphia during spring and summer (detected means an average pf >0.9). No Unclassified Orders are presented unless the phylum would have been excluded from the list.

Phylum	Class	Order	Families	OTUs
Acidobacteria	Acidobacteria	Acidobacteriales	1	20
Actinobacteria	Actinobacteria	Acidimicrobiales	2	2
Actinobacteria	Actinobacteria	Actinomycetales	20	102
Actinobacteria	Actinobacteria	Bifidobacteriales	1	7
Actinobacteria	Actinobacteria	Coriobacteriales	1	1
Actinobacteria	Actinobacteria	Rubrobacterales	1	4
AD3	Unclassified	Unclassified	1	1
Aquificae	Aquificae	Aquificales	1	1
Bacteroidetes	Bacteroidetes	Bacteroidales	3	8
Bacteroidetes	Flavobacteria	Flavobacteriales	2	3
Bacteroidetes	Sphingobacteria	Sphingobacteriales	3	6
BRC1	Unclassified	Unclassified	1	2
Caldithrix	Unclassified	Caldithrales	1	2
Chlamydiae	Chlamydiae	Chlamydiales	1	1
Chlorobi	Chlorobia	Chlorobiales	1	1
Chloroflexi	Anaerolineae	Chloroflexi-1a	1	1
Chloroflexi	Anaerolineae	Chloroflexi-1f	1	1
Cyanobacteria	Cyanobacteria	Chloroplasts	1	13
Cyanobacteria	Cyanobacteria	Chroococcales	1	1
Cyanobacteria	Cyanobacteria	Nostocales	1	1
Cyanobacteria	Cyanobacteria	Oscillatoriales	1	1
Cyanobacteria	Cyanobacteria	Plectonema	1	1
Cyanobacteria	Cyanobacteria	Prochlorales	1	1
Deinococcus-Thermus	Unclassified	Unclassified	1	2
DSS1	Unclassified	Unclassified	1	1
Firmicutes	Bacilli	Bacillales	8	136
Firmicutes	Bacilli	Lactobacillales	6	38
Firmicutes	Clostridia	Clostridiales	6	130
Firmicutes	Mollicutes	Anaeroplasmatales	1	4
Firmicutes	Symbiobacteria	Symbiobacterales	1	2
Gemmatimonadetes	Unclassified	Unclassified	1	5
Lentisphaerae	Unclassified	Unclassified	1	3
marine group A	mgA-2	Unclassified	1	1
Natronoanaerobium	Unclassified	Unclassified	1	3
NC10	Unclassified	Unclassified	1	1
Nitrospira	Nitrospira	Nitrospirales	1	1
OD1	OP11-5	Unclassified	1	1
OP10	CH21 cluster	Unclassified	1	1
OP3	Unclassified	Unclassified	1	2
OP8	Unclassified	Unclassified	1	1
OP9/JS1	OP9	Unclassified	1	1
Planctomycetes	Planctomycetacia	Planctomycetales	3	7
Proteobacteria	Alphaproteobacteria	Azospirillales	1	3
Proteobacteria	Alphaproteobacteria	Bradyrhizobiales	3	8

Table 18.2. contd

Proteobacteria	Alphaproteobacteria	Caulobacterales	1	4
Proteobacteria	Alphaproteobacteria	Consistiales	1	2
Proteobacteria	Alphaproteobacteria	Ellin329/Riz1046	1	3
Proteobacteria	Alphaproteobacteria	Rhizobiales	4	18
Proteobacteria	Alphaproteobacteria	Rhodobacterales	1	7
Proteobacteria	Alphaproteobacteria	Rickettsiales	1	2
Proteobacteria	Alphaproteobacteria	Sphingomonadales	1	14
Proteobacteria	Betaproteobacteria	Burkholderiales	3	5
Proteobacteria	Betaproteobacteria	Nitrosomonadales	1	2
Proteobacteria	Betaproteobacteria	Rhodocyclales	1	1
Proteobacteria	Deltaproteobacteria	AMD clone group	1	5
Proteobacteria	Deltaproteobacteria	Bdellovibrionales	1	1
Proteobacteria	Deltaproteobacteria	dechlorinating clone group	1	1
Proteobacteria	Deltaproteobacteria	Desulfobacterales	4	15
Proteobacteria	Deltaproteobacteria	Desulfovibrionales	2	8
Proteobacteria	Deltaproteobacteria	Desulfuromonadales	1	1
Proteobacteria	Deltaproteobacteria	Myxococcales	1	6
Proteobacteria	Deltaproteobacteria	Syntrophobacterales	2	6
Proteobacteria	Epsilonproteobacteria	Campylobacterales	2	20
Proteobacteria	Gammaproteobacteria	Alteromonadales	1	2
Proteobacteria	Gammaproteobacteria	Chromatiales	1	2
Proteobacteria	Gammaproteobacteria	Ellin307/WD2124	1	1
Proteobacteria	Gammaproteobacteria	Enterobacteriales	1	4
Proteobacteria	Gammaproteobacteria	GAO cluster	1	1
Proteobacteria	Gammaproteobacteria	Legionellales	1	4
Proteobacteria	Gammaproteobacteria	Oceanospirillales	1	2
Proteobacteria	Gammaproteobacteria	Pseudomonadales	2	3
Proteobacteria	Gammaproteobacteria	Symbionts	1	2
Proteobacteria	Gammaproteobacteria	Thiotrichales	2	5
Proteobacteria	Gammaproteobacteria	uranium waste clones	1	1
Proteobacteria	Gammaproteobacteria	Xanthomonadales	1	1
SPAM	Unclassified	Unclassified	1	2
Spirochaetes	Spirochaetes	Spirochaetales	1	1
Synergistes	Unclassified	Unclassified	1	2
TM7	TM7-3	Unclassified	1	3
Verrucomicrobia	Verrucomicrobiae	Verrucomicrobiales	3	5
WS3	Unclassified	Unclassified	1	1
WS5	Unclassified	Unclassified	1	1

The unexpectedly high level of diversity of common organisms that is found in Table 18.2 suggests either that urban areas share numerous and equivalent types of reservoirs for dispersal of bacteria into the air, or that bacteria are transported for long distances to effectively homogenise bacterial types in the atmosphere. Regardless of the reason for the large number of shared OTUs for these two cities, the variety of organisms identified demonstrates how little is known about their fate and transport in urban environments. We have previously used split samples to compare results of the microbial composition of bioaerosols from our phylogenetic microarray with the results from DNA sequences of 16S library clones and found that both methods agree on the types of organisms present with additional classes of organisms found with the array method. We have verified the presence of the additional organisms picked up by the array by designing primers to specifically amplify the selected sequence from the sample. One of the reasons that additional organisms are identified by the phylogenetic array may be that a pool containing all of the amplicons from the sample is hybridised to the array, while sequencing even several hundred clones will miss many of the less abundant amplified products.

The results shown in Table 18.2 highlight the power of microarrays to elucidate the bacterial diversity present in samples. A second important factor for these kinds of studies is having meaningful taxonomic specificity in the probe sets. Individual bacterial types, as defined by OTU, had different distributions in the samples from different seasons or cities. One of the advantages of the phylogenetic microarray is the ability to monitor specific OTUs for their presence or absence in different samples. Fig. 18.3 shows an example of an OTU containing marine/aquatic *Agrobacterium* sequences that was found to be present in a Detroit sample based on the fluorescence signal of the perfect match (PM) probes being higher than the fluorescence signal of the corresponding mismatch (MM) probes for 23 of the 25 probe pairs, giving a positive fraction score of 0.92. The two probe-pairs that were not scored positive were numbers 1 and 25. In contrast, the Philadelphia sample for the corresponding week had six probe-pairs where the MM probe had an equal or greater fluorescence signal than the PM probe (positive fraction = 0.76), and did not meet our criteria for presence. Although a number of probe-pairs from the Philadelphia sample had a positive response and a strong florescence signal, most likely due to cross-reaction with a similar 16S rDNA sequence, the redundancy of requiring multiple probes increases our confidence in the results. The 16S rDNA sequences that were used to create this OTU came from very closely related species of *Agrobacterium, Stappia* and *Crassostrea,* and have been found to have no relation to terrestrial *Agrobacterium* (Uchino *et al.*, 1998).

Knowing the distribution of bacteria, especially potential human or plant pathogens, is important for biosurveillance efforts and also for general scientific knowledge. For example, an OTU containing only *Mycetocola* sequences was present in one Detroit spring sample and two Detroit summer samples, but not in any Philadelphia samples. Bacteria in this genus have been isolated from fruiting bodies of cultivated mushrooms, particularly those with brown blotch disease. *Morganella morganii*, which is found commonly in the environment, yet is also an opportunistic human pathogen, was detected only in the Detroit summer samples. One other opportunistic pathogen, *Rahnella* sp., was detected in Detroit summer samples and one Philadelphia summer sample. Bacteria in the Family Rhizobiaceae (likely to be *Sinorhizobium* and/or *Rhizobium)* were detected in all city-seasons. *Xanthomonas* and *Lysobacter* were detected in all samples except two (both being Philadelphia spring samples).

As stated before, we are still learning about the natural distribution of bacteria, and bacteria with 'known' distributions may be found in many other locations/niches as more studies are completed. It may simply be that microbial ecologists have not yet fully defined a bacterium's natural habitat and so we are now finding it in 'surprising' places. Aerosols contain bacteria carried from a variety of places and so it is plausible to find soil, water and animal-associated bacteria in air samples. As an example, several OTUs detected only in the Detroit summer samples comprised marine or aquatic bacteria (GenBank sequences from isolates or amplicons). It is possible that this result occurred because there are several bodies of water nearby (Lake Huron, Lake Erie, Lake St. Clair, and the Detroit River) and that air masses may have entrained some of the bacteria from these sources. It is also possible that the air masses had carried the bacteria from much more distant locations. Of course, it is possible that the natural distribution of these bacteria is more cosmopolitan than one could guess from the habitats they have been recovered from to date. This is undoubtedly the case for many bacteria, and molecular detection methods are expected to help reveal a more comprehensive view of bacterial distributions. We can no longer afford to remain unaware of the types of microorganisms that we are routinely breathing in and their impact upon our health and the health of the global environment. Bacteria may persist for long periods in the atmosphere under ideal conditions and pose hazards, either directly as pathogens, or indirectly as allergenic agents. In addition to this, human activities, such as the increasing desertification of certain areas, are making it increasingly important to develop baseline information of types of organisms that are observed in bioaerosols. Only by repeated monitoring can we determine if the microbial composition is remaining relatively constant or if certain activities will measurably alter the composition.

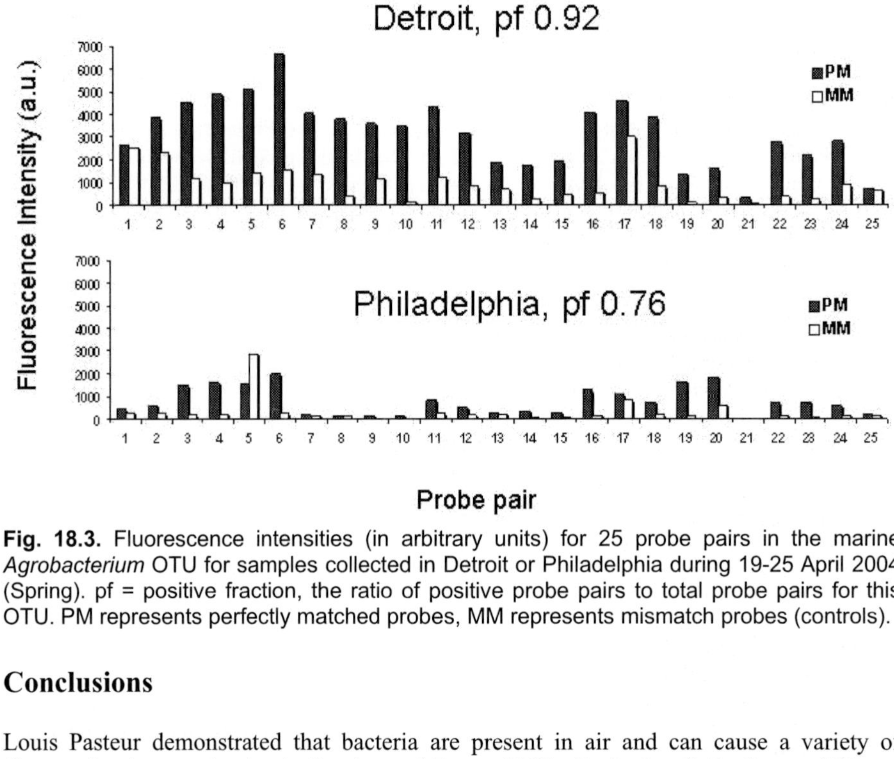

Fig. 18.3. Fluorescence intensities (in arbitrary units) for 25 probe pairs in the marine *Agrobacterium* OTU for samples collected in Detroit or Philadelphia during 19-25 April 2004 (Spring). pf = positive fraction, the ratio of positive probe pairs to total probe pairs for this OTU. PM represents perfectly matched probes, MM represents mismatch probes (controls).

Conclusions

Louis Pasteur demonstrated that bacteria are present in air and can cause a variety of diseases in plants and animals (Levine and Evers, 1999). Zuckerkandl, Pauling and Woese provided a way to classify and learn about these organisms. Technological advances have provided tools to elucidate microbial interactions even further, rapidly and with replication, to start to understand the variability that exists in these interactions and to discern trends. These advances will continue to increase the taxonomic resolution attainable on high-density DNA microarrays, in turn leading to a better understanding of immigration, emigration and dynamic population changes of bacteria due to growth, senescence or dormancy. As complex as natural systems are, no one method will provide all the information we desire and yet a tool as powerful as microarrays provides a strength of analysis not available in other formats. DNA microarrays are expected to help us track bacterial and fungal populations over great distances and over time where similar studies using other methods, such as cloning and sequencing, would be time, labour, and/or cost prohibitive. Both phylogeny and function may be combined on one array in the future, providing even more information about ecosystem processes. We will undoubtedly continue to learn about completely new organisms through cloning and sequencing, as microarrays are limited to sequences we know something about, yet the value of being able to analyse replicates for community-wide trends or changes over time cannot be overstated.

How processes like bacterial immigration and emigration affect ecosystems can be studied more comprehensibly with microarrays than with alternative techniques. This may become important for understanding worldwide circulation patterns of plant and animal diseases, as well as understanding more localised population dynamics. Supplementing, not supplanting, culture-based studies with molecular studies is expected to open new areas of research, even for what have been recalcitrant issues.

References

Buttner, M.P., Willeke, K. and Grinshpun, S.A. (2002) Sampling and analysis of airborne microorganisms. In: Hurst, C.J., Crawford, R.L., Knudsen, G.R., McInerney, M.J. and Stetzenbach, L.D. (eds) *Manual of Environmental Microbiology*, 2nd edition, ASM Press, Washington, USA. pp. 814-826.

Castiglioni, B., Rizzi, E., Frosini, A., Sivonen, K., Rajaniemi, P., Rantala, A., Mugnai, M.A., Ventura, S., Wilmotte, A., Boutte, C., Grubisic, S., Balthasart, P., Consolandi, C., Bordoni, R., Mezzelani, A., Battaglia, C. and DeBellis, G. (2004) Development of a universal microarray based on the ligation detection reaction and 16S rRNA gene polymorphism to target diversity of Cyanobacteria. *Applied and Environmental Microbiology* 70, 7161-7172.

DeSantis, T.Z., Dubosarskiy, I., Murray, S.R. and Andersen, G.L. (2003) Comprehensive aligned sequence construction for automated design of effective probes (CASCADE-P) using 16S rDNA. *Bioinformatics* 19, 1461-1468.

DeSantis, T.Z., Stone, C.E., Murray, S.R., Moberg, J.P. and Andersen, G.L. (2005) Rapid quantification and taxonomic classification of environmental DNA from both prokaryotic and eukaryotic origins using a microarray. *FEMS Microbiology Letters*, 245, 271-278.

Dojka, M.A., Hugenholtz, P., Haack, S.K. and Pace, N.R. (1998) Microbial diversity in a hydrocarbon- and chlorinated-solvent-contaminated aquifer undergoing intrinsic bioremediation. *Applied and Environmental Microbiology* 64, 3869-3877.

Dunbar, J., Takala, S. Barns, S.M., Davis, J.A. and Kuske, C.R. (1999) Levels of bacterial community diversity in four arid soils compared by cultivation and 16S rRNA gene cloning. *Applied and Environmental Microbiology* 65, 1662-1669.

Griffin, D.W., Garrison, V.H., Herman, J.R. and Shinn, E.A. (2001) African desert dust in the Caribbean atmosphere: microbiology and public health. *Aerobiologia* 17, 203-213.

Hirano, S.S. and Upper, C.D. (1989) Diel variation in population size and ice nucleation activity of *Pseudomonas syringae* on snap bean leaflets. *Applied and Environmental Microbiology* 55, 623-630.

Hirano, S.S. and Upper, C.D. (2000) Bacteria in the leaf ecosystem with emphasis on *Pseudomonas syringae* – a pathogen, ice nucleus, and epiphyte. *Microbiology and Molecular Biology Reviews* 64, 624-653.

Huber, T., Faulkner, G. and Hugenholtz, P. (2004) Bellerophon: a program to detect chimeric sequences in multiple sequence alignments. *Bioinformatics* 20, 2317-2319.

Hugenholtz, P. and Huber, T. (2003) Chimeric 16S rDNA sequences of diverse origin are accumulating in the public databases. *International Journal of Systematic and Evolutionary Microbiology* 53, 289-293.

Kellogg, C.A., Griffin, D.W., Garrison, V.H., Peak, K.K., Royall, N., Smith, R.R. and Shinn, E.A. (2004) Characterization of aerosolized bacteria and fungi from desert dust events in Mali, West Africa. *Aerobiologia* 20, 99-110.

Kelly, J.J., Siripong, S., McCormack, J., Janus, L.R., Urakawa, H., El Fantroussi, S., Noble, P.A., Sappelsa, L., Rittmann, B.E. and Stahl, D.A. (2005) DNA microarray detection of nitrifying bacterial 16S rRNA in wastewater treatment plant samples. *Water Research* 39, 3229-3238.

Lehner, A., Loy, A., Behr, T., Gaenge, H., Ludwig, W., Wagner, M. and Schleifer, K-H. (2005) Oligonucleotide microarray for identification of Enterococcus species. *FEMS Microbiology Letters* 246, 133-142.

Levetin, E. (2002) Bioaerosols in agricultural and outdoor settings. In: Bitton, G. (ed.) *Encyclopedia of Environmental Microbiology*, Vol. 1 John Wiley & Sons, New York, USA. pp. 404-416.

Levine, R. and Evers, C. (1999) *The Slow Death of Spontaneous Generation.* Access Excellence/National Health Museum [Accessed 2005] Available at http://www.accessexcellence.org/RC/AB/BC/Spontaneous_Generation/

Lighthart, B. and Shaffer, B.T. (1994) Bacterial flux from chaparral into the atmosphere in mid-summer at a high desert location. *Atmospheric Environment* 28, 1267-1274.

Lighthart, B. and Shaffer, B.T. (1995) Airborne bacteria in the atmospheric surface layer: temporal distribution above a grass seed field. *Applied and Environmental Microbiology* 61, 1492-1496.

Lindemann, J. and Upper, C.D. (1985) Aerial dispersal of epiphytic bacteria over bean plants. *Applied and Environmental Microbiology* 50, 1229-1232.

Lindemann, J., Constantinidou, H.A., Barchet, W.R. and Upper, C.D. (1982) Plants as sources of airborne bacteria, including ice nucleation-active bacteria. *Applied and Environmental Microbiology* 44, 1059-1063.

Lindow, S.E. and Andersen, G.L. (1996) Influence of immigration on epiphytic bacterial populations on navel orange leaves. *Applied and Environmental Microbiology* 62, 2978-2987.

Liu, W-T. and Stahl, D.A. (2002) Molecular approaches for the measurement of density, diversity, and phylogeny. In: Hurst, C.J., Crawford, R.L., Knudsen, G.R., McInerney, M.J. and Stetzenbach, L.D. (eds) *Manual of Environmental Microbiology*, 2nd edition. ASM Press, Washington, USA. pp. 114-134.

McCarthy, M. (2001) Dust clouds implicated in spread of infection. *The Lancet* 358, 478.

Mohr, A.J. (2002) Fate and transport of microorganisms in air. In: Hurst, C.J., Crawford, R.L., Knudsen, G.R., McInerney, M.J. and Stetzenbach, L.D. (eds) *Manual of Environmental Microbiology*, 2nd edition. ASM Press, Washington, USA. pp. 827-838.

Pace, N.R. (1997) A molecular view of microbial diversity and the biosphere. *Science* 276, 734-740.

Piceno, Y.M. and Davis, G.A. (2002) Sampling techniques for environmental microbiology. In: Bitton, G. (ed.) *Encyclopedia of Environmental Microbiology.*, Vol. 5. John Wiley & Sons, New York, USA. pp. 2809-2822.

Radosevich, J.L., Wilson, W.J., Shinn, J.H., DeSantis, T.Z. and Andersen, G.L. (2002) Development of a high-volume aerosol collection system for the identification of air-borne micro-organisms. *Letters in Applied Microbiology* 34, 162-167.

Shaffer, B.T. and Lighthart, B. (1997) Survey of culturable airborne bacteria at four diverse locations in Oregon: urban, rural, forest, and coastal. *Microbial Ecology* 34, 167-177.

Sheeran, T.J. (2002) Bioterrorism. In: Bitton, G. (ed.) *Encyclopedia of Environmental Microbiology*. Vol. 2. John Wiley & Sons, New York, USA. pp. 771-782.

Staley, J.T. and Konopka, A. (1985) Measurements of *in situ* activities of nonphotosynthetic microorganisms in aquatic and terrestrial habitats. *Annual Review of Microbiology* 39, 321-346.

Stites, F.L. (2002) Bioaerosols: Transport and fate. In: Bitton, G. (ed.) *Encyclopedia of Environmental Microbiology*. Vol. 1. John Wiley & Sons, New York, USA. pp. 425-434.

Sultan, B., Labadi, K, Guegan, J-F. and Janicot, S. (2005) Climate drives the meningitis epidemics onset in West Africa. *Plos Medicine* 43-49. Available at http://www.plosmedicine.org/

Torsvik, V., Goksoyr, J. and Daae, F.L. (1990) High diversity in DNA of soil bacteria. *Applied and Environmental Microbiology* 56, 782-787.

Uchino, Y., Hirata, A., Yokota, A. and Sugiyama, J. (1998) Reclassification of marine *Agrobacterium* species: proposals of *Stappia stellulata* gen.nov., *Stappia aggregata* sp. nov., nom. rev., *Ruegeria altantica* gen. nov., comb. nov., *Ruegeria gelatinovora* comb. nov., and *Ahrensia kieliense* gen. nov., sp. nov., nom. rev. *The Journal of General and Applied Microbiology* 44, 201-210.

Upper, C.D. and Hirano, S.S. (2002) Revisiting the roles of immigration and growth in the development of populations of *Pseudomonas syringae* in the phyllosphere. In: Lindow, S.E.,

Hecht-Poinar and E.I., Elliott, V.J. (eds) *Phyllosphere Microbiology*. American Phytopathological Society, St. Paul, USA. pp. 69-79.

Urakawa, H., Noble, P.A., El Fantroussi, S., Kelly, J.J. and Stahl, D.A. (2002) Single-base-pair discrimination of terminal mismatches by using oligonucleotide microarrays and neural network analyses. *Applied and Environmental Microbiology* 68, 235-244.

Warsen, A.E., Krug, M.J., LaFrentz, S., Stanek, D.R., Loge, F.J. and Call, D.R. (2004) Simultaneous discrimination between 15 fish pathogens by using 16S ribosomal DNA PCR and DNA microarrays. *Applied and Environmental Microbiology* 70, 4216-4221.

Weir-Brush, J.R., Garrison, V.H., Smith, G.W. and Shinn, E.A. (2004) The relationship between gorgonian coral (Cnidaria: Gorgonacea) diseases and African dust storms. *Aerobiologia* 20, 119-126.

Wilson, K.H., Wilson, W.J., Radosevich, J.L., DeSantis, T.Z., Viswanathan, V.S., Kuczmarski, T.A.,and Andersen, G.L. (2002) High-Density microarray of small subunit ribosomal DNA probes. *Applied and Environmental Microbiology* 68, 2535-2541.

Yang, C-H., Crowley, D.E., Borneman, J. and Keen, N.T. (2001) Microbial phyllosphere populations are more complex than previously realized. *Proceedings of the National Academy of Sciences*, USA. 98, 3889-3894.

Zuckerkandl, E. and Pauling, L. (1965) Molecules as documents of evolutionary history. *Journal of Theoretical Biology* 8, 357-366.

Chapter 19

Human Pathogens and the Health Threat of the Phyllosphere

M.T. Brandl
Produce Safety and Microbiology, USDA/ARS, WRRC, 800 Buchanan Str., Albany, CA 94710, USA
mbrandl@pw.usda.gov

Introduction

Human enteric illness has been associated traditionally with the consumption of contaminated meat products or water. However, foodborne illness linked to the contamination of fresh fruits and vegetables has been increasing dramatically since the 1970s (Sivapalasingam *et al.*, 2004). This increase may be explained in part by a noticeable shift in dietary habits in the USA between 1970-1979 and 1997. During these periods, the consumption of fresh fruits and fresh vegetables alone increased 34% and 26%, respectively (Anonymous, 2001). This trend is mainly attributable to the introduction of salad bars in restaurants and supermarkets, and of fresh-cut fruits and vegetables, pre-packaged salads and exotic and specialty produce in supermarkets (Anonymous, 2001). Along with this development, the intensification of agricultural production in industrialised countries has resulted in the disposal of increasing amounts of animal waste on agricultural fields and the frequent use of effluent or reclaimed water to irrigate crops. Application of contaminated irrigation water, or animal waste as raw or improperly composted manure, may introduce enteric pathogens in the field. Other causes for the increase in the incidence of enteric illness linked to fresh produce may include: (i) changes in processing and packaging methods, particularly the introduction of ready-to-eat fruits and vegetables on the market; (ii) the expansion of global trade, which provides fresh produce out of season in local markets; and (iii) the enhanced surveillance of foodborne illness by regulatory agencies.

The continued rise in the number of cases and outbreaks of foodborne illness associated with fresh produce is of great concern to the fresh produce industry and to public health agencies. This has fuelled a new wave of studies on the ecology of human pathogens on plants. Although still in its infancy, this field of research already has yielded new insights into human pathogen behaviour on plant surfaces. This chapter presents some of the information that is arising from this emerging field.

Epidemiology of Enteric Illness Linked to Produce

Incidence of enteric pathogens on produce

Large international epidemics originating from almonds (Isaacs *et al.*, 2005) and from sprouts grown from contaminated seeds (Mahon *et al.*, 1997) have occurred. In the USA, the number of outbreaks per year that are linked to produce doubled between the periods of 1973-1987 and 1988-1992 (Bean *et al.*, 1997; Mead *et al.*, 1999). Outbreaks have originated from lettuce (Ackers *et al.*, 1998; Hilborn *et al.*, 1999; Takkinen *et al.*, 2005), cilantro (Campbell *et al.*, 2001), parsley, tomato (Hedberg *et al.*, 1999; Cummings *et al.*, 2001; Anonymous, 2005), cantaloupe (Mohle-Boetani *et al.*, 1999; Anonymous, 2002a), and unpasteurised orange juice (Cook *et al.*, 1998; Anonymous, 1999a) and apple juice (Besser *et al.*, 1993), to name only a few. According to a report by the Center for Science in the Public Interest, produce was the second most important food vehicle of outbreaks in the USA between 1990 and 2002, and accounted for 20% of the total number of illness cases reported (Anonymous, 2002b). The food categories included in these data were fruits, vegetables and produce dishes. Table 19.1 lists the number of outbreaks by type of produce as reported to the Center for Disease Control and Prevention (CDC) in the USA from 1973 to 1997. A high proportion of the outbreaks were associated with salads, lettuce, juice, melon and sprouts. In other countries where surveillance of foodborne illness is extensive, a significant proportion of outbreaks has been attributed to fresh produce also. In England and Wales, salad, vegetables and fruit caused 6.4% and 10.1% of all outbreaks with a known food vehicle in the periods of 1993-1998 (Anonymous, 2000) and 1999-2000 (Anonymous, 2003), respectively.

Table 19.1. Produce items implicated in foodborne illness outbreaks in the USA, 1973-1997.

Items	No. of outbreaks
Multiple produce items	105
Salad	76
Mixed fruit	22
Mixed vegetables	7
Single produce items	85
Lettuce	25
Melon	13
Seed sprout	11
Apple and orange juice	11
Berry	9
Tomato	3
Green onion	3
Carrot	2
Apple	1
Pear	1
Pineapple	1
Basil	1
Celery	1
Cucumber	1
Fresh elderberry juice	1
Fresh-squeezed lemonade	1

Sivapalasingam *et al.* (2004). Reprinted with permission from the *Journal of Food Protection*. Copyright held by the International Association for Food Protection, Des Moines, Iowa, USA.

Although some of the outbreaks linked epidemiologically to produce may have originated from cross-contamination by water or another food item during food preparation or processing, the data presented above raise great concerns about the safety of raw produce consumption. Several surveys conducted by individual research groups revealed the occurrence of contaminated fresh fruits and vegetables at the marketplace (Ercolani, 1976; Garcia-Villanova Ruiz *et al.*, 1987; Wells and Butterfield, 1997). Ercolani (1976) reported that 68% and 72% of lettuce and fennel samples, respectively, purchased from retail markets in Italy over a period of two years, harboured one or more of six *Salmonella* serotypes. A survey conducted in Spain between 1981 and 1983 showed that 7.4% of the samples from a variety of vegetables were contaminated with *Salmonella* (Garcia-Villanova Ruiz *et al.*, 1987). More recently, a large survey of produce by the US Food and Drug Administration in 2000-2001 revealed that 1.6% and 4.4% of the samples from domestic and import distribution markets, respectively, were contaminated with human pathogens (www.cfsan.fda.gov). In addition, an ongoing survey by the Microbial Data Program of the US Department of Agriculture, showed that in 2002, 0.62% of the samples were contaminated with *E. coli* strains harbouring virulence factors (www.ams.usda.gov/science/mpo/MDPSumm02.pdf). The types of produce sampled in this survey were: cantaloupe, celery, leaf lettuce, romaine lettuce and tomatoes. Overall, these studies provide evidence that outbreaks may not necessarily occur according to a cross-contamination scenario, and suggest that plant contamination with human pathogens can occur in a pre-harvest fashion in the field or during post-harvest transit before distribution in the marketplace.

Sources of Contamination

Unlike meat and fresh-cut produce, which are regularly tested in the packing plant, most produce is not inspected for the presence of foodborne pathogens. The short shelf life of fresh fruits and vegetables makes it difficult to identify the exact source of the contamination when an outbreak occurs, and epidemiological studies must be used to trace back to the contamination source. Although good bookkeeping at the farm and throughout the distribution chain may help to link a particular contaminated product to a specific farm or field, the very source of the inoculum usually remains unidentified. For example, in an outbreak of illness caused by *E. coli* O157:H7-contaminated lettuce, manure piled in the vicinity of the lettuce fields, roaming free-range chickens in the cow pasture as well as in the lettuce fields, bird activity in the processing shed, and contaminated wash water that was used to clean produce, were all potential sources of inoculum and risk factors identified at the farm implicated in the outbreak (Hilborn *et al.*, 1999).

In general, application of improperly composted manure, contaminated water used for irrigation or for application of pesticides, and animal activity, which leaves droppings on the ground, are considered as the primary causes of contamination in the field. Poor sanitation facilities for farm workers in the field may account for a high percentage of illness transmitted via human faeces, such as shigellosis. In addition, insects may be an important epidemiological factor. In particular, flies, which rely on faecal material and rotten fruit as a protein source to develop eggs (Lauzon, 2003), may act as vectors of faeces-borne pathogens. Both, the vinegar fly (*Drosophila melanogaster*) and the Mediterranean fruit fly (*Ceratitis capitata*) were shown to transmit *E. coli* to wounded and intact apples, respectively (Janisiewicz *et al.*, 1999; Sela *et al.*, 2005).

While airborne transmission is recognised as an important factor in the dispersal of pathogens among livestock (Dowd *et al.*, 2004), the role of bioaerosols as a source of bacterial contamination of produce has remained unexplored. Bioaerosols are a probable source of inoculum in field crops since the presence of high quantities of animal faeces on the ground and of large manure piles at animal production facilities may lead to the transport of enteric pathogens via air currents and their deposition onto crops in the field.

Enteric Pathogens Linked to Foodborne Illness from Produce

A variety of bacterial enteric pathogens have been isolated from fresh fruits and vegetables and have caused produce-linked outbreaks. They include *S. enterica*, *Listeria monocytogenes*, *E. coli*, *Shigella* spp., *Campylobacter* spp., *Yersinia* spp. and *Staphylococcus aureus*.

S. enterica

S. enterica is the enteric pathogen most commonly recovered from produce and the most frequent bacterial causal agent in outbreaks associated with produce (De Roever, 1998; Castillo and Rodriguez-Garcia, 2004). In the USA, between 1973 and 1997, 48% of the outbreaks from produce that had a known bacterial etiological agent, were caused by *Salmonella* (Sivapalasingam *et al.*, 2004). Most alarming are the recent large outbreaks caused by multi-drug resistant (MDR) *S. enterica* Typhimurium DT104 strains that were associated with the consumption of lettuce in the UK (Horby *et al.*, 2003) and in Finland (Takkinen *et al.*, 2005). *S.* Typhimurium phage type DT104 is an emerging pathogen that is resistant to at least five antibiotics (Threlfall *et al.*, 1994). Multi-drug resistance of pathogens interferes with treatment of infections and puts patients at higher risk of death. Consequently, the emergence of MDR human pathogenic strains associated with fresh produce is of great concern to public health agencies.

L. monocytogenes

Although not implicated in outbreaks as frequently as *S. enterica*, *L. monocytogenes* is one of the deadliest foodborne pathogens associated with produce. The case-fatality rate of *L. monocytogenes* is 20% and is highest among pregnant women, neonates and immunocompromised adults (Mead *et al.*, 1999). Because of its lethality, the USA and the UK have established a "zero tolerance" level of *L. monocytogenes* per 25-g food sample, and consequently, the pathogen has caused the highest number of recalls by the food industry. These recalls include not only meat products, but also red peppers, sprouts and lettuce (Gorski, 2003). In 1981, cabbage processed as coleslaw was the first food and vegetable reported to be associated with an epidemic of listeriosis; the implicated cabbage was suspected to have been fertilised with sheep manure, which was the most probable source of contamination (Schlech *et al.*, 1983). In a recent study in Italy, Caggia *et al.* (2004) detected the presence of *L. monocytogenes* in green table olives, even after thermal treatment. This reflects the enhanced ability of this species to resist heat under low water activity conditions. *L. monocytogenes* is a versatile foodborne pathogen that has the ability to grow at a wide range of temperatures and at low pH (Montville and Matthews, 2005), characteristics that make it more difficult to control in food, including produce.

E. coli

While various pathogenic *E. coli* strains have caused produce-related outbreaks, *E. coli* O157:H7 is the causal agent of the most publicised outbreaks linked to fresh fruits and vegetables. O157:H7 is a dangerous serotype of *E. coli* that causes haemolytic-uremic syndrome, which can lead to death, particularly in children, the elderly and those with compromised immune systems. In 1990-1999, 18 out of 55 produce-related outbreaks were caused by *E. coli* O157:H7 (http://www.cspinet.org/new/prodhark.html). Many of these outbreaks were traced back to the consumption of lettuce (Sivapalasingam *et al.*, 2004), a popular fresh vegetable that is difficult to wash thoroughly and is mostly consumed raw. While the Food and Drug Administration in the USA has approved the use of irradiation to decontaminate ground beef, a main food-vehicle of *E. coli* O157:H7 outbreaks, the decontamination of produce remains a significant challenge.

Shigella spp.

Shigella is a foodborne pathogen for which primates, including humans, are the only reservoir. It has a low infectious dose and therefore, can be transmitted easily through the faecal-oral route (Montville and Matthews, 2005). The major cause of contamination of foods with *Shigella* is poor personal hygiene of food handlers. Thus, fresh produce can be contaminated when sanitary facilities for farm workers are lacking or are inconveniently located in the field. Foodborne outbreaks of shigellosis have been associated with several types of raw produce, including green onions, iceberg lettuce and uncooked baby maize (Anonymous, 1999b). In particular, in 1998, seven outbreaks of *Shigella sonnei* within a 2-month period in the USA and Canada were attributed to fresh parsley (Anonymous, 1999b), and another outbreak of a strain with the same PFGE subtype that sickened 300 people in California was linked to cilantro shortly after (Tsang *et al.*, 1999). These outbreaks, along with enterotoxigenic *E. coli* infections linked to parsley (Naimi *et al.*, 2003), were traced back to the same farm, revealing that unsafe agricultural practices have widespread public health consequences.

Campylobacter spp.

In contrast to *Salmonella enterica* and pathogenic *E. coli*, which cause frequent foodborne outbreaks, *Campylobacter* infections appear to occur mainly as sporadic illnesses (Friedman *et al.*, 2000). While poultry products and contaminated water are considered to be major sources of *Campylobacter* illness, outbreaks of campylobacteriosis linked to contaminated fruits, vegetables or other produce-related products have been reported (Beuchat, 1996; Anonymous, 1999b; Jacobs-Reitsma, 2000). In a retrospective cohort study conducted in the UK, Evans *et al.* (2003) observed that salad vegetables were the second highest risk factor in relation to the number of individual cases of *Campylobacter* infection. Similarly, a review of the epidemiology of campylobacteriosis outbreaks and cases between 1990 and 1999 in the USA revealed that produce was associated with more cases of *Campylobacter* illness than any other food source during this period, and was second only to dairy products in the total number of outbreaks (Mandrell and Brandl, 2004). Although cross-contamination from other foods may help explain the unexpectedly high number of produce-related outbreaks compared to other food sources, it is noteworthy that several studies have reported the presence of *Campylobacter* species on produce sampled at the marketplace (Doyle and Schoeni, 1986; Park and Sanders, 1992; Kumar *et al.*, 2001), in ready-to-eat (Federighi *et al.*, 1999), and in modified atmosphere packaged (MAP) (Phillips, 1998) vegetables.

Other Pathogens

Y. enterocolitica, *Y. pseudotuberculosis* and *S. aureus* are involved in comparatively fewer outbreaks than the pathogens described above and cause mostly sporadic illness. However, yersiniosis is often misdiagnosed as acute appendicitis (Montville and Matthews, 2005) and outbreaks of this foodborne disease may have occurred from contaminated produce but remained unrecognised. For example, in an outbreak of *Y. pseudotuberculosis* from iceberg lettuce in Finland in 1998, five of 45 case patients underwent appendectomies (Nuorti *et al.*, 2004). The outbreak was nevertheless identified because of routine surveillance and of subtyping of clinical isolates on a broad geographical scale (Tauxe, 2004). As surveillance intensifies, and as fruits and vegetables become increasingly investigated as a likely vehicle of foodborne disease, other pathogens may join the list of potential etiological agents linked to outbreaks from produce.

Ecology of Pathogens on Produce

Most studies on the behaviour of human pathogens on plants have been performed on cut plants or plant tissue in order to simulate the contamination of produce during post-harvest handling and food processing. While the results of these studies do not necessarily reflect the dynamics of pre-harvest crop colonisation by enteric pathogens, they provide clues about the behaviour of pathogens on plants in a simple system with controlled variables. On the other hand, several studies have assessed the persistence of enteric pathogens on crops in the field over prolonged periods of time. The following section presents experimental evidence for the ability of enteric pathogens to attach, grow and survive on plant surfaces in pre- and post-harvest environments.

Attachment of Enteric Pathogens to Plant Tissue

Localisation

Studies of *E. coli* O157:H7 on store-bought lettuce indicate that the bacterial cells attach to leaves in a relatively short time period, and that not all cells are removed by vigorous washing (Seo and Frank, 1999; Takeuchi and Frank, 2000; Wachtel and Charkowski, 2002). Similar observations have been reported with other enteric pathogens and plant species (Reina *et al.*, 2002; Ukuku and Fett, 2002). In addition, enteric pathogens appear to attach preferentially to specific sites on plant tissue. After immersion of lettuce pieces in a suspension of *E. coli* O157:H7, the pathogen was observed predominantly on the cut edges of the leaves, whereas fewer cells attached to the intact cuticle region. Cells that attached to the undamaged leaf surface were located near stomates, on trichomes and on veins (Seo and Frank, 1999). Likewise, *S. enterica* was observed in the vein region of cilantro leaves after its inoculation onto potted plants (Brandl and Mandrell, 2002), and epiphytic bacteria are frequently located near trichomes and on the veins of leaves (Leben, 1988; Monier and Lindow, 2004). It remains unclear whether this common localisation of human pathogenic bacteria and epiphytic bacteria on specific sites on leaves is caused by the enhanced wettability of these sites or by the presence of plant factors that favour bacterial attachment.

Although similarities in the pattern of attachment of human pathogens and epiphytic bacteria to plants appear to exist, clear differences also have been observed. Takeuchi *et al.* (2000) evaluated strains of *E. coli* O157:H7, *Pseudomonas fluorescens*, *S. enterica* and *L. monocytogenes* for their attachment to the intact and cut region of lettuce leaf pieces. *P.*

fluorescens attached preferentially and in the greatest number to the intact cuticle of the leaf, *E. coli* O157:H7 and *L. monocytogenes* attached better to the cut edges, and *S. enterica* bound equally to both sites, suggesting that the plant epiphyte may be better adapted to the phyllosphere than the human pathogens tested in the study. Additional evidence of such differential behaviour among bacterial species in adhesion to plant tissue has been presented. Barak *et al.* (2002) showed that four non-pathogenic *E. coli* strains isolated from field-grown cabbage attached to alfalfa sprouts (cotyledons, hypocotyls and roots) better than *E. coli* O157:H7, and that *S. enterica* attached in 10- to 1000-fold greater numbers than *E. coli* O157:H7 strains. However, in a separate study, strains of *E. coli* O157:H7 implicated in produce outbreaks attached to lettuce roots at ca. 10-fold higher levels than did two of five non-pathogenic *E. coli* strains (Wachtel *et al.*, 2002). Such differences in attachment have been observed also with strains of *L. monocytogenes*. While minimal differences between attachment of seven strains of *L. monocytogenes* were detected on cut radish tissue (Gorski *et al.*, 2003), 100- to 1000-fold differences were reported in the attachment and colonisation of different *L. monocytogenes* strains to alfalfa sprouts (Gorski *et al.*, 2004). These differences in the strength of attachment at the bacterial species and strain level, as well as with various plant species, suggest that specific factors may be involved in the attachment of enteric pathogens to plant tissue.

Attachment Factors

It is likely that some of the plant attachment mechanisms in enteric pathogens are similar to those described in plant-associated bacteria. These attachment factors may include lectins, polysaccharides, flagella, fimbriae, cellulose fibrils and nonfibrillar adhesins. At present, few fundamental studies have been initiated to identify plant attachment factors in human pathogens and their corresponding moiety in plants. Gorski *et al.* (2003) screened a mutant library of *L. monocytogenes* for attachment to radish slices at 30°C and identified three mutants that were reduced in attachment by at least 10-fold compared to the parental strain. Two mutations were in genes of unknown function within an operon encoding flagellar biosynthesis, but only one of these caused a lack of motility. The third mutant carried an insertion in a gene necessary for the transport of arabitol. The motility mutant was impaired in attachment also at 10 and 20°C, whereas the arabitol transport mutant was decreased in attachment also at 10°C. Interestingly, no difference in attachment between the mutants and the parental strain was observed when the assay was performed at 37°C. The effect of temperature in this system suggests that *L. monocytogenes* may express different plant attachment factors under different environmental conditions.

S. enterica serovar Thompson was observed in the cilantro phyllosphere by immunoelectron microscopy after its inoculation and incubation on cilantro plants (Brandl and Monier, 2006). The high resolution of the electron microscope combined with antibodies specific to the flagella of serovar Thompson allowed observation of the flagella of the pathogen cells located on the leaf cuticle. The flagella appeared to anchor the bacterial cells to the leaf surface, suggesting that they may serve for attachment to plants. Flagella have been described to have a role in the adhesion of *Azospirillum brasilense* to wheat roots based on experiments with a flagellar mutant as well as with purified flagella (Croes *et al.*, 1993). Besides flagella, other adhesins such as pili and fimbriae also are involved in the attachment of several bacterial species to plants surfaces (Korhonen *et al.*, 1986; Vesper and Bauer, 1986; Vesper, 1987; Romantschuk *et al.*, 2002). Recently, several *S. enterica* mutants were identified that have reduced adhesion to alfalfa sprouts (Barak *et al.*, 2005). In addition to mutants that had insertions in previously uncharacterised genes, one mutant was inactivated in curli production. Curli are thin aggregative fimbriae produced by *E. coli* (Olsen *et al.*, 1989) and *S. enterica* (Romling *et al.*, 1998), which mediate binding to and invasion of animal cells (Sukupolvi *et al.*, 1997; Uhlich *et al.*, 2002). It is possible therefore, that curli fimbriae have a role in the attachment of *S. enterica* to both animal and plant tissue.

In an attempt to identify plant factors that affect the attachment of *S. enterica* to cilantro leaves, polymers were purified from leaf surface extracts and tested individually for attachment by *S. enterica*. One of the few compounds to which *S. enterica* attachment was significantly superior to the control, was identified by mass spectrometry as stigmasterol (Mandrell *et al.*, 2006), a sterol detected also in other leaf extracts (Esmelindro *et al.*, 2004). It is unclear whether such particular components of the cuticle layer of leaves interact specifically with *S. enterica* cells upon their arrival in the phyllosphere, or if they mediate bacterial adhesion through electrostatic and hydrophobic bonds. Hassan and Frank (2003) have probed the nature of the forces that drive the adhesion of *E. coli* O157:H7 to the leaf surface of lettuce pieces. Their results indicated that highly hydrophobic surfactants were the most effective in removing *E. coli* O157:H7 cells from the intact portion of the leaf cuticle, and thus, that hydrophobic interactions are involved in the attachment of this pathogen to leaves. In a study on the attachment of several foodborne pathogens to the outer surface of whole cantaloupe melons, Ukuku and Fett (2002) demonstrated a significant linear correlation between the ability of the various bacterial species to attach to the cantaloupe surface and their relative cell hydrophobicity and surface charge. Therefore, it is likely that passive mechanisms of attachment that are dictated by the chemical nature of bacterial cell surfaces and plant surfaces, including electrostatic and hydrophobic interactions, are involved in the adhesion of human pathogens to plants, in addition to more specific factors (e.g. adhesins). This model has been proposed previously for the attachment of plant pathogenic bacteria and fungi to plants (Romantschuk *et al.*, 1996).

Fitness of Enteric Pathogens on Plants

Growth

It is the ability of immigrant foodborne pathogenic bacteria, not only to attach, but also to grow and/or survive in the plant environment that determines their fate in that habitat and their potential to cause foodborne illness. Few studies have investigated the ability of enteric pathogens to multiply on plant surfaces. Brandl and Mandrell (2002) developed a model with cilantro and a clinical isolate of *S. enterica* serovar Thompson that was linked to an outbreak from cilantro in California (Campbell *et al.*, 2001), to assess the role of various biotic and abiotic factors in the fitness of this pathogen in the phyllosphere. Bacterial population dynamics revealed that *S. enterica* reached lower population sizes than *Pseudomonas chlororaphis* and *Pantoea agglomerans*, two common epiphytic bacterial colonisers, after their inoculation onto the leaves of cilantro plants incubated under humid conditions in the laboratory. Despite this lower apparent fitness, microcolonies of GFP-labelled *S. enterica* cells, located mainly in the vein area of the leaf, were observed as soon as 2 days after inoculation. Larger colonies and high density aggregates were visualised on the veins of old leaves after 9 days of incubation under conditions conducive to the presence of free water on the leaves (Plate 4, see colour plate section). Leaf surface areas between the veins also hosted cells of the pathogen, but at much lower density, following a distribution pattern that has been observed previously with epiphytic bacterial species (Leben, 1988; Monier and Lindow, 2004).

S. enterica grows optimally at 37°C. When inoculated cilantro plants were incubated under high humidity at 30 or 37°C, *S. enterica* achieved higher growth rates and population sizes in the cilantro phyllosphere than at 24°C (Fig. 19.1) (Brandl and Mandrell, 2002). This suggested that under growth-conducive conditions, in this case warm temperature and high water availability, *S. enterica* is capable of multiplying rapidly in the phyllosphere and thus, reaching significant population sizes. It is noteworthy that the ratio of *S. enterica* to resident bacteria population sizes decreased after an initial increase during the first day following inoculation. A probable scenario explaining this change in the competitiveness of *S. enterica* over time is that the high growth rate of the pathogen at warm temperatures allows it to utilise simple nutrient sources that are present on the leaf surface efficiently. However, once that nutrients of this type are depleted, the pathogen may be less adapted to derive much energy from the remaining nutrients and is outcompeted by the resident microbiota. Evidence from studies by O'Brien and Lindow (1989) also supports the hypothesis that *S. enterica* can colonise the phyllosphere under conditions of high water availability on the plant surface. In their studies, performed in a growth chamber at 24°C, *S.* Typhimurium and *E. coli* colonised the wet leaves of maize and bean plants to population levels similar to those of three strains of *P. syringae*. This level of fitness on plants has not been observed with all enteric pathogens. For example, *C. jejuni*, a foodborne pathogen with a microaerophilic and thermophilic lifestyle, was unable to grow and survived poorly in the phyllosphere and rhizosphere of several plant species (Brandl *et al.*, 2004), indicating a lack of microsites on leaves and roots that are suitable for colonisation by fastidious human pathogens.

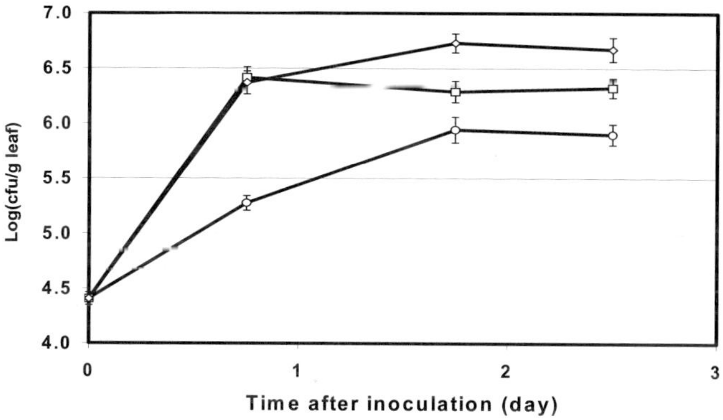

Fig. 19.1. Population dynamics of *S. enterica* in the cilantro phyllosphere after its inoculation onto plants and incubation under high humidity at 24°C (○), 30°C (□) and 37°C (◊).

Despite its epiphytic growth under optimal conditions, *S. enterica* did not multiply in the cilantro phyllosphere under dry conditions (Brandl and Mandrell, 2002). This is unlike *P. syringae*, a common coloniser of plant surfaces, which is able to grow on bean leaves in the laboratory even under conditions of low water availability (Wilson and Lindow, 1992). Similar differences between *P. syringae* strains, and *S. enterica* and *E. coli* on dry plant surfaces were reported by O'Brien and Lindow (1989). It is noteworthy, however, that after an initial phase of overall decline, which has been observed also with epiphytic bacteria, the population size of *S. enterica* stabilised on the dry leaf surface of cilantro plants maintained at low relative humidity (Fig. 19.2). Most importantly, the human pathogen recovered well to reach high population levels under subsequent wet conditions on the leaves. This implies

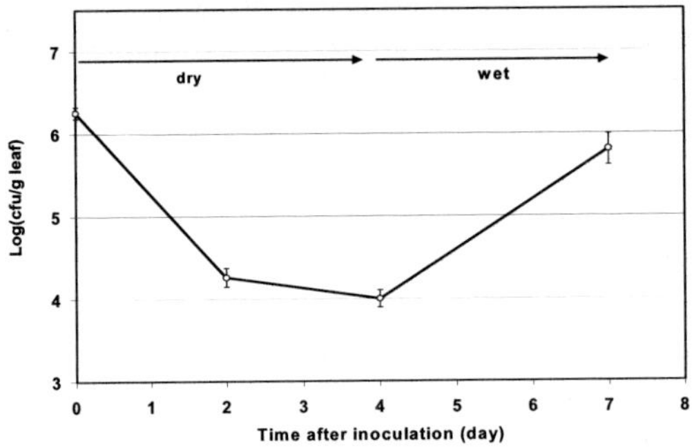

Fig. 19. 2. Population sizes of *S. enterica* on cilantro leaves after its inoculation onto plants incubated under dry conditions at 50% RH, and then under conditions allowing for the presence of water on the plant surface.

that even small populations of the pathogen that survived desiccation stress on crops in the field or during storage may increase to infectious dose levels during growth-conducive conditions before the produce is consumed. It has been reported that *S. enterica* forms large heterogeneous aggregates on cilantro leaves (Brandl and Mandrell, 2002), and that up to 54% of the pathogen population on leaves reside in large aggregates (Brandl *et al.*, 2005). The presence of such aggregates may have important implications for the survival of *S. enterica* under stressful conditions on plants, as was demonstrated for *P. syringae* in the bean phyllosphere (Monier and Lindow, 2003).

Survival

The frequent detection of enteric pathogens on produce, even before it is distributed to retailers, suggests that pre-harvest contamination occurs and that the pathogens have the ability to survive on crops after their arrival in that habitat. Consequently, there is a great interest in determining the survival of enteric pathogens on plants in order to better assess the risks associated with contamination of produce in the field. The survival of enteric pathogens on plants has been documented in several soil microcosm-based studies in the laboratory. Wachtel *et al.* (2002) investigated the association of *E. coli* O157:H7 with lettuce seedlings after their growth from seeds in soil that was irrigated with an aqueous suspension of the pathogen. *E. coli* O157:H7 cells were detected on the hypocotyls and cotyledons, as well as on the roots of the seedlings, where they formed small colonies.

Evidence of growth of E. coli O157:H7 on lettuce was demonstrated also when plants were spray-irrigated with water containing low concentrations of the pathogen; following this growth period, the population of E. coli O157:H7 on the lettuce leaves declined, but the pathogen was still detectable 30 days after inoculation at plant maturity (Solomon, 2003). Similar results were obtained when spinach seeds inoculated with *E. coli* were planted in soil microcosms, and the upper part of the plants tested for the presence of *E. coli* at harvest (Warriner *et al.*, 2003). In growth chamber experiments using soil cores and simulated seasonal environmental conditions, *S.* Typhimurium survived on the leaves of arugula plants until harvest when inoculated manure was applied to the soil in the "summer" and the crop harvested in the "autumn", but was undetectable at harvest when the manure was applied in the "spring" (average daily temperature <10ºC) and crops harvested in the "summer" (Natvig *et al.*, 2002). This suggested a critical role for warm temperatures in the population sizes that *S. enterica* can achieve on leaf surfaces, as reported by Brandl and Mandrell (2002).

Contamination of the entire plant, including the seeds, was reported in a study where *Arabidopsis thaliana* was grown in soil irrigated with *S. enterica* or *E. coli* O157:H7 (Cooley *et al.*, 2003). In this model system, movement of *S. enterica* up the plant to the seeds was eliminated when nonmotile mutants were used. Seed contamination has a direct impact on the microbial safety of sprouts, which are among the top food vehicles of outbreaks associated with produce (Sivapalasingam *et al.*, 2004). Due to factors inherent to their mode of production, such as warm temperature, ample water and high levels of nutrients exuded during seed germination, sprouts offer an ideal environment for the amplification of human pathogens. Furthermore, as is recognised already for plant pathogens, the presence of human pathogens on seeds has important implications for the contamination cycle on the farm, since the planting of contaminated seeds may spread or reintroduce the pathogen into the field.

Field studies have demonstrated that enteric pathogens are able to survive for extended periods of time on vegetables and in soil inoculated by applying contaminated water to the crops or manure to the soil in the field. Ercolani (1979) provided the first evidence that enteric pathogens can persist on crops under agricultural conditions. After inoculation of young lettuce plants, *S.* Typhi was recovered from the plants at harvest, whether the crop was grown in the field during the winter or summer season. More than 20 years later, with a heightened concern regarding the microbial safety of produce, similar field investigations were conducted with an avirulent mutant of *S.* Typhimurium (Islam *et al.*, 2004b) and a nontoxigenic mutant of *E. coli* O157:H7 (Islam *et al.*, 2004a). Irrigation with contaminated water or application of manure to the soil, at the seedling stage, resulted in the persistence of both enteric species for at least 3 and 6 months in the lettuce and parsley phyllosphere, respectively. It is significant that the pathogens survived in the soil at higher rates when the crops were in the ground than after harvest. This observation suggests that the aerial part of leafy plants, such as lettuce and parsley, may contribute to the persistence of enteric pathogens in the field, either by providing a continuous source of inoculum or by creating physicochemical conditions that are conducive to the survival of the pathogens in soil or on roots. The effect of roots on the enhanced survival of *E. coli* O157:H7 in soil microcosms has been proposed (Gagliardi and Karns, 2002), and may be part of a plethora of factors that affect the fitness of enteric pathogens in the agricultural environment.

Conclusion

The prevailing dogma about the fitness of human pathogens in the plant environment has been that enteric pathogens lack the ability to grow and survive on plants, and thus should be of little threat to the microbial safety of fresh produce. Emerging outbreaks are challenging this concept by providing evidence that foodborne illness can occur from contaminated produce. This is supported by recent studies that demonstrated that human pathogens have the ability to colonise and/or survive on plant surfaces. Although not evolutionarily as well adapted to colonise plants as plant-associated or plant pathogenic bacteria, enteric pathogens such as *S. enterica* and *E. coli* appear to be fit enough on plants to grow or persist in that environment. It is still unclear, however, whether this fitness is sufficient to explain their common association with enteric illness from the consumption of fresh fruits and vegetables, or if post-harvest conditions that promote the growth of these human pathogens are required to achieve the infectious doses necessary to cause disease in their host.

It is clear that the prevention and control of crop contamination with enteric pathogens in the pre- and post-harvest environments will require an increased understanding of their ecology in those habitats. At present, there is a great lack of fundamental knowledge of the bacterial and plant determinants that enable foodborne pathogens to multiply and survive on plant surfaces. Genomic information about most common enteric pathogens is now available and should be valuable for the investigation of the molecular events that drive the fitness of these pathogens on plants, and their interaction with the resident plant microbiota. Such knowledge may help identify critical control points for intervention strategies or treatments for the decontamination of produce. A better understanding of the biology of enteric pathogens in the plant environment also will provide insight into their minimum infectious dose when present in the plant-food matrix, and help assess the risks associated with the consumption of contaminated produce. Present epidemiological data suggests that enteric pathogens should not be discounted as important transient members of plant microbial communities.

Acknowledgements

Work in my laboratory is funded by the US Department of Agriculture, Agriculture Research Service CRIS projects 201-5325-210-44 and 201-5325-210-45.

References

Ackers, M.L., Mahon, B.E., Leahy, E., Goode, B., Damrow, T., Hayes, P.S., Bibb, W.F., Rice, D.H., Barrett, T.J., Hutwagner, L., Griffin, P.M. and Slutsker, L. (1998) An outbreak of *Escherichia coli* O157:H7 infections associated with leaf lettuce consumption. *Journal of Infectious Diseases* 177, 1588-1593.

Anonymous (1999a) Outbreak of *Salmonella* serotype Muenchen infections associated with unpasteurized orange juice-United States and Canada. *Centers for Disease Control and Prevention Morbidity and Mortality Weekly Report* 48, 582-585.

Anonymous (1999b) Outbreaks of *Shigella sonnei* infection associated with eating fresh parsley - United States and Canada, July-August 1998. *Centers for Disease Control and Prevention Morbidity and Mortality Weekly Report* 48, 285-289.

Anonymous (2000) 7th Report of WHO surveillance programme for control of foodborne infections and intoxications in Europe. *The FAO/WHO Collaborating Centre for Research and Training in Food Hygiene and Zoonoses.*

Anonymous (2001) Agriculture fact book. *US Department of Agriculture, Office of Communications.*

Anonymous (2002a) Multistate outbreaks of *Salmonella* serotype Poona infections associated with eating cantaloupe from Mexico-United States and Canada, 2000-2002. *Centers for Disease Control and Prevention Morbidity and Mortality Weekly Report* 51, 1044-1047.

Anonymous (2002b) Outbreak Alert: Closing the gaps in our federal food safety net. Washington, DC: Report from *Center for Science in the Public Interest*, pp. 58.

Anonymous (2003) 8th Report of WHO surveillance programme for control of foodborne infections and intoxications in Europe. *The FAO/WHO Collaborating Centre for Research and Training in Food Hygiene and Zoonoses.*

Anonymous (2005) Outbreaks of *Salmonella* infections associated with eating Roma tomatoes - United States and Canada, 2004. *CDC Morbidity and Mortality Weekly Report* 54, 325-328.

Barak, J.D., Whitehand, L.C. and Charkowski, A.O. (2002) Differences in attachment of *Salmonella enterica* serovars and *Escherichia coli* O157:H7 to alfalfa sprouts. *Applied and Environmental Microbiology* 68, 4758-4763.

Barak, J.D., Gorski, L., Naraghi-Arani, P. and Charkowski, A.O. (2005) *Salmonella enterica* virulence genes are required for bacterial attachment to plant tissue. *Applied and Environmental Microbiology* 71, 5685-5691.

Bean, N.H., Goulding, J.S., Daniels, M.T. and Angulo, F.J. (1997) Surveillance of foood-borne disease outbreaks: United States, 1988-1992. *Journal of Food Protection* 60, 1265-1286.

Besser, R.E., Lett, S.M., Weber, J.T., Doyle, M.P., Barrett, T.J., Wells, J.G. and Griffin, P.M. (1993) An outbreak of diarrhea and hemolytic uremic syndrome from *Escherichia coli* O157:H7 in fresh-pressed apple cider. *Journal of the American Medical Association* 269, 2217-2220.

Beuchat, L.R. (1996) Pathogenic microorganisms associated with fresh produce. *Journal of Food Protection* 59, 204-216.

Brandl, M.T. and Mandrell, R.E. (2002) Fitness of *Salmonella enterica* serovar Thompson in the cilantro phyllosphere. *Applied and Environmental Microbiology* 68, 3614-3621.

Brandl, M.T., Haxo, A.F., Bates, A.H. and Mandrell, R.E. (2004) Comparison of survival of *Campylobacter jejuni* in the phyllosphere with that in the rhizosphere of spinach and radish plants. *Applied and Environmental Microbiology* 70, 1182-1189.

Brandl, M.T., Miller, W.G., Bates, A.H. and Mandrell, R.E. (2005) Production of Autoinducer 2 in *Salmonella enterica* Serovar Thompson contributes to its fitness in chickens but not on Cilantro leaf surfaces. *Applied and Environmental Microbiology* 71, 2653-2662.

Brandl, M.T. and Monier, J.-M. (2006) Methods in microscopy for the visualization of bacteria and their behavior on plants. In: Sapers, G.M., Gorny, J.R. and Yousef, A.E. (eds) *Microbiology of Fruits and Vegetables,* CRC Press LLC, Boca Raton, FL, USA. pp. 595-619.

Caggia, C., Randazzo, C.L., Di Salvo, M., Romeo, F. and Giudici, P. (2004) Occurrence of *Listeria monocytogenes* in green table olives. *Journal of Food Protection* 67, 2189-2194.

Campbell, V.J., Mohle-Boetani, J., Reporter, R., Abbott, S., Farrar, J., Brandl, M.T., Mandrell, R.E. and Werner, S.B. (2001) An outbreak of *Salmonella* serotype Thompson associated with fresh cilantro. *Journal of Infectious Diseases* 183, 984-987.

Castillo, A. and Rodriguez-Garcia, M.O. (2004) Bacterial hazards in fresh and fresh-cut produce: sources and controls. In: Beier, R.C., Pillai, S.D., Phillips, T.D. and Ziprin, R.L. (eds) *Preharvest and Postharvest Food Safety*. Blackwell Publishing, Ames, IA, USA.

Cook, K.A., Dobbs, T.E., Hlady, W.G. and al., (1998) Outbreak of *Salmonella* serotype Hartford infections associated with pasteurized orange juice. *Journal of the American Medical Association* 280, 1504-1509.

Cooley, M., Miller, W. and Mandrell, R. (2003) Colonization of *Arabidopsis thaliana* with *Salmonella enterica* or enterohemorrhagic *Escherichia coli* O157:H7 and competition by an *Enterobacter asburiae*. *Applied and Environmental Microbiology* 69, 4915-4926.

Croes, C.L., Moens, S., Van Bastelaere, E., Vanderleyden, J. and Michiels, K.W. (1993) The polar flagellum mediates *Azospirillum brasilense* adsorption to wheat roots. *Journal of General Microbiology* 139, 2261-2269.

Cummings, K., Barrett, E., Mohle-Boetani, J.C., Brooks, J.T., Farrar, J., Hunt, T., Fiore, A., Komatsu, K., Werner, S.B. and Slutsker, L. (2001) A multistate outbreak of *Salmonella enterica* serotype Baildon associated with domestic raw tomatoes. *Emerging Infectious Diseases* 7, 1046-1048.

De Roever, C. (1998) Microbiological safety evaluations and recommendations on fresh produce. *Food Control* 9, 321-347.

Dowd, S.E., Thurston-Enriquez, J.A. and Brashears, M. (2004) Environmental reservoirs and transmission of foodborne pathogens. In: Beier, R.C., Pillai, S.D., Philips, T.D. and Ziprin, R.L. (eds) *Preharvest and Postharvest Food Safety: Contemporary Issues and Future Directions*. Blackwell Publishing Professional, Ames, IA, USA. pp. 161-171.

Doyle, M.P. and Schoeni, J.L. (1986) Isolation of *Campylobacter jejuni* from retail mushrooms. *Applied and Environmental Microbiology* 51, 449-450.

Ercolani, G.L. (1976) Bacteriological quality assessment of fresh marketed lettuce and fennel. *Applied and Environmental Microbiology* 31, 847-852.

Ercolani, G.L. (1979) Differential survival of *Salmonella typhi*, *Escherichia coli*, and *Enterobacter aerogenes* on lettuce in the field. *Zentralblatt Bakteriology Naturwissenschaft* 134, 402-411.

Esmelindro, A.A., Girardi Jdos, S., Mossi, A., Jacques, R.A. and Dariva, C. (2004) Influence of agronomic variables on the composition of mate tea leaves (*Ilex paraguariensis*) extracts obtained from CO_2 extraction at 30°C and 175 bar. *Journal of Agricultural Food and Chemistry* 52, 1990-1995.

Evans, M., Ribeiro, C. and Salmon, R. (2003) Hazards of healthy living: bottled water and salad vegetables as risk factors for *Campylobacter* infection. *Emerging Infectious Diseases* 9, 1219-1225.

Federighi, M., Magras, C., Pilet, M.F., Woodward, D., Johnson, W., Jugiau, F. and Jouve, J.L. (1999) Incidence of thermotolerant *Campylobacter* in foods assessed by NF ISO 10272 standard: Results of a two-year study. *Food Microbiology (London)* 16, 195-204.

Friedman, C.R., Neimann, J., Wegener, H.C. and Tauxe, R.V. (2000) Epidemiology of *Campylobacter jejuni* infections in the United States and other industrialized countries. In: Blaser, M.J. (ed.). *Campylobacter*. ASM Press, Washington, DC, USA. pp. 121-138.

Gagliardi, J.V. and Karns, J.S. (2002) Persistence of *Escherichia coli* O157:H7 in soil and on plant roots. *Environmental Microbiology* 4, 89-96.

Garcia-Villanova Ruiz, B., Galvez Vargas, R. and Garcia-Villanova, R. (1987) Contamination of fresh vegetables during cultivation and marketing. *International Journal of Food Microbiology* 4, 285-291.

Gorski, L., Palumbo, J.D. and Mandrell, R.E. (2003) Attachment of *Listeria monocytogenes* to radish tissue is dependent upon temperature and flagellar motility. *Applied and Environmental Microbiology* 69, 258-266.

Gorski, L., Palumbo, J.D. and Nguyen, K.D. (2004) Strain-specific differences in the attachment of *Listeria monocytogenes* to alfalfa sprouts. *Journal of Food Protection* 67, 2488-2495.

Hassan, A.N. and Frank, J.F. (2003) Influence of surfactant hydrophobicity on the detachment of *Escherichia coli* O157:H7 from lettuce. *International Journal of Food Microbiology* 87, 145-152.

Hedberg, C.W., Angulo, F.J., White, K.E., Langkop, C.W., Schell, W.L., Stobierski, M.G., Schuchat, A., Besser, J.M., Dietrich, S., Helsel, L., Griffin, P.M., McFarland, J.W. and Osterholm, M.T. (1999) Outbreaks of salmonellosis associated with eating uncooked tomatoes: implications for public health. *Epidemiology and Infection* 122, 385-393.

Hilborn, E.D., Mermin, J.H., Mshar, P.A., Hadler, J.L., Voetsch, A., Wojtkunski, C., Swartz, M., Mshar, R., Lambert-Fair, M.A., Farrar, J.A., Glynn, M.K. and Slutsker, L. (1999) A multistate outbreak of *Escherichia coli* O157:H7 infections associated with consumption of mesclun lettuce. *Archives of Internal Medecine* 159, 1758-1764.

Horby, P.W., O'Brien, S.J., Adak, G.K., Graham, C., Hawker, J.I., Hunter, P., Lane, C., Lawson, A.J., Mitchell, R.T., Reacher, M.H., Threlfall, E.J. and Ward, L.R. (2003) A national outbreak of multi-resistant *Salmonella enterica* serovar Typhimurium definitive phage type (DT) 104 associated with consumption of lettuce. *Epidemiology and Infection* 130, 169-178.

Isaacs, S., Aramini, J., Ciebin, B., Farrar, J.A., Ahmed, R., Middleton, D., Chandran, A.U., Harris, L.J., Howes, M., Chan, E., Pichette, A.S., Campbell, K., Gupta, A., Lior, L.Y., Pearce, M., Clark, C., Rodgers, F., Jamieson, F., Brophy, I. and Ellis, A. (2005) An international outbreak of salmonellosis associated with raw almonds contaminated with a rare phage type of *Salmonella enteritidis*. *Journal of Food Protection* 68, 191-198.

Islam, M., Doyle, M.P., Phatak, S.C., Millner, P. and Jiang, X. (2004a) Persistence of enterohemorrhagic *Escherichia coli* O157:H7 in soil and on leaf lettuce and parsley grown in fields treated with contaminated manure composts or irrigation water. *Journal of Food Protection* 67, 1365-1370.

Islam, M., Morgan, J., Doyle, M.P., Phatak, S.C., Millner, P. and Jiang, X. (2004b) Persistence of *Salmonella enterica* serovar *Typhimurium* on lettuce and parsley and in soils on which they were grown in fields treated with contaminated manure composts or irrigation water. *Foodborne Pathogens and Disease* 1, 27-35.

Jacobs-Reitsma, W. (2000) *Campylobacter* in the food supply. In: Blaser, M.J. (ed.). *Campylobacter*. ASM Press, Washington, D.C., USA. pp. 467-481.

Janisiewicz, W.J., Conway, W.S., Brown, M.W., Sapers, G.M., Fratamico, P. and Buchanan, R.L. (1999) Fate of *Escherichia coli* O157:H7 on fresh-cut apple tissue and its potential for transmission by fruit flies. *Applied and Environmental Microbiology* 65, 1-5.

Korhonen, T.K., Haahtela, K., Romantschuk, M. and Bamford, D. (1986) Role of fimbriae and pili in the attachment of *Klebsiella, Enterobacter*, and *Pseudomonas* to plant surfaces. In: Lugtenberg, B. (ed.) *Recognition in Microbe-Plant Symbiotic and Pathogenic Interactions*. Vol. H4. Springer-Verlag, Heidelberg, Germany. pp. 229-241.

Kumar, A., Agarwal, R.K., Bhilegaonkar, K.N., Shome, B.R. and Bachhil, V.N. (2001) Occurrence of *Campylobacter jejuni* in vegetables. *International Journal of Food Microbiology* 67, 153-155.

Lauzon, C.R. (2003) Symbiotic relationships of tephritids. In: Bourtzis, K. and Miller, T.A. (eds) *Insect Symbiosis*. CRC Press, Boca Raton, FL, USA.

Leben, C. (1988) Relative humidity and the survival of epiphytic bacteria with buds and leaves of cucumber plants. *Phytopathology* 78, 179-185.

Mahon, B.E., Ponka, A., Hall, W.N., Komatsu, K., Dietrich, S.E., Siitonen, A., Cage, G., Hayes, P.S., Lambert-Fair, M.A., Bean, N.H., Griffin, P.M. and Slutsker, L. (1997) An international outbreak of *Salmonella* infections caused by alfalfa sprouts grown from contaminated seeds. *Journal of Infectious Diseases* 175, 876-882.

Mandrell, R.E. and Brandl, M.T. (2004) *Campylobacter* species and fresh produce: Outbreaks, incidence and biology. In: Beier, R., Pillai, S., Philips, T. and Ziprin, R. (eds) *Pre-harvest and Post-harvest Food Safety: Contemporary Issues and Future Directions*. Blackwell Publishing Professional. Ames, IA, USA. pp. 59-72.

Mandrell, R.E., Gorski, L. and Brandl, M.T. (2006) Attachment of microorgamisms to fresh produce. In: Sapers, G.M., Gorny, J.R. and Yousef, A.E. (eds) *Microbiology of Fruits and Vegetables* CRC Press LLC, Boca Raton, FL, USA. pp. 34-61.

Mead, P.S., Slutsker, L., Dietz, V., McCaig, L.F., Bresee, J.S., Shapiro, C., Griffin, P.M. and Tauxe, R.V. (1999) Food-related illness and death in the United States. *Emerging and Infectious Diseases* 5, 607-625.

Mohle-Boetani, J., Reporter, R. and Werner, S.B. (1999) An outbreak of *Salmonella* serogroup Saphra due to cantaloupes from Mexico. *Journal of Infectious Diseases* 180, 1361-1364.

Monier, J.M. and Lindow, S.E. (2003) Differential survival of solitary and aggregated bacterial cells promotes aggregate formation on leaf surfaces. *Proceedings of the Nationa Academy of Science USA* 100, 15977-15982.

Monier, J.M. and Lindow, S.E. (2004) Frequency, size, and localization of bacterial aggregates on bean leaf surfaces. *Applied and Environmental Microbiology* 70, 346-355.

Montville, T.J. and Matthews, K.R. (2005) *Food Microbiology, An Introduction*. ASM Press, Herdon, VA, USA.

Naimi, T.S., Wicklund, J.H., Olsen, S.J., Krause, G., Wells, J.G., Bartkus, J.M., Boxrud, D.J., Sullivan, M., Kassenborg, H., Besser, J.M., Mintz, E.D., Osterholm, M.T. and Hedberg, C.W. (2003) Concurrent outbreaks of *Shigella sonnei* and enterotoxigenic *Escherichia coli* infections associated with parsley: implications for surveillance and control of foodborne illness. *Journal of Food Protection* 66, 535-541.

Natvig, E.E., Ingham, S.C., Ingham, B.H., Cooperband, L.R. and Roper, T.R. (2002) *Salmonella enterica* serovar *Typhimurium* and *Escherichia coli* contamination of root and leaf vegetables grown in soils with incorporated bovine manure. *Applied and Environmental Microbiology* 68, 2737-2744.

Nuorti, J.P., Niskanen, T., Hallanvuo, S., Mikkola, J., Kela, E., Hatakka, M., Fredriksson-Ahomaa, M., Lyytikainen, O., Siitonen, A., Korkeala, H. and Ruutu, P. (2004) A widespread outbreak of *Yersinia pseudotuberculosis* O:3 infection from iceberg lettuce. *Journal of Infectious Diseases* 189, 766-774.

O'Brien, R.D. and Lindow, S.E. (1989) Effect of plant species and environmental conditions on epiphytic population sizes of *Pseudomonas syringae* and other bacteria. *Phytopathology* 79, 619-627.

Olsen, A., Jonsson, A. and Normark, S. (1989) Fibronectin binding mediated by a novel class of surface organelles on *Escherichia coli*. *Nature* 338, 652-655.

Park, C.E. and Sanders, G.W. (1992) Occurrence of thermotolerant campylobacters in fresh vegetables sold at farmers' outdoor markets and supermarkets. *Canadian Journal of Microbiology* 38, 313-316.

Phillips, C.A. (1998) The isolation of *Campylobacter* spp. from modified atmosphere packaged foods. *International Journal of Environmental Health Research* 8, 215-221.

Reina, L.D., Fleming, H.P. and Breidt, F., Jr. (2002) Bacterial contamination of cucumber fruit through adhesion. *Journal of Food Protection.* 65, 1881-1887.

Romantschuk, M., Roine, E., Bjorklof, K., Ojanen, T., Nurmiaho-Lassila, E.-L. and Haahtela, K. (1996) Microbial attachment to plant aerial surfaces. In: Morris, C.E., Nicot, P.C. and Nguyyen-The, C. (eds) *Aerial Plant Surface Microbiology* Plenum Press, New York, USA.

Romantschuk, M., Boureau, T., Roine, E., Haapalainen, M. and Taira, S. (2002) The role of pili and flagella in leaf colonization by *Pseudomnas syringae*. In: Lindow, S.E., Hecht-Poinar, E.I. and Elliott, V.J. (eds) *Phyllosphere Microbiology* APS Press, St. Paul, Minnesota, USA. pp. 101-114.

Romling, U., Bian, Z., Hammar, M., Sierralta, W.D. and Normark, S. (1998) Curli fibers are highly conserved between *Salmonella typhimurium* and *Escherichia coli* with respect to operon structure and regulation. *Journal of Bacteriology* 180, 722-731.

Schlech, W.F., 3rd, Lavigne, P.M., Bortolussi, R.A., Allen, A.C., Haldane, E.V., Wort, A.J., Hightower, A.W., Johnson, S.E., King, S.H., Nicholls, E.S. and Broome, C.V. (1983) Epidemic listeriosis--evidence for transmission by food. *New England Journal of Medicine* 308, 203-206.

Sela, S., Nestel, D., Pinto, R., Nemny-Lavy, E. and Bar-Joseph, M. (2005) Mediterranean fruit fly as a potential vector of bacterial pathogens. *Applied and Environmental Microbiology* 71, 4052-4056.

Seo, K.H. and Frank, J.F. (1999) Attachment of *Escherichia coli* O157:H7 to lettuce leaf surface and bacterial viability in response to chlorine treatment as demonstrated by using confocal scanning laser microscopy. *Journal of Food Protection* 62, 3-9.

Sivapalasingam, S., Friedman, C.R., Cohen, L. and Tauxe, R.V. (2004) Fresh produce: a growing cause of outbreaks of foodborne illness in the United States, 1973 through 1997. *Journal of Food Protection* 67, 2342-2353.

Solomon, E.B., H-J. Pang, and K.R. Matthews (2003) Persistence of *Escherichia coli* O157:H7 on lettuce plants following spray irrigation with contaminated water. *Journal of Food Protection* 66, 2198-2202.

Sukupolvi, S., Lorenz, R.G., Gordon, J.I., Bian, Z., Pfeifer, J.D., Normark, S.J. and Rhen, M. (1997) Expression of thin aggregative fimbriae promotes interaction of *Salmonella typhimurium* SR-11 with mouse small intestinal epithelial cells. *Infection and Immunity* 65, 5320-5325.

Takeuchi, K. and Frank, J.F. (2000) Penetration of *Escherichia coli* O157:H7 into lettuce tissues as affected by inoculum size and temperature and the effect of chlorine treatment on cell viability. *Journal of Food Protection* 63, 434-440.

Takkinen, J., Nakari, U.-M., Johansson, T., Niskanen, T., Siitonen, A. and Kuusi, M. (2005) A nationwide outbreak of multiresistant *Salmonella Typhimurium* va. *Copenhagen* DT104B infection in Finland due to contaminated lettuce from Spain. *Eurosurveillance Weekly www.eurosurveillance.org* 10.

Tauxe, R.V. (2004) Salad and pseudoappendicitis: *Yersinia pseudotuberculosis* as a foodborne pathogen. *Journal of Infectious Diseases* 189, 761-763.

Threlfall, E.J., Frost, J.A., Ward, L.R. and Rowe, B. (1994) Epidemic in cattle and humans of *Salmonella typhimurium* DT 104 with chromosomally integrated multiple drug resistance. *The Veterinary Record* 134, 577.

Tsang, T.H.F., Brawley, L., Abbott, S., Werner, S.B. and Vugia, D. (1999) Large *Shigella sonnei* outbreak linked to salsa in California. In *Abstracts of the 48th Epidemic Intelligence Service Conference*: Centers for Disease Prevention and Control, Atlanta, Georgia, USA.

Uhlich, G.A., Keen, J.E. and Elder, R.O. (2002) Variations in the *csgD* promoter of *Escherichia coli* O157:H7 associated with increased virulence in mice and increased invasion of HEp-2 cells. *Infection and Immunity* 70, 395-399.

Ukuku, D.O. and Fett, W. (2002) Behavior of *Listeria monocytogenes* inoculated on cantaloupe surfaces and efficacy of washing treatments to reduce transfer from rind to fresh-cut pieces. *Journal of Food Protection.* 65, 924-930.

Vesper, S.J. and Bauer, W.D. (1986) Role of pili (fimbriae) in attachment of *Bradyrhizobium japonicum* to soybean roots. *Applied and Environmental Microbiology* 52, 134-141.

Vesper, S.J. (1987) Production of pili (fimbriae) by *Pseudomonas fluorescens* and correlation with attachment to corn roots. *Applied and Environmental Microbiology* 53, 1397-1405.

Wachtel, M. and Charkowski, A. (2002) Cross-contamination of lettuce with *Escherichia coli* O157:H7. *Journal of Food Protection* 65, 465-470.

Wachtel, M.R., Whitehand, L.C. and Mandrell, R.E. (2002) Association of *Escherichia coli* O157:H7 with preharvest leaf lettuce upon exposure to contaminated irrigation water. *Journal of Food Protection* 65, 18-25.

Warriner, K., Ibrahim, F., Dickinson, M., Wright, C. and Waites, W.M. (2003) Interaction of *Escherichia coli* with growing salad spinach plants. *Journal of Food Protection* 66, 1790-1797.

Wells, J.M. and Butterfield, J.E. (1997) *Salmonella* contamination associated with bacterial soft-rot of fresh fruits and vegetables in the marketplace. *Plant Disease* 81, 867-872.

Wilson, M. and Lindow, S.E. (1992) Relationship of total viable and culturable cells in epiphytic populations of *Pseudomonas syringae*. *Applied and Environmental Microbiology* 58, 3908-3913.

Chapter 20

Post-harvest Spoilage of Wheat Grains: Malodour Formation and the Infection Process

P.T.N. Spencer-Phillips[1], A. Wallington[1], W. Köckenberger[2], H.E. Gunson[1] and N.M. Ratcliffe[1]
[1]School of Biosciences, Faculty of Applied Sciences, University of the West of England, Coldharbour Lane, Bristol, BS16 1QY, UK; [2]Sir Peter Mansfield Magnetic Resonance Centre, School of Physics and Astronomy, University of Nottingham, University Park, Nottingham, NG7 2RD, UK
peter.spencer-phillips@uwe.ac.uk

Introduction

Global populations are rising, resulting in increased human energy requirements. These could be met by increased food production and also by reduction of losses, both when crops are growing in the field and after harvest. World food supply relies on 15 key species of plants; cereals comprise more than half of this group, with wheat and rice providing 60% of food energy intake (Gallagher, 1984). Fungi are the most significant cause of spoilage of stored cereal grain, and therefore reduction of this problem would result in substantial economic and social benefits.

Heads of wheat grain have a diverse microbiota, with more than 150 species of fungi associated with the carposphere (Flannigan, 1970, 1978 and references therein). Most are field fungi, such as species of *Alternaria, Cladosporium* and *Fusarium* (Pitt and Hocking, 1997). The majority do not cause problems post-harvest, with a few notable exceptions. For example, *Tilletia tritici* (also known as *T. caries*) causes a disease known as bunt or stinking smut, and produces teliospores (Fig. 20.1) with a fishy odour due to the production of trimethyl-amine (Neergaard, 1988); grain contaminated by these spores is usually rejected by processors due to the distinctive unpleasant odour. Storage fungi, including *Aspergillus* and *Penicillium* spp., are also widespread although not dominant on plant surfaces in the field (Flannigan, 1978), but may develop to dominate the ecosystem of stored grains if environmental conditions are favourable (Lacey and Magan, 1991). Key factors include temperature and moisture content (Cairns-Fuller *et al*., 2005), with post-harvest spoilage likely to occur if the latter is above 14% (Harris, 1997), although the distribution of water within grains and its role in regulating colonisation processes is not clear.

Post-harvest spoilage of wheat grain causes a number of very significant problems, including: reduced germination, significant for the seed industry and in the malting process; caking, where the grain becomes a solid mass and impedes handling operations; heating, which in extreme cases has been known to cause combustion and explosion in silos. For food manufacture, major problems are discolouration and the production of malodours, which are potential causes of consumer dissatisfaction. Of greater significance, however, is the formation of mycotoxins, which are thought to include some of the most carcinogenic of all naturally occurring compounds (Skudamore, 1994). Mycotoxins within grain and other seeds such as groundnuts are of major health concern, and the resulting mycotoxicoses often lead to death in both livestock and humans (Hussein and Brasel, 2001). As well as the many well documented historic incidences in Europe and elsewhere, the problem continues today particularly in the developing world. In the poorest regions, such as parts of Africa, recent data on dietary intake of mycotoxins and their effects on human health indicate the severity of the problem (Egal *et al.*, 2005).

Fig. 20.1. Teliospores of *Tilletia tritici* observed by environmental scanning electron microscopy on the surface of a commercial wheat grain sample with a fishy odour.

In the UK, total annual yield of the major grain crops (mainly wheat, barley, oats and oilseed rape) amounts to approximately 25 million tonnes (Hook, 2004). A modest estimate of spoilage between harvest and processing at £1 per tonne suggests that total losses may approximate to £25 million each year, although precise data are not available. More specifically, rejection of wheat grain with malodours and taints by processors in the UK is estimated to cost up to £1 million annually (Roger Williams, HGCA, pers. comm.), with fungal contamination a significant factor. Therefore, control of fungal spoilage is a priority for store managers. Whilst the profuse production of sporophores (e.g. of *Aspergillus* and *Penicillium* spp. on cereal grain, see Figs 20.2 and 20.3) can be recognised macroscopically, less subtle spoilage cannot always be detected even with rigorous examination in the grain intake laboratories of major cereal processors in developed countries. Indeed, the present method for detection is limited to visual inspection and assessment of samples by the human nose for musty and mouldy odours. The health risks to workers from inhalation of spores include allergies and lung infections such as aspergillosis, which may lead to invasive infection with 60-90% mortality (Hoskin, 1999). Airborne dust and spores may also contain mycotoxins (Palmgren and Lee, 1986; Skaug *et al.*, 2001), which can result in pulmonary mycotoxicosis if inhaled (Selim *et al.*, 1998). The problem is compounded because odours may not be detectable before the final processed product. Cooking contaminated grain during processing releases odours that were previously not detectable, perhaps resulting in a prolonged down-time and cleaning procedures that add to the cost of lost production and reduce profit margins. Further, mycotoxins may still enter the product.

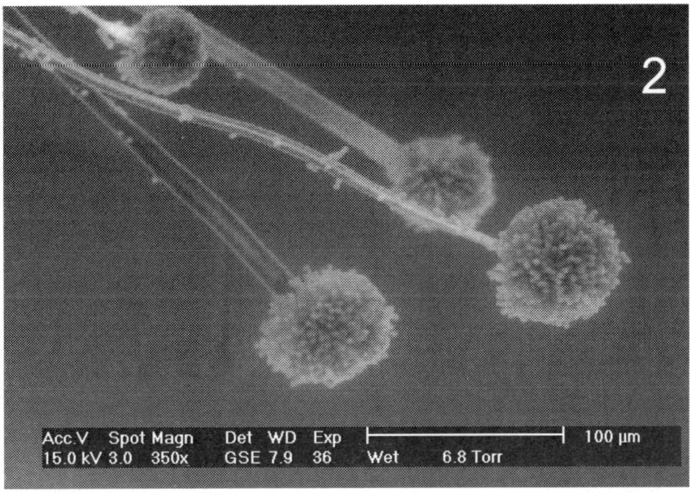

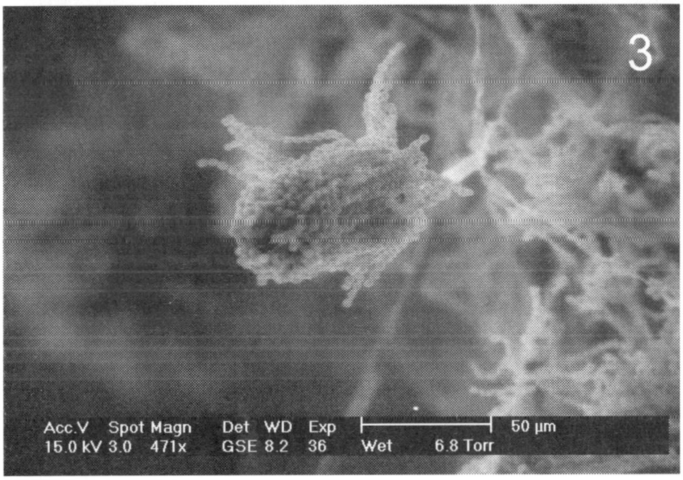

Figs 20.2 and 20.3. Conidiophores of *Aspergillus* (Fig. 20.2) and *Penicillium* (Fig. 20.3) observed by environmental scanning electron microscopy growing from the surface of commercial wheat grain samples with musty odours.

Thus a more reliable and sensitive method is needed for early detection of the growth of spoilage organisms and the odours produced, both to avoid economic loss and hazards to health. Volatile organic compounds (VOCs) emitted during the colonisation process provide suitable markers of post-harvest spoilage (Magan and Evans, 2000). Prototype electronic noses, based on sensors that detect indicator VOCs at lower concentrations than the human nose, have been developed recently for stored produce such as potato tubers (de

Lacy *et al.,* 2000) and wheat grain (de Lacy *et al.,* 2003). These promise to provide more safe, reliable and standardised methods for detection of the VOCs that indicate microbial spoilage.

Selection of appropriate sensors first required a method for routine and reliable infection of grain by specific spoilage organisms. These methods enabled identification of the characteristic VOCs by gas chromatography-mass spectrometry (GC-MS), and then the rigorous testing of the selected sensors against grain samples with malodours that could be replicated reliably in the laboratory (Brown *et al.,* 2002). Experiments to select a sterilisation method to ensure that grain became infected by the introduced organism and the effect of the selected method on grain viability and development of malodours, in comparison with 25 commercial wheat grain samples collected in the UK, are described here.

An inoculation technique for individual grains was also devised so that the progress of infection could be related to the production of VOCs. Methods to study the infection and colonisation processes previously have included light and electron microscopy, but recent technological advances now permit new approaches. Environmental scanning electron microscopy (ESEM) enables hydrated cereal grains and spoilage fungi to be imaged in the living condition. Nuclear magnetic resonance (NMR) microscopy, a variant of the medical body scanning technique used in hospitals, has sufficient resolution to image very small structures to a cellular resolution (Köckenberger 2001a, b). In conjunction with conventional light microscopy and ESEM, this technique has been used to investigate how the spoilage fungi *Aspergillus niger*, *Fusarium culmorum* and *Penicillium aurantiogriseum* colonise wheat grains.

Infection of Grain and Malodour Formation

Surface Sterilisation of Wheat Grains

A range of treatments was assessed for surface sterilising grains of *Triticum aestivum* cv. Hereward. The methods used the combinations of Tween 80, sodium hypochlorite, ethanol and Plant Preservative Mixture (Plant Cell Technology Inc., USA) listed in Table 20.1. Approximately 100 wheat grains (4.4 g) were exposed to each of the 14 different treatments, and then three replicate batches of five grains from each sample were plated onto potato dextrose agar (PDA). Microbial contamination was assessed visually after incubation of the plates in the dark for 7 days at 19°C.

All plates showed microbial growth from all grains following the treatments listed in Table 20.1, indicating that the grains remained colonised by viable microorganisms. Thus none of the surface sterilisation methods was effective. It was noted, however, that contamination was slower to appear in treated samples than in the control samples (1, 2 and 14 in Table 20.1). Contamination was more apparent as mycelial fungal colonies in treated samples than as either yeasts or bacteria, whilst colonies arising from control grains were a mixture of mycelial and unicellular fungi, as well as bacteria. In contrast, irradiated grains plated onto PDA showed no microbial growth after 7 days' incubation, as described below.

Grain Germination Test for Effect of Irradiation and Odour

Twenty-five commercial samples of wheat grains collected in the UK from a variety of unknown cultivars (Table 20.2), and grains of cv. Hereward that either had been treated with a 12 kGy dose of irradiation or were untreated, were supplied by the Campden and

Chorleywood Food Research Association (CCFRA; Chipping Campden, UK). Three replicates of 100 grains were prepared to assess the germination rate of each sample. One hundred grains were laid out in a grid pattern on a damp paper towel, then covered by another towel, and the paper and grain sandwich rolled up into a cylinder. The rolls of paper towel were placed in a penicillin assay dish, covered with the lid and incubated in the dark at 19°C for 10 days. The experiment with irradiated grain was repeated twice. Microbial contamination was assessed as before, and the odours of grain samples were determined by a panel of assessors as described in de Lacy Costello *et al.* (2003).

Normal germination was considered to have occurred when the grain produced a seedling with an intact shoot and at least two main roots. In the experiment to assess the effect of irradiation on grain viability, non-irradiated control samples showed a mean germination of 86 and 94% in two separate experiments. No grains germinated normally with the irradiated samples, and where germination did occur, either the root and shoots stopped growing after they were about 1 cm in length or the grain only produced a root or a shoot but not both. This was termed irregular germination.

Table 20.1. Methods of surface sterilisation of *Triticum aestivum* cv. Hereward grains (SDW = sterile distilled water; NaOCl = sodium hypochlorite; EtOH = ethanol; PPM = Plant Preservative Mixture; Tween 80 at 0.1% v/v).

	Pre-treatment	Main treatment	Post-treatment
1.	None	None	None
2.	None	3 x SDW wash	None
3.	None	6% NaOCl (10 min)	3 x SDW wash
4.	None	6% NaOCl + Tween 80 (10 min)	3 x SDW wash
5.	None	70% EtOH dip/5% NaOCl (10 min)	3 x SDW wash
6.	None	70% EtOH dip/5% NaOCl + Tween 80 (10 min)	3 x SDW wash
7.	None	0.5% PPM (1 h)	3 x SDW wash
8.	SDW + Tween 80 (1 h)	6% NaOCl (10 min)	3 x SDW wash
9.	SDW + Tween 80 (2 h)	6% NaOCl (10 min)	3 x SDW wash
10.	70% EtOH dip/ SDW + Tween 80 (1 h)	6% NaOCl (10 min)	3 x SDW wash
11.	70% EtOH dip/ SDW + Tween 80 (2 h)	6% NaOCl (10 min)	3 x SDW wash
12.	SDW + Tween 80 (1 h)	70% EtOH dip/6% NaOCl (10 min)	3 x SDW wash
13.	SDW + Tween 80 (2 h)	70% EtOH dip/6% NaOCl (10 min)	3 x SDW wash
14.	SDW + Tween 80 (2 h)	None	None

Grain samples with distinctive odours (Table 20.2) varied in their ability to germinate according to the type of odour. Grain with a normal control, "wheat", odour produced a shoot greater than 2.5 cm long and roots greater than a total length of 4.5 cm in 0 to 94% of the grains. Two control samples (Fc/38242/001 and 025) showed a similar pattern of germination to previous results for non-irradiated control grain. In contrast, control samples Fc/38242/007, 002 and 022 had 75%, 94% and 96%, respectively, of their grains not germinating at all, with only 007 and 002 showing a very low percentage of normal germination.

Seed with a musty or mouldy component to the odour was always severely damaged in its ability to germinate, with the greatest normal germination at 21% in the musty/fishy sample 006 (Table 20.2). The grains did not germinate at all in six of the musty or mouldy samples, and close inspection showed that most were supporting the growth of discrete fungal colonies after the 10-day incubation period. Non-germinated grains of other musty samples also often had fungal colonies associated especially with the embryo end of the grain (e.g. as in Plate 5 a and b, see colour plate section). Where germination had occurred, it often was classed as irregular. This was defined as before, but with the additional requirement that shoots and roots were less than 2.5 cm and 4.5 cm long, respectively.

Table 20.2. Odour and germination of wheat grains from 25 samples collected in the UK and ascribed an odour by CCFRA. Germination was assessed by incubating three replicate samples of 100 grains in damp tissue paper at 19°C for 10 days. Normal germination was considered to have occurred when the grain produced a seedling with an intact shoot and at least two main roots; irregular germination was defined as when shoots and roots were less than 2.5 cm and 4.5 cm long, respectively. Grain with a normal wheat odour is described as control.

Grain sample	Odour	Germination (%, mean of 3 replicates)		
		None	Irregular	Normal
Fc/38242/025	Control	3	3	94
Fc/38242/001	Control	4	12	84
Fc/38242/007	Control	75	11	14
Fc/38242/002	Control	94	4	2
Fc/38242/022	Control	96	4	0
Fc/38242/003	Musty	100	0	0
Fc/38242/024	Musty	35	65	0
Fc/38242/035	Musty	54	38	8
Fc/38242/008	Musty/Mouldy	100	0	0
Fc/38242/009	Mouldy	100	0	0
Fc/38242/004	Musty/Sour	100	0	0
Fc/38242/010	Musty/Sour	100	0	0
Fc/38242/016	Musty/Green	100	0	0
Fc/38242/018	Musty/Green	82	18	0
Fc/38242/019	Musty/Green	73	25	2
Fc/38242/006	Musty/Fishy	73	6	21
Fc/38242/034	Sour/Control	11	12	77
Fc/38242/021	Sour	4	7	89
Fc/38242/017	Sour	55	17	28
Fc/38242/012	Green/Sour	5	3	92
Fc/38242/020	Green	4	10	86
Fc/38242/023	Green	34	17	49
Fc/38242/013	Fishy	5	3	92
Fc/38242/014	Fishy	2	5	93
Fc/38242/015	Fishy	18	27	55

Grain samples with either a green (freshly mown grass) or a sour component to the odour generally had higher rates of germination than those with musty or mouldy odours.

Some of these, and two of the three samples with a fishy odour, showed germination close to the highest rate found for control grain.

Examination with the ESEM showed that sporophores of *Aspergillus* (Fig. 20.2) and *Penicillium* (Fig. 20.3) spp. were typically associated with musty and mouldy grain. All samples of grain with a fishy odour had teliospores of *Tilletia tritici* (the bunt or stinking smut fungus) attached to the grain surface (Fig. 20.1).

Inoculation of Grain with Spoilage Microorganisms and Development of Malodours

Initial experiments with non-sterile grain used *Triticum aestivum* cv. Consort, whilst cv. Hereward was used for subsequent work with irradiated grain. Seed quality grain of cv. Consort was obtained from the Plant Breeding Institute, Cambridge (UK). It had not received any chemical seed treatments and had a 92% germination rate. The moisture content was 15%, as determined with a Protimeter (UK) Digital Grainmaster, and was conditioned following the CCFRA protocol (method ref: TES-CM-0001, and described in de Lacy *et al.,* 2003) to give 18% moisture before inoculation. The amount of grain inoculated was either 250 g or 500 g, and was incubated in 500 ml or 1000 ml glass Duran media bottles, respectively, at 22°C in the dark. *Streptomyces griseus* and *Fusarium culmorum* were cultured on V8 juice (Campbells, UK) agar at 22°C for 5-10 days and the spores harvested in maximum recovery diluent (MRD; Brown *et al.,* 2002), then adjusted to the required density. Inoculated grain samples were incubated at 22°C in the dark, and odours were assessed at various intervals.

Unconditioned grain of wheat cv. Hereward had a moisture content of 12% and 13% for the non-irradiated and irradiated samples, respectively, and was conditioned as before to give 18% moisture for the non-irradiated grain and 25% for the irradiated grain. The conditioned grain was inoculated using batches of 500 g incubated in the 1000 ml glass media bottles as described above. Spores from V8 juice cultures of *Aspergillus flavus, Aspergillus niger, F. culmorum, Penicillium aurantiogriseum, Penicillium vulpinum, Penicillium verrucosum* and *S. griseus* were harvested in either MRD or sterile distilled water (SDW), depending on which was used for suspending spores to inoculate the grain.

After 36 days following inoculation of non-irradiated grains of cv. Consort with either MRD, *F. culmorum* or *S. griseus,* musty odours were detected. However, these odours were not limited to any one particular grain treatment (Table 20.3). Other odours were present but not detected consistently. In a second experiment with this grain, no strongly musty odours were detected until 27 days after inoculation. The control wheat samples inoculated with MRD developed an obvious musty odour, whilst the *F. culmorum* and *S. griscus* infected grain developed an odour described as "damp" in all but one *F. culmorum* replicate (Table 20.4). Other odours were apparent, but the sensory panel did not agree consistent descriptors, so the descriptions given by various panel members are listed in footnotes to Tables 20.3 and 20.4.

With grains of cv. Hereward, malodours were detected after 12 days in several of the non-irradiated samples (Table 20.5), but were not linked to a specific grain treatment. Where no consensus was reached, odours were detected at very low levels and descriptors were varied. After 36 days, all assessors agreed that malodours were present in all the grain samples tested. A consensus of opinion confirmed a musty component to the odours, but there was no consensus on the descriptors of the other odours present. It was noted that the odours seemed to peak at 22-29 days, and decline in intensity at 36 days. Microscopic examination of non-irradiated grain confirmed the presence of a variety of microorganisms, often including *F. culmorum*, as well as growth of the organism originally introduced as inoculum.

Irradiated grain of cv. Hereward developed musty odours in three out of the four samples tested after 8 days, in all four samples tested after 15 and 22 days and in all eight samples after 29 days following inoculation with *P. aurantiogriseum* at both spore densities (10^4 and 10^6 spores/ml) tested. There was a very strong odour and complete consensus that it was the same for all samples. After 29 days the musty odour was still dominant, but other odours were noted although no consensus on descriptors was reached. The grain in all samples had a blue-green coloured growth visible macroscopically after 8 days, with the embryo becoming colonised preferentially giving the typical "blue eye" symptoms of infection by *P. aurantiogriseum* (Plate 5 c and d, see colour plate section). No significant differences were noted between the grains inoculated with the different spore densities, nor with MRD or with SDW as diluent. Some samples were only tested at 29 days post-inoculation to ensure there was no chance of extraneous microbial contamination of the bulk inoculated grain.

Table 20.3. Odour assessment of non-irradiated grains of wheat cv. Consort inoculated with maximum recovery diluent (MRD) alone, or with MRD containing spores of *Fusarium culmorum* or *Streptomyces griseus* at a density of 6 x 10^4 spores/ml (2.7 x 10^5 spores per 250 g grain), at 18% moisture content and after 36 days incubation at 22°C in the dark (three replicates of each treatment).

Treatment	Odour
1. MRD	Musty
2. MRD	Musty
3. MRD	Other[1]
4. *F. culmorum*	Other[2]
5. *F. culmorum*	Musty/other[3]
6. *F. culmorum*	Musty/other[4]
7. *S. griseus*	Other[4]
8. *S. griseus*	Musty/other[4]
9. *S. griseus*	Musty/damp

[1] Woody; damp grass
[2] Damp; chlorine; stale sock
[3] Stale sock; damp
[4] Damp; slightly malty

The same inoculation technique was then applied to a range of other spoilage organisms, *A. flavus, A. niger, F. culmorum, P. aurantiogriseum, P. vulpinum, P. verrucosum* and *S. griseus*, at two different spore densities. At a spore density of 5 x 10^5 spores/ml, musty odours were apparent 2 days after inoculation for all organisms and one of the SDW controls. The difference between the two control samples may be because one was tested blind, whilst the other was used as the known control to prime the sensory panel. The latter sample also had air passed through it for the duration of the odour assessment for the first three sampling periods, hence a significant amount of the detectable odour may have been flushed out. At the subsequent sampling times the known control was sampled at the beginning and halfway through the sampling. Both control samples developed odours, although visually they appeared uncontaminated. In contrast, the inoculated grain samples showed discoloration characteristic of each organism by day 14.

Table 20.4. Odour assessment of non-irradiated grains of wheat cv. Consort inoculated as in Table 20.3, but at a density of 1 x 10^4 spores/ml (9 x 10^4 spores per 500 g grain) and over a 33-day period. Odours in addition to those described as damp, sour, yeasty, wheaty, musty and green are listed in the footnote.

Treatment	Odour, days post-inoculation				
	5 days	12 days	19 days	27 days	33 days
1. MRD	Sour	Yeasty	Wheaty	Musty [7]	Musty [1]
2. MRD	Sour	Wheaty	Wheaty [2]	Musty [7]	Wheaty/Green [9]
3. MRD	Sour	Wheaty	Wheaty [2]	Musty [7]	Wheaty/Musty/ Green [9]
4. *F. culmorum*	Sour	Wheaty	Wheaty/Musty/ Other [3]	Musty	Musty [10]
5. *F. culmorum*	Sour	Other [1]	Wheaty [6]	Damp [8]	Other [9,10]
6. *F. culmorum*	Sour	Other [1]	Wheaty/Other [4]	Damp [8]	Other [9,10]
7. *S. griseus*	Sour	None	Wheaty [5]	Damp [8]	Wheaty/Musty/ Green [9,10]
8. *S. griseus*	Sour	Wheaty	Wheaty [4]	Damp [7,8]	Musty [9,10]
9. *S. griseus*	Sour	Wheaty	Wheaty [6]	Damp [7]	Other [8,9]

[1] Damp chlorine
[2] Dry chlorine; woody; mushroom
[3] Grassy; mushroom; dry chlorine; wet soil
[4] Chlorine; mushroom soup; amine
[5] Chlorine, mushroom soup
[6] Woody
[7] Stale sock
[8] Slight musty
[9] Damp; chlorine; metallic; dusty
[10] Fragrant; mealy

There was a general trend for the odours to increase in intensity up to day 14, where after some declined in intensity. *P. vulpinum* demonstrated the most notable decrease in odour between days 14 to 28, with little difference in odour from the control by day 28. It was also noted that the odour changed during the assessment event; for example with *S. griseus* the odours detected at day 21 were initially described as earthy/musty, but after a few minutes had changed to fishy.

The description "musty" gives no indication of the variety of odours produced by these organisms. Whilst recognising that odour descriptors are subjective, it was possible to ascribe different characteristic odours to the different organisms following their inoculation onto irradiated wheat. All of the *Penicillium* isolates produced a very characteristic Brazil nut-type odour, with *P. vulpinum* having earthy undertones and *P. verrucosum* having socky/damp undertones. *A. niger* developed a fragrant, earthy and nutty odour, whilst *A.*

flavus had a more damp, socky and savoury odour. *S. griseus* produced the musty/earthy odour characteristic of *Streptomyces* spp., whilst *F. culmorum* developed an odour which was hard to describe but had earthy, sweet and bitter components to its mustiness. One sample each of *A. flavus*, *S. griseus* and *P. verrucosum* developed fishy odours at various times.

All the organisms produced musty odours at the greater inoculum density (10^6 spores ml^{-1}) that were not significantly different (as detected by human nose) from those in previous experiments. This last experiment at the higher spore density was performed with scrupulous care to avoid contamination of the SDW controls. They were assessed first, and always remote from samples inoculated with microorganisms. It is particularly significant that neither of the SDW inoculated samples developed musty odours over the 28-day incubation period.

The Colonisation Process

Light Microscopy and Environmental Scanning Electron Microscopy

Irradiated wheat grains of cv. Hereward were conditioned to 25% moisture content prior to inoculation with a suspension 5 x 10^5 spores/ml of *A. niger*, *F. culmorum* and *P. aurantiogriseum* by either bulk or local methods, before incubation at 22°C in the dark. Bulk inoculation was more representative of conditions in storage silos than local inoculation, but the latter enabled precise delivery of spores to a single point on the grain surface. It was then possible to observe whether sporulation occurred at the point of inoculation, or whether the grain was colonised first by a mycelium that sporulated elsewhere.

For bulk inoculation, grain was inoculated with 4.5 x 10^5 spores per 50 g batch in 500 ml Schott bottles. For local inoculation, individual seeds were inoculated either on the embryo or at the mid-point of the grain with a 10 µl droplet of spore suspension, and incubated on microscope slides over damp filter paper in Petri dishes. For both, the inoculated grain was incubated at 20°C in the dark for 30 days. Progress of infection was followed using light microscopy. A dissecting microscope was used to follow development of surface colonisation. Internal colonisation was examined in grains embedded in wax, then sectioned to give 5 µm thick sections, which were stained with Toluidine blue (1%, pH 5.6) before examination with bright field optics using a Nikon Optiphot compound microscope.

The external surfaces and internal colonisation of wheat grains were also examined using a Philips XL30 environmental scanning electron microscope (ESEM). Hydrated grains were fractured with a razor blade and mounted directly with no pre-treatment on 9 mm diameter self-adhesive carbon discs (Agar Scientific Ltd, Cambridge) attached to stainless steel stubs before being placed on the Peltier stage of the ESEM at 5°C. The grain surface was examined at an accelerating voltage of 150 kV and a vapour pressure of between 6.8 and 7.4 Torr, whilst fracture faces were examined at 20 kV and between 4.7 and 5.7 Torr.

A multi-directional, surface mycelium was established by *A. niger*, followed at 3 days post-inoculation (dpi) by sparse formation of the characteristic black conidiophores at the point of inoculation. At 13 dpi, there was preferential colonisation of the embryo, with 50% of the grains showing sporulation mainly from the embryo following inoculation at either point (Plate 5 a, see colour plate section). These conidiophores were also much larger than those formed elsewhere on the grain, and typically formed a cluster arising from the extreme tip of the grain. Existing sporulation at the embryo was profuse at 30 dpi, and the grains had a slightly shrivelled appearance with clusters of conidiophores also forming elsewhere (Plate 5 b, see colour plate section). When sectioned, the endosperm of some of the infected grains had a yellow hue radiating from the embryo, with a few hyphae visible in the endosperm.

Table 20.5. Odour assessment of non-irradiated grains of wheat cv. Hereward inoculated as in Table 20.4, but including *Penicillium aurantiogriseum.*

Treatment	Odour, days post-inoculation			
	12 days	22 days	29 days	36 days
1. MRD	No consensus	Green	Musty/Green	Musty
2. MRD	Musty	Wheaty/Sour	No consensus	Musty
3. MRD	Wheaty	Wheaty	Not tested	Not tested
4. *F. culmorum*	No consensus	No consensus	Musty	Musty
5. *F. culmorum*	No consensus	Wheaty	No consensus	Musty
6. *F. culmorum*	Green	Not tested	Not tested	Not tested
7. *S. griseus*	Wheaty	Musty	Musty	Musty
8. *S. griseus*	Wheaty/Musty	Wheaty	Musty	Musty
9. *S. griseus*	Wheaty	Not tested	Not tested	Not tested
10. *P. aurantiogriseum*	Wheaty	No consensus	Musty	Musty
11. *P. aurantiogriseum*	Musty	Wheaty	Musty	Musty
12. *P. aurantiogriseum*	Musty	Not tested	Not tested	Not tested

P. aurantiogriseum had formed a mycelium on the grain surface at the point of inoculation by 3 dpi. Characteristic "blue eye" symptoms were observed by 13 dpi due to sporulation from the embryo in 50% of the grains inoculated at the mid-point, and nearly 100% of the grains inoculated at the embryo end (Plate 5 c and d, see colour plate section). At 13 to 21 dpi, further sporulation erupted from other points on the seed coat (Plate 6 a and b, see colour plate section) and by 30 dpi grains had an irregular, bulging appearance. The latter was due to conidiophore formation between the endosperm and the seed coat (Plate 6 c, see colour plate section). This internal sporulation occurred as early as 7 dpi, but was not seen with the other fungi. Hyphae were also visible in the endosperm of sectioned grains, especially beneath points of sporulation. Colonised areas were sometimes visible in fresh grains cut in half by a scalpel blade and viewed with a dissecting microscope (Plate 6 b, see colour plate section). Examination of Toluidine blue stained sections of wax embedded grains confirmed the presence of hyphae beneath the seed coat and within the endosperm in these areas (Plate 8 b and c, see colour plate section).

F. culmorum had formed branching hyphae radiating out from the point of inoculation over the surface of the seed coat at 7 dpi. By 14 dpi, a diffuse web of hyphae had formed between the seed surface and the surfaces of the Petri dish where moisture had accumulated. No spore bearing structures were visible with the dissecting microscope and, whilst the progression of infection was slightly more advanced in seeds inoculated at the embryo end, both points yielded a mycelium with the inoculation point at the centre. By 21 dpi, the seed was no longer visible beneath a dense white, mycelium (Plate 6 d, see colour plate section). The pattern of internal colonisation was not as consistent as for the other two fungi. For example, the endosperm of some grains sectioned at 14 dpi contained hyphae, whilst no internal hyphae were visible in other grains sectioned at 28 dpi.

As with locally inoculated grain, both *A. niger* and *P. aurantiogriseum* introduced by bulk inoculation preferentially colonised the embryo, and its cellular structure was destroyed by 21 dpi to give a collapsed appearance (compare Plate 7 a with b and c, see colour plate section). Sparse hyphal growth was observed radiating from the embryo into the endosperm with both fungi. By 29 dpi with *P. aurantiogriseum*, hyphae growing beneath the seed coat had penetrated through the aleurone layer (compare Plate 8 a and b, see colour plate section) to colonise the endosperm to form pockets of dense mycelium (Plates 6 b and 8 c, see colour plate section). In contrast, whilst *A. niger* hyphae penetrated the aleurone cells, subsequent colonisation of the endosperm was more uniform with fewer hyphae observed. There was no obvious pattern of infection for *F. culmorum*, which gave the same external appearance as with local inoculation. Structural integrity of the embryo was retained for the duration of observations (to 28 dpi), despite extensive colonisation.

ESEM proved to be an excellent technique for imaging the internal structure of hydrated grain that had not been modified by the fixation and dehydration regimes required for conventional scanning electron microscopy. It was also very rapid, with images such as those in Figs 20.4 and 20.5 obtained within 10 min of fracturing hydrated grain. ESEM revealed that starch grains showed an altered appearance in areas colonised by hyphae of all three fungi. Large lenticular grains becamc pitted, giving a "swiss cheese" appearance (Figs 20.4 and 20.5). Compound grains also appeared to disintegrate to give smaller spherical grains. Starch grains were not penetrated by hyphae, even though they were closely associated with the grain surface.

Nuclear Magnetic Resonance Micro-imaging

The moisture distribution in grain inoculated with *A. niger* and *P. auratiogriseum* was examined using nuclear magnetic resonance (NMR) microscopy. Visualisation of structures within the grains was difficult due to their low moisture content and the correspondingly short relaxation time of water protons, which resulted in low intensity NMR signals from the water within the grains. Therefore, images were acquired using a spin-echo experiment with a very short echo time (2.5 ms). Usually, a 128 x 128 image matrix was obtained with a field-of-view of 7 mm and a slice thickness of 1 mm. NMR microscopy was performed with a 4 mm resonator in a 9.4 T vertical magnet. The repetition time was set to 0.8 s and usually 30 averages were acquired, thus resulting in a total experimental time of 52 min.

In the NMR images obtained, areas of relatively higher water content were resolved clearly enough to match structures observed by conventional microscopy. Thus embryos in uninfected control grains appeared relatively large and intact (Plates 7 a and 9 a, see colour plate section), whilst those following infection were collapsed and reduced in overall size (Plates 7 c and 9 b, see colour plate section). Whilst the crease side of uninfected grains in longitudinal scans appeared to have a greater signal intensity than the opposite side (Plate 9 a, see colour plate section), this was shown by transverse scans (Plate 9 e, see colour plate section) to be due to the relatively high intensity of the seed coat in the crease, and the intensity of the endosperm to be uniform throughout. Within infected grains, a zone of endosperm immediately adjacent to the embryo showed a marked reduction in signal intensity (Plate 9 b and c, see colour plate section), corresponding to a reduction in either moisture content or water mobility. This reduced signal intensity was most marked in grains infected by *A. niger* (Plate 9 b, see colour plate section), where the tips of conidiophores were also visible due to their high intensity. In *P. aurantiogriseum* infected

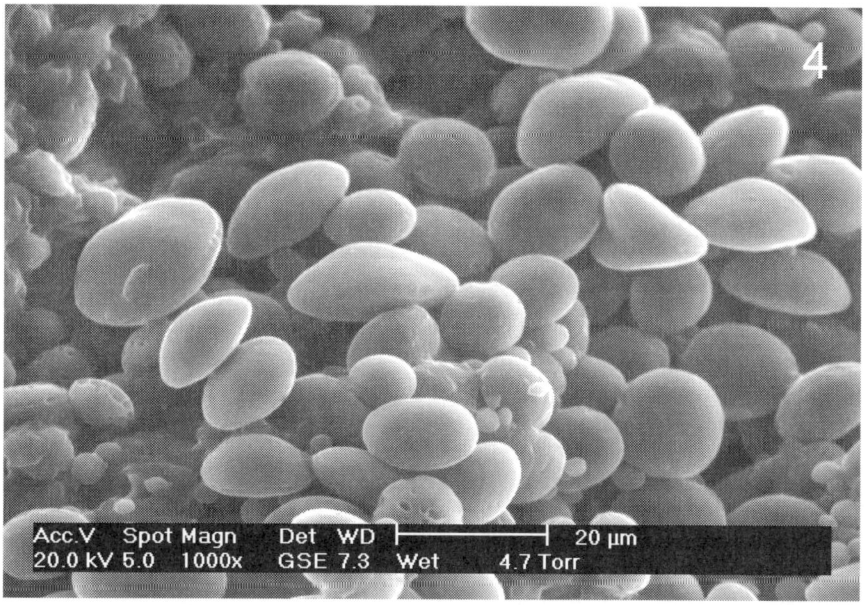

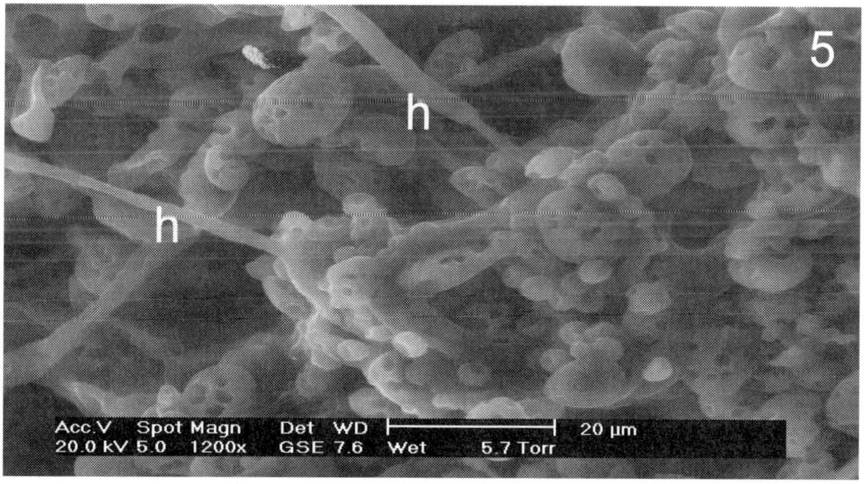

Figs 20.4 and 20.5. Environmental scanning electron microscope images of fractured, hydrated endosperm of grains infected by *A. niger* (27 dpi). Starch grains adjacent to areas of colonisation appear intact (Fig. 20.4), whilst those permeated by hyphae (h in Fig. 20.5) are degraded with a pitted appearance.

grains, the areas of reduced signal were less clear, but zones could be distinguished around all areas of fungal sporulation, such as from the embryo (Plate 9 c and d, see colour plate section) and in the crease and nucellar projection regions (Plate 9 d and f, see colour plate section).

Discussion and Conclusions

None of the surface sterilisation treatments proved effective, which suggests that contaminating microorganisms were protected from the treatments used. The observation that contamination was slower to appear in treated than in control samples supports this notion. It is likely that the crease and embryo regions of the grain are areas affording special protection. Some of the microorganisms may also be endophytic (i.e. growing within the internal tissues), which is why irradiation alone was able to inactivate them.

The predominance of mycelial fungi following surface sterilisation probably was not due to dehydrated fungal spores being resistant to the sterilising chemicals. The pre-treatments which involved wetting the grains with Tween 80 prior to the main treatments (8, 9, 12 and 13 in Table 20.1) are likely to have allowed fungal spores to hydrate and germinate and therefore be more vulnerable to the sterilising agents.

The germination experiments clearly demonstrated that the irradiated grain was severely damaged in its ability to germinate. Although many grains germinated with both shoot and root initiation occurring, seedlings did not develop any further than this stage, with none developing to the same extent as non-irradiated control grains. Therefore irradiation is likely to modify nutrients available to both the natural microbiota and introduced microorganisms, including those causing spoilage. This may influence the predominance of certain microbial species and the progress of infection, and hence odour production. Despite this reservation, irradiation proved to be the only method that allowed routine and reliable infection by specific microorganisms.

There appears to be a correlation between type of odour and grain viability, although this conclusion is based on a relatively small number of commercial samples. Typically, musty odours were associated with non-viable grain. However, it is not clear whether the grains were non-viable as a result of infection, or were more readily infected by spoilage fungi because they were non-viable. Microscopy certainly indicated that *A. niger* and *P. aurantiogriseum* preferentially colonised embryos which became necrotic. The alternative scenario is significant because reduced viability may be a consequence of over-heating grain during the drying process after harvest, and merits investigation.

In this regard, it was interesting to observe that some samples with control (wheaty) odours had greatly reduced viability (Fc/38242/007, 002 and 022 in Table 20.2), which was most probably caused by either heat damage or some other post-harvest effect. A reduction in seed viability may be a significant factor in the establishment of infections leading to musty odours in stored wheat grains. If this is the case, it emphasizes the importance of ensuring that grain remains viable when it is dried following a damp harvest. In these conditions the abundance of spoilage organisms in the carposphere is likely to be enhanced, and the amount of inoculum within grain handling equipment increased as a consequence. Thus damp conditions plus increased inoculum and the greater susceptibility of heat killed grain to infection may explain the prevalence of major spoilage problems associated with harvests in wet summers in the UK.

Fishy, sour and green odours were associated with grain with a greater viability than musty odours. Irradiated grain with reduced viability therefore may not be an appropriate substrate for reproducing these odours in the laboratory. The presence of *T. tritici* spores on the fishy samples provides conclusive evidence that in these samples the odour was not a consequence of fungal infection whilst in storage.

With non-irradiated grain inoculated in the laboratory, there seemed to be no treatment-specific effects: raising grain moisture content to 18% and adding MRD either alone or with spores of the organisms generally gave off-odours. This may indicate that at an inoculum density of 10^4 spores/ml, the introduced organism does not consistently become established and predominate. Off-odours are therefore a cocktail resulting from the activities of the natural microbiota, with the odours developing not always consistent between different batches of the same sample of grain.

Microscopy showed that *A. niger* rapidly produced conidiophores which emanated preferentially from the embryo. Following dense colonisation of the embryo, hyphae penetrated aleurone cells to give uniform, but sparse colonisation of the endosperm. Therefore, it appears that *A. niger* hyphae initially colonise the embryo, and only subsequently spread to the endosperm. The grain appeared to become desiccated as infection progressed, with NMR micro-imaging indicating that fungal infection reduced moisture content in the endosperm adjacent to the embryo. Similar results were obtained with *P. aurantiogriseum*. This localised effect may restrict further colonisation of the endosperm as water availability is a limiting factor for hyphal growth in spoilage fungi (e.g. *P. verrucosum*; Pardo *et al.*, 2006). Water stress may also result in altered production of secondary metabolites such as VOCs and mycotoxins. Results of the experiments reported here indicate that the characteristic musty odours peak at the onset of abundant sporulation from the embryo, but then decline as water availability reduces. This notion is supported by GC-MS data on the key VOCs emitted (Brown *et al.*, 2002).

P. aurantiogriseum appeared to infect via the embryo and defects in the seed coat to produce individual pockets of mycelia, culminating in heavy sporulation erupting through the surface. The embryo of most seeds was colonised and resulted in profuse sporulation, which frequently also occurred between the endosperm and seed coat. Sporulation also typically occurred in the base of the crease, adjacent to the nucellar projection. In contrast, whilst *F. culmorum* produced a dense surface mycelium, the pattern of internal colonisation was inconsistent and embryos retained the same structure as in uninfected grains.

The effects on the starch grains caused by all fungi presumably were due to the release of extracellular enzymes, for example the glucoamylase secreted by hyphal tips of *A. niger* (Vinck *et al.*, 2005), and illustrate how ESEM can provide information on the extent of degradation rapidly and without artefacts caused by the preparation techniques of conventional SEM. An alternative explanation is that the altered appearance of starch grains was due to dehydration, but this is unlikely as NMR microscopy showed the dehydrated zone to extend beyond areas colonised by hyphae, into areas which ESEM showed the grains to have a normal appearance. The difference in the effects on structural integrity of the embryos suggests that *F. culmorum* produces different enzymes and utilises different substrates to the other two species.

NMR micro-imaging has been used previously to investigate the development of rice caryopses (Horigane *et al.,* 2001) and to study water movement within a single wheat grain (Jenner *et al.*, 1988). The present study shows for the first time that this technique can be applied to monitor the infection of wheat grain by fungi. A major advantage for such studies is the non-invasive and non-destructive nature of NMR microscopy. This will make it possible to follow colonisation processes in real-time for individual cereal grains. The preliminary work presented here illustrates the potential of this novel approach for elucidating the relationship between localised changes in water availability and fungal spoilage. Refinement of the technique is likely to enable mapping of moisture gradients within grains, and relate this to progression of infection and the production of malodours and mycotoxins.

Acknowledgements

We are grateful to the British Mycological Society for the award of a vacation bursary to AW, the Home-Grown Cereals Authority for funding, collaborators at CCFRA, and to Dr Dave Patton for assistance with the ESEM. WK is grateful to the Royal Society for support provided during the time these experiments were performed.

References

Brown, H.M., Ratcliffe, N., Voysey, P.A., Williams, J.S., Spencer-Phillips, P., de Lacy Costello, B.P.J., Beales, N., Salmon, S., Gunson, H., Sivanand, P.S. and Ewen, R.J. (2002) Development and testing of a sensor to detect microbiological spoilage in grain. Project Report No. 295, Home-Grown Cereals Authority, London, UK.

Cairns-Fuller, V., Aldred, D. and Magan, N. (2005) Water, temperature and gas composition interactions affect growth and ochratoxin A production by isolates of *Penicillium verrucosum* on wheat grain. *Journal of Applied Microbiology* 99, 1215-1221.

de Lacy Costello, B.P.J., Ewen, R.J., Gunson, H.E., Ratcliffe, N.M. and Spencer-Phillips, P.T.N. (2000) The development of a sensor system for the early detection of soft rot in stored potato tubers. *Measurement Science and Technology* 11, 1685-1691.

de Lacy Costello, B.P.J., Ewen, R.J., Gunson, H.E., Ratcliffe, N.M., Sivanand, P.S. and Spencer-Phillips, P.T.N. (2003) A prototype sensor system for the early detection of microbially linked spoilage in stored wheat grain. *Measurement Science and Technology* 14, 397-409.

Egal, S., Hounsa, A., Gong, Y.Y., Turner, P.C., Wild, C.P., Hall, A.J., Hell, K. and Cardwell, K.F. (2005) Dietary exposure to aflatoxin from maize and groundnut in young children from Benin and Togo, West Africa. *International Journal of Food Microbiology* 104, 215-224.

Flannigan, B. (1970) Comparison of seed-borne mycofloras of barley, oats and wheat. *Transactions of the British Mycological Society* 55, 267-276.

Flannigan, B. (1978) Primary contamination of barley and wheat grain by storage fungi. *Transactions of the British Mycological Society* 71, 37-42.

Gallagher, E.J. (1984) *Cereal Production*. Butterworth & Co., London, UK.

Harris, B. (1997) Minimizing mycotoxin problems. *Feed Management* 48, 27-28.

Hook, S.C.W. (2004) A national grain sampling and analysis system for improved food marketing and safety. Project Report No. 349, Home-Grown Cereals Authority, London, UK.

Horigane A.K., Egelaar H.G., Maruyama S., Yoshida M., Okubo A. and Nagata T. (2001) Visualisation of moisture distribution during development of rice caryopses (*Oryza sativa* L.) by nuclear magnetic resonance microimaging. *Journal of Cereal Science* 33, 105-114.

Hoskin, S. (1999) Fungi as animal pathogens. In: Oliver, R. and Schweizer, M. (eds) *Molecular Fungal Biology*. Cambridge University Press, Cambridge, UK. pp. 322-340.

Hussein, H.S. and Brasel, J.M. (2001) Toxicity, metabolism, and impact of mycotoxins on humans and animals. *Toxicology* 167, 101-134.

Jenner C.F., Xia Y., Eccles C.D. and Callaghan P.T. (1988) Circulation of water within wheat grain revealed by nuclear magnetic resonance micro-imaging. *Nature* 336, 399-402.

Köckenberger W. (2001a) Nuclear magnetic resonance micro-imaging in the investigation of plant cell metabolism. *Journal of Experimental Botany* 52, 641-652.

Köckenberger W. (2001b) Functional imaging of plants by magnetic resonance experiments. *Trends in Plant Science* 6, 286-291.

Lacey, J. and Magan, N. (1991) Fungi in cereal grains: their occurrence and water and temperature relationships. In: Chelkowski, J. (ed.) *Cereal Grain – Mycotoxins, Fungi and Quality in Drying and Storage*. Elsevier, Amsterdam, The Netherlands. pp. 77-118.

Magan, N. and Evans, P. (2000) Volatiles as an indicator of fungal activity and differentiation between species, and the potential use of electronic nose technology for early detection of grain spoilage. *Journal of Stored Products Research* 36, 319-340.

Neergaard, P. (1998) *Tilletia caries* (DC.) Tul. and *Tilletia foetida* (Wallr.) Liro. In: Smith, I.M., Dunez, J., Lelliot, R.A., Phillips, D.H. and Archer, S.A. (eds) *European Handbook of Plant Diseases*. Blackwell Scientific Publications, Oxford, UK. pp. 465-466.

Palmgren, M.S. and Lee, L.S. (1986) Separation of mycotoxin-containing sources in grain dust and determination of their mycotoxin potential. *Environmental Health Perspectives* 66, 105-108.

Pardo, E., Malet, M., Marín, S., Sanchis, V. and Ramos, A.J. (2006) Effects of water activity and temperature on germination and growth profiles of ochratoxigenic *Penicillium verrucosum* isolates on barley meal extract agar. *International Journal of Food Microbiology*, 106, 25-31.

Pitt, J.I. and Hocking, A.D. (1997) *Fungi and Food Spoilage*, 2nd edn. Blackie Academic and Professional, London, UK.

Scudamore, K.A. (1994) *Aspergillus* toxins in food and animal feedstuffs. In: Powell, K.A., Renwick, A. and Peberdy, J.F. (eds) *The Genus* Aspergillus: *From Taxonomy and Genetics to Industrial Application*. Plenum Press, New York, USA. pp. 59-71.

Selim, M.I., Juchems, A.M. and Popendorf, W. (1998) Assessing airborne aflatoxin B_1 during on-farm grain handling activities. *American Industrial Hygiene Association Journal* 59, 252-256.

Skaug, M.A., Eduard, W. and Størmer, F.C. (2001) Ochratoxin A in airborne dust and fungal conidia. *Mycopathologia* 151, 93-98.

Vinck, A., Terlou, M., Pestman, W.R., Martens, E.P., Ram, A.F., van den Hondel, C.A.M.J.J. and Wösten, H.A.B. (2005) Hyphal differentiation in the exploring mycelium of *Aspergillus niger*. *Molecular Microbiology* 58, 693-699.

Chapter 21

Atmospheric Composition and the Phyllosphere: the Role of Foliar Surfaces in Regulating Biogeochemical Cycles

D. Fowler[1], J.N. Cape,[1] J. Muller[1], M. Coyle[1], M.A. Sutton[1], R. Smith[1] and C. Jeffree[2]
[1]Centre for Ecology and Hydrology, Bush Estate, Penicuik, Midlothian, EH26 0QB, UK; [2]Institute of Molecular Plant Sciences, University of Edinburgh, King's Buildings, Mayfield Road, Edinburgh, EH9 3JH, UK
dfo@ceh.ac.uk

The focus of scientific interest in plant-atmosphere exchange of reactive gases has been in stomatal uptake and the effects of the absorbed gas on physiological processes. Current work on the effects of ozone on vegetation are focused on the link between stomatal uptake and yield loss. However, for the main reactive gases present in the atmosphere, including O_3, SO_2, HNO_3 and HCl, deposition of the trace gas is primarily to the external surfaces of vegetation, which provide the primary control over rates of deposition. The list can be extended for semi-natural vegetation which is generally a sink rather than a source to NH_3, avoiding the vegetation which commonly exhibit compensation points in excess of ambient concentrations.

For O_3, long-term measurements show that external surfaces are the sink for two-thirds of the deposited ozone. The processes of destruction of ozone at the surface are poorly known, but appear consistent with thermal decomposition at the surface, with activation energies of typically 35 kJmole^{-1}. In the presence of surface water on vegetation, rates of O_3 deposition are smaller, consistent with this sparingly soluble gas, but again there is a clear dependence with surface temperature. Furthermore, growing evidence in the literature implies an additional solar radiation linked process accelerating the destruction of ozone at the surface, through photochemical processes.

For SO_2, the leaf surface has been shown to be the site of control over the deposition process, that the surface chemistry is the prime regulator and that surface water plays a crucial role, providing the medium within which most of the chemistry takes place. The surface water pH is primarily regulated in rural areas by the relative concentrations of SO_2 and NH_3. The consequence of this mechanism has been a gradual increase in the deposition velocity for SO_2 over Europe during the last 20 years.

For HNO_3, the surface is considered a perfect sink and deposition rates are regulated by the aerodynamic roughness of vegetation. However, few data have been obtained to test the hypothesis and even small surface resistances would make a substantial difference to the rates of deposition, either way surface processes are at the heart of the control mechanism.

For small particles, short vegetation has been shown to be a modest sink for aerosols, but aerodynamically rough surfaces, including forests are efficient sinks for small particles, yet the underlying control process remains obscure.

Processes on foliar surfaces therefore regulate surface-atmosphere exchange of most of the reactive atmospheric trace gases. However, interactions with leaf microbiota remain largely unknown and represent an important research focus.

Introduction

Vegetation plays a central role in regulating the fluxes of water and energy between the atmosphere and terrestrial surfaces. The momentum transfer due to frictional drag at the surface is influenced strongly by the aerodynamic roughness of the vegetation and the partitioning of net radiation at the surface between sensible and latent heat is strongly regulated by vegetation through the control of stomatal conductance (Monteith and Unsworth, 1990; Jarvis *et al*., 1997). Vegetation also regulates the uptake and assimilation of CO_2 by terrestrial surfaces (Monteith and Unsworth, 1990; Jarvis *et al*., 1997). Over regional scales, the recycling of water through evapotranspiration helps maintain entire ecosystems, as in the case of tropical forest in the Amazon Basin (Shuttleworth *et al*., 1991). While the role of vegetation in these important processes is clear, the role of the phylloplane is less central. Certainly foliar surfaces intercept incoming short wave radiation, but the control over leaf temperatures is regulated by latent heat flux and hence by stomatal aperture (Monteith and Unsworth, 1990).

Phylloplane surfaces play almost no role in regulating the net exchange of carbon between the surface and the atmosphere. Thus for the carbon cycle there has been relatively little research interest in the phylloplane. Similarly for other radiatively active gases with long residence times in the atmosphere, including nitrous oxide (N_2O) and methane (CH_4), foliar surfaces are of little importance in regulating either the sources or sinks of these gases (Graedel and Crutzen, 1992). However, in the case of reactive gases, including ozone (O_3), sulphur dioxide (SO_2), nitric acid (HNO_3) and ammonia (NH_3), reactions on external surfaces of vegetation are important in regulating or modifying the net surface-atmosphere exchange. The chemical reactivity of these gases reduces their atmospheric lifetimes relative to radiatively active gases from decade to century timescales to hours or days. The reactions to remove these gases from the atmosphere occur both in the free atmosphere, through gas phase (homogeneous) processes and on aerosols, cloud and rain droplets as well as with terrestrial surfaces. Thus processes at the surface compete with those within the atmosphere in regulating their lifetime and hence travel distance. The relationship between atmospheric lifetime and travel distance for a wide range of gaseous and particulate species is illustrated in Fig. 21.1, showing that for the gases of interest in this chapter, the lifetimes range from a few hours to a few days and the corresponding travel distances range from a few km to 1000 km.

The exchange of CO_2 with vegetation is focused on stomatal exchange, as external foliar surfaces have little involvement in the process. In the case of the reactive gases, both stomata and external foliar surfaces are involved significantly in the exchange, yet the range of processes involved is large. The objective of this chapter is to outline the importance of surfaces of vegetation and the phylloplane in regulating the exchange of a range of different trace gases in the atmosphere, differing in reactivity. The gases chosen for consideration are all prominent components of current environmental problems. Aerosols are also considered briefly as most of the trace gases are transformed within the atmosphere to aerosols.

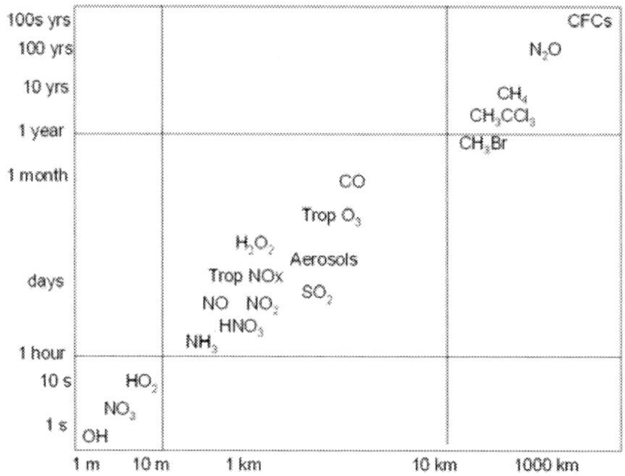

Fig. 21.1 Approximate atmospheric lifetimes and travel distances for a range of trace gases and aerosols.

There are three very different spatial scales relevant to the exchange of trace gases at terrestrial surfaces, first the micro-scale, at which the chemical and biological interactions occur (Plate 10, see colour plate section). Second is the spatial scale at which most of the current measurement and interpretation takes place, which is the field scale (10^3 to 10^5 m^2) for measurements using micrometeorological methods. Lastly, the application of current understanding of the surface exchange process is primarily at regional to continental scales to characterise the fluxes and budgets within chemical transport models (CTM), so that net fluxes averaged over large areas of the landscape are provided.

It is now appropriate to use a series of examples of the role of the phyllosphere in regulating the surface-atmosphere exchange of trace gases.

Ozone

Concern over ozone in the atmosphere centres around two issues which differ in the underlying cause, the part of the atmosphere concerned and the current trend. Stratospheric ozone concentrations have declined during the last three decades as a consequence of depletion through catalytic destruction of ozone on polar stratospheric cloud by breakdown products of halocarbon compounds (Wayne, 1990). Removal of ozone at terrestrial surfaces plays only a minor role in the stratospheric ozone budget and will not be considered further here. The other ozone problem is too much rather than too little ozone in the troposphere, and more especially at the surface, where average concentrations have doubled over substantial areas of the northern mid latitudes, and especially in the industrialised countries since the start of the 20th century (Volz and Kley, 1988). In Europe for example, summer daytime concentrations frequently exceed thresholds for effects on human health and sensitive vegetation, despite substantial reductions in emissions of the ozone precursor gases in Europe (Fuhrer *et al.*, 1997). The removal of ozone from the atmosphere occurs both through chemical processes in the atmosphere and through deposition to terrestrial surfaces. For boundary layer ozone, removal at the surface by dry deposition is the major removal mechanism over land. Typical deposition velocities for ozone of 10 mm s^{-1} would limit the lifetime of ozone in the boundary layer to about a day. The deposition to vegetation comprises two different sinks, stomatal and leaf surface uptake, illustrated in Plate 11 (see colour plate section).

Measurements of the deposition flux may be made with an uncertainty of typically 10% to 30% averaged over 30 to 60 minutes at a field site, providing a horizontal uniform fetch of 200 m to 500 m. The overall transfer resistance between atmosphere and surface may be estimated directly from the measured flux and concentration using methods described by Duyzer and Fowler (1994), and the atmospheric component of the deposition pathway may be estimated from the measured turbulence and removed from the total resistance. Any residual from this process represents a measure of the affinity of the surfaces for ozone uptake. If the surface absorbs ozone molecules at the rate they are delivered to the surface by the atmospheric transfer processes, then the residual is zero and the surface is termed a 'perfect sink'. Few gases are absorbed at such large rates, and ozone invariably exhibits a residual or surface resistance. The uptake at the surface comprises two terms, stomatal and external or cuticular uptake, illustrated in Plate 11 (see colour plate section) and described in more detail by Fowler *et al.* (2001). The relative magnitudes of the stomatal and cuticular sinks in vegetation are similar during the day, but at night, as stomata close, the importance of cuticular uptake increases. Over the year cuticular uptake is the dominant deposition process. Thus the average deposition rate of ozone is strongly dependent on external processes on foliar surfaces. While recent publications illustrate the importance of leaf surface processes in regulating ozone deposition, the underlying chemical, physical and biological mechanisms involved are unknown. The most revealing analysis shows a dependence of the cuticular resistance on irradiance of the surface and surface temperature, which are of course related. Some observers have suggested that the dependence on irradiance provides evidence of a photochemical surface process which destroys ozone. However, the reduction in cuticular resistance with increasing irradiance is also consistent with the thermal destruction of ozone on surfaces of foliage, which increases with surface temperature. In the case of thermal destruction of ozone at the surface, the effect can be characterised quantitatively through an Arrhenius analysis, in which the activation energy of the reaction is given by the relationship between the deposition rate and 1/RT (Fowler *et al.*, 2001) as shown in Fig. 21.2.

While such an analysis does not reveal the underlying molecular process, the resulting activation energy (of 36 kJ $mole^{-1}$) for the analysis is consistent with thermal destruction on the external surfaces of vegetation. Regardless of the underlying mechanism, the reduction in surface resistance with increasing surface temperature has some important consequences. First, the surface removal would increase with increasing global temperatures and therefore provide negative feedback into the ozone budget in the lower troposphere, in which surface removal is an important controlling process. Second, the partitioning into stomatal and non-stomatal fluxes would be expected to vary with latitude, being most important in the dry tropics.

The ozone absorbed by plants through stomata is clearly the important component for effects on photosynthesis (Fuhrer *et al.*, 1997). However, when integrated over annual time scales, the non-stomatal flux to terrestrial surfaces exceeds stomatal uptake in the few data series which extend over sufficiently long periods to estimate the overall annual deposition of ozone and its partitioning at the surface. Thus it is necessary to understand ozone uptake by stomata to quantify the effects on crop productivity, and the total flux and hence non-stomatal uptake to quantify the global distribution of ozone concentration at the surface and its change with time, especially as climate changes.

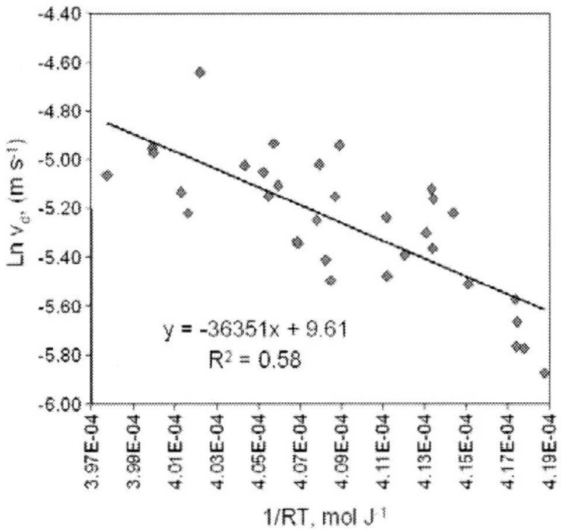

Fig. 21.2. The relationship between deposition velocity (plotted as Ln v_d) and 1/RT, showing the temperature dependence of ozone deposition process and the activation energy of the surface reaction.

Reactive Nitrogen Gases

HONO

Concentrations of HONO in the atmosphere are small, and generally in the ppt (parts in 10^{12} by volume) range, especially during the day, when it is readily photolysed to OH and NO (Plate 12, see colour plate section). This gas is important despite the small concentrations present as it is a significant source of the very powerful atmospheric oxidant OH. Substantially larger concentrations have been reported widely in both urban and rural atmospheres at night (Logan *et al.*, 1981), and the processes forming these large concentrations of HONO are poorly known. It has been suggested that surface chemistry on terrestrial surfaces and on aerosols is responsible for some of the observed HONO (Harrison *et al.*, 1994). The field observations are consistent with a surface source of HONO and reactions on the surfaces of vegetation, being the dominant surface, in the area at the site of the measurements. Thus processes at the phylloplane appear central to HONO production, yet the processes remain obscure.

HNO_3

It is assumed widely that the deposition of nitric acid to vegetation and soil is regulated entirely by the turbulent transfer resistance through the boundary layer, or that the surface may be considered a 'perfect sink' in relation to nitric acid, in which the effective concentration at the surface is zero. The limited field data by micrometeorological methods show that the atmospheric resistances (r_a and r_b) are the dominant terms (Hubert *et al.*, 1988), however considerable uncertainty in the measurements is present and a small non-zero surface resistance would be very important in constraining the deposition flux, especially over aerodynamically rough surfaces. As in the case of HONO and O_3, it is the processes at the surface of vegetation that pose the greatest uncertainty in current understanding, and to make progress it is necessary to identify the chemical reactions at the surfaces of vegetation involving HNO_3 and their rate coefficients.

NH_3 and SO_2

Ammonia, like nitric acid is deposited rapidly onto the cuticular surfaces of vegetation and, as a very soluble species, it is taken up rapidly by water on external surfaces of plants (Sutton *et al.*, 1995). However, the external surfaces of plants present a sink for NH_3 from both atmospheric and surface sources, because NH_3 is present in plant tissues, and specifically in the apoplast, which is directly connected to the atmosphere through stomatal apertures as illustrated schematically in Plate 13 (see colour plate section).

Thus vegetation may be a sink for atmospheric NH_3 when the net flux is onto foliar surfaces. Vegetation may also be a source of NH_3 when the equilibrium NH_3 is larger than ambient concentrations, i.e. below the compensation point. The apoplast NH_4^+ is generally larger for agricultural crops due to the fertiliser nitrogen provided, and smaller for semi-natural vegetation. Thus in the countryside, crops are the source areas while the semi-natural vegetation are the sinks in the landscape. The spatial patterns of emission and deposition are therefore complex and vary over distances of a few 10/s of metres, from sinks to sources. The spatial patterns are further complicated by the presence of livestock units, which are substantial sources of NH_3 in the landscape. To illustrate the link between nitrogen supply to crops and the net exchange of NH_3, data from a field measurement programme are shown in Fig. 21.3, in which the net exchange of NH_3 above grassland in the Scottish lowlands was measured by micrometeorological methods during the summer of 1998.

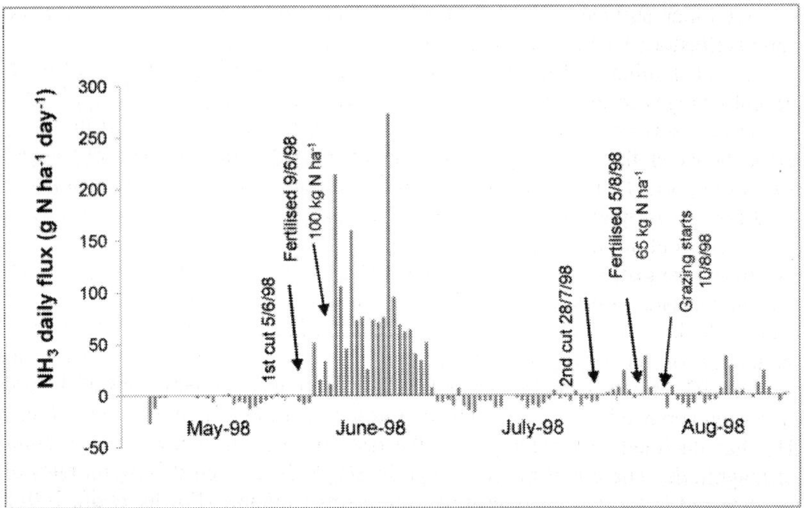

Fig. 21.3. Ammonia exchange, mean daily NH_3 flux as g N $ha^{-1}day^{-1}$ (from Milford *et al.*, 2001).

The measurements show small deposition (-ve) fluxes during May, with values in the range -5 to -10 g N $ha^{-1}day^{-1}$ followed by a large emission, reaching values of +270 g N $ha^{-1}day^{-1}$. Following a month of significant emissions, the fluxes are again dominated by deposition to the grassland as the foliar N concentration declined, to values small enough to be below the compensation point for the grassland with respect to ambient NH_3 concentrations at the site. The small, mainly deposition fluxes are maintained for the remainder of the summer, reflecting the generally cool and wet conditions, and the fact that the vegetation was not fertilised again during the monitoring period. These measurements show that even grassland, supplied with fertiliser, is for much of the time a sink for NH_3. The characteristics of the external surfaces of vegetation which regulate the deposition of NH_3 are known in only a very general sense. The deposition rates to external surfaces increase with surface wetness. For UK conditions, the surface is effectively wet most of the time due to the presence of deliquescent salts on foliar surfaces (Burkhardt and Eiden, 1994), which are associated with liquid water at humidities in excess of approximately 70%. Only during the warmest parts of dry days do humidities at the leaf surface fall below this value in moist temperate regions such as northern Europe. The presence of a liquid film on vegetation also influences the exchange of other soluble gases such as SO_2.

The reason that SO_2 is also considered in this section is that the exchange of this gas is influenced strongly by the pH of surface water on vegetation, which in turn is influenced by NH_3. Early literature on the deposition of SO_2 to vegetation concentrated on stomatal uptake, and suggested that this was a major deposition pathway (Garland, 1977; Fowler and Unsworth, 1979). However, it was shown that the majority of SO_2 is deposited to the external surfaces of vegetation as is the case for O_3. The deposited SO_2 following uptake into surface water, is oxidised, first to SO_3 and then to SO_4, an oxidation pathway which releases acidity (H^+) into the liquid film. The acidification of the surface film inhibits SO_2 uptake, reducing the deposition rate. However, the presence of NH_4^+ in the liquid film maintains a high solution pH, and maintains a rapid uptake of SO_2. The chemistry of the very small amount of water present on foliar surface is therefore the regulator of deposition rates for some of the most important reactive species. A simplified treatment of the solution chemistry on foliar surfaces was developed by Flechard *et al.* (1999) as a mechanistic model, and is illustrated schematically in Fig. 21.4.

The model is initialised from measured rainfall composition and is able to simulate the temporal changes in the surface-atmosphere exchange of SO_2 and NH_3 with changing atmospheric and surface conditions and exchange through the cuticular membrane. The interactions between these trace gases in regulating surface-atmosphere exchange rates represents an important advance in understanding. In particular it suggested that deposition rates of these gases may change with time entirely as a consequence of changes in atmospheric concentrations of another trace gas. In the case of SO_2 and NH_3, relative emissions have changed considerably over the last 20 years. Emissions of SO_2 were at their maximum in the late 1960s in most of Europe, and have declined by about an order of magnitude in the last 20 years. By contrast, emissions of NH_3 have changed relatively little, with reductions of the order of 10% in the UK. The only country to make marked reductions in NH_3 emissions is the Netherlands, where emission reductions of the order of 40% have been achieved. Thus over most of Europe the ratio of ambient concentrations of SO_2:NH_3 has declined substantially over the last 20 years, in many areas by almost an order of magnitude. The effect of the change in SO_2:NH_3 ratio on deposition rates of SO_2 has been detected in the few long term measurements available (Fowler *et al.*, 2001). The surface resistance for SO_2 deposition declined from 120 s m^{-1} in 1996 to 50 s m^{-1} in 2003, resulting in the increase in deposition rate by a factor of two over the 7 years. The consequence is therefore substantial and has had the effect of reducing the atmospheric lifetime and travel distance of SO_2. In long-range transport model runs, when the effect of SO_2:NH_3 changes on deposition rates of SO_2 were ignored, the model-measurement comparison showed a steady decline, with the model indicating much larger concentrations than those observed. The model-measurement comparison was brought into satisfactory agreement by increasing the deposition rate for SO_2 as indicated below.

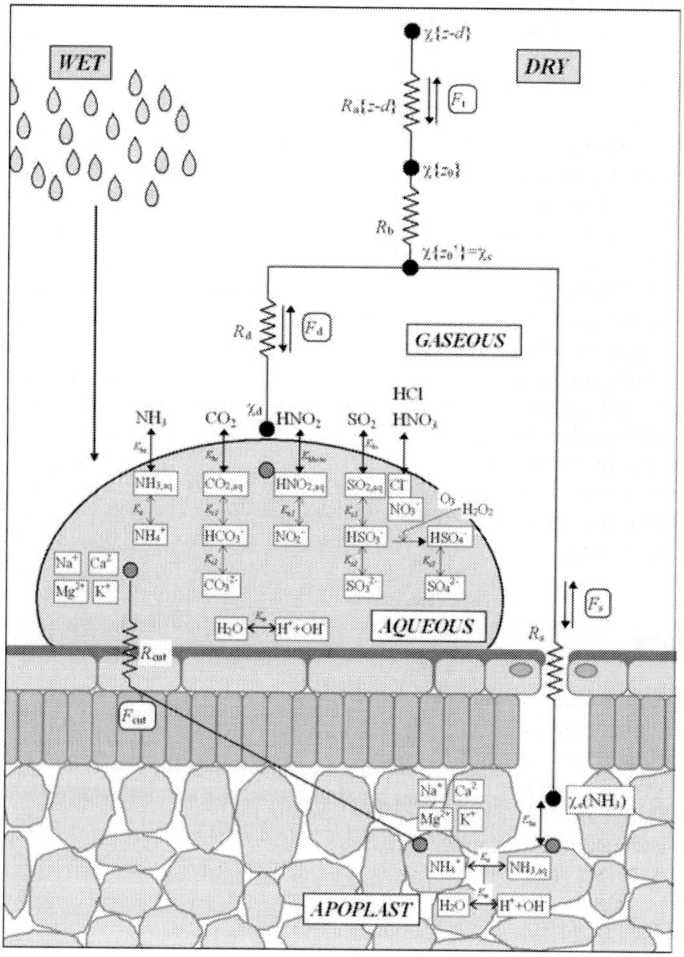

Fig. 21.4. The surface chemistry of SO_2 and NH_3 interactions in moist films on vegetation (from Flechard *et al.*, 1999).

Particles

Atmospheric particles are at the heart of some of the most intractable atmospheric issues including air pollution and human health, and the radiative forcing of climate. In the case of air pollution and human health, atmospheric concentrations of particles with less than 10 μm diameter (PM_{10}) have been shown to be responsible for premature mortality. However, to date the specific composition responsible for the effect remains unknown. The role of particles in the issues underlying climate change are complex, but include both direct effects of the particles on the radiative budget of the earth, through increasing the albedo, and indirect through changing the optical properties of clouds (IPCC, 2001).

The deposition of particles to terrestrial surfaces varies strongly with particle size, and for the bulk of the sub-micron particles, the deposition to vegetation is very inefficient. These particles are mainly removed from the atmosphere by precipitation. However, for aerodynamically rough vegetation, such as woodland, hedgerows and other tall vegetation, deposition rates are somewhat larger. The effect of trees on deposition rates of particles is now clear (Gallagher *et al.*, 2002), and is illustrated in Plate 14 (see colour plate section), in which the peak in deposition rates on woodland for 0.5 μm particles is about an order of magnitude larger than values simulated by the most widely accepted theoretical treatment (Slinn, 1982). The large deposition rates of particles on woodland is a potentially valuable property which could be exploited to remove particulate material from urban atmospheres (Fowler *et al.*, 2004). However, the scientific problem is that the physical mechanism underlying the cause of large particle deposition rates for woodland remains unknown.

The complexity of leaf surfaces, and especially the structure of epicuticular waxes, present very large areas for the capture of particles, but similar complexity is present in both short and tall vegetation. What is required is a mechanism for transporting particles across the leaf boundary layer, in which turbulence is suppressed. One possibility for tall vegetation is an electrophoretic effect, in which charged particles are transported across the leaf boundary layer by differences in the potential of the leaf surface and the particle. This, at present remains only a possible mechanism, and needs to be challenged by field and laboratory experiment.

Conclusions

The examples of trace gases and aerosols considered in this chapter are all major atmospheric contaminants, and all show processes at the phylloplane to be those which regulate the rates of surface-atmosphere exchange. While the underlying scientific understanding has progressed quite rapidly, the direction of progress has been towards provision of large area average rates of exchange. This progress has not been matched by progress in understanding the processes at the surface, on the phylloplane. An exception is the chemistry of SO_2 and NH_3 in moisture on leaf surfaces. But for the other examples, including O_3, HNO_3, HONO, NO_2 and particles, the physical and chemical processes operating at the leaf surface are very poorly understood.

The examples chosen here are primarily from anthropogenic sources of gases and particles. The list of gases and particles of natural origin exchanged at foliar surfaces is extensive. Volatile organic gases emitted by plants and responsible for O_3 production through photochemical mechanisms in the atmosphere in the presence of NO and NO_2 are important for the atmospheric oxidant cycle. The production, release and deposition of spores, pollen, bacteria and virus particles provided the early understanding of particle deposition and remain an important research field (Deacon, 1997).

The improvements in understanding have followed a need for estimates of sources and fate of compounds of interest and, importantly, new measurement techniques. The development of reliable flux measurement techniques for micrometeorological methods has advanced our understanding of long-term CO_2 exchange for a wide range of ecosystems.

To unpick the chemistry of interactions on foliar surfaces will require a combination of controlled environment and field measurement studies, at concentrations in the range of ambient values. It will also require methods capable of observing the chemistry and physics involved at the fine scale of the architecture of leaf wax structure.

Acknowledgements

The authors gratefully acknowledge the support of the Department for Environment, Food and Rural Affairs (Defra) for support of the measurements of SO_2 and NH_3 exchange under the contract Acid Deposition Processes RMP2258.

References

Burkhadt, J. and Eiden, R. (1994) Thin water films on coniferous needles. *Atmospheric Environment* 28, 2001-2011.

Deacon, J.W. (1997) *Modern Mycology* 3rd edition. Blackwell Science, Oxford, UK.

Duyzer, J. and Fowler, D. (1994) Modelling land atmosphere exchange of gaseous oxides of nitrogen in Europe. *Tellus* 46 B, 353-372.

Flechard, C.R., Fowler, D., Sutton, M.A. and Cape, J.N. (1999) A dynamic chemical model of bi-directional ammonia exchange between semi-natural vegetation and the atmosphere. *Quarterly Journal of the Royal Meteorological Society* 125, 1-33.

Fowler, D. and Unsworth, M.H. (1979) Turbulent transfer of sulphur dioxide to a wheat crop. *Quarterly Journal of the Royal Meteorological Society* 105, 767-783.

Fowler, D., Flechard, C., Cape, J.N., Storeton-West, R.L. and Coyle, M. (2001) Measurements of ozone deposition to vegetation quantifying the flux, the stomatal and non-stomatal components. *Water, Air and Soil Pollution* 130, 63-74.

Fowler, D., Skiba, U., Nemitz, E., Choubedar, F., Branford, D., Donovan, R. and Rowland, P. (2004) Measuring aerosol and heavy metal deposition on urban woodland and grass using inventories of ^{210}Pb and metal concentrations in soil. *Water, Air and Soil Pollution – Focus* 4, 483-499.

Fuhrer, J., Skarby, L. and Ashmore, M.R. (1997) Critical levels for ozone effects on vegetation in Europe. *Environmental Pollution* 97, 91-106.

Gallagher, M.W., Nemitz, E., Dorsey, J.R., Fowler, D., Sutton, M.A., Flynn, M. and Duyzer, J.H. (2002) Measurements and parameterizations of small aerosol deposition velocities to grassland, arable crops, and forest: influence of surface roughness length on deposition. *Journal of Geophysical Research* 107, no. 4154.

Garland, J.A. (1977) Dry deposition of sulphur dioxide to land and water surfaces. *Proceedings of the Royal Society of London Proceedings Series A* 354, 245-268.

Greadel, T.E. and Crutzen, P.J. (1992) *Atmospheric Change: An Earth System Perspective*. W.H. Freeman and Company, New York, USA.

Harrison, R.M. and Kitto, A.M.N. (1994) Evidence for a surface source of atmospheric nitrous-acid. *Atmospheric Environment* 28, 1089-1094.

Huebert, B.J., Luke, W.T., Delaney, A.C. and Brost, R.A. (1988) Measurements of concentrations and dry surface fluxes of atmospheric nitrates in the presence of ammonia. *Journal of Geophysical Research - Atmospheres* 93, 7127-7136.

IPCC (2001) *Climate Change 2001: the Scientific Basis*. Intergovernmental Panel on Climate Change, http://www.ipcc.ch/

Jarvis, P.G., Massheder, J.M., Hale, S.E., Moncrieff, J.B., Rayment, M. and Scott, S.L. (1997) Seasonal variation of carbon dioxide, water vapor, and energy exchanges of a boreal black spruce forest. *Journal of Geophysical Research - Atmospheres* 102, 28953-28966.

Logan, J. I., Prather M. J., Wofsy S. C. and McElroy, M. B. (1981) Tropospheric chemistry: a global perspective. *Journal of Geophysical Research* 86, 7210–7254.

Milford, C., Theobald, M.R., Nemitz, E. and Sutton, M.A. (2001) Dynamics of ammonia exchange in response to cutting and fertilising in an intensively-managed grassland. *Water, Air and Soil Pollution - Focus* 1, 167-176.

Monteith, J.L. and Unsworth, M. (1990) *Principles of Environmental Physics* 2nd edition. Butterworth-Heinemann, Oxford, UK.

Nemitz, N., Milford, C. and Sutton, M.A. (2001) A two-layer compensation point model for describing bi-directional biosphere-atmosphere exchange of ammonia. *Quarterly Journal of the Royal Meteorological Society* 125, 815-833.

PORG (UK Photochemical Oxidants Group) (1998) *Ozone in the United Kingdom.* Fourth Report of the Photochemical Oxidants Review Group. The Department of the Environment Transport and the Regions/CEH Edinburgh.

Shuttleworth, W.J., Gash, J.H.C., Roberts, J.M., Nobre, C.A., Molion, L.C.B. and Ribeiro, M.D.G. (1991) Post-deforestation Amazionian climate - Anglo Brazilian research to improve prediction. *Journal of Hydrology* 129, 71-85.

Slinn, W.G.N. (1982) Predictions for particle deposition to vegetative canopies. *Atmospheric Environment* 16, 1785-1794.

Sutton, M.A., Schjørring, J.K. and Wyers, G.P. (1995) Plant-atmosphere exchange of ammonia. *Philosophical Transactions of the Royal Society London Series A* 351, 261-278.

Volz, A. and Kley, D. (1988) Evaluation of the Montsouris series of ozone measurements made in the 19th century. *Nature* 332, 240-242.

Wayne, R.P. (1990) Atmospheric chemistry. *Science Progress* 74, 379-409.

INDEX

Acremonium zeae 211
Aggregates *see* Biofilms
Air pollution 311, 312
Alternaria 287
Ammonia 306
  deposition 310–311
  levels 310–311
Antibiosis 213, 214–215, 218
Antibiotics 135–136, 216–218
  brefeldin A 214–215
  lateritin 216
  mevinolin 216
  phomopsin A 216
Aphid 56–57, 59
Aspergillosis 288
Aspergillus
  flavus
    wheat grain 290, 293–296
  niger
    colonisation of seed 296–298, 300, 301
    wheat grain 290, 293–296
  stress response 150
  wheat grain 287, 288
    odour 293
Atmosphere
  ammonia 306, 310–311
  carbon dioxide 305, 306
  nitric acid 305, 306, 309–310
  nitrous acid 309
  ozone 305, 306, 307–308
  particles 312, 313
  sulphur dioxide 305, 306, 311–312
Aureobasidium pullulans 10, 55, 56, 61
  control 141
  distribution 240, 244
  secondary metabolites 241
Auxin 12, 92, 203

Bacillus
  mycoides 138, 140, 141, 142
  subtilis 138, 139, 143
Bean 4, 90
Beech 57, 59
Biofilms
  cellulose 122–124, 125–126
  formation 67, 69, 71–73, 74-75, 76, 84–85, 126, 202
  role 67–68, 69, 77–78, 96, 97, 121–122, 127–129
Biological control 149
  commercial utilisation 133, 157–158, 161–162, 165, 172, 174, 175–176
    strategies 165–166
  inoculum 149, 154, 157–159, 162
    see also compost teas
  integrated control 140
  mode of action 134
    antibiosis 135–136
    competition 134–135
    parasitism 136–137
    systemic induced resistance 137–140
Biosurfactants 12
Botrytis
  cinerea 134, 136
  control 133, 140–141, 143, 170, 171
Brevibacillus brevis 136

Campylobacter 273
Candida sake 159
Canola 112
Carbon dioxide
  elevated levels 61
  uptake 306
Cercospora
  control 138, 139, 140
Cladosporium 61
  biological control agent 137, 139–140
wheat grain 287
Claviceps purpurea 211
Climate change 51, 55, 306, 307–308, 312–313
Colletotrichum 211
Colony development 74–75, 245–248
  assemblages 84–85, 87–89, 93–94, 97–100
  exopolymeric matrix 87, 89, 98
  leaf surface 252
  seed 296–298
  variant formation 76
Colpoda cucullus 12–13

Competition 45–46, 215, 277
Compost teas 166, 167, 176
    disease control 168–169
    production 167–168, 170–174
        aeration 168, 169–170, 173
Cotton 109
Cryptococcus 55
Curtobacterium flaccumfaciens 45

Dispersal 74, 252–253
    diurnal cycles 253
    seasonal 259–260
Distribution 5, 11, 40, 248
    host architecture 244
    leaf topography 93, 94, 239–240,245, 274–275
    spatial distribution 90, 95, 239–240, 244
Diversity 6, 25–29
Denaturing gradient gel electrophoresis DGGE 224, 225, 254
    direct amplification 41
    leaf age 24
    microarrays 255–264
        analysis of 258–259
        applications 259
        development 255–258
    population
        bacteria 9, 10, 21, 27, 93, 96, 256, 259–260, 263
        fungal 26–27, 33, 47, 209
        host genetics 113, 115, 223 228–229, 234, 235–236
    ribosomal RNA 23–24, 25, 29, 32–33, 254, 262
    sampling 39, 40–41, 225, 242, 251–252

Endophytes 37–39, 42–43, 108
    broad-leaved crops 210
    grasses 210, 218–219
    secondary metabolites 210, 211, 217–219
Enteric pathogens *see* Human pathogens
Epicoccum nigrum 61, 154, 156–157
Erwinia
    amylovora
        control 134, 135
        hrp 184
    carotovora subsp. *atroseptica*
        hrp 193, 195
        genome 191–194
        pathogenicity 194, 195, 199
        type III secretion system 191 192, 195–197
    carotovora pv. *betavasculorum*
        control 139
    herbicola 5, 10, 12
Escherichia coli 73, 122, 123, 173–174, 272–273
    attachment 73, 122, 123, 275–276
    colony development 173–174, 277
    distribution on leaf 274
    movement on plant 279
    occurrence 271
    survival 278, 279

Fusarium
    control 169
    culmorum
        colonisation of seed 296–298, 301
        wheat grain 290, 293–296
    verticillioides 211
        wheat grain 287

Gene expression
    biofilm 67, 69–70, 72–73, 74–75
    transcription factor 194, 196
Gene regulation 72, 74–75, 182–183
    host 185–186

hrp cluster
    expression 181–183, 184–185, 196–197
    host response 184–185, 186
    regulation 181–183
Human pathogens 269
    attachment 275–276
    distribution on leaf 274–275
    diversity 272
    fresh produce 271–274

Human pathogens *contd.*
sources of 269–270, 271–274
seeds 279
survival 278
Hypoxylon 217–218
fragiforme 215

Inhibition 114
Insect 56

Leaf
surface 2
topography 239–240, 243, 245
trichomes 90, 91–92, 96, 100
veins 90, 242, 245, 276
Lipase 73
Listeria monocytogenes 272, 274–275
attachment 275
Lupins 216

Metarhizium anisopliae
stress tolerance 150, 151, 154
Microscopy 86–87, 88–89, 290, 296
GFP 87, 93, 182, 242
nuclear magnetic resonance 298, 300, 301
Monilinia 133
Mortality 77
Movement
airborne 251–252, 253, 259–260, 262, 263, 271–272
insect 8, 22, 271
motile bacteria 279
rain splash 8, 22, 95–96, 248
sporulation 44
seeds 44, 279
Mycotoxins 288, 289, 301

Neurospora crassa 14
Nitric acid
deposition 309–310
uptake 305, 306
Nitrogen fixation 46
Nitrous acid 309
Nutrients 3–5, 89, 154–157, 171–172, 240
availability 14, 15 96, 240, 245, 246
Nutrients *contd.*
carbon 3–4, 32, 51, 156–157
cycling 51
carbon 52, 54, 306
nitrogen 52, 54
honeydew 56–57, 59, 60
nitrogen 4
sugars 4, 152–153, 156

Oak, 57, 59
Oilseed rape *see* Canola
Ozone
deposition 307–308
levels 307
uptake 305, 306

Pantoea agglomerans 91, 92, 93–94, 96
antibiotic production 135
biological control agent 135, 142–143, 152
colony development 276
water stress 154
Pathogenicity
type III secretion system 191, 193 195–197, 203, 204, 206
gene expression 204
Pea 25–32
Penicillium
aurantiogiseum
colonisation of seed 296–298, 300, 301
wheat grain 290, 293–296
control 133, 140, 159
expansum 134
verrucosum 293–296
vulpinum 293–296
wheat grain 287, 288
Phenyalanine ammonia lyase (PAL) 185
Phoma medicaginis 212–215
Phomopsis leptostromiformis 216
Photosynthesis 46, 308
Phytophthora infestans 171
Pichia anomala 154, 158, 160
Plant
genetic traits 107–109, 110, 111, 115
secondary metabolites 110–111, 113, 114

Plasmid 98
Population 6
density 14–15, 23, 31–32, 60, 90, 92, 202, 206, 241
host genetics 202–203
ribosomal RNA 23–24
immigration 7, 8–10, 22, 91, 95–96, 240, 252–253
island biogeography 8–9
stability 31–32
see also Diversity
Populus 109–111, 112–113, 114
Postharvest spoilage
colonisation of seed 296–298
detection 288–289
electronic noses 289–290
odour 289, 291–296, 300–301
economic cost 288
environmental conditions
moisture 287, 299–300, 301
temperature 287
mycotoxins 288, 301
Powdery mildew
control 136, 137, 142, 143, 172
Pseudomonas 31, 121–126, 231
aeruginosa 66, 69, 73, 74, 76, 78
biofilm 122, 124
chlororaphis 276
fluorescens 274–275
antibiotic production 135–136
biofilm 76–77, 122, 124–126
biological control agent 135–136, 142–143
colony development 10, 22–23
control 134
distribution on leaf 4, 93
putida 66, 70, 124–126
syringae 2, 5, 9, 10, 11, 12, 13, 14–15, 70 90, 91, 93–94, 98
biofilm 122, 126, 128, 202, 205
colony development 9, 10, 13, 277
control 13, 134, 142
distribution 5, 90, 91, 93–94
epiphytic colonisation 11,14–15
hrp 183, 187
pathogenicity 12, 201, 203
stress 2, 70, 98
Pseudomonas contd.
tabaci
hrp 184
Pseudoxyma flocculosa 142
Pulmonary mycotoxicosis288
Pythium
control 168, 169, 171–172, 175

Quorum sensing 69–72, 74, 77, 193, 197, 205, 248

Ralstonia solanacearum 182–183
Rhizobium 109
Rhizoctonia solani 140
Rice 112
Rust 136–137

Sampling
methods
culture 42, 242, 251–252 254
molecular 254–255
strategies 39, 259–260
Salmonella
enterica 123, 173–174, 272
attachment 274–276
colony development 276–278
movement on plant 279
systemic induced resistance 138, 139–140
nutrients 277
survival 278
temperature 277, 278
water availability 277–278
occurrence 271
typhimurium
colony development 277
survival 278, 279
Scots pine 54–55
Secondary metabolites 210
Serratia marcescens 70, 71–75, 77–78
Shigella 273
Sphingomonas 231
elodea 128

Sporobolomycs roseus 55, 56

Spruce 54, 57, 59
Staphylococcus aureus 274
Stilbella erythrocephala 218
Stomata 90
  conductance 306
  gaseous absorption 308, 311
Streptomyces griseus
  wheat grain
    inoculation 293
    odour 293–296
Stress 51, 149
  environmental 206
  pollution 55
  survival 96, 98, 150, 151, 152–153
  water availability 121, 150, 151 152–153, 202
Strobilurin 218
Sugarbeet 25–32
Sulphur dioxide
  deposition 311–312
  levels 311
  uptake 305, 306
Systemic induced resistance 137–140
  elicitors 185–186
  hrp 185

Tilletia caries
  odour 293, 301
  wheat grain 287
Tilletia tritici see T. caries
Tomato
  bacterial spot *see Xanthomonas campestris* pv. *vesicatoria*
Trichoderma
  biological control agent 136, 140–141, 143
  secondary metabolites 218
  stress tolerance 156
Trichomes 90, 91–92, 96, 100

Ulocladium atrum 135, 151, 153, 158
UV 2–3, 55, 56, 202, 223, 224 225–233, 235–236

Vibrio cholerae 76, 77–78

Water
  availability 3, 5, 89, 127–128
Wheat
  diversity 25–32, 112
  grain infection 290
    Alternaria 287
    Aspergillus 287, 293–296
    Cladosporium 287
    Fusarium 287, 293, 296
    odour 291–294
    Penicillium 287, 293–296
    postharvest losses 288
    Streptomyces griseus 293–296
    Tilletia caries 287

Xanthomonas
  campestris 128
  campestris pv. *vesicatoria*
    gene regulation 182–183
    host response 184–185
    hrc genes 181
    hrp genes 181–183
    pathogenicity 182–183
    population size 203
Xylaria 38, 40, 44, 45
  secondary metabolites 217–218
Xylella fastidiosa 45, 128

Yersinia 274

Zea mays (maize) 224, 235–236